COHERENT SYSTEMS

STUDIES IN LOGIC

AND

PRACTICAL REASONING

VOLUME 2

ELSEVIER

AMSTERDAM • BOSTON • HEIDELBERG • LONDON • NEW YORK • OXFORD • PARIS
SAN DIEGO • SAN FRANCISCO • SINGAPORE • SYDNEY • TOKYO

COHERENT SYSTEMS

Karl SCHLECHTA
Université de Provence and
Laboratoire d'Informatique Fondamentale (CNRS UMR 6166)
Marseille, France

ELSEVIER

2004

AMSTERDAM • BOSTON • HEIDELBERG • LONDON • NEW YORK • OXFORD • PARIS
SAN DIEGO • SAN FRANCISCO • SINGAPORE • SYDNEY • TOKYO

ELSEVIER B.V.
Sara Burgerhartstraat 25
P.O. Box 211, 1000 AE
Amsterdam, The Netherlands

ELSEVIER Inc.
525 B Street
Suite 1900, San Diego
CA 92101-4495, USA

ELSEVIER Ltd.
The Boulevard
Langford Lane, Kidlington,
Oxford OX5 1GB, UK

ELSEVIER Ltd.
84 Theobalds Road
London WC1X 8RR
UK

First edition 2004

Library of Congress Cataloging in Publication Data
A catalog record is available from the Library of Congress.

British Library Cataloguing in Publication Data
A catalogue record is available from the British Library.

ISBN: 0-444-51789-8
ISSN: 1570-2464

The paper used in this publication meets the requirements of ANSI/NISO Z39.48-1992 (Permanence of Paper).
Printed in The Netherlands.

Foreword (by David Makinson)

This book contains the deepest and most comprehensive study yet undertaken of preferential and related models for nonmonotonic inference.

It is based on a series of papers and monographs by the author, published over the last fifteen years. It goes beyond them, bringing separate results into a coherent overall framework, accompanied by discussion of the general concepts underlying the field. At the same time, the author solves new problems that rose to the surface as he carried out this conceptual organization.

The subject is logic, specifically nonmonotonic logic. The focus is on representation theorems for such systems, particularly in the tradition of preferential models. These were introduced by Shoham, and brought to wide attention when Kraus, Lehmann and Magidor obtained the first deep results in a classic paper. Whereas classical logic takes the conclusions obtainable from a set of premises to be those formulas that are true in all the models satisfying them, the driving idea of the preferential tradition is to permit as conclusions all those formulas that are true in all the models that are minimal, under a given ordering, among the models satisfying the premises. A simple idea — but one that opened the door to the exploration of new worlds, as well as revealing links with investigations in several other areas of logic such as belief change, update, counterfactual conditionals, and conditional directives.

The author's approach is resolutely semantic, in the sense that as much work as possible is done on the level of the models themselves, independently of the logical superstructure. The representation problem is thus transformed into one of determining when certain kinds of model using selection functions can have that selection function determined by other devices such as minimalisation under a relation. The results obtained on the model-theoretic level are then applied to yield, as corollaries, representation theorems about nonmonotonic inference relations, as well as about belief revision and update.

Thus while the enterprise is motivated by logical considerations, most of the hard work is carried out on a pre-linguistic level. One might call it pre-logic. It is, in effect, an algebraic analysis of the inter-relations between selection operations, neighborhoods, qualitative notions of size and of distance, and above all ordering relations on arbitrary sets. The analysis is then applied to the particular case of sets of valuations of logical formulas, and thus ultimately to notions such as that of nonmonotonic consequence over the formulas.

Even in the finite case, when a propositional language contains only finitely many elementary letters, representation needs care. In particular, if formulas are composed used only Boolean connectives, we need "copies" of classical valuations to allow two different states that satisfy exactly the same elementary letters to occupy different positions in the ordering of the model. This issue does not arise for completeness theorems in, say, modal logic, whose models also involve states with positions under an ordering. For there the modal operators of the object language already provide a means to differentiate states even when they satisfy the same elementary letters. But when the formulas of the object language are all Boolean, as they are in most treatments of nonmonotonic logic and also in this book until its last two chapters, then such copies must be provided as part of the construction in a representation theorem.

Nevertheless, proofs are usually relatively straightforward in the finite case. But they are beset with snags when we go infinite. This is essentially because when we have infinitely many elementary letters in the language, there are sets of Boolean valuations that are not definable by any set of formulas. For such a set of valuations, there is always another valuation outside the set that satisfies all the formulas that are satisfied by all valuations in the set. This creates difficulties as soon as we allow our nonmonotonic consequence operations to take infinite sets of formulas, as well as finite ones, as premises. Compactness properties, so useful in many areas of logic, become scarce and new methods are needed.

The author has developed his own distinctive techniques for getting around these difficulties and thereby constructing representative structures in the infinite case. His techniques are subtle, abstract and sometimes complex. But they are also robust and versatile: he shows how, with suitable modifications, they can be deployed successfully in many different contexts. Because of their potential, they will interest researchers in the area as much as the results that they have been used to obtain.

In some cases, however, representation theorems are not forthcoming. Faced with such recalcitrant situations, researchers usually give up and pass on to other cases where positive results are obtainable. One of the features of this book is its passage from impasse to proof of impossibility. The author invents techniques to show, for certain natural classes of model structure, that no representation theorems, of a very broad kind, are possible.

This is not an easy book; the reader will need mathematical maturity and perseverance to get through it. But the effort is well repaid. It provides the most advanced account presently available of representation for selection structures and nonmonotonic inference operations in the infinite case, indispensable for all those seeking to understand the area; and it will not be displaced from that position quickly. Its level of mathematical sophistication and penetration will engage the interest of professional model theorists and mathematicians, who up to now have tended to leave the study of nonmonotonic reasoning to computer scientists and philosophical logicians. That in turn should lead to further advances in our understanding of the area.

David Makinson

King's College, London

Contents

Summary

All logic can be based on very few general principles.

Procrustes of Attika,
around 800 BC

This book has three distinct, but strongly connected, directions:

First, we base several types of human reasoning on a small number of basic semantical notions. This is a back and forth procedure, as we take simple common-sense notions like "size" or "distance" to clarify human reasoning and, in turn vary these notions to obtain different forms of reasoning. This is a rather philosophical enterprise.

Second, we establish a number of representation theorems for such reasoning, showing how properties of the basic semantical notions are reflected by the logical properties. This is a logical, or even algebraic, enterprise. In more abstract terms, the basic semantical notions result in coherence conditions in corresponding model choice functions, which carry over (often almost one-to-one) to logical properties.

Third, we sketch a unified treatment of several such types of reasoning in a uniform modal framework, with several layers of abstraction which, we hope, gives some of the richness and variability of human reasoning. This is almost an engineering effort, and intended as an approach to a flexible system of argumentation. In this part, we will indicate problems and solutions in rough outline only.

These three parts are interrelated. The second part formalizes the first, it establishes a 1-1 correspondence between semantics and reasoning based on it, or demonstrates the impossibility to do so. Without the ideas in the first part (at least in rough outline), there would not have been anything in the second part.

The first and the third parts are related: We not only obtain a principled way of reasoning and arguing, but we can also refine our arguments by going down to the basic notions themselves.

Finally, the second part gives the necessary technical rigour to construct a sound system of argumentation. Conversely, the third part gave motivation for some of the results in the second part.

In the second and main part (completeness and incompleteness constructions), we have three main directions of thrust. First, we want to show that, for preferential and related structures, a basic (and provably most general) idea can be used and modified to give a number of results for semantics with various strengths. Second, we insist on stressing the role of domain closure conditions, which seem to have been mostly neglected so far. In particular, we will show that some results depend on such seemingly innocent closure conditions. But, we will also show how to avoid such nice closure conditions, and what is the price to pay for it (more complicated conditions elsewhere). Third, we will show that several, at first sight very simple, semantics do not allow a finite characterization, or even no characterization at all by usual means. Again, the third and the second point are related, as finite representation can be possible, provided the domain is sufficiently rich.

We will usually decompose representation problems into two steps: as we work with model sets, the basic semantical notions, like a relation between models, give rise to functions on model sets. Thus separating the algebraic from the logical problem brings them better to light, makes their solution easier, and we can re-use results for the (usually more difficult) algebraic part in various logical contexts.

Finally, reformulating the logics as "generalized modal logics", expressing the core concepts in the object language has many advantages, like increased flexibility and expressivity, and also results in increased "quality" of the logic, permitting, e.g. contraposition.

The reader will probably have guessed that the quotation attributed to Procrustes is a hoax. Procrustes was not so much concerned with a systematization of logic (by perhaps unorthodox and slightly violent means), but rather with a standardization of people (also by unorthodox and slightly violent means).

Acknowledgements

Scientifically — and beyond —, I owe very much to my co-authors D. Lehmann, M. Magidor, D. Makinson. Others have contributed more indirectly, through discussions, or, even just by some remarks, which have left traces that widened over time.

A number of the formal results presented in this text had been developed over the years, but I needed time to take a step back and put things into perspective — and attack new problems.

The French CNRS financed a two years stay at the Institut des Sciences Cognitives at Lyon, where I was free to think in whatever direction I chose. M. Jeannerod, its director, gave me a very warm and generous welcome, and provided the ample but inspiring framework I needed. Without CNRS's financial support and M. Jeannerod's patience and generosity, I would probably not have been able to take the step back and write these pages. I hope that both will not regret having made this possible.

I would also like to thank my co-authors for their permission to use joint results in this book.

The editors and publishers of various journals kindly gave me permission to include in the book material published before in their journals:

- Oxford University Press for Journal of Logic and Computation, see [Sch92], [Sch95-1], [Sch96-1], [Sch99], [Sch00-2],
- Hermes, Paris, for Journal of Applied Non-Classical Logics, see [SM94],
- Kluwer Academic Publishers for Journal of Logic, Language and Information, see [Sch96-3],
- IOS Press, Amsterdam, for Fundamenta Informaticae, see [SD01], [SGMRT00],

- The Association for Symbolic Logic for The Journal of Symbolic Logic, see [LMS01], [Sch00-1],

- Notre Dame Journal of Formal Logic, see [Sch91-1].

The parts concerned are described in detail in Section 1.2.

Two referees have helped with their comments to transform this text from a pile of notes to something more readable. Their comments were very valuable.

Finally, I would like to thank my Marseille collegues for their patience with me while I was busy writing the book.

Chapter 1

Introduction

1.1 The main topics of the book

The book is organized into the following chapters:

(1) Introduction

(2) Concepts

(3) Preferences

(4) Distances

(5) Definability preservation

(6) Sums

(7) Size

(8) Integration

(9) Conclusion and outlook

Chapter 1 contains — apart from the general introduction — some basic definitions and results.

Chapter 2 presents and analyzes the concepts we use.

Chapters 3 – 7 contain most of the technical results. They are the hard core of the book.

Chapter 8 shows a way to integrate the different formalisms we have discussed previously.

Chapter 9 summarizes in very abstract terms what we consider the main ideas of the book, those discussed or used, but also those which need much further elaboration in future research.

1.1.1 Conceptual analysis

We argue that an important number of common-sense reasoning logics can be based on a small number of semantical notions, which, in some cases, are interdefinable. These semantical concepts express in some qualitative way (un)certainty, using notions of preference, distance, or size. They motivate or even impose coherence conditions on the semantics, and thus on the logics.

For instance, nonmonotonic logics can be based on preference, we consider only those cases preferred by their normality, or they can be based on size, we neglect "small" or "unimportant" sets of exceptions. Theory revision can be based on distance, we revise one theory with another by considering only those models of the second one, which are closest to the models of the first, thus coding minimal change into closeness. Counterfactual conditionals also have a distance based semantics, given by Lewis and Stalnaker.

We will look at various such concepts, in various forms, e.g. a relation of preference might be an arbitrary binary relation, it might be transitive, free from cycles, ranked, etc. Such additional properties ususally result in additional properties of the resulting logics, or, on a still more elementary, algebraic, level, of the resulting model choice functions. E.g. a distance generates in the theory revision approach a function f, which associates to two (model) sets X and Y, a subset of Y, i.e. those elements of Y which are closest to X. The examination of the properties of such model choice functions will be at the core of our formal completeness and incompleteness results.

Conversely, interesting logical and algebraic properties of the model choice functions can motivate powerful new concepts for the semantical structures. For instance, cumulativity has motivated the concept of smoothness for relations.

In some cases, we can transform in a natural way one concept (e.g. size) to another one (e.g. ranking of sets). Thus, we will not only look at these concepts individiually, but also at their interplay.

We will examine the concepts in Chapter 2, which has a somewhat more philosophical character. We will not and cannot discuss the question whether such reductions are cognitively adequate, in the sense that people really reason this way. This problem has to be examined by cognitive psychologists.

The reader will not find a ready made receipt here, but should be prepared to accept a multitude of alternative suggestions. The author thinks that this is due to the subject, and not to his inability to come to a decision. Common-sense reasoning has to be adequate to its subject, and there are many subjects, so we should not expect to find very few approaches. The systematization is in the common basic notions, not in the way they are used. In the formal part, we will follow some, but not all, of these alternatives, and leave the rest for the interested reader, giving him, we hope, some useful techniques to tackle the problems.

Common to all logics we consider is that they work with somehow defined sets of classical models. This leads to the concept of a generalized modal logic, preparing at the same time an integrating framework within which we can construct a system incorporating several common-sense reasoning mechanisms. We present this now in outline.

1.1.2 Generalized modal logic and integration

When we look at logics like modal logic, the logic of counterfactual conditionals, preferential reasoning, etc., but also the classical operators $\wedge$, $\vee$, $\neg$, etc., we see a common point: all work with sets of models, e.g. in the case of preferential reasoning: $\alpha \mid\sim \beta$ iff β holds in the set of preferred models of α, $M(\phi \wedge \psi) = M(\phi) \cap M(\psi)$ for classical $\wedge$, etc., where $M(\phi)$ is the set of models of ϕ.

It is therefore natural to take a more general look at logics defined via some set or sets of models chosen one way or the other. So, in a first abstraction, we have some choice function f for model sets, and the logic is defined via f, i.e. by $T \mid\sim \phi$ iff $f(M(T)) \models \phi$, where $\models$ is classical validity. In a more complicated case, we have perhaps a family of M_i, and f chooses again some family $(f\{M_i : i \in I\})_j$ — theory revision is a case of several M_i. As the logics will then be defined by the formulas true in some set of classical models, it seems appropriate to call such logics "generalized modal logics".

Of course, such model choice by some f should be made in some principled way, we should choose a "good" set in some sense, perhaps guessing a little bit, but not too much by some standard, giving a reasonable approximation. The present book is, if you like, an essay on such "good" choice.

We will see that this framework will also allow us to interpret notions like "certainty" (via some kind of neighborhood relation), and more. So, we can and will, go beyond the aim of presenting a unified framework for several logics.

Studying a notion of this generality (arbitrary f) is usually not very interesting. It will become more so when we base the choice functions on some underlying structure — as already hinted at above — like distance (e.g. for choosing a neighborhood), size (of sets or of elements, choosing a "big" subset), relations between elements (preference or reachability). Of course, there has to be some intuitive connection between the notion motivating the choice function (e.g. nonmonotonic reasoning), and the notion we try to base it on (e.g. preference) — this is a philosophical problem.

But we also face, sometimes not very trivial, mathematical problems like:

- How are structures based on different notions (e.g. size and preference) related?
- Can we sometimes transform one notion into the other?
- What are properties common to all choice functions generated by a fixed structure with certain properties (like choice functions of modal logics for reflexive relations, etc.)?

Once we see this general framework, we can try to put many things into the object language, which are otherwise often left outside the language, e.g. in the form of operators like $\mid\!\sim$ for nonmonotonic logics, or $*$ and $\vdash$ for theory revision. As this transformation will be largely obvious, we will do it in most cases only implicitly. E.g., $\alpha \mid\!\sim \beta$ will become $\vdash N(\alpha) \rightarrow \beta$ — in the normal cases of α, β always (classical implication) holds. This has several advantages: In our example, normality has now become much more powerful, as we can negate it, and speak, e.g. about the abnormal $\alpha-$case $(\alpha \wedge \neg N(\alpha))$. We also have contraposition, $\neg\beta \rightarrow \neg N(\alpha)$, for $\alpha \mid\!\sim \beta$, and this is perfectly reasonable. "Classical" nonmonotonic logics do not have contraposition as they hide the complexity in the formalism: $\alpha \mid\!\sim \beta$ and $\neg\beta$ means: either α does not hold, or some of the stuff we put into $\mid\!\sim$ is not true — i.e. we are not in a normal $\alpha-$case. But in our approach, it is explicit, and we can reason about it; this is obviously much better! Every time we have to backtrack, we have a case of contraposition, and the author thinks these things should be expressed clearly.

Thus, we see that the strategy of putting as many notions as possible into the object language has many advantages:

- The formalism itself becomes more expressive, we can reason about the notions concerned, not only with them
  - this usually results in a notion of consistency,
  - this can give contraposition, and thus an increased quality of the logic, as argued above,
  - we have nestedness, boolean combinations, etc.,
  - we have relativization in the following sense: Sometimes, we might not be sure about the normal cases of ϕ, but, we know that if x a normal $\phi-$case, then it will be a normal $\psi-$case — this can now be expressed, and used, e.g. in an argument or in dynamic reasoning.
- The "grammatical" role or abstract framework of a notion has to be made explicit:
  - is it an operator like $\wedge$?
  - is it a relation, and if so, between which notions?
  - or is it something more complicated?
  - does it impose conditions on other operators?
  - which are its arguments, which its results?
- The approach is more flexible:
  - detailed properties may be neglected for the moment, and added later directly to the object language,
  - sometimes, it seems difficult or even impossible to describe the details for all possible situations (we have not a constant, but a variable), again the details can be added "on the fly" in the object language.
- We have a clear separation between the general (grammatical) role of the notion, and its perhaps complicated details, and can pursue the often very useful strategy of "divide and conquer".
- The cooperation with other notions has to be made explicit.

If we put not only notions like "normal" into object language, but also the generating semantical concept, like preference, distance, size, etc., we can (and have to) bind the two concepts together by an object language analogue of a soundness and completeness theorem. The important point is that we win this way multi-level description and reasoning, where unnecessary complexity can be hidden from upper levels (here, from normality), and is visible only in the lower levels (here, e.g. preference of one situation over another), whenever this is necessary. For instance, we may know that in all minimal $\alpha-$models β holds, without knowing exactly the relation of normality. Often, this will be sufficient, but there might be situations where we need the more precise information, e.g. to convince an opponent in argumentation. If we go further and put uncertainty directly into the language, as an object, as the amount of cases we neglected, we can speak about arguments: they are paths or graphs, where the "gaps" are annotated by the size of the set of the neglected cases. We can say, e.g. to our opponent in argumentation: Look, your gaps are much bigger than mine, so your argument is weaker. Thus, we can put a suitable theory of argumentation in the language, and reason with and about it.

The price we have to pay is not very high, as we have big and small sets ($N(X)$ are the big subsets of X, $X - N(X)$ the small ones), etc. already in our object language. We can now put, e.g. also the relation into the language, add sufficiently many axioms to make it work, and give reasons why one set is big, and the other is not. This is somewhat elaborated in Chapter 8.

Limitations of our approach

Representing reasoning this way as generalized modal logics is certainly an idealization, as human reasoners (and computers) cannot always calculate all consequences — not even up to classical propositional equivalence. But it is a reasonable approximation.

On the other hand, it is evident that some reasoning will never fall into this category. All reasoning about aims, actions, desires, intentions is outside: If my aim is to become a rich man, then it is in all but pathological cases not true that I want to become a rich man or to commit suicide, or that I want truth (in the logical sense) to hold. These notions are incompatible with deductive closure, at least in such a primitive way. This is plausible, as they speak about states of the world where something holds which does not hold already, but, e.g. truth will already hold. It is another question whether they can be caught by something like $D := Th(X) - Th(Y)$, where X is some desired state, and Y some actual state.

1.1.3 Formal results

Coherence properties:

A central subject, implicit and explicit, in our discussion will be "coherence properties". They describe answers to the following question: Given $f(M(T))$, can we still choose freely $f(M(T'))$, or is there some coherence between $f(X)$ and $f(Y)$ (given some relation connecting X and Y)? For instance, in preferential structures, if $X \subseteq Y$, then $f(Y) \cap X \subseteq f(X)$ will always hold. Our uniform approach as generalized modal logics will help to bring such properties to light in various logics, and, as a matter of fact, we can consider such "algebraic" properties of the choice functions f as the core properties to describe in characterizations, and thus in completeness results. In other terms, if we work in the generalized modal logics framework, (almost) all that is left are coherence properties. We will see that there are often natural ways to generate f by underlying structures, like preference or reachability relations, or distances between models, and that such structures determine sometimes very strong coherence properties.

Such properties are all the more interesting as the logics considered might be quite "wild" and any property taming them is very welcome. So coherence is not just a purely academic subject, but an eminently practical one. At least when the algebraic properties are themselves generated by an underlying structure, the resulting logical properties can be considered a "byproduct". But, in some situations, they are at the very core of considerations, this is the case, e.g. in analogical or inductive reasoning. Analogy says that T is in some way similar to T', and induction says that the big set behaves as does a small subset. Here, coherence is the hypothesis, a fact about the world, and not the byproduct of a logic (more precisely, of the semantics), and we should not expect one case of coherence to be very similar to the other.

As said in the Summary, we will usually decompose representation problems into two steps: the basic semantical notions, like a relation between models, give rise to functions on model sets. We can thus separate the algebraic from the logical problem. This brings them better to light, makes their solution easier, and we can re-use results for the (usually more difficult) algebraic part in various logical contexts — as we see, e.g. for plausibility logic in Section 3.7.

We will consider in detail representation results based on preference, distance, size, and sums.

Preference:

Our results on preference are the most developed ones. We apply there very general techniques (suitable choice functions to code our ignorance, trees to code transitivity, "hulls" to keep away from certain sets) to various situations. Particular emphasis is made on domain closure properties (see, e.g. again the section on plausibility logic 3.7), which seem to have been quite neglected so far, and definability preservation. We will also show that important classes of the limit variant are equivalent to the conceptually and technically much simpler minimal variant. In all cases, we describe problems which can arise when these conditions are not satisfied, but sketch also solutions which circumvent them. We will also show that topological constructions can give interesting counterexamples, but can open new ways to achieve properties like cumulativity, too. Often, and not only for preferential structures, such counterexamples can be obtained by using in the structure implicitly a topology different from, even against, the natural topology of propositional logic. We examine the ranked case in more detail, giving not only a representation result for the limit case, but demonstrate again its equivalence with the usual minimal case in an important class of problems. The question of copies finds an answer in a different approach, by total orders instead of partial orders, and we can show a corresponding representation result in the finite case. This Section 3.9 stands a little apart. On the one hand, it is motivated by philosophical considerations about completeness proofs and the role of copies. On the other hand, it has its natural place in the formal part, as it gives a formal answer to these philosophical questions.

Distance:

We then turn to distance based formalisms, in particular to theory revision and counterfactual conditionals. We prove several completeness results for the first, but also demonstrate that finite characterization is impossible, unless the domain is sufficiently rich. In an application of a similar idea we show that counterfactuals based on one single metric are essentially the same as counterfactuals based on several metrics. The argument is, just as for the lack of finite characterizations for revision, that close elements can hide those farther away.

Definability preservation:

The model set operators are usually assumed to preserve definability, i.e. $f(M(T))$ is supposed to be exactly $M(T')$ for some other theory T'. Many results (not only by the author) hold only under this assumption. We show a uniform technique to avoid this assumption (approximation), but show also that, at least in some cases, there is no characterization in the usual form possible, as we have to admit arbitrarily big (in cardinality) sets of exceptions, which we cannot describe by usual logical means. This negative result applies also to the general limit version of preferential structures.

Sums:

Formalisms based on sums, like update or other types of reasoning about sequences, have a uniform representation criterion, the possibility to solve certain systems of inequalities. In these cases, we are interested in those sequences which are minimal by some criterion. As minimality is calculated component-wise, we are interested in minimal sums, and have therefore to determine whether systems of inequalities between sums have a solution. This can be verified by an old algorithm, due to Farkas. Often again, we can show that essentially simpler, finite, characterizing conditions are impossible.

Size:

We then incorporate abstract size into a first order framework, in the form of a generalized quantifier, expressing "almost all". This is done in an isolated fashion, and with very weak conditions, to be able to extend it in any direction desired. We also show how to base theory revision on model size, revealing a general technique which we consider interesting. Finally, we compare several coherent systems of size.

1.1.4 The role of semantics

Semantics are in the center of this book, so the author should explain why he thinks that semantics deserve this place and, more generally, what their role is in his opinion.

Semantics seem to the author the binding link between the world (or better, our intuitive reasoning about it, our basic concepts like "possible", "normal") and logic.

We should not try to construct logics directly, too many attempts have led to inconsistent logics, and futile work. It seems very difficult to find good logical systems without having (at least implicitly) a semantics. Logics seem too difficult to understand, as their concise and constructive nature (construct all theorems from a very small set of axioms) can hide what they are about. Formal semantics, in the author's view, try to capture an aspect of the world, they are like a simplifying painting of nature, and it is usually quite clear what they capture, and where their limits are. Their adequacy is a philosophical question, and one of utility: does this aspect of the world interest us or not.

Logics are then the second step: reasoning about (the objects of) the formal semantics. The question of their adequacy is now a mathematical problem: soundness and completeness. In particular, the semantics are now the gauge by which logics are to be measured, and logics are adequate or not for certain semantics.

A logic is, of course, necessary if we want to do formal reasoning about the aspects of the world that interest us. We need both, logics and semantics, but the right strategy seems to start with the semantics.

Working with a semantics thus splits the difficult problem of constructing a suitable logic in two parts: a more philosophical part, which discusses the adequacy of the formalisation (does the formal semantics really capture the essence of what we want to speak about in our logic?), and a purely mathematical part — soundness and completeness — which shows that the logic corresponds one-to-one to the semantics we found. This mathematical part is per se interesting from an academic point of view, as it shows the equivalence of two different constructions: an axiomatization of the syntactic side, and a semantics. But it also has eminent philosophical and even practical interest, as it binds reasoning via semantics to (an aspect of) the world.

If we adopt this view of semantics and logics, we should neither demand too little from semantics, nor too much.

For instance, to say that certain extensions or fixpoints of a formal system are the semantics of this formal system, is demanding too little. When we look back to classical logics (in an analogical reasoning), the set of all theorems would then be their semantics. This does not seem the right notion of semantics. In this approach, we would construct the semantics from the logics, so we can never measure logics independently by the semantics. Semantics would have lost its normative power. Of course, this does not mean that extensions, etc. are not interesting: they elucidate the formal system, but they do not seem to be a true semantics to the author.

On the other hand, if we demand from our semantics to be one for (correct) common-sense reasoning, this is demanding too much, as the semantics should then be a more or less complete description of the actual world around us. This seems a much too difficult problem.

To summarize: semantics should formalize one or several aspects of the world and our reasoning about it, like "possible", "normal", "causes", etc., in a clear and intuitive way. We should then, in a second step, try to see which logics correspond to our semantics, and construct them, if necessary.

The picture is perhaps a little blurred by the situation in common-sense reasoning. Common-sense reasoning seems to proceed often by rules, which have proved successful in certain areas. "If bird, then fly." For this reason, rule systems like Reiter defaults, or inheritance networks are so attractive. Thus, in common-sense reasoning, logics (or rules) preceed semantics. The research task is then a back and forth movement. First, we have to try and find out for what type of situation (semantics) those rules are adequate, construct from this vague description of situations a formal semantics, and look back at the rules whether they are (now in a precise, mathematical sense) adequate to the formal semantics or not. If necessary, we have to construct new logics (rules, etc.) to really fit the formal semantics in a rigourous, mathematical sense.

1.1.5 Various remarks

The different kinds of contents in the book will be reflected by different styles of exposition. Whereas the first part, discussing concepts, will mostly be a more informal discussion, commentaries in the second part, which show completeness and incompleteness results, will mostly be restricted to an explanation of the formal development, with motivation given already in the first part. Finally, the third part will largely be a partial summary and use of the first two parts, with indications how to fuse the various concepts together.

Naturally, part 1 is addressed more to the more philosophically interested reader. Part 2 is more for the reader with formal interests, and a certain mathematical maturity is perhaps sometimes needed to follow the arguments in detail. There is, of course, no harm (at least in a first reading) to skip proofs and lemmas, and concentrate on the main definitions and results. The reader who wants to do his own completeness proofs for related cases, will, we hope, find here ideas and quite general techniques which are useful to his interests. Finally, part 3 is destined for the "bricoleur" of reasoning systems, giving him some philosophical ideas and mathematical nuts and bolts to work with.

1.2 Historical remarks

Theory revision as presented in Section 2.2.10 is, of course due to Alchourron, Gärdenfors, and Makinson. The semantical approach to counterfactual conditionals via distance in Section 4.3 is, again of course, due to Stalnaker/Lewis. The approach to update in Section 6.4 is due to Boutilier, Friedman, and Halpern (BFH for short). For shorter presentations of other people's results, we have given references locally.

In the sequel, "new" will abbreviate "new to the author's knowledge".

The basic results on preferential structures (Sections 3.2 and 3.3) were published in [Sch92], [Sch96-1], [Sch00-1]. The basic results on plausibility logic (Section 3.7) were published in [Sch96-3], the representation result presented here is new. The counterexample to the KLM result (Section 3.5) was published in [Sch92], the nonsmooth model of cumulativity (Section 3.6) appeared in [Sch99]. The approach to preferential structures via total orders (Section 3.9) was published in [SGMRT00]. Some basic material on ranked structures (Section 3.10) was published in [Sch96-1], the examples and representation results given here are new (though some are close to the ones given in [LM92]), as well as the discussion of the limit variant.

Section 4.2 on theory revision (with the exception of Sections 4.2.4 and 4.2.5) was written with D.Lehmann and M.Magidor, and was published in [LMS01] and [SLM96]. The other results are essentially my own work. The results on the limit variant for revision (i.e. without closest elements, Section 4.2.5) are new. Section 4.3 on counterfactuals was written with D.Makinson, and is published in [SM94].

The lack of finite characterizations for theory revision (Section 4.2.4), for update by minimal sums (Section 6.3.4), for Markov developments (Section 6.4.3), and for "between" and "behind" (Section 6.5) are new. The other results on BFH style update were written down in [SFBMS00], but never published.

Some of the results on representation without definability preservation (part of Section 5.2) were published in [Sch00-2].

Update by minimal sums (Section 6.3) was published in [DS99] and [SD01].

Weak filters (Section 7.2) were introduced in [Sch95-1], see also [Sch96-2]. The filter and order systems (Section 7.3) were compared in [Sch97-4]. The approach to theory revision by model size (Section 7.4) was published in [Sch91-1] and [Sch91-3].

The rest of the material is unpublished.

In summary, and to help the reader to orient himself, the following results and approaches are new to the author's knowledge, and not published before:

On the more conceptual side:

(1) A systematic reduction of a certain number of reasoning mechanisms to a small number of basic semantical concepts (Chapter 2).

- The principled generation of a robust ranking from model size (Section 7.4.2.2).

(2) The unification of several logics of common-sense reasoning into one system of generalized modal logic (Chapters 2 and 8).

- The analysis of a problem of this fusion and its (somewhat brutal) solution.
- The resulting system of argumentation based on clear semantical principles, and with different levels of abstraction.

More on the formal side:

(1) The systematic treatment of problems of domain closure and definability preservation for representation problems (at various places).

(2) The detailed analysis of ranked preferential structures (Section 3.10).

(3) The representation result for smooth preferential structures without closure of the domain under finite union (Section 3.7.2.3).

(4) Representation results without definability preservation for

- smooth preferential structures (Section 5.2),
- distance based theory revision (Section 5.3).

(5) The reduction of the limit variant or version of preferential and distance based structures to the much simpler minimal variant in many important cases:

- for general preferential structures (Section 3.4.1),
- for ranked structures (Section 3.10.3),
- for distance based theory revision and update (Section 4.2.5 and 4.3.2.3).

(6) The impossibility of representation

- By finite descriptions for
  - theory revision based on distances (Section 4.2.4),
  - theory update based on minimal sums (Section 6.3.4),
  - theory update for Markov developments (Section 6.4.3),
  - "behind" and "between" (Section 6.5).
- By any "normal" description for preferential structures without definability preservation and the general limit variant, in the general case, the ranked case, and for distance based theory revision (Section 5.2.3).

(7) The characterization of update by Markov developments (Section 6.4.2.2).

(8) The short analysis of X-logics (Section 3.2.4).

Personal appreciation of the results published here for the first time

Of the formal part, I like the results on the limit versions, on the lack of (finite or even infinite) representations, and on the not definability preserving cases best. Conceptually, but still formally, the investigation of domain closure properties for representation results was very interesting for me, as I had encountered the problem before, but just seen it as a nuisance, not worth while a more detailed discussion. Of the more analytical part, it was interesting to see how much can be done with so few basic concepts. It was at least as interesting to go from single bits of reasoning to a uniform framework of generalized modal logics, and beyond, to meta-reasoning (and argumentation), with much increased flexibility as a result. This also gave sense to the extremely fruitful idea (e.g. in computer network architecture) of complexity hiding in the context of logic. (In computer networks, different levels of communication have clearly separated tasks, the lower levels work with details the upper levels need not and should not see. For instance, when you send an email, you need not know which path the message takes. Intermediate levels will find a path for you automatically, but might not know whether the message goes via optical fiber, or wire, but low level procedures will have to know that.)

The parts on basic concepts are, of course, addressed in the first place at more philosophically minded people. Readers working on completeness problems might be most interested in the parts which treat lack of finite characterization, and domain closure problems. They might also like to find reductions of the more difficult limit case to the simpler minimal one, and might find the negative results about impossibility of characterization useful. We also hope that persons interested in applications and building systems, perhaps of argumentation, might find the remarks on multi layered systems and combinations of several operators stimulating. Those parts contain, of course, also remarks for the more conceptually oriented readers. A combination of more philosophically oriented reflection with "hard" formal results are to be found in Section 3.9.

1.3 Organization of the book

The basic concepts like size and distance, as well as their connections are discussed in Chapter 2.

Chapters 3 – 7 contain most of the formal results: Preferential structures in Chapter 3, distance based approaches in theory revision and counterfactuals in Chapter 4, sums in Chapter 6, and size in Chapter 7, definability preservation problems and their solution in Chapter 5. Absence of finite representation is dicussed in various places in Sections 4.2.4, 6.3.4, 6.4.3, 6.5. Absence of any usual characterization is discussed in Section 5.2.3. Problems with domain closure are discussed in Sections 3.4, 3.5, 3.7, 3.10.3, 4.2.4, 4.2.5, 5, 6.3.3. A somewhat analogous problem is to be found in Section 8.3.3.

In Chapter 8, we address the uniform object language, and indicate very briefly how to put the various pieces together.

1.4 Overview of the chapters

The author is aware that the present Section 1.4 is probably difficult to understand before knowing the results of the book. It is not really intended as an introduction, rather as a companion, while reading the formal parts, helping the reader to orient himself or herself. It might be a good idea to read it first superficially, trying to guess more than to understand, and come back from time to time while reading the formal parts seriously.

1.4.1 The conceptual part (Chapter 2)

This chapter contains a mostly informal analysis of several forms of common sense reasoning, and bases them to a large extent on a small number of semantical notions, size, preference, and distance. We do not argue that this reduction is totally exhaustive, only that it seems to capture astonishingly many aspects of common sense reasoning with less than a handful of concepts. We think that such concepts like distance and size are given (in a naive way), but feel free to change their properties as seems necessary and helpful. We shall also see that the different types of common-sense reasoning are interdependent. E.g. a notion of certainty can be used to define nonmonotonic reasoning or theory revision, and conversely, given, e.g. a theory revision operator, we can recover a notion of certainty — this is the well known equivalence of AGM-style theory revision with epistemic entrenchment. (We use "AGM" as abbreviation for the article [AGM85], its authors Alchourron, Gärdenfors, and Makinson, and their approach.) Preference can also be seen as distance from a fixed ideal point, and we shall see how to generate in a natural and robust way a ranking of sets from a notion of size of elements.

In particular, it is very desirable to base as many reasoning types as possible on a few basic semantical notions, when we try to create integrated systems of common-sense logics and reasoning. (In Chapter 8, we will sketch such an enterprise in rough outline.)

This Section 1.4 also serves to illustrate that the approaches to common-sense reasoning usually discussed in the literature are by no means the only possible ones. So, the (poor) reader will be presented a multitude of alternative suggestions, but, will also see that they can often again be based on our basic semantical notions, so multitude is partly a superficial phenomenon. The alternatives presented can also be seen as suggested research programs, which, we think, should often be quite straightforward, using the machinery developed in subsequent chapters.

Thus, it seems that the basic concepts are few, and that it is the way they are used and combined, which generates the richness of common-sense reasoning.

1.4.2 The formal part (Chapters 3–7)

First, an overview by subjects treated.

In the sections on formal results, we will treat various representation results for

- preferential structures,
- distance based revision and distance based counterfactuals,
- concepts based on minimal sums (update, "between"/"behind", revision sequences),
- abstract size.

(Chapter 5 is discussed in the paragraphs on preference and distance.)

Preference

More precisely, we show rather general representation results for the minimal version of definability preserving

- general preferential structures,
- general transitive preferential structures (the transitive and the not necessarily transitive case satisfy the same conditions),
- smooth preferential structures,
- smooth and transitive preferential structures, (again, the transitive and the not necessarily transitive case satisfy the same conditions),
- two systems of plausibility logic, for one we also prove a negative result, due to lack of domain closure,
- ranked preferential structures,
- ranked and smooth preferential structures.

We also show representation results for the limit variant or version of preferential structures, in particular, we show equivalence of the limit and the minimal version in important classes of problems. This is an interesting result, reducing the limit version in many cases to the conceptually and technically much simpler minimal variant. The essential property and purpose of the limit version is that it does not "degenerate" when we have no minimal elements: In the minimal version, we are interested in the formulas ϕ that hold in the minimal models of T. In the limit version, we are interested in those formulas ϕ, which finally hold in the $T-$models, when we are sufficiently low. In the ranked case, this amounts to the following: there is a level of $T-$models, below which ϕ holds in all $T-$models. This is a natural extension of the minimal version to the case where we have not necessarily minimal models.

Finally, we treat the nondefinability preserving cases for the general, general transitive, the smooth and the smooth transitive versions of preferential structures. A, we think, important negative result is that a characterization in the usual form is impossible in these general cases.

The investigation of total orders as basic entities of preferential reasoning stands a little apart in this series. It gives an answer to the problem of copies of models, and discusses the difference between various types of completeness results. It is more in the spirit of classical logic, where models have maximal information. A completeness result for the finite case is given. The use of copies of models, or noninjective labelling functions in a different, but equivalent terminology, has probably historical origins in modal logic. We give here a different justification via disjoint unions of total orders. For details, the reader is referred to Section 3.9.

In all cases, we show first purely algebraic characterizations of the associated choice functions, which we translate then into logical properties. This translation is trivial under the caveat of definability preservation, we use classical soundness and completeness.

Distance

We show representation results for two minimal versions of distance based definability preserving revision (for symmetric and not necessarily symmetric revision), first the purely algebraic, then the logics version.

We then prove representation results for the limit variant or version of distance based revision. In particular, we show again equivalence of the limit and the minimal version in an important class of problems. Recall that the limit version of distance defined revision is the analogue of the limit version of preferential structures. $T * \phi$ will be the set of those formulas, which hold in all those $\phi-$models which are sufficiently close to the $T-$models. So, even if there are no $\phi-$models closest to the set of $T-$models, this definition does not collaps, as does the minimal variant.

We treat the nondefinability preserving case of distance based revision, which uses the same ideas and techniques as the analogous case for preferential structures.

We also prove that distance based revision has no finite characterizations, roughly, by presenting arbitrarily complex positive and negative examples. This result is again related to domain closure problems, as a sufficently rich domain would sometimes make it possible to summarize much information into just one piece of information, and the problem with arbitrarily big amounts of information would not be there.

To give one example: In revision, the problem is with transitivity, as we cannot always "see" the intermediary steps, they are literally hidden. So we cannot argue that $a \leq b \leq c \leq d$ implies $a \leq d$, because, roughly, we do not necessarily see $a \leq c$, but only the result. More precisely, we may be able to observe that $a \leq b \leq c \leq d$ holds, but not necessarily that, e.g. $a \leq c$ holds, as there might be closer elements in the way. So we cannot summarize $a \leq b \leq c$ to $a \leq c$, but need all three bits of information. The trick for the formal proof is then to construct arbitrarily complicated situations, where changing just one element of information transforms a bad case to a good one. As we cannot summarize the bad case in a small amount of information, there is no way to have a finite description distinguishing the good from the bad cases.

We then show a converse, that we can fuse many different local metrics to one global metric in the counterfactual semantics.

Sums

We show a general technique for representation results for semantics based on sums, using an old algorithm, due to Farkas, to determine whether systems of inequalities have a solution. We also show in all cases discussed that no finite characterization is possible, again by presenting arbitrarily complex positive and negative examples.

Size

We first interpret "normal" or "almost all" as a generalized quantifier in first order logic. The semantics is a weak filter, and we give a sound and complete axiomatization. This is really a bare bones system, and we intend it as such: this way, it is easy to make extensions in whatever direction we choose. In particular, we give no coherence whatever, this can be added in trivial ways by just writing down the corresponding conditions for semantics and proof theory.

We then discuss various abstract coherent systems based on size, due to Ben-David/Ben-Eliyahu, Friedman/Halpern, and the author. The first and the third are based on filter systems, the second is a relation comparing sizes. All three are roughly equivalent, and reveal that essentially the same intuition is expressed in various forms.

Finally in this section, we show how to base revision on model size. In hindsight, this was done by a general principle, which gives a robust and intuitive way to construct a ranking of sets from size of elements. This ranking is essentially stable under intersections — which is its robustness, and makes it compatible with other operations, like revision.

1.4.3 Integration (Chapter 8)

We first discuss in this chapter the, in our opinion, important advantages of formulating concepts in object language rather than as rules. We see the main advantages in increased expressiveness, clarification of the role of the operators, and a better "quality" of the logic, which shows, e.g. in contraposition which is some form of revision: If $\alpha \mid\!\sim \beta$ fails, as we see α and $\neg\beta$, we do not know what to do. If we put normality into the language, we will have $N(\alpha) \to \beta$, α, and $\neg\beta$, and this is perfectly reasonable, we can revise the assumption $N(\alpha)$ to $\alpha \wedge \neg N(\alpha)$.

We then discuss the several levels of reasoning, the usual level, the level below, where the underlying structure is, and where notions like preference or size can be represented, and the higher level, where we can reason about our own arguments. We argue that a rich system should allow all levels, to be able to fall back on reasons, and assess qualities of arguments.

We present almost in table form the types of common-sense reasoning and their basic concepts we have found in this book. This is intended as an overview and reference.

Finally, we discuss formal problems and questions when integrating several formalisms into one system. We will first present in anecdotical fashion some problems of classical (modal) logic. We then discuss a subtle problem, very similar to that of definability preservation, when putting several formalisms and their semantics together. To conclude, we show that, under some caveats, putting things together is trivial.

We turn to an overview of the main problems, ideas and techniques

1.4.4 Problems, ideas, and techniques

We hope that these chapters are not only interesting for their results, but also for the problems they treat, their ideas and techniques they use for solutions. Therefore, we will now enumerate the main problems, ideas, and techniques discovered and used in the present book.

We see "isolated" and "recurrent" problems (and their solutions, if any).

Isolated problems are specific representation issues, recurrent ones those which present themselves at several moments in our discussion. Among these are definability preservation, domain closure, and finite or even more general representation questions.

Before we discuss them, we turn to the isolated problems.

In most cases, we split the representation problem into a purely algebraic and a purely logical part. This has several advantages. First, we can recycle algebraic results in other contexts, as we have done several times (preferential logics, plausibility logic, preferences among developments). Second, we see more clearly the different problems, e.g. with sets logically definable on the logics side, properties of the basic choice functions on the algebraic side.

The first such problem is to code our ignorance about minimization in general preferential structures — we do not know which elements minimize some given element. We solve this problem by making copies for all choice functions in a suitable cartesian product. This is the best possible solution, the relation defined will, however, be the worst possible solution, again justified by our ignorance (Section 3.2.1).

The next problem is to make this construction transitive. We see that indexing by the trees of successors is again the right and most general technique to obtain full control about successors (Section 3.2.2). (In the simple, most general case, we can do with a much simpler technique, see Proposition 3.2.8.)

We then turn to smooth structures. Our construction will be a successive repair process of all smoothness we might have destroyed in previous steps. Some form of hull $H(U)$ around U to avoid U sufficiently generously seems to be the thing to do (Section 3.3.1).

For smooth and transitive structures, we essentially combine the ideas from the general transitive and the smooth case.

The author was asked if it were not possible to use complete theories for the construction of representation results, as is done usually for modal logics. This might well be possible, but, we think, the constructions presented here are the natural ones. Modifying and adapting them to other constructions would only blur the picture, we think.

We then show that important classes of the limit variant or version (transitive structures with either cofinally many definable closed minimizing subsets (or minimizing initial segments, MISE), or considering only formulas on the left of $\mid\sim$ for the resulting logic) are equivalent to the much simpler minimal variant (Section 3.4.1).

The techniques for ranked structures are quite different, due to the strength of the rankedness condition. In particular, the question of copies has a simple answer — mostly, one copy suffices (Lemma 3.10.4). On the other hand, there is a multitude of natural and quite similar conditions, which, however, might differ in somewhat exotic cases. We discuss these, and obtain several completeness results (Section 3.10.2).

We then discuss the limit variant or version of ranked structures, obtain a general technique to treat the limit version (also applied for distance based revision), and show that for an important class of problems, the minimal and the limit version are equivalent (Section 3.10.3). This is due to domain closure properties, and has again its analogue in distance based revision. The main idea is to show that the systems of closed minimizing sets have (after some modification, essentially) the properties of the minimal variant.

In many examples for ranked structures we use essentially topological considerations to show negative results. Such techniques are also used to obtain cumulativity in the absence of smoothness (Section 3.6).

Domain closure properties are also used to show a counterexample to an extension of the original KLM characterization (Section 3.5).

We can apply our techniques and results immediately to plausibility logic in Section 3.7 to obtain a representation result by general preferential logic. The cumulative version proved much more interesting. A counterexample (Section 3.7.2.2) shows that the lack of closure of the domain under finite union can be a serious problem for the smooth case. This was the first time we saw the real importance of such closure conditions of the domain. We now present the main building block of a solution also for the smooth case, but with more complicated conditions (Section 3.7.2.3).

We turn to distance based revision. Again, we give first an algebraic characterization, using an arbitrarily big, but finite, loop condition (Section 4.2.2, in particular Proposition 4.2.2). An at first sight innocent example (Example 4.2.1) about lack of information (closer elements hide more distant ones) turns out to give the crucial idea later on (Section 4.2.4) to show that there is no finite characterization possible. To this purpose, we construct arbitrarily big positive and negative examples, which differ only by one bit of information we can obtain.

For distance based revision, too, we discuss the limit variant or version (Section 4.2.5), getting rid of the necessity of closest elements, and, in particular, show that both versions are equivalent in an important class of problems (Section 4.2.5.3).

We then drop the definability preservation condition, by admitting "small" sets of exceptions to our conditions (Chapter 5). The ideas and techniques are the same for preferential structures and distance based revision. The reader should keep in mind that we see a problem similar to definability preservation in the last Section 8.3.3, where we will combine several semantical structures. There again, we have to be careful, when, on a finer level, elements are missing, which are not visible on a coarser level. In the same chapter, Section 5.2.3, we show that such general preferential structures cannot be characterized in the usual way. The same applies to general ranked structures and to distance based theory revision — see again Section 5.2.3.

The fact that closer elements hide elements farther away is used in a positive way for counterfactuals (Section 4.3). We can construct in essentially independent ways the neighborhoods of all worlds, arranging the recurrent distances in a way that the unwanted neighborhoods are hidden by closer ones. This technique might prove useful in other situations, too.

Representation problems for situations defined by minimal sums have a uniform solution via a suitable algorithm, due to Farkas, see Section 6.2, determining whether systems of inequalities have a solution. It depends on the richness of the domain whether we can let the algorithm work directly on the domain and its information, or whether it has to turn outside as a black box. This is essentially the same problem as the one which causes absence of finite representation, and, as a matter of fact, we show that all cases considered will have no finite representation (Sections 6.3.4, 6.4.3, 6.5.1).

The, in the author's opinion, most important recurrent problems we have treated or begun to treat are perhaps:

- domain closure problems and their relation to possible characterization, e.g. the existence or absence of finite characterizations,
- the importance of definability preservation, and approximation as a solution (see Chapter 5), and analogous problems (see Section 8.3.3),
- size of exception sets and general incompleteness results (see Chapter 5).

These subjects should be treated in a more systematic way in future research. We have only scratched the surface here.

Despite all these shortcomings and problems left open, there are some methods the author thinks are useful, and merit attention. Among these are:

- to split representation into an algebraic and a logics part,

- to use choice functions for preferential structures, and trees to encode transitivity, to use suitable hulls $H(U)$ — sets to be to be avoided — in the construction of smooth structures (e.g. Sections 3.2.1, 3.2.2, 3.3.1, 3.3.2),
- to use topological constructions to obtain positive and negative results for preferential (and similar) structures (e.g. Sections 3.10.2, 3.10.3),
- to use approximations to solve definability preservation problems (Chapter 5),
- to reduce the more complicated limit variant or version to the much more simple minimal version (Sections 3.4.1, 3.10.3, 4.2.5),
- to use incompleteness of information about behavior to show lack of certain representations (Sections 4.2.4, 6.3.4, 6.4.3, 6.5.1),
- to deduce from the arbitrarily big size of exception sets in definability problems general incompleteness results (Section 5.2.3),
- to construct negative examples sufficiently close to positive ones by making them well-behaved up to a certain cardinality, but not beyond (Section 5.2.3).

1.5 Specific remarks on propositional logic

A natural distance between (finite) propositional models

In many cases we will do as if the set of models were an arbitrary set and work with additional information or structure, like a relation, a distance, etc. But sometimes it is very useful to consider the natural structure of the set of models — especially if this is all we have — and, thus (partially) answer the question where such additional information comes from. At the same time, this will give us a concrete example to refer to.

We will concentrate here on the finite propositional case, but some transfer is possible, e.g. to the first order case: we can work in a fixed countable universe, giving each element a name to fix things, and introduce a new predicate to describe the limits of the universe if it is finite.

First, there is a natural distance between models, the Hamming distance defined by: $d(m, m') :=$ the number of propositional variables for which they differ.

E.g., if the language consists of p, q, r and $m \models p \wedge q \wedge \neg r$, $m' \models p \wedge \neg q \wedge r$, then $d(m, m') = 2$. In the first order case, we would count for all elements all the differences (predicates, values of functions, etc.).

Second, each formula has a natural decomposition into a disjunction of conjunctions (which need not be unique, consider the example $p \wedge q$, $p \wedge \neg q$, $\neg p \wedge q$, but we can take the decompositions into a minimal number of conjunctions).

Decomposition and Hamming distance cooperate:

Define a set X convex wrt. to distance d, iff for all $x, x' \in X$ and all x'' $d(x, x'') + d(x'', x') = d(x, x')$ implies $x'' \in X$. (If we have no addition, we can define "between x and y" by $d(z, x) \leq d(x, y) \wedge d(z, y) \leq d(x, y)$.)

If X is a model set described by a pure conjunction, then it is convex wrt. the Hamming distance: Let A be the set of propositional variables not determined in X, then X is everything between the model deciding positively on A (and outside as X dictates) and the model deciding negatively on A.

Conversely, each set X convex wrt. the Hamming distance can be described by a pure conjunction: Let A be the set of propositional variables s.t. for $a \in A$ there are $x, x' \in X$ and x decides a positively, x' negatively. Suppose there is $A' \subset A$ maximal s.t. there is $x \in X$ with x decides all $a \in A'$ negatively. If $a' \in A - A'$, then there is $x' \in X$ which decides a' negatively. Now, x'', which is like x, but decides a' negatively, is between x and x', so in X, contradiction. In the same way, there is y deciding all $a \in A$ positively, and X is the convex set between x and y. So X can be described by the conjunction of all $y \notin A$, or their negation, on which all $x \in X$ agree.

All kinds of modifications of the Hamming distance are possible, we can give different weights to different variables, work with formulas instead of variables, consider only subsets of all variables, etc. The reader is referred to [Sch95-2] for a more detailed discussion of various distances.

Natural model sets

Given any distance, e.g. above Hamming distance, convex model sets are particularly simple and natural ones.

Given a finite theory, or a formula, if we write it as a disjunction of conjunctions, these conjunctions define a granularity. If $\phi = (p \wedge q) \vee (p \wedge s)$, and the language has also the variable t, then we do not see down to t, as t and $\neg t$ will always be treated together. Thus, ϕ is coarser than, e.g. $p \wedge q \wedge s \wedge t$. We can either say that the granularity is $p \wedge q \wedge s$, or $p \wedge q$ and $p \wedge s$ — the case at hand will tell us which is the better solution. It can now be reasonable (e.g. in theory revision) not to go below the granularity given by the theory, e.g. either we choose all of $p \wedge q \wedge s$ to be part of the result, or not, but will not put $p \wedge q \wedge s \wedge t$ in, and $p \wedge q \wedge s \wedge \neg t$ out.

Note that some of the ideas of R. Parikh and his co-authors are in the same spirit — see, e.g. Chopra/Parikh [CP00].

Propositional formulas as approximations

A propositional formula can be seen as a conjunction of disjunctions, and thus as an approximation from above (to its set of models), or as a disjunction of conjunctions, and thus as an approximation from below.

In particular, if, e.g. $\phi = \phi' \wedge \phi''$, and $m \not\models \phi$, we may nonetheless say that ϕ holds "almost" in m, if it holds in many of the disjunctions making up the formula, here in at least one of ϕ' or ϕ''. Conversely, if $\phi = \phi' \vee \phi''$, and $m \models \phi$, we may say that ϕ strongly holds in m iff many of the conjunctions making up ϕ have m as a model, here, iff $m \models \phi' \wedge \phi''$.

Thus, in some cases, we have already a natural notion of graded validity of a formula in a model.

1.6 Basic definitions

1.6.1 The algebraic part

The reader will find in this Section 1.6.1 some basic algebraic definitions, in particular those of (weak) filters and (weak) ideals. They are collected together, so it will be easier to find them when necessary.

We make ample and tacit use of the Axiom of Choice.

Definition 1.6.1

We use $\mathcal{P}$ to denote the power set operator, $\Pi\{X_i : i \in I\} := \{g : g : I \rightarrow \bigcup\{X_i : i \in I\}, \forall i \in I.g(i) \in X_i\}$ is the general cartesian product, $card(X)$ shall denote the cardinality of X, and V the set-theoretic universe we work in — the class of all sets. Given a set of pairs $\mathcal{X}$, and a set X, we denote by $\mathcal{X}\lceil X := \{< x, i > \in \mathcal{X} : x \in X\}$.

$A \subseteq B$ will denote that A is a subset of B or equal to B, and $A \subset B$ that A is a proper subset of B, likewise for $A \supseteq B$ and $A \supset B$.

Given some fixed set U we work in, and $X \subseteq U$, then $\boldsymbol{C}(X) := U-$X.

$\prec^*$ will denote the transitive closure of the relation $\prec$. If a relation $<$, $\prec$, or similar is given, $a \perp b$ will express that a and b are $<$ – (or $\prec$ –) incomparable — context will tell.

A child (or successor) of an element x in a tree t will be a direct child in t. A child of a child, etc. will be called an indirect child. Trees will be supposed to grow downwards, so the root is the top element.

A subsequence $\sigma_i : i \in I \subseteq \mu$ of a sequence $\sigma_i : i \in \mu$ is called cofinal, iff for all $i \in \mu$ there is $i' \in I$ $i \leq i'$.

Given two sequences σ_i and τ_i of the same length, then their Hamming distance is the quantity of i where they differ.

We recall or introduce the definitions of a filter, weak filter, ideal, weak ideal. Intuitively, a filter (dually an ideal), describes the "big" ("small") subsets of a set X. What is in the filter, is big, just as in your coffee machine, what is in the ideal, is small. In both definitions, the first two conditions should hold if the notions should have anything to do with usual intuition, and there are reasons to consider only the weaker, less idealistic, version of the third.

Definition 1.6.2

Fix a base set X.

A (weak) filter on or over X is a set $\mathcal{F} \subseteq \mathcal{P}(X)$, s.t. (F1)–(F3) ((F1), (F2), (F3') respectively) hold:

(F1) $X \in \mathcal{F}$

(F2) $A \subseteq B \subseteq X$, $A \in \mathcal{F}$ imply $B \in \mathcal{F}$

(F3) $A, B \in \mathcal{F}$ imply $A \cap B \in \mathcal{F}$

(F3') $A, B \in \mathcal{F}$ imply $A \cap B \neq \emptyset$.

So a weak filter satisfies (F3') instead of (F3).

A filter is called a principal filter iff there is $X' \subseteq X$ s.t. $\mathcal{F} = \{A : X' \subseteq A \subseteq X\}$.

An (weak) ideal on or over X is a set $\mathcal{I} \subseteq \mathcal{P}(X)$, s.t. (I1)–(I3) ((I1), (I2), (I3') respectively) hold:

(I1) $\emptyset \in \mathcal{I}$

(I2) $A \subseteq B \subseteq X$, $B \in \mathcal{I}$ imply $A \in \mathcal{I}$

(I3) $A, B \in \mathcal{I}$ imply $A \cup B \in \mathcal{I}$

(I3') $A, B \in \mathcal{I}$ imply $A \cup B \neq X$.

So a weak ideal satisfies (I3') instead of (I3).

A filter is an abstract notion of size, elements of a filter on X are called big subsets of X, their complements are called small, and the rest have medium size. The dual applies to ideals, this is justified by the following trivial fact:

Fact 1.6.1

If $\mathcal{F}$ is a (weak) filter on X, then $\mathcal{I} := \{X - A : A \in \mathcal{F}\}$ is a (weak) ideal on X, if $\mathcal{I}$ is a (weak) ideal on X, then $\mathcal{F} := \{X - A : A \in \mathcal{F}\}$ is a (weak) filter on X.

Definition 1.6.3

(1) We use the usual interval notation for subsets of the reals: $(a, b) := \{x \in \Re : a < x < b\}$, $(a, \infty) := \{x \in \Re : a < x\}$, $[a, b) := (a, b) \cup \{a\}$, etc.

(2) For two functions $f : X \to Y$, $g : Y \to Z$, let $g \circ f : X \to Z$ be defined by $(g \circ f)(x) := g(f(x))$. For $f : X \to Y$, $A \subseteq X$, let $f[A] := \{f(x) : x \in A\}$, and $ran(f) := range(f) := f[X]$.

1.6.2 The logical part

We collect here, just as for the algebraic side in Section 1.6.1, basic logical definitions and properties, which will be used many times throughout the text. In particular, Definition 1.6.5 will contain many of the logical and corresponding algebraic conditions which we discuss in various contexts. The reader will probably use it continually for reference.

Unless said otherwise, we always work in propositional logic. A notable exception is Section 7.2.

Notation 1.6.1

We use sometimes FOL as abbreviation for first order logic, and NML for nonmonotonic logic. To avoid Latex complications in bigger expressions, we replace $\widetilde{xxxxx}$ by $\overbrace{xxxxx}$.

Definition 1.6.4

If $\mathcal{L}$ is a propositional language, $v(\mathcal{L})$ will be the set of its variables, $M_{\mathcal{L}}$ the set of its classical models, ϕ, etc. shall denote formulas, T, etc. theories in $\mathcal{L}$, and $M(T)$ or $M_T \subseteq M_{\mathcal{L}}$ the models of T, likewise for ϕ.

(A theory will just be an arbitrary set of formulas, without any closure conditions.)

For any classical model m, let $Th(m)$ be the set of formulas valid in m, likewise $Th(M) := \{\phi : m \models \phi$ for all $m \in M\}$, if M is a set of classical models. $\models$ is the sign of classical validity. For two theories T and T', let $T \vee T' := \{\phi \vee \psi : \phi \in T, \psi \in T'\}$. $\bot$ stands for falsity, and $\boldsymbol{T}$ for truth.

$\overline{T} \subseteq \mathcal{L}$ will denote the closure of T under classical logic, and $\vdash$ the classical consequence relation, thus $\overline{T} := \{\phi : T \vdash \phi\}$. Given some other logic $\mid\!\sim$, $\overline{\overline{T}}$ will denote the set of consequences of T under that logic, i.e. $\overline{\overline{T}} := \{\phi : T \mid\!\sim \phi\}$.

$Con(T)$ will say that T is classically consistent, likewise $Con(\phi)$, etc.

Note that the double bar notation does not really conflict with the single bar notation: closing twice under classical logic makes no sense from a pragmatic point of view, as the classical consequence operator is idempotent.

$\boldsymbol{D}_{\mathcal{L}} \subseteq \mathcal{P}(M_{\mathcal{L}})$ shall be the set of definable subsets of $M_{\mathcal{L}}$, i.e. $A \in \boldsymbol{D}_{\mathcal{L}}$ iff there is some $T \subseteq \mathcal{L}$ s.t. $A = M_T$. If the context is clear, we omit the subscript $\mathcal{L}$ from $\boldsymbol{D}_{\mathcal{L}}$.

For $X \subseteq \mathcal{P}(M_{\mathcal{L}})$, a function $\mu : X \to \mathcal{P}(M_{\mathcal{L}})$ will be called definability preserving, iff $\mu(Y) \in \boldsymbol{D}_{\mathcal{L}}$ for all $Y \in \boldsymbol{D}_{\mathcal{L}} \cap X$. If $\boldsymbol{D}_{\mathcal{L}} \subseteq X$, then $\mu : X \to \mathcal{P}(M_{\mathcal{L}})$ defines a logic $T \mapsto T^{\mu}$, $T^{\mu} := \mu(T)$ on $\mathcal{L}$ by $T^{\mu} := \{\phi: \forall m \in \mu(M_T).m \models \phi\} = Th(\mu(M_T))$.

Note that $\mu(M_T) \subseteq M_{T^{\mu}}$ always holds, but not necessarily $\mu(M_T) = M_{T^{\mu}}$, the latter only iff f is definability preserving, as $M_{T^{\mu}} = M(Th(\mu(M_T)))$, and $X \subseteq M(Th(X))$ will always hold, but not always the converse, as we see in the following Fact.

We recall the following basic facts about definable sets. The reader should be familiar with such properties.

Fact 1.6.2

1. $\emptyset$, $M_{\mathcal{L}} \in \boldsymbol{D}_{\mathcal{L}}$.

2. $\boldsymbol{D}_{\mathcal{L}}$ contains all singletons, is closed under arbitrary intersections and finite unions.

3. If $v(\mathcal{L})$ is infinite, and m any model for $\mathcal{L}$, then $M := M_{\mathcal{L}} - \{m\}$ is not definable by any theory T. (Proof: Suppose it were, and let ϕ hold in M', but not in m, so in m $\neg\phi$ holds, but as ϕ is finite, there is a model m' in M' which coincides on all propositional variables of ϕ with m, so in m' $\neg\phi$ holds, too, a contradiction.)

4. If $v(\mathcal{L})$ is infinite, then $\boldsymbol{D}_{\mathcal{L}} \neq \mathcal{P}(M_{\mathcal{L}})$.

□

We recollect and note:

Fact 1.6.3

Let $\mathcal{L}$ be a fixed propositional language, $\boldsymbol{D}_{\mathcal{L}} \subseteq X$, $\mu : X \to \mathcal{P}(M_{\mathcal{L}})$, for a $\mathcal{L}-$theory T $\overline{\overline{T}} := Th(\mu(M_T))$, let T, T' be arbitrary theories, then:

(1) $\mu(M_T) \subseteq M_{\overline{\overline{T}}}$,

(2) $M_T \cup M_{T'} = M_{T \vee T'}$ and $M_{T \cup T'} = M_T \cap M_{T'}$,

(3) $\mu(M_T) = \emptyset \leftrightarrow \bot \in \overline{\overline{T}}$.

If μ is definability preserving or $\mu(M_T)$ is finite, then the following also hold:

(4) $\mu(M_T) = M_{\overline{\overline{T}}}$,

(5) $T' \vdash \overline{\overline{T}} \leftrightarrow M_{T'} \subseteq \mu(M_T)$,

(6) $\mu(M_T) = M_{T'} \leftrightarrow \overline{T'} = \overline{\overline{T}}$. □

We add the following example, useful for some cases:

Example 1.6.1

(A theory with countably many models)

We give here the example of a theory in a language with countably many propositional variables, which has countably many models.

We code a tree of height ω, with ω many cofinal branches, where we "open" a new branch at every level. At level 0, we fix p_0. At level 1, we branch for p_1, i.e. we permit p_1 and $\neg p_1$. At level 2, we will continue just one branch in two ways. Thus, we continue the branch $< p_0, p_1 >$ just by p_2, but we continue the branch $< p_0, \neg p_1 >$ both ways, i.e. by p_2 and by $\neg p_2$. At level 3, we have already three branches $< p_0, p_1, p_2 >$, $< p_0, \neg p_1, p_2 >$, $< p_0, \neg p_1, \neg p_2 >$, which we continue as follows: $< p_0, p_1, p_2, p_3 >$, $< p_0, \neg p_1, p_2, p_3 >$, $< p_0, \neg p_1, \neg p_2, p_3 >$, $< p_0, \neg p_1, \neg p_2, \neg p_3 >$, branching just for $\neg p_2$. Coding this into logic, we have $\phi_0 := p_0$, $\phi_1 := p_1 \vee \neg p_1$, $\phi_2 := p_2 \vee \neg p_1$, $\phi_3 := p_3 \vee (\neg p_1 \wedge \neg p_2)$, etc.

All models of T will make p_0 true. Starting at p_1, once they make any p_i, $1 \leq i$ true, they will have to make all p_j, $i < j$ true. Making all p_i false is also a model. So T has the models $\{m_i : 0 < i < \omega\} \cup \{m_\omega\}$, where

$m_i \models p_0$, and for $1 \leq j < \omega$ $m_i \models \neg p_j$ iff $1 \leq j \leq i$, and $m_\omega \models \neg p_i$ for all $i \in \omega$. □

We also note en passant:

Fact 1.6.4

Let T, T' be s.t. $M(T) \cup M(T) = M_{\mathcal{L}}$, and $M(T) \cap M(T') = \emptyset$. Then T and T' are equivalent to single formulas ϕ and ϕ'.

Proof:

By $M(T) \cap M(T') = \emptyset$, $\neg Con(T, T')$, so there are finite subsets $T_0 \subseteq T$, $T_0' \subseteq T'$ s.t. $\neg Con(T_0, T_0')$. But then $M(T) \subseteq M(T_0)$, $M(T_0) \cap M(T') \subseteq M(T_0) \cap M(T_0') = \emptyset$, so $M(T) = M(T_0)$, likewise $M(T') = M(T_0')$. Take now $\phi := \bigwedge T_0$, etc. □

We collect now for easier reference some conditions on logics and choice functions we will often see in the course of development. We will, however, repeat the conditions in the main results, to make reading easier. The main definitions for preferential structures are in Section 2.3.1, and for theory revision in Section 2.2.10.2.

We show, wherever adequate, in parallel the versions for a formula on the left of $\mid\!\sim$ in the left column, for a full theory on the left of $\mid\!\sim$ in the middle column, and the semantical or algebraic counterpart in the right column. The algebraic counterpart gives conditions for a function $f : \mathcal{Y} \rightarrow \mathcal{P}(U)$, where U is some set, and $\mathcal{Y} \subseteq \mathcal{P}(U)$.

Definition 1.6.5

(AND) $\phi \mid\!\sim \psi, \phi \mid\!\sim \psi' \Rightarrow$ $\phi \mid\!\sim \psi \wedge \psi'$	(AND) $T \mid\!\sim \psi, T \mid\!\sim \psi' \Rightarrow$ $T \mid\!\sim \psi \wedge \psi'$	
(OR) $\phi \mid\!\sim \psi, \phi' \mid\!\sim \psi \Rightarrow$ $\phi \vee \phi' \mid\!\sim \psi$	(OR) $T \mid\!\sim \psi, T' \mid\!\sim \psi \Rightarrow$ $T \vee T' \mid\!\sim \psi$	$(\mu \cup w)$ — w for weak $f(A \cup B) \subseteq f(A) \cup f(B)$
(LLE) or Left Logical Equivalence $\vdash \phi \leftrightarrow \phi', \phi \mid\!\sim \psi \Rightarrow$ $\phi' \mid\!\sim \psi$	(LLE) $\overline{T} = \overline{T'} \Rightarrow \overline{\overline{T}} = \overline{\overline{T'}}$	
(RW) or Right Weakening $\phi \mid\!\sim \psi, \vdash \psi \rightarrow \psi' \Rightarrow$ $\phi \mid\!\sim \psi'$	(RW) $T \mid\!\sim \psi, \vdash \psi \rightarrow \psi' \Rightarrow$ $T \mid\!\sim \psi'$	
	(CCL) or Classical Closure $\overline{\overline{T}}$ is classically closed	
(SC) or Supraclassicality $\phi \vdash \psi \Rightarrow \phi \mid\!\sim \psi$	(SC) $\overline{T} \subseteq \overline{\overline{T}}$	$(\mu \subseteq)$ $f(X) \subseteq X$
(CP) or Consistency Preservation $\phi \mid\!\sim \bot \Rightarrow \phi \vdash \bot$	(CP) $T \mid\!\sim \bot \Rightarrow T \vdash \bot$	$(\mu\emptyset)$ $f(X) = \emptyset \Rightarrow X = \emptyset$
(RM) or Rational Monotony $\phi \mid\!\sim \psi, \phi \not\mid\!\sim \neg\psi' \Rightarrow$ $\phi \wedge \psi' \mid\!\sim \psi$	(RM) $T \mid\!\sim \psi, T \not\mid\!\sim \neg\psi' \Rightarrow$ $T \cup \{\psi'\} \mid\!\sim \psi$	$(\mu =)$ $X \subseteq Y, Y \cap f(X) \neq \emptyset \Rightarrow$ $f(X) = f(Y) \cap X$
(CM) or Cautious Monotony $\phi \mid\!\sim \psi, \phi \mid\!\sim \psi' \Rightarrow$ $\phi \wedge \psi \mid\!\sim \psi'$	(CM) $T \subseteq \overline{T'} \subseteq \overline{\overline{T}} \Rightarrow$ $\overline{\overline{T}} \subseteq \overline{\overline{T'}}$	(μCM) $f(X) \subseteq Y \subseteq X \Rightarrow$ $f(Y) \subseteq f(X)$
(CUM) or Cumulativity $\phi \mid\!\sim \psi \Rightarrow$ $(\phi \mid\!\sim \psi' \Leftrightarrow \phi \wedge \psi \mid\!\sim \psi')$	(CUM) $T \subseteq \overline{T'} \subseteq \overline{\overline{T}} \Rightarrow$ $\overline{\overline{T}} = \overline{\overline{T'}}$	(μCUM) $f(X) \subseteq Y \subseteq X \Rightarrow$ $f(Y) = f(X)$
(PR) $\overline{\overline{\phi \wedge \phi'}} \subseteq \overline{\overline{\overline{\phi}} \cup \{\phi'\}}$	(PR) $\overline{\overline{T \cup T'}} \subseteq \overline{\overline{\overline{T}} \cup T'}$	(μPR) $X \subseteq Y \Rightarrow$ $f(Y) \cap X \subseteq f(X)$

(PR) is also called infinite conditionalization — we choose the name for its central role in preferential structures. Note that in the presence of $(\mu \subseteq)$, and if $\mathcal{Y}$ is closed under finite intersections, (μPR) is equivalent to

$(\mu PR')$ $f(X) \cap Y \subseteq f(X \cap Y)$.

The system of rules (AND), (OR), (LLE), (RW), (SC), (CP), (CM), (CUM) is also called system P (for preferential), adding (RM) gives the system R

(for rationality or rankedness).

We will see that limit preferential structures separate the finitary from the infinitary versions of (CUM) and (PR), see Section 3.4.1.

1.6.2.1 General incompleteness results

We will show in this book a number of negative representation results, i.e. that (finite or even infinite) representation is not possible. It is important to make the problem precise, and, for this purpose, we have to define possible candidates for a representation. Of course, we will choose them as generalizations of the descriptions we normally use, and as they are collected in above Definition 1.6.5. This might therefore be the adequate place to discuss the general framework of negative representation results.

The logical rules and conditions of above table are all composed of $\overline{T}$, $\overline{\overline{T}}$, $\overline{\phi}$, $\overline{\overline{\phi}}$, with perhaps some constants like $\bot$, $\emptyset$, $\boldsymbol{T}$, etc., where the T's and ϕ's, etc. can be combined by the usual logical operators, sets of formulas can be combined with the usual set operators like $\cup$, $\cap$, set difference, etc., expressions are formed with $\subseteq$ and $=$, etc., and we can combine expressions again with boolean connectives like $\overline{T} = \overline{T'} \Rightarrow \overline{\overline{T}} = \overline{\overline{T'}}$, etc.

These expressions are implicitly universally quantified.

In different contexts, the expressions will be formed somewhat differently, e.g. for theory revision, we will have binary operators on formulas or model sets, for "between" we will work with triples like $< a, b, c >$, expressing that b is between a and c, but the principle stays the same.

In particular, we will not allow to introduce new, arbitrary predicates, as this would allow cheating: we could then introduce a new predicate, which expresses exactly the property we examine, and if it holds, the system is representable by the type of structure we look at, if not, this is not the case. Such "dirty tricks" are certainly not intended.

We leave, however, sometimes open the possibility to form infinite expressions up to a certain length which we make precise beforehand, going beyond the expressive power of standard logic.

To summarize: the conditions we admit have the form

$\forall x_1, \ldots, \forall x_\kappa(\phi(x_1, \ldots, x_\kappa))$,

where the x_i can range over formulas, theories, elements, sets, or whatever the context allows, and ϕ is quantifier free, but perhaps infinite, though bounded by some fixed infinite cardinal κ.

We have now two classes of situations, $\mathcal{M}$ and $\mathcal{M}'$, with $\mathcal{M} \subseteq \mathcal{M}'$, where $\mathcal{M}$ is the subclass of those situations which can be generated by some structure. For instance, $\mathcal{M}'$ is the class of all consequence relations, and $\mathcal{M}$ the class of those consequence relations which can be generated by a preferential structure.

A characterization is now a formula ϕ (or a set of formulas Φ) in the meta-language used to formulate the logical properties, which separates $\mathcal{M}$ from $\mathcal{M}' - \mathcal{M}$, i.e. ϕ holds in all $M \in \mathcal{M}$, but fails in all $M' \in \mathcal{M}' - \mathcal{M}$ (or, respectively, all $\phi \in \Phi$ hold in all $M \in \mathcal{M}$, but for all $M' \in \mathcal{M}' - \mathcal{M}$, there is at least one $\phi \in \Phi$ which fails in M'). In the cases we consider, we can always assume that we have one single formula, either we look at finite characterizations, or we admit infinitary conjunctions.

As ϕ is universally quantified, any possible instance of ϕ has to hold in all $M \in \mathcal{M}$, but at least one instance has to fail in any $M' \in \mathcal{M}' - \mathcal{M}$.

We call such characterizations normal ones.

We will show in our negative results that we just cannot use enough information to separate the positive cases (in $\mathcal{M}$) from the negative ones (in $\mathcal{M}' - \mathcal{M}$). Take for instance finite characterization. If there is one, it has some fixed length, say n. So, it cannot speak about more than n instantiations, e.g. only about $T_1, \ldots, T_n$. We consider now a negative case M', where we need $n+1$ bits ("bit" in the nontechnical sense) of information to see that it is really a negative one. More precisely, we show that for all sets S of n bits of information, there is a positive case $M_S \in \mathcal{M}$, which agrees exactly with M' on S. If there is a characterization of length n, which holds in all instances in all positive cases M, but fails in all negative cases M' in at least one instance, we have a contradiction: For one set S of size n, it has to fail in M', but it has to hold in M_S, yet all elements from S (and thus all expressions using only elements from S) evaluate to the same truth value in M_S and M'.

A simple example may help to illustrate the problem and the solution. Suppose we want to discern centipedes from millipeds. Both species have a head and a tail segment, and in between 50 or 500 leg segments. If we have information of size at most 50, we cannot tell the difference. We see, e.g. a head segment, a tail segment, and 48 leg segments. This can be a centipede or a milliped. If we have information of size 51, we can tell, because we might see 51 leg segments, and this cannot be a centipede. But, suppose that there is for every 10 leg segments one control center to coordinate leg movements, then information of size 6 will suffice to distinguish, as seeing six control centers allows to conclude for a milliped. This is exactly what may or may not happen with the transitivity problems described below:

transitivity allows to bundle much information into one piece of information (a control center in above picture), but we have to be able to see it, and this is not always possible if the domain is not sufficiently rich. Of course, we have to prevent cheating. It will not be allowed to glue a label with "m" on the back of millipeds, and a "c" on the back of centipedes, allowing information of size one to distinguish — we will not be permitted to add arbitrary new predicates to the language.

In more abstract terms, to prove our results, we will construct logics which are locally (up to size 50 in above picture, finitely or up to size κ in our results) compatible with logics generated by a structure under examination, but not globally. So we have to take the structures, and model them closely on the local level with a new logic, which we will make globally incompatible with the structures. "Globally" may mean in the infinite case, or for a bigger infinite cardinal.

We consider briefly some examples.

(1) In Section 4.2.4, we show that there is no finite characterization of distance definable revision. The set of all revision relations (sets of various $\phi * \psi \mid\sim \sigma$, etc.) is in above notation $\mathcal{M}'$, the subset of those which are distance definable is $\mathcal{M}$. We construct "hamster wheels" of arbitrary size, which contain a loop in the revision results, and are thus not distance representable. Yet any true subset of information has its counterpart in a distance definable revision structure, where it has the same truth value as in the negative situation. (The case is a little more complicated, as we use sets of models, but we see easily that only sets of cardinality 2 are interesting.)

We consider here also possible complications caused by nestedness of the revision operator, of the form $(\phi * \psi) * \sigma$, etc. Since we have to stay finite, this might, in principle, cause problems. But, as we will see, the picture is the same as in the flat case.

(2) In Section 6.5.1, we show that there is no finite characterization of distance definable "between" and "behind". We use essentially the same technique. We show that there are arbitrarily big situations, tupels of $< a, b, c >$, meaning that b is between a and c, which are not distance definable, but any true subset has the same truth value in a distance definable structure.

(3) In Section 5.2.3, we show (among analogous results for ranked structures and for distance defined revision) that there is no characterization at all (i.e. not even an infinite one of fixed maximal cardinality) of consequence relations generated by arbitrary, not necessarily definability preserving, preferential structures. As the general limit variant or version gives in some special cases (which we consider here) the same results, we show at the same

time that the general limit variant has no characterization either.

Given any cardinal κ, we construct a consequence relation which is not generated by a preferential structure ("but almost"). By prerequisite, there has to be a formula of size $\leq \kappa$, one of whose instances fails in this structure. We can show that there were only $\leq \kappa$ model pairs involved in determining the values of all $\leq \kappa$ $\overline{\overline{T}}$'s, etc. (see Section 5.2.3 for details), and that there is a true preferential structure (constructed from just those $\leq \kappa$ model pairs), which produces the same values for all those $\overline{\overline{T}}$'s, so the formula fails in the positive case, too, contradicting the assumption that it holds for all instances in all preferential structures. As a matter of fact, we show more: We not only show that exactly those $\leq \kappa$ T's give the same results, but that all T's, which use only the $\leq \kappa$ pairs to define their consequences give the same results.

Of course, nestedness could appear here, too, e.g. as something like $\overline{\overline{\overline{\overline{T}} \cup T'}}$, but this will also be up to a certain depth only, and as we work with arbitrarily big cardinals, we have plenty of room to move, and this will not cause any problems.

This negative result might seem in contradiction with the representation result in Sections 5.2.1 and 5.2.2, but a closer inspection reveals that we use there arbitrary unions, which are not bounded by any cardinality.

Thus, our negative results concern the finite and the arbitrary case. The finite case results from the fact that transitivity concerns arbitrarily long, but finite chains. The arbitrary case results from the fact that logically small sets may have arbitrary size — depending on the language. We do not know whether there are interesting and natural cases in between. If normality is defined via, e.g. co-countable sets, in a natural way, this might perhaps be a candidate to consider. We can also use the minimal cardinal, if this exists, of a characterization as a measure of distance of a semantics from logic. The first case above will then have distance 1, and the second one infinite distance, i.e. not be bound by any cardinal. As we will see in the discussions, such values are not absolute, as sufficient domain closure conditions can reduce this measure of distance.

Chapter 2

Concepts

2.1 Introduction

In this chapter, we will first consider a list of different types of common-sense reasoning. We will discuss them, and base them loosely on what we consider basic semantical concepts, like distance, size, etc.

We think that these concepts are present in human thinking, in some form or the other. But we are also prepared to modify and generalize them, as it suits our purposes. Thus, what we call a distance will not necessarily be the same concept as the distance from Bachenpfuhl to Eberswutz, or from Marseille to Pernes-les-Fontaines. We still feel that our notions of distance have sufficient in common with usual distance to merit their name. Our approach is then a bootstrap procedure. We begin with a vague and naive concept, and generalize or refine it as it seems necessary. We are not only guided by the application to logic — e.g. the syntactic property of cumulativity leads naturally to the semantic notion of a smooth relation — but also by abstract considerations like generalization, etc.

We do not pretend to have given all possible or reasonable interpretations of common-sense reasoning types in an exhaustive study, and are conscious of the essay character of these pages. In a way, it is a stroll through a lush rain forest of logics and problems, with a lineean project in the back of our mind.

A moment's reflection will often show that there are many alternative interpretations of our notions possible. For instance, when we consider the notion of center, and base it on distance, we can define "center" by the distance to other elements of the set considered, or to those outside the set — see Section 2.2.1.1. It seems impossible to exhaust all possibilities, so we often just indicate alternatives, to give the reader some suggestions what one can do. What should be done depends on the concrete case at hand, and cannot be decided beforehand. We strongly insist here that this multitude of possibilities is not due to an incapacity of the author to decide one way or the other, but is imposed by the possibilities of reasoning and determined by the adequacy to the problem. The multitude of approaches comes from the multitude of situations one might wish to treat. The systematisation is in the fact that we often find the same basic notions, but not in the way they are used or composed. Yet, finding over and over the same basic concepts is already very useful for a systematic and formal treatment of several cases. So, the reader should not feel confused by the multitude of possibilities, but rather see the basic concepts behind, and, use whatever seems adequate to him for his problem. We will pursue some approaches in great detail, others not as far, or not at all, but hope that the reader will find in the techniques explained in detail also ideas how to treat the other approaches in more detail. So, this part of the book can be read on different levels. Either as an introduction to what could be done, and how we can analyse some types of common-sense reasoning, or, as headlines whose chapters have been written to various depths, and incitations to the (advanced) reader to fill in the missing details for the other chapters.

We then describe the basic concepts we saw in more detail in Section 2.3, in particular, we show some natural transformations from one to the other. E.g., we will discuss how a size (with addition) for elements can be transformed in a natural way to a ranking of sets. We will re-emphasize the ubiquity of our basic semantical notions of human reasoning by describing the situation we saw in the first part from above, now as seen from those basic concepts. The fact that there are some (also formal) connections between the notions should not obscure the fact that basic intuitions seem to diverge sometimes.

We then turn to another basic concept in reasoning, coherence. Abstractly, this is the "transfer" of conclusions from situation T to situation T'. For instance, in classical, monotonic, logic, we have that $T \vdash \phi$ implies $T' \vdash \phi$, if $T \subseteq T'$. Such transfer operations allow us to do relevant thinking, we can transfer conclusions from one situation to another one, and need not start anew every time. Such transfer is also possible in nonmonotonic logic, e.g. when the logic is cumulative. We describe now the situations we have already seen from the perspective of coherence. For instance, we look at the coherence conditions imposed by a preference relation on the models. We will mostly investigate here the algebraic side of the picture, e.g. the properties imposed on the model choice functions by a preference relation, or a distance.

This perspective gives us a unifying view, but we do not go as far as to formulate a theory of possible or reasonable coherence properties. (We would probably enter there the domain of analogical reasoning in more detail than we wish to do. For instance, it might be a reasonable assumption that, if we transfer ϕ from T to T', and T'' is between T and T' (by some measure), then we should also be able to transfer ϕ from T to T''.) Moreover, as coherence is (usually more than) half of the logic, in particular, coherence properties are often the key to representation results, this perspective gives us a good starting point for the more technical Chapters 3–7 to come.

It is somewhat difficult to begin our analysis. We will see that most notions we look at reveal themselves as interdependent, so it is somewhat arbitrary where we enter the mesh-work. Nonmonotonic reasoning can sometimes be seen as choosing the "best" elements (and reasoning with them), certainty is sometimes based on the same choice, counterfactuals (in the Lewis/Stalnaker interpretation) work with closest (or best, seen from our point of view) elements, likewise theory revision (in a slightly modified approach), etc. We will just begin somewhere.

We conclude with a remark on terminology: D. Makinson, in [Mak03] has made the difference between completeness and representation results, we just merrily use both in naive terminology, and go even farther: we sometime also speak about characterization. But in all cases, it will be clear what we mean (at least we hope so).

2.2 Reasoning types

Common to all formalisms we consider is that they (usually) work with model sets other than those of classical logic. Instead of considering $M(T)$, the set of classical models of T, we work with some other set of models, say M'. Nonmonotonic logics consider normal cases, leave exceptions aside, theory revision considers only closest cases, as do counterfactual conditionals, certainty looks how well one formula is embedded in another, etc. Often, we "forget" in a controlled way some models, or, we "forget" the exact limits of the model set, and look how far we can go without running into disaster.

2.2.1 Traditional nonmonotonic logics

Being nonmonotonic is a property, not a name for a unique logic. In all likelihood, the following "logic" is nonmotonic: $\phi \mid\sim \psi$ iff the last two digits of the Gödel numbers of ϕ and ψ are the same. Any formal, abstract approach to reasoning with information of differing quality will be nonmonotonic: better information can override weaker information.

But traditionally, the term nonmonotonic is used for logics stronger than classical logic in the sense of (SC) — see Definition 1.6.5. This is due to the motivation behind their introduction: to create a logic able to conjecture beyond certain knowledge (formalized by classical logic). Nonmonotonic logics, NML for short, were designed to formalize aspects of reasoning about the "normal", "interesting", "important", "useful" cases, the "majority", or the like. This is underspecified as it corresponds to somewhat different intuitions, and is one of the reasons why there is a multitude of NML's: after all, if they are adequate, they have to code the intuition somehow into the formalism, so different intuitions will usually generate different formalisms. Note that such logics about normality, etc. will also be naturally nonmonotonic, as any fact, e.g. about the normal case need not be plausible any more once we add the information that the case at hand is not normal.

We now discuss some of these intuitions or interpretations.

We follow the tradition of being stronger than classical logic, so all nonmonotonic logics are represented by a model choice function f s.t. $f(X) \subseteq X$, property $(\mu \subseteq)$ in Definition 1.6.5.

Here are some reasonable interpretations of a nonmonotonic consequence relation $\alpha \mid\sim \beta$:

(1) in the normal, important or interesting $\alpha-$cases, β holds, too,

(2) in the prototypical or ideal $\alpha-$cases, β holds, too (e.g. "center" of $\alpha-$cases),

(3) in the majority of $\alpha-$cases, β holds, too,

(4) in as many cases where α holds, and where it is consistently possible, β will also hold (Reiter defaults).

Note that all properties are relative to a base set (here the $\alpha-$ cases), and a priori there is no connection between those cases singled out for α and those for α' (except if $\alpha \leftrightarrow \alpha'$). Coherence properties describe just such connections.

"Important", "interesting", "normal", "prototypical", "useful", and "ideal" are properties of individuals (or of pairs of individuals when we compare them for importance, etc.), whereas "many", "majority", etc. are properties of sets. "Important" admits perhaps more substitution than "normal" in the sense that we can substitute one important element by three less important ones. But even if this is not your intuition it will not harm to do as if, and examine such substituitions. "Prototypical cases" are perhaps rarer than normal ones, and close to ideal ones, which might not even exist, but be mere idealizations.

Ideal cases may be special normal cases, maximally normal cases, or they may be formulated in a simpler language, where we forget about irrelevant properties, this would be outside our framework.

2.2.1.1 Normal, important, or interesting cases

We do not distinguish the notions "normal", "important", "interesting", etc., we treat them together, and, for brevity, just speak about "normal" cases.

We work first in

Propositional logic

(1) Abstract normality:

First, and in the simplest case, "normal" is just a predicate, an arbitrary subset $X' \subseteq X$ which contains exactly the normal cases of X. X' can be anything. In particular, for different X and Y, we can choose X' and Y' in a totally independent manner. In more abstract terminology, there is no coherence in the choices for the different X and Y. We then reason — unless we have information to the contrary — with X' instead of with X. As the choice of X' is independent from the choice of Y' for Y, this kind of reasoning will usually be nonmonotonic. Obviously, (AND) holds, i.e., if we conclude from $Th(X)$ that ϕ is the case — because it holds in all $x \in X'$ — and that ϕ' holds (for the same reason), then we will also conclude that $\phi \wedge \phi'$ holds, by the rules of classical validity for $\wedge$.

(2) Normality defined by a binary relation:

Much more interesting is the situation where normality is defined by a relation, which expresses that x is more normal than x', $x \prec x'$. (We follow tradition that the more normal elements are $\prec$ −smaller. Thus, until you are used to it, better read: less abnormal.) The normal elements of X are then those of maximal normality, i.e. those $x \in X$ s.t. there is no $x' \in X$ $x' \prec x$. Note that this is relative to X, i.e. there might well be $x' \prec x$, but for $x' \notin X$. Now, we have some coherence: the property of being normal is downward absolute. If x is normal in X, $Y \subseteq X$, $x \in Y$, then x must also be normal in Y. This is a trivial consequence of the definition. We shall see in Chapter 3 that this is the essential characterizing property of normality defined by a binary relation.

(3) Normality defined by a distance:

Another definition of normality might be given with distance: The normal cases X' of X are those elements which are in the center of X. They are the least marginal cases. Of course, it is not clear how "center" should be defined, even given a distance:

(1) The center of X might be the set of elements which have maximal mean distance to elements outside of X.

(2) It might be the set of elements which have minimal mean distance to the other elements of X. (This is not the same as variant (1), as easy examples show.)

(3) It might be the set of those elements of X, whose minimal distance to the elements outside of X is maximal: $x \in X'$ iff $\forall x' \in X(d(x, U - X) \geq d(x', U - X))$.

And there may be other reasonable definitions of "center".

It is easy to see that the coherence property of normality determined by a binary relation is usually not satisfied: Take the natural numbers with standard distance, and consider $X := \{2, 3, 4, 5, 6\}$, $Y := \{2, 3, 4\}$. Then the center of X will be $\{4\}$, but the center of Y will be $\{3\}$. On the other hand, the notion might seem so weak that there are no coherence properties at all. This is not the case, as the following example shows for variant 2 (at least as long as we work without copies — you will understand the remark once you have read Section 4.3 on counterfactual conditionals):

Consider $X := \{a, b, c, d\}$ and a symmetric distance. If, for all three-element subsets $X' := \{x, y, z\}$ $C(X') = X'$, then we have $d(x, y) + d(x, z) = d(y, x) + d(y, z) = d(z, x) + d(z, y)$, so all these three distances inside X' are the same, but then, as it holds for all such X', all distances are the same, and $C(X) = X$ has to hold, too.

We can also take a dual approach with a relation or a distance: instead of considering the "best" — i.e. most normal, or in the center — elements as done above, we take away only the worst, consider all but the least normal, all but the most excentric ones, etc.

There are certainly still other ways to determine normality than by arbitrary choice or by a binary relation, or by a distance. We could, for instance, single out globally the normal elements, say $U' \subseteq U$, if U is the universe, and then choose X' as $X \cap U'$ — logically, this would just amount to adding $Th(U')$ in classical logic, so this is not very interesting, but it has strong coherence properties: If $X \subseteq Y$, then $X' = Y' \cap X$.

The definition of normality by a binary normality relation between elements is the most developed one, and we return to it now. First, one can impose additional properties on the relation. Such properties sometimes have very interesting repercussions on the resulting logic. For instance, "smoothness" (see below in Section 2.3.1) results in cumulativity, (CUM), "rankedness" (see Section 2.3.1, too) in rational monotony, (RM), etc. Second, and on a more basic level, one can work with "copies" of models, (in an equivalent language: noninjective labelling functions, see Definition 2.3.1 below), as is done in modal logic, or without copies (or injective labelling functions). An object is then called minimal, iff at least one copy is minimal. Third, if there are no most normal elements, our definition collapses, as X' will be empty, so falsity will be a consequence. For this situation, we need a different, and more complicated approch: the "limit" version of normality (see again Definition 2.3.1 below). In this variant, ϕ is a consequence of T iff, "from a certain degree of normality onward", ϕ holds whenever T holds, i.e. when we approach the limit of normality, ϕ will finally always hold when T does. For instance, suppose that the models m_i, $i \in \omega$, $m_i \models T$, get ever more normal with increasing index, $m_i \succ m_j$ for $j > i$, and from some n onward, ϕ will always hold, we then say that $T \mid\sim \phi$ in our structure.

Note that these remarks also apply to distances, in particular, the limit variant is very useful if we have infinite approaching chains of elements.

We turn to

First order logic

(1) Normality within the models or over the model set?

We have two possibilities for a normality predicate (whether it is an abstract predicate, or generated, e.g. by a binary relation): We can define it on the set of all models, or within each model (or perhaps a combination of both). The latter was not possible in the propositional case. Within one propositional model, a formula holds, or it does not hold. There is no graduation, there is no approximation. (This can, of course, be different when we have more than two truth values.) Essentially, in the first case (meta-) quantifiers of normality range not over all models, but only over part of them, in the second case, the object language quantifiers will be restricted to normal cases. For instance, "normally $\phi(x)$ holds", will signify in the first case: in all normal models, $\phi(x)$ holds everywhere, i.e. $\forall x \phi(x)$ holds in all normal models, and in the second: in all models, $\phi(x)$ holds for all normal elements of the model. We can capture — at least in some cases — the first interpretation by the second: we just choose the "local" normality predicate as $\emptyset$ for the nonnormal models, and as everything for the others.

We do not know whether one interpretation should generally be preferred over the other, the most important is perhaps that we are aware of the two possibilities. We have to make a caveat about the second variant: We could, for instance, choose in models where $\phi(x)$ fails almost everywhere, normality in such a way that still, normally $\phi(x)$ holds: normal elements are just those very few where $\phi(x)$ holds. It is doubtful whether this has still much to do with intuition.

(2) Normality and maximal extension of one predicate

There is an important variant of the first interpretation, in which normality of models cooperates with maximal extensions of certain predicates. This idea seems to correspond rather well to intuition, and we find it also with Reiter defaults. (We consider for simplicity only formulas with one free variable, which we do not write down systematically.)

It seems natural to say that in a structure $\mathcal{M}$ (a set of models, with a normality predicate or relation $\prec$ on the set), "normally ϕ" holds, if those models M in $\mathcal{M}$ are normal, where $\phi(x)$ has maximal extension (perhaps for a given universe, perhaps restricted by some cases where $\neg\phi(a)$ has to hold, etc.). "Maximal extension" can be either by set inclusion, or by cardinality. We consider here only maximality by set inclusion. This can be more complicated when the universe changes: we may have added one $\phi-$case, but also many $\neg\phi-$cases. So a more careful comparison is necessary — we will not go into details.

(3) Normality and maximal extension of several predicates

When we consider a set of such formulas $\Phi(x)$, we can distribute in two ways: Either we say $\Phi(x)$ holds in such a structure iff all $\phi(x) \in \Phi(x)$ individually hold normally, or, iff in the normal models, for each $\phi(x) \in \Phi(x)$, we could increase the extension of $\phi(x)$ only at the cost of another $\phi'(x)$, i.e. there might be $\phi-$better models, but they will be $\phi'-$worse for some $\phi, \phi' \in \Phi$. The two variants are then, using $[\phi]_M$ for the extension of $\phi(x)$ in the classical FOL model M :

(a) $\mathcal{M} \models_a \Phi$ iff: if M is a $\prec$ −minimal model in $\mathcal{M}$, then there is no $M' \in \mathcal{M}$ s.t. for some $\phi \in \Phi$ $[\phi]_{M'} \supset [\phi]_M$,

(b) $\mathcal{M} \models_b \Phi$ iff: if M is a $\prec$ −minimal model in $\mathcal{M}$, then there is no $M' \in \mathcal{M}$ s.t. for all $\phi \in \Phi$ $[\phi]_{M'} \supseteq [\phi]_M$ and for some $\phi \in \Phi$ $[\phi]_{M'} \supset [\phi]_M$.

We can then pose the question: What are the rules for the consequence relations $\Phi \models_a \psi$ and $\Phi \models_b \psi$, where this is defined as follows: $\Phi \models_a \psi$ iff for all $\mathcal{M}$ $\mathcal{M} \models_a \Phi$ implies $\mathcal{M} \models_a \Phi \cup \{\phi\}$.

For the second variant, suppose that ϕ and ψ are partially contradictory, i.e. there are elements, where either one may hold, but not both together. One might want to see here some equilibrium, like, in half of the cases ϕ holds, in the other half ψ holds. But this seems difficult to achieve. For instance, if we know that there must be more ϕ−cases than ψ−cases, shall we try to follow this proportion also for the contradictory cases?

(4) Normality and maximal extension with prerequisites

When we extend the approach to sentences like: "if, $\phi(x)$, then normally $\psi(x)$", something like a default with prerequisites, we see new questions which need to be answered. As long as we have just this sentence, we will probably make the same approach as above, only relativized to $[\phi]_M$ in each classical model M. But, if we also have "normally $\phi(x)$", then we cannot simply fix $[\phi]_M$, but we have to vary it, too. And if we make $[\phi]$ bigger, but let $[\psi]$ constant, we seem to have decreased the validity of "if, $\phi(x)$, then normally $\psi(x)$". Thus, such sentences merit a more detailed treatment, in the line of the above variant (b).

2.2.1.2 The majority of cases

We definite now $T \mid\sim \phi$ iff ϕ holds in a majority of T−models, or in a big subset of the T−models.

Note that the alternative definition: "holds in all big subsets of the set of T−models" is degenerate: If majority, "big", etc. are defined reasonably, the whole set of T−models will be big, so $\mid\sim$ would then just be $\vdash$.

The first question is how we determine "majority" and related notions.

There are at least three natural possibilities:

(1) By counting.

(2) By an abstract measure in the sense of mathematical measure theory. Note that here, size of singletons does not necessarily determine size

of sets, e.g. in the standard measure on the reals, all countable sets have size 0, but there are sets of arbitrarily big size.

(3) Still more abstractly by a (weak) filter, where the filter contains the big subsets (see Definition 1.6.2).

It is easy to see that counting results in problems in the finite case: If the only majority is the whole set, then $\mid\!\sim$ is $\vdash$. Otherwise, as each x is definable by a formula ϕ_x, and each $X - \{x\}$ is a majority, $\neg\phi_x$ is a consequence for each x. But the conjunction contradicts T. Thus, $\mid\!\sim$ is contradictory, when we admit arbitrary conjunctions (this applies to weak filters, too).

An interesting approach is to consider sets of size 1 for an abstract probability measure. This system is closed under countable intersections (by the properties of a measure), and thus even stronger than a filter, which is closed under finite intersections only. We will, however, concentrate on the usual filter approach. (The main reason for its popularity is probably that it cooperates well with usual logics, which admit arbitrarily long finite formulas, but no infinite ones, so, in particular no infinite $\bigwedge$.)

By the central filter property (F3) (recall Definition 1.6.2), (AND) holds trivially in true filters (and not necessarily in weak filters). (RW), right weakening, will always hold by (F2).

We have seen above that, e.g. normality defined by a binary relation imposes coherence properties. This is not the case with filters. In principle, we can choose all filters for each subset independently. Yet, there are natural coherence conditions for such filter systems. The most natural one is probably the analogue to (F2): If $A \in \mathcal{F}(B)$, and $A \subseteq C \subseteq B$, then $A \in \mathcal{F}(C)$. (Often, intuition is more clearly expressed by considering the dual notion of an ideal, e.g. if X is a small subset of Y, and $Y \subseteq Z$, then X is small in Z. We will change from filters to ideals and back liberally.) An immediate consequence on the logical side is cautious monotony (CM): If $T \mid\!\sim \phi$ and $T \mid\!\sim \psi$, then by (F3) $T \mid\!\sim \phi \wedge \psi$, so there is a big subset $X \subseteq M(T)$ s.t. $X \subseteq M(T \cup \{\phi, \psi\})$, but as $X \subseteq M(T \cup \{\phi\}) \subseteq M(T)$, X is big in $M(T \cup \{\phi\})$, so $T \cup \{\phi\} \mid\!\sim \psi$. In the case of weak filters, we will generally only have the weaker property $T \mid\!\sim \phi \wedge \psi \Rightarrow T \cup \{\phi\} \mid\!\sim \psi$.

A principal filter (i.e. if there is $A \in \mathcal{F}$ s.t. for all $B \in \mathcal{F}$ $A \subseteq B$), can now be seen as the set of supersets of the normal cases, with A the set of normal cases, and for many purposes it suffices then to consider the generator $\bigcap \mathcal{F}$.

In the FOL case similar considerations as above for normality apply. We can use a filter (or a system of filters) on the set of models, or inside each model. The latter is investigated in detail in Section 7.2.

2.2.1.3 As many as possible (Reiter defaults)

We will treat defaults only as far as they reveal or contrast some basic ideas we also see in formalisms which we have treated more in depth. In particular, we will only discuss normal defaults with or without prerequisites here and in Section 7.2. We will also neglect the problem of consistency of default theories: In our opinion, a default theory $\{: \phi,: \neg\phi\}$ with the meaning: "if consistent, assume ϕ" and "if consistent, assume $\neg\phi$ ", is simply inconsistent, and has no meaning — see Section 7.2 for a more detailed discussion. Our argument about Reiter defaults is, in essence, pessimistic: there seems to be no global solution. One can often find implausible special cases of generally plausible rules.

We first look at propositional logic.

In propositional logic, when we say "by default α", we can only exclude $\neg\alpha-$models as abnormal. This is at first sight nothing more than normality as already discussed. There is a possible difference, when we consider subsets X of the universe. As long as there is one $\alpha-$case in X, i.e. X is consistent with α, Reiter defaults impose the normal $X-$cases to be $\alpha-$cases — no matter how intuitively bizarre X might be. Consider such examples for X like: A big set of birds, all penguins, emus, etc., and just one sparrow. The default in its usual use will tell us that birds in this set normally fly, which goes against (my) intuition. We see the same behavior in ranked structures (see Section 2.3.1): one minimal model suffices to "kill" all the rest. This property may be desirable or not, it seems to depend on the choice of X.

Human reasoning does not treat arbitrary predicates like $\phi \vee \psi \wedge \neg\sigma$, natural kinds like ravens or tables, and predicates like white or big in the same way, as classical logic does. And if they are not treated the same way, they must also be different things for any logic that pretends to be close to human reasoning. We should not be blinded by classical logic, which was created for other purposes.

The idea is then to prevent situations where $X \cap [\phi]$ is very small, perhaps even "artificial", to force $f(X) = X \cap [\phi] \neq \emptyset$. To address the problem, we may single out "admissible" predicates, which are meant to be "independent", "irrelevant" for the default. For instance, for the default "birds fly", the color of the bird will probably be irrelevant. This prevents artificial examples like the one above. Can we say anything about such predicates? We do not think we should admit arbitrary intersections, as we might well construct counterexamples from reasonable predicates this way — sets can get arbitrarily small this way, and we want to be closer to simple common-sense reasoning than operations with small sets are. Neither should arbitrary unions be admitted, as we could use the same positive case (one flying bird) again and again, to obtain a case similar to the one above. But the following operation seems possible: If X and X' are admissible, and $X \cap X' = \emptyset$, then $X \cup X'$ is admissible, too.

We turn to several defaults. This situation seems to be the same as the FOL case of normality discussed above in Section 2.2.1.1. It seems reasonable to make as many conclusions of as many defaults as possible true, to abbreviate, we just say that we make as many defaults as possible true. It also seems reasonable not to sacrifice one default to make another true. The difference to usual preferential models is that even subideal cases are still differentiated: without explicit coding, preference says nothing about $\alpha \wedge \neg\beta-$cases, when $\alpha \mid\sim \beta$ — here we save as much as possible, and this seems to correspond often better to intuition. (This question is also known as the blond Swede problem: normally, Swedes are tall and blonde. But even not blonde Swedes should be tall — unless we suspect a common mechanism behind being tall and blonde, then failure of one makes failure of the second probable. We see again that there seems to be no universal solution, at least not on this level of abstraction.) The behavior for subideal cases can also be described as "greediness" of defaults: we make as many cases as possible hold.

We turn to first order logic.

Again, the same ideas as for normality, discussed in Section 2.2.1.1, seem to be valid, as well as the considerations for predicates X as above, since we may — and perhaps should — choose not only $[\phi]$ maximal, but also $[\phi] \cap X$ for suitable X. This might be contradictory if we have information in the background theory which says that more $X \cap [\phi]-$cases will generate less $[\phi]-$cases globally. We have to decide what to do then.

2.2.2 Prototypical and ideal cases

We mention this case here for completeness' sake only, but will not, and are not able to, go into detail. In the author's opinion, prototypical reasoning is the domain of (experimental) cognitive psychology. The notion had fallen there into discredit, probably as it was conceived too narrow, and seems to be reviving slowly. The author is simply incompetent in this domain.

Reasoning with prototypes is an example of semantical reasoning. If we see prototypical reasoning as a form of efficient reasoning, it is natural that one class can have several prototypes — depending on the kind of answer we try to give. This is, as far as we know, confirmed by psychologists. A zoologist will probably work with a different prototype for a chicken than a chef de cuisine. We also conjecture that prototypes are dynamical, we start with a coarse and very simple prototype, and refine if insufficient.

We will thus have several operations on prototypes:

- choice, if there are several candidates,
- modifying (revising) prototypes:
  - correcting,
  - refining,
- combining several prototypes: if we think about chicken in a garden, we will probably not have a ready made prototype for this situation, but somehow combine a chicken with a garden prototype,
- learning prototypes.

We should also be able to measure the quality of a prototype — the natural criterion being the question whether it does what it should do, and to what degree.

One, but certainly not the only formal interpretation of prototypical elements of a set is to take those "at the center" of the set — an operation we saw already above. Often, a prototype will not be a member of the set of cases considered, but an idealization, which combines the "salient" properties — even if this combination does not exist in reality.

2.2.3 Extreme cases and interpolation

Extreme cases are somewhat the opposite to prototypical and normal cases. They may be the worst elements in some normality order, or excentric elements wrt. some distance. Extreme cases can be very interesting for verifying: Suppose we do quick reasoning with some prototypical elements, and check the result with extreme cases. The reason behind is that reasoning with complete theories, perhaps in a simplified language, is fast, and we think that results somehow "respect" the between relation (between extremes). More precisely, if ϕ holds at a, and at b, we hope that it will hold at every element between a and b, too. Of course, there are counterexamples: take two extreme models m, m', and their defining formulas $\phi(m)$, $\phi(m')$ (in the finite case), then $\phi(m) \vee \phi(m')$ will hold there, and nowhere else. The utility of such reasoning depends on the kind of question (and distance).

2.2.4 Clustering

Clusters are "natural" subsets (or sets of subsets) of the domain, where elements from one cluster are supposed to have more in common with each other than with elements from other clusters.

It seems probable that human reasoning works, for simplification, with such clusters, perhaps choosing one (or several) prototype(s) for each cluster. We can conjecture that natural kinds form clusters.

If we form clusters inside a small set, we can expect clusters to be finer that when we form them inside a big set: a, b may be in the same cluster in a big set, but in different ones in a small set. For instance, when we look at all animals, dogs and cats may form one cluster each, but when looking at domestic animals, we might differentiate between different kinds of dogs. Abstractly, we would expect clusters in a smaller set to form subsets of those in a bigger one — borders in the bigger set will be respected.

There are several ways clusters can be formed:

- Via some distance:
  - Given some fixed distance d, all elements connected by paths with step length $\leq d$ form a cluster. Adding points can fuse old clusters by adding "stepping stones" between the old clusters. Fusing clusters will necessarily contain additional elements.
  - Varying the maximal distance: clusters are formed as above, but d is chosen, e.g. as $1/2$ the minimal distance needed to form one big cluster containing everything. If clusters fuse, we need at least one element more distant than the old elements.
- Via "simple" sets as discussed briefly in Section 2.2.6. In this case, different clusters may have nonempty intersections. We may treat the intersection in the way of defeasible inheritance.
- By some other equivalence relation, e.g. by considering the levels of a ranked structure as clusters. In this case, clusters will be absolute, the size of the set in which we form them is not important. This is done, e.g. in the model size based approach to theory revision, see Section 7.4.

2.2.5 Certainty

We discuss now something which is perhaps best seen as a re-interpretation of the notion of epistemic entrenchment of theory revision: at least when both are defined via distance, certainty corresponds exactly to epistemic entrenchment: Given information T, with $T \vdash \phi$, the more $M(\neg\phi)$ is distant from $M(T)$, the more ϕ is certain, the more it is entrenched.

The notion of certainty has a number of more or less different uses.

First, we can firmly believe in some classical information T, but, just in case T is wrong, we can pose the question which parts of T we still consider most certain. In this case, we go beyond classical logic in certainty: If we weaken (i.e. increase) $X := M(T)$ to X', then $T \mid\!\sim \phi$ iff $X' \models \phi$. Thus, $\mid\!\sim$ is more certain than classical logic.

Second, we may know that T is wrong, but would like to "save" some of T — this is theory revision.

Third, we can doubt information T, and would like to have some degree of certainty we might still give it.

Fourth, when we go beyond classical logic, e.g. to normal cases, we can ask the question how much certainty we have lost.

In the first and second case, we have to weaken T to a certain degree, this corresponds to looking for a bigger (model) set than X. In general, the more we weaken the set, the safer we are — with truth the limit. The problem is how far to go. The choice can be made by simplicity: we look for a simple set including X. If we are given some distance, we can take an area of diameter S surrounding X, etc. These alternatives are discussed in more detail below.

In the third case, degree of certainty, a good idea might be to first find a weakening of T to some T', and then measure somehow the difference between the two. In the case of weakening by distance, we have a natural candidate, a simple possibility is to take just the set difference (of the model sets) — see also below in the paragraph on approximation.

In the forth case, we can again measure the loss of certainty by the set difference (i.e. the set of cases we have omitted) between X and $\mu(X)$. This, however, will neglect any graduation in importance we may have between the various elements. In more definite cases (ranked structures of preferential reasoning, see below), we can use the normality difference of the ranks as a measure — the natural distance between the best and the worst elements of X.

We discuss now in some more detail the second case.

There are various reasons why information can be uncertain. (1) The communication channel can be noisy, the source of information may be unreliable in various ways, (2) the source may be interested not to tell the truth and thus distort in a certain direction, (3) the source may have been under the influence of alcohol, (4) visibility may have been poor, etc.

Without details about the coding, it may be difficult to draw any conclusions in the first case. We are not familiar with drunk witnesses, so we leave this question to the competent.

If we suspect an information source to lie, we will try to determine the direction of its lies, what it may want to hide, or what it may want to make us believe. We might then — in a first approach — "shift" the information given by the source in the opposite direction. The set we might accept will be "drawn" into one direction, i.e. we will make the distance to a point or another set smaller.

If visibility was the problem, we cannot be sure about details, but the general picture was perhaps correct. The car discribed as black might have been blue, but it probably was not a bicycle. We might be willing to "weaken" the information to a certain degree, i.e. $X \subseteq f(X)$, where the "certain degree" might be coded by an area around X, covering a certain distance from X, i.e. $f(X) := \{y : d(y, X) \leq k\}$ for some k — however this may be defined. We might also think of some "natural" neighborhood of X, e.g. the convex hull of X wrt. to some distance d. The more we consider the information reliable, the less we will weaken it. The more we weaken it, the more we can be sure about the information, with truth the limiting case. Of course, we can, in a subsequent step or in parallel, also take "normality" (and thus perhaps a preference relation) into account.

In some cases, we will not be ready to give a definite boundary until which we are prepared to go, but can still do relative reasoning: if we are prepared to accept x, then a fortiori we will have to accept y.

Instead of working with distance, we can also work with size, taking bigger supersets of X to be more certain. There is no universal solution, it depends on the structures we have, and what seems to fit best.

2.2.6 Quality of an answer, approximation, and complexity

Suppose we ask the question ϕ?, and get as answer ψ. If $\psi \to \phi$, or $\psi \to \neg\phi$, (and we know it) the question is answered, even "over-answered" if the implication is not an equivalence, we have more precision than we wanted. The interesting case is where neither holds. Suppose we cannot do any more questioning for the moment, and have to do with the answer. We then have to decide for ϕ or $\neg\phi$, and also may want to have some idea of the quality of the answer, e.g. in order to decide for further inquiries later on.

We think that the decision is best based on a comparison of the quality of the argument ψ for or against ϕ.

For instance, if $\psi \wedge \phi$ is a big subset of ϕ, and $\psi \wedge \neg\phi$ is a small subset of $\neg\phi$, ψ seems a good argument for ϕ, and a bad one for $\neg\phi$, and we can measure the quality, e.g. by the pair (big, small), or directly by the pair of sets $(\psi \wedge \phi, \psi \wedge \neg\phi)$, if we want more information.

But the situation can be more complicated: The sizes we discuss may also depend strongly on what we want to do with the information ϕ. If we decide for an action involving high risks in case of error, even a small probability of assuming ϕ, when really $\neg\phi$ is the case, will weigh heavily. So size or probability should not be an absolute notion.

Approximation is (partly) a related problem. It has two parts: first, how to find an approximation, second, how to judge its quality. Sometimes, not all sets will be admitted as candidates for an approximation to a set X, due to various reasons. (E.g. only "simple" sets might be chosen, where, e.g. a set convex wrt. some distance might be a simple set. In this case, the best approximation from above is the smallest convex set containing it, and from below any of the biggest convex sets contained in it.) Approximation from above can also be seen as nonmonotonic reasoning applied to the complement — choosing a suitable subset.

The quality of an approximation can be determined similarly to the quality of a reply. We can, e.g. measure the size of the set difference between the approximated set X, and the approximation A. We might also work with distances, and determine the quality of an approximation A to X from above by the biggest distance of a point $a \in A - X$ from X, and if A is an approximation from below, we might take the biggest distance of a point $a \in X - A$ from A.

Complexity

We have already said repeatedly that a distance relation — in particular the Hamming distance — and its associated notion of convexity can be used to describe simple sets. We can now define the complexity of a set, $c(X)$ by counting the minimal number of convex sets it decomposes to. In a next step, we can define the complexity of a logic described by a model set function f as the maximal (or average) difference $c(f(X)) - c(X)$. Note that this is not a complexity of computation, but of argument vs. result.

2.2.7 Useful reasoning

We consider here situations like the following: a physician has several drugs d, d', etc. at his disposal, he is confronted with a patient, but has only limited diagnostic possibilities, he may make an error in his diagnosis, and consequently give the wrong drug. The right drug has beneficial effects, the wrong one can do harm. So he has to weigh the benefits and costs, and the likeliness of error. Essentially, what he does here, is some kind of rough integration, or rough multiplication, in the following sense: "The likeliness of error is small, but if I err, the consequences are very serious. On the other hand, with this diagnosis, I am less sure, but if I am mistaken, the drug may not help but at least will not do much harm, what shall I do? Is $small * big$ better than $medium * small$?" There will not be any global answer to this, but we can probably still give a reasonable framework. In particular, it might be important to be able to do relative reasoning here. Essentially, the question is how abstract measures inherit from sets to their products (like $small * small = (very)\ small$?).

It seems reasonable to decompose the problem into two sub-problems:

(1) What is the utility (or harm) of assuming ϕ in model m?

(2) What is the overall utility for a set of models?

On (1):

It seems that we can make only very weak reasonable assumptions: Diagnosing truth has probably value 0, as we know beforehand that truth holds, diagnosing for a model m $Th(m)$ (its complete description) will probably have maximal value for that model — one cannot get better — and logically equivalent formulas have the same utility.

Unfortunately, this seems about all we can do. Suppose, e.g. $m \models \phi \wedge \psi$. Diagnosing correctly $\phi \wedge \psi$ may lead to an efficient treatment, but diagnosing correctly ϕ, without realizing at the same time ψ may be a catastrophe. Now, the diagnosing person (or apparatus) might first say "ϕ", stop a moment, and go on "ψ". So he has really said "$\phi \wedge \psi$". Consequently, what we mean by diagnosis is something like: "ϕ, ψ, ...— and this is all I know", or, equivalently, the conjunction as final answer. So the diagnosis is a single formula (and, if you like, all its logical consequences). Consequently, in above case, the (correct but incomplete) diagnoses ϕ and ψ may have very negative value, the (correct and perhaps still incomplete) diagnosis $\phi \wedge \psi$ may have a very positive value. Conversely, if $m \models \phi \wedge \psi \wedge \sigma$, diagnosing ϕ and ψ (separately) may have positive value, but diagnosing $\phi \wedge \psi$ (and failing to see σ) may have very negative value. As the values depend on the world as it is, and the world can be bizarre, it seems that any attribution of values is possible.

Fortunately, the world is not as bizarre as it could be, so there will be regularities. Often, utility will none the less be additive, i.e. realizing $\phi \wedge \psi$ will have the sum of the utilities of ϕ and ψ.

On (2):

We are here in a situation where we have some information, and make a diagnosis. As we have only some information, we cannot be sure which actual case we are in, say we only know that we are in set X (e.g. the patient is conscious, middle-aged, ...). So the utility of the diagnosis is the sum of the individual utilities of this diagnosis for all $x \in X$. Note that, even if we think that ϕ is the best diagnosis by symptoms, we may still prefer to work with a weaker (or different) diagnosis, because the danger in case of error may be smaller.

Usually, we will not calculate a detailed sum, but will do an estimation. This can either be a rough sum as discussed above (e.g. a mortal risk for a patient will carry it all), or a rough product: If, in most cases, the diagnosis is very beneficial, and in a few cases slightly harmful, we will probably consider the overall utility positive.

If a rare occurrence of an extreme value can carry it all, it seems that we look more at an extreme form of sum than at a multiplication. In that case, in a way, one of the coordinates is stronger than the other. In the other cases, we probably often just calculate the minimum of both values: $small * anything$ is small, $medium * anything$ is medium or small (if "anything" is small), etc.

2.2.8 Inheritance and argumentation

Inheritance and arguments have a realistic feeling: they look like real reasoning. It might be their combination of relatively simple nonmonotonicity and analogical reasoning which makes them attractive. In our context, a theory of argumentation will compare arguments, and does not tell how to generate them.

We will discuss in this section:

(1) Why we think that it is at least hard to find a satisfactory theory of argumentation — and we have certainly none to offer — and a way out of this difficulty.

(2) Present in anecdotal fashion elements to consider in this framework. They will fit into the general considerations presented in (1).

The problems with theories of argumentation and a way out

We think it might be difficult to find a good and universal theory of argumentation.

First, an anecdotal (meta) argument. Defeasible inheritance can be seen as a simple case of argumentation — the language is poor, so is the logic. Yet there are a big number of different approaches (upward vs. downward concatenation, choice of admissible reference classes, etc.), and we have never seen a convincing (meta) argument why to prefer one approach over the other. Note also that being cautious in one case might mean being bold in the other: the argument we were cautious about could have served to oppose the one we end up bold about.

Second, as a side remark, note that the existence of a perfect theory of argumentation would lead to a funny situation:

As long as a theory of argumentation is deficient, we can find situations where it gives the intuitively wrong answer: it says argument a is better than b, where the converse intuitively seems true.

But once we have found an optimal and universal theory TA of argumentation, what do we see? Suppose we have another theory TA'. If TA is optimal, and TA' not, we have to be able to see this, so there must be good arguments in favor of TA and against TA'. As TA is optimal and universal, it has to be able to recognize them as such. They are instances of the situations TA describes as good arguments. So the quality of TA is based on its own instances — and not on anything more: if there were other reasons to consider TA optimal, by optimality and universality, TA would be able to recognize them.

So any optimal TA must be some kind of fixed point — you cannot improve, but the ground you stand on is shaky, as it is a cycle, or else TA has to be of limited scope.

Third, forms of argumentation which may be bizarre in one context can be justified in others. Usually, it is bad argumentation to give more weight to components of an argument, because we like the result. A very bad example can often be seen in politics, where people deny facts to embellish their favorite political monstrosity (Hitler and Auschwitz, Stalin and the Gulag, BenLaden and the destruction of the World Trade Center). But, in tentative scientific reasoning, making correct predictions, or solving some difficult problem, can give credit to a new scientific theory (if we are not hard core Popperians). Despite the differences in these cases, the underlying question is whether we can justify an argumentation by its results.

Thus, the quality of an argument seems to depend on the context, and perhaps we cannot do more than enumerate some cases, questions, and problems, without any hope for an exhaustive treatment.

One such question is whether one argument can influence the quality of another one. An example is specificity in defeasible inheritance. Establishing that $B \to A$ may allow to give precedence to a path from C via B to one from C via A — as more specific information is considered more reliable.

The general situation can probably be described as follows: we have "bits of logic", i.e. perfectly logical inferences (in whatever logic), and "shaky bridges" between these inferences to make them fit for concatenation, which really ARE the argument (and which are weaker than justified by the logic we work with). These bridges can be seen as the strengthening of the first inference, or of the second, or of something in between. They may be temporary and local constructions, just for this one argument, or more global and durable strengthenings of the respective inferences. An argument can thus be considered a mesh of inferences in some logic, with some patchwork to fill the gaps.

A comparison of two arguments will then be the comparison of the quality of these "bridges" — plus the comparison of the different underlying logics when they differ. The question how we measure the quality, and whether it is fixed once and for all, or itself subject to reasoning, is left open.

We have then the following situation: there might be no universal and optimal theory of argumentation. The quality of an argument is determined by the quality of the underlying logic (classical logic, or, even better, something stronger than classical logic, some form of nonmonotonic logic, etc.) and the size of the "gaps" between the inferences. In our approach, the quality of an argument is the size, importance, or whatever, of the cases we have neglected. In the simplest form, this is just the set of cases neglected. In a more subtle approach, this set might itself be reasoned about, we may consider it big, important, negligeable, etc., and this size (or importance, etc.) need not be fixed from the outset, but might be reasoned and argued about. If the theory of argumentation itself is not a constant, we treat it as a variable, and put it into object language, too.

We then have to put the necessary tools for this type of reasoning into the language.

For the moment, we see two things we need: We have to speak about (relative) size, and about (relative) distance. Distance might be used here in two variants: first, an elementwise distance between sets as in the revision semantics, second as a measure of similarity between sets (where, e.g. one might be a subset of the other), e.g. for analogical reasoning.

If we work in the finite framework, we then add for instance a new operator $<$ to the language (with, of course, precise meaning), where $\phi < \psi$ means that the $\phi-$models form a small subset of the $\psi-$models. The connection to the N-operator is obvious: $\phi < \psi$ iff $\phi \rightarrow \neg N(\psi)$ — provided we speak about the same notion of size, but this will not always be the case. In argumentation as described above, a neglected set of cases might still be much bigger than the complement of N, we will go beyond what usual nonmonotonicity allows us to do. We can relativize further by, e.g. $\phi < \psi \rightarrow \phi' < \psi'$. Similarly, we may express $d(\phi, \psi) < d(\phi', \psi')$, a quaternary relation. Once we have these operators, we can express counterfactuals, revision, nonmonotonicity, etc., by using sufficiently many of such operators. Soundness and completeness themselves migrate (essentially) to the object language.

Further elements to consider in a theory of argumentation

In defeasible inheritance, a diagram like $A \rightarrow B \rightarrow C$ (which allows for exceptions) is, we think, best read as: normal A's are B's, and normal B's are C's, and by default, normal A's behave like normal B's, so they are C's, too. The problem is, of course, the default reasoning. Contradictions are (usually) resolved by specificity. So, we have to look at specificity. The default reasoning has to be weaker than normality in above example: If N expresses normality, and N' the default, then the diagram $A \rightarrow B \rightarrow C$ says: $N(A) \subseteq B$, $N(B) \subseteq C$, $N'(N(A) \subseteq B \rightarrow N(A) \subseteq N(B))$ — we prefer situations where normal A's are normal B's in the language of preference. This preference can be seen as preference between whole structures, or as "the most normal elements of A are (most, if we want to concatenate further) normal elements of B".

We turn to a justification of the specificity criterion.

Reasoning by specificity may best be described as a special case of analogical reasoning. Specificity says that a more specific reference class is the better one in case of conflict. But, this corresponds well to the simple notion of distance between subsets: If $X \subseteq Y \subseteq Z$, then X is closer to Y than to Z, so it seems reasonable to take the closer one as reference class.

There are a few complications:

(1) In the case of conflict, we will not abandon all information of the losing reference class, but only that which is in conflict with the better one. So, this behavior is the one of "subideal defaults" seen above, or, alternatively, we have a case of revision: we fuse the information of the superclasses, giving precendence in case of conflict to the more specific one — we preserve as much as possible of each superclass, and solve conflicts by specificity. If there is no such criterion, we drop both pieces of information (direct scepticism). At the same time, conflicts are thus isolated: we have no EFQ (ex falso quodlibet, from $\perp$, we can conclude anything in classical logic), but we solve conflicts locally by revision. To do so, we use the "place" of the information, the point in the diagram, which is a criterion for its scope (effective only at it and below) and strength (by specificity).

(2) The reference class itself is not really the better candidate, but the set of its normal elements is — as an example, see the Tweety diagram: Tweety is a penguin, and a bird, peguins normally cannot fly, birds can, and penguins are birds. The latter gives the specificity criterion: the information about penguins has priority.

(3) Which reference classes do we admit? This is not so trivial as it might seem, as the difference between split validity and total validity preclusion shows, see Section 6.1 in [Sch97-2].

We can find a second argument for following specificity: if we did not take information from the nearest neighbor upward, we would have to make another change going down the inheritance chain: If Tweety were to fly, we change from "fly" at bird, to "not fly" at penguin, again to "fly" at Tweety, one change more than by having Tweety fly. And we should make changes only when we are forced to.

Note that it is easy to see that independent arguments for the same result give additional strength: the sets we neglect (and which determine certainty) are different — this leads to a natural generalization of the measure of certainty of an argument (or a proof). (Un)Certainty is the amount of risk we take. Independent arguments give different risks. The more different (and individually smaller) risks we take, the less likely is an error. A bigger risk corresponds to (AND), independent risks to (OR).

In realistic reasoning, inheritance networks will often be complemented by analogical reasoning: cases that are close to those in a known inheritance network will be treated in an analogous way — thus the accepted paths of a network will have a certain "width of channel". Consequently, an argument system has a certain "fit" to a class of situations, and a suitable system will then be close to a given situation.

2.2.9 Dynamic systems

We call reasoning systems dynamic when they are decomposed into a sequence of reasoning steps, whose results are evaluated, and, if they are found insufficient by some criterion, reasoning continues.

In a general form, at each step some question is posed, and the system returns an answer after some time.

Before posing a question, we should evaluate the following:

- how good (reliable, precise, etc.) is the information given to the system to solve the question?
- do we know something about the system's competence to answer the question based on the information it is given?
  - do we have prior experience?
  - do we have several systems at our disposal?
  - can we estimate precision, reliability, etc.?
- do we know how long the system will probably take to return an answer?
- do we know the overall probable cost of obtaining the answer (time, other resources)?
- how useful will the probable answer(s) be to solve the original problem?

Once we have received an answer, we have to evaluate it and the method used to answer it (reliability, precision, usefulness, see above) wrt. the question posed, but also wrt. the original problem (which need not coincide with the question posed). An improbable answer (unlikely cases) can also lead to a revision of the prerequisites the answer is based on.

If we receive several, perhaps contradictory answers, from perhaps several reasoning systems, we have again a problem of theory revision.

So, in general, the procedure is a cycle: evaluate which question to pose, pose the question, evaluate the answer, until we are satisfied. A good strategy might be to mix fast, bold guessing (perhaps based on prototypes or few cases) with more careful verification.

We think it is important that the start of the procedure is not so crucial for the final outcome: we should be able to start with a very bad guess, but still converge to something reasonable — there must be sufficient correction and control in the approximation.

A formalization of such reasoning will certainly contain meta-reasoning about:

- quality of conclusion,
- normalities and coherences used (and thus the logical system: preferential/rational, etc.),
- reasoning itself,
- choice of arguments.

Note that such a procedure will also provide some hiding of complexity: We may have small modules, and a "mastermind" which consults the small modules, posing questions. The answers to these questions can be simpler than the apparatus inside the modules needed to answer the questions. The answers are then composed to yield an overall answer. Inconsistencies can be treated on the level of answers. The whole system is close to defeasible inheritance, composing nodes of information in a more complex picture.

It is desirable to have progressively finer reasoning in the following sense: we can first reason on the level of, e.g. a preferential structure, and only if this is not fine enough, we will reason about the structure, i.e. the relation itself, provided we put it into the object language. Progressive reasoning is then a kind of argumentation, refining its methods, if we are not satisfied with a conclusion, we want the argument leading to it, revealing the underlying complexity — i.e. going from more to less abstraction. Formally, this can be achieved by admitting progressively more elements of the language into argumentation.

2.2.10 Theory revision

2.2.10.1 General discussion

Theory revision (TR for short) is the problem of "fusing" contradictory information to obtain consistent information. The now classical, and most influential, approach is the one by Alchorron, Gärdenfors, Makinson, AGM. (Recall that we use "AGM" indiscriminately for the article [AGM85], for its three authors, and for their approach.) We will present it below in very rough outline, but will first take a more general point of view, and discuss shortly some different cases and basic ideas of the problem. We will not, or only marginally, present more complicated approaches to theory revision, where for instance the theory itself contains information about revision.

Of course, there are situations where a contradiction is so coarse that both informations will just be discarded. If, e.g. one witness says to have seen a bicycle, the other an airplane, without further information, we will probably exclude both, and not conclude that there was a means of transportation. This does not interest us here — the operation is trivial (but it is not trivial to decide whether this is the adequate procedure).

Let us first state that the problem of theory revision is underspecified, so, just as for nonmonotonic logics, there are different reasonable solutions for different situations. Consider for instance the following cases:

- Two witnesses in court tell different versions of a nighttime accident with poor visibility. We will probably conclude on their reliability from the difference of their testimonies. So the outcome will probably be some "haze" around the OR of the two stories.
- Theories are deontic statements, the old law was as reliable as the new one, but the new one shall nonetheless have priority.
- We have contradictory information from two sources, but have good assumptions where each source might err.

Speaking semantically, we are given (at least) two sets X, Y, and look for a suitable choice function $f(X, Y)$, which captures some of the ideas of revision.

There are the following two basic approaches (with their reasons):

- we choose a subset of the union of models, if we do not take any reliability into account,

- we consider the contradictions a sign of lack of precision, and do not really believe any of the sources, so we choose some set not included in the union.

Can we find some postulates for a problem, which is posed in so general terms? It seems so:

(1) If we have no good reason for the contrary, each bit of information should have at least potential influence.

(2) We should not throw the baby with the bathwater:

  - The result should not be overly strong, i.e. it should be consistent, if possible (but there might be good reason to have, e.g. $f(\emptyset, \emptyset) = \emptyset$).
  - The result should not be overly weak (i.e.U), if possible.

  This point is usually summarized by the informal postulate of "minimal change" — ubiquous in common-sense reasoning (counterfactuals, theory update)! But this smells distance: given situation A, we look for situation B, which is minimally different from A, given some conditions (which usually exclude A itself), and some criterion of difference. If the "distance" between A and B is the amount of change, the cloud of smell concretizes to a formal definition: B is the one among a set of candidates $\mathcal{B}$, which is closest to A. And that is exactly what we will do later on in Section 4.2. We can also base revision on a notion of size of models, considering the biggest models as most important. This is developed in more detail in Section 7.4, and also shows a connection between size and ranking.

A further major distinction between different approaches to theory revision is whether each bit of information has the same weight, or whether some have a priviledged position. We call the first variant symmetrical revision, the second asymmetrical revision. Traditional (AGM) theory revision is asymmetrical.

We have then the following semantical situation: we have two sets, X and Y, perhaps disjoint, and Y should be given more weight than X. The approach by AGM is extremist in the following sense: $f(X, Y)$ will always be a subset of Y, so the influence of Y is very very strong. In particular, we do not doubt the reliability of Y. The problem is how to choose this subset of Y, recalling that some influence of X should nonetheless be felt. Given a distance, the more we go away from X, the less likely a point y will be, seen from X. So it is a natural idea to take those points in Y, which seem most likely, seen from X, i.e. which are closest to X.

We can be less radical, and still be asymmetrical in a variant of the symmetrical approach (putting a doubt on the precision of Y): Given a distance d, we do not only take all points of Y, but also all those between (wrt. d) X and Y, up to the middle, e.g.: $Y \cup \{z : d(z,X) + d(z,Y) = d(X,Y) \wedge d(z,X) \geq d(z,Y)\}$. (If we have no addition, we can define "between x and y" by $d(x,z) \leq d(x,y) \wedge d(y,z) \leq d(x,y)$.) We might also take all points around Y, up to the distance $d(X,Y)$. Again, one can imagine a number of possible approaches, playing around with distances.

But we can also work with structural information: Let, e.g. $X = X_1 \cap X_2$, and let $X \cap Y = \emptyset$, but $X_1 \cap Y \neq \emptyset$. Then $Y \cap X_1$ is a possible candidate for revision of X by Y, as X has a certain influence on the result, but the result is a sharpening of Y. If there are several candidates, we can take the "best" (e.g. biggest by number of models), or, if this is impossible, take the union of the results, etc. If X is a union of intersections, we can work with each component of X separately, then take the union of the results, or work with the "biggest" component, etc. There are many ways to play around with such ideas. This generalizes naturally to many sets X, Y, Z, as we can first form unions, they have the same form (unions of pure intersections). Instead of working with intersections, we can also work with convex sets wrt. some distance.

Other revision ideas are possible, too, e.g., we might choose for $X * Y$ a "useful" subset of Y.

Again, we have seen the basic semantical notions of distance and size appear in various possible interpretations of revision.

Two remarks on the Ramsey test

The problem of the Ramsey test has found much interest in the field of theory revision, we address it very briefly.

The idea behind the Ramsey test is that the theory K itself fully expresses all its possible revisions: $\phi > \psi \in K$ iff $\psi \in K * \phi$ — whatever "$>$ " may be.

Note that the deduction theorem and the Ramsey test have the same form: they tell how logical operators migrate through $\models$ or $\vdash$. In the case of the deduction theorem, $\wedge$ on the left changes to $\rightarrow$ on the right, and vice versa: $\phi \wedge \psi \models \sigma$ iff $\phi \models \psi \rightarrow \sigma$. The Ramsey test would like to have s.t. like $\phi * \psi \models \sigma$ iff $\phi \models \psi > \sigma$ for some kind of operator $>$. We now have the (perhaps embarrassing) result that if $M(\phi) \subseteq M(\phi')$, then $\phi' * \psi \models \sigma$ implies $\phi' \models \psi > \sigma$, so $\phi \models \psi > \sigma$, and $\phi * \psi \models \sigma$, and this is quite a restriction under normal circumstances (and is heavily violated in usual revision). It is a property of individual evaluation (like counterfactuals or classical modal logic), and it seems accepted that theory revision does not follow this property.

There are several ways out:

(1) first, we go from K first to some $f(K)$ (like in nonmonotonic reasoning), and evaluate the Ramsey test there: $\phi > \psi \in f(K)$.

(2) We accept monotonicity, but put it on a higher level, i.e. "$>$ " is not a symbol of the base language, so we can have low level nonmonotonicity. (Take, e.g. distance based revision for propositional logic, and just add as new formulas all $\phi > \psi$ corresponding to the Ramsey test. Evaluate classical formulas as usual, and Ramsey formulas by the Ramsey test. This does at least the trick for the case without iteration.)

We will finally put $\phi * \psi$ into the object language. The formula $\phi * \psi$ will evaluate true in those models of ψ, which are closest to $\phi-$models (thus, $\phi * \psi$ cannot evaluate true at the same time as ϕ if ϕ and ψ are inconsistent). "$\sigma \in \phi * \psi$" is now replaced by $\vdash \phi * \psi \rightarrow \sigma$, and we have made ϕ explicit as starting point of the revision, and not implicit as in the Ramsey test.

We now address in more formal detail the AGM approach to theory revision.

2.2.10.2 The AGM approach

We give here in very very rough outline the essentials of the AGM approach. Definitions and results are due to AGM. We present in parallel the logical and the semantic (or purely algebraic) side. For the latter, we work in some fixed universe U, and the intuition is $U = M_{\mathcal{L}}$, $X = M(K)$, etc., so, e.g. $A \in K$ becomes $X \subseteq B$, etc. The translation is trivial, but worth while being written down, otherwise one always has to translate on the fly, and risks to make mistakes. (For reasons of readability, we omit all caveats about definability.)

Definition 2.2.1

$K_\perp$ will denote the inconsistent theory, contraction will be written $-$ $(\ominus)$, revision $*$ $(|)$ on the language (algebraic) side.

The following Definitions 2.2.2 and 2.2.3 contain the principal concepts of AGM theory revision. Propositions 2.2.2 and 2.2.3 show the core equivalences between revision, contraction, and epistemic entrenchment. The reader interested in the subject should become familiar with these definitions and facts.

Definition 2.2.2

We consider two functions, $-$ and $*$, taking a deductively closed theory and a formula as arguments, and returning a (deductively closed) theory on the logics side. The algebraic counterparts work on definable model sets. It is obvious that $(K-1)$, $(K*1)$, $(K-6)$, $(K*6)$ have vacuously true counterparts on the semantical side. Note that K (X) will never change, everything is relative to fixed K (X). $K*\phi$ is the result of revising K with ϕ. $K-\phi$ is the result of subtracting enough from K to be able to add $\neg\phi$ in a reasonable way.

If they satisfy the following "rationality postulates" for -:

$(K-1)$	$K-A$ is deductively closed		
$(K-2)$	$K-A\subseteq K$	$(X\ominus 2)$	$X\subseteq X\ominus A$
$(K-3)$	$A\notin K\Rightarrow K-A=K$	$(X\ominus 3)$	$X\not\subseteq A\Rightarrow X\ominus A=X$
$(K-4)$	$\nvdash A\Rightarrow A\notin K-A$	$(X\ominus 4)$	$A\neq U\Rightarrow X\ominus A\not\subseteq A$
$(K-5)$	$K\subseteq\overline{(K-A)\cup\{A\}}$	$(X\ominus 5)$	$(X\ominus A)\cap A\subseteq X$
$(K-6)$	$\vdash A\leftrightarrow B\Rightarrow K-A=K-B$		
$(K-7)$	$(K-A)\cap(K-B)\subseteq K-(A\wedge B)$	$(X\ominus 7)$	$X\ominus(A\cap B)\subseteq(X\ominus A)\cup(X\ominus B)$
$(K-8)$	$A\notin K-(A\wedge B)\Rightarrow K-(A\wedge B)\subseteq K-A$	$(X\ominus 8)$	$X\ominus(A\cap B)\not\subseteq A\Rightarrow X\ominus A\subseteq X\ominus(A\cap B)$

and for $*$

$(K*1)$	$K*A$ is deductively closed		
$(K*2)$	$A\in K*A$	$(X\mid 2)$	$X\mid A\subseteq A$
$(K*3)$	$K*A\subseteq\overline{K\cup\{A\}}$	$(X\mid 3)$	$X\cap A\subseteq X\mid A$
$(K*4)$	$\neg A\notin K\Rightarrow\overline{K\cup\{A\}}\subseteq K*A$	$(X\mid 4)$	$X\cap A\neq\emptyset\Rightarrow X\mid A\subseteq X\cap A$
$(K*5)$	$K*A=K_\perp\Rightarrow\vdash\neg A$	$(X\mid 5)$	$X\mid A=\emptyset\Rightarrow A=\emptyset$
$(K*6)$	$\vdash A\leftrightarrow B\Rightarrow K*A=K*B$		
$(K*7)$	$K*(A\wedge B)\subseteq\overline{(K*A)\cup\{B\}}$	$(X\mid 7)$	$(X\mid A)\cap B\subseteq X\mid(A\cap B)$
$(K*8)$	$\neg B\notin K*A\Rightarrow\overline{(K*A)\cup\{B\}}\subseteq K*(A\wedge B)$	$(X\mid 8)$	$(X\mid A)\cap B\neq\emptyset\Rightarrow X\mid(A\cap B)\subseteq(X\mid A)\cap B$

they are called a (syntactical or semantical) contraction and revision function respectively.

Remark 2.2.1

(1) Note that $(X \mid 7)$ and $(X \mid 8)$ express a central condition for ranked structures, see Section 3.10: If we note $X \mid .$ by $f_X(.)$, we then have: $f_X(A) \cap B \neq \emptyset \Rightarrow f_X(A \cap B) = f_X(A) \cap B$.

(2) It is trivial to see that AGM revision cannot be defined by an individual distance (see Definition 2.3.5 below): Suppose $X \mid Y := \{y \in Y : \exists x_y \in X(\forall y' \in Y.d(x_y, y) \leq d(x_y, y'))\}$. Consider a, b, c. $\{a, b\} \mid \{b, c\} = \{b\}$ by $(X \mid 3)$ and $(X \mid 4)$, so $d(a, b) < d(a, c)$. But on the other hand $\{a, c\} \mid \{b, c\} = \{c\}$, so $d(a, b) > d(a, c)$, contradiction.

Proposition 2.2.2

Both notions are interdefinable by the following equations:

$$\begin{array}{rl} K * A := \overline{(K - \neg A) \cup \{A\}} & X \mid A := (X \ominus \boldsymbol{C}A) \cap A \\ K - A := K \cap (K * \neg A) & X \ominus A := X \cup (X \mid \boldsymbol{C}A) \end{array}$$

i.e., if the defining side has the respective properties, so will the defined side. □

Definition 2.2.3

Let $\leq_K$ be a relation on the formulas relative to a deductively closed theory K on the formulas of $\mathcal{L}$, and $\leq_X$ a relation on $\mathcal{P}(U)$ or a suitable subset of $\mathcal{P}(U)$ relative to fixed X s.t.

(EE1)	$\leq_K$ is transitive	(EE1)	$\leq_X$ is transitive
(EE2)	$A \vdash B \Rightarrow A \leq_K B$	(EE2)	$A \subseteq B \Rightarrow A \leq_X B$
(EE3)	$\forall A, B$ $(A \leq_K A \wedge B$ or $B \leq_K A \wedge B)$	(EE3)	$\forall A, B$ $(A \leq_X A \cap B$ or $B \leq_X A \cap B)$
(EE4)	$K \neq K_\perp \Rightarrow$ $(A \notin K$ iff $\forall B.A \leq_K B)$	(EE4)	$X \neq \emptyset \Rightarrow$ $(X \not\subseteq A$ iff $\forall B.A \leq_X B)$
(EE5)	$\forall B.B \leq_K A \Rightarrow \vdash A$	(EE5)	$\forall B.B \leq_X A \Rightarrow A = U$

We then call $\leq_K$ $(\leq_X)$ a relation of epistemic entrenchment for K (X). When the context is clear, we simply write $\leq$.

A remark on intuition:

The idea of epistemic entrenchment is that ϕ is more entrenched than ψ (relative to K) iff $M(\neg\psi)$ is closer to $M(K)$ than $M(\neg\phi)$ is to $M(K)$. In shorthand, the more we can twiggle K without reaching $\neg\phi$, the more ϕ is entrenched. Truth is maximally entrenched — no twiggling whatever will reach falsity. Another word for entrenchment is certainty, see Section 2.2.5. The more ϕ is entrenched, the more we are certain about it. Seen this way, the properties of epistemic entrenchment relations are very natural (and trivial): As only the closest points of $M(\neg\phi)$ count (seen from $M(K)$), ϕ or ψ will be as entrenched as $\phi \wedge \psi$, and there is a logically strongest ϕ' which is as entrenched as ϕ — this is just the sphere around $M(K)$ with radius $d(M(K), M(\neg\phi))$.

Again, we have an interdefinability result:

Proposition 2.2.3

The function $K-$ $(X\ominus)$ and the ordering $\leq_K$ $(\leq_X)$ are interdefinable in the following sense:

Define $K - A$ by $B \in K - A :\leftrightarrow B \in K$ and $(A <_K A \vee B$ or $\vdash A)$ $(A <_K B$ means: $A \leq_K B$, and not $B \leq_K A)$

and

$X \ominus A := X$ iff $A = U$, and $\bigcap\{B : X \subseteq B \subseteq U,\ A <_X A \cup B\}$ otherwise.

Define $A \leq_K B$ by $A \leq_K B :\leftrightarrow A \notin K - (A \wedge B)$ or $\vdash A \wedge B$

and

$A \leq_X B :\leftrightarrow A, B = U$ or $X \ominus (A \cap B) \not\subseteq A$.

Then, if the defining side has the respective properties, so will the defined side. □

Taking up the intuition behind epistemic entrenchment, $K * \phi$ is (the theory defined by) that part (of the models) of ϕ, which is closest to K (or its models). $K - \phi$ is (the theory defined by) $M(K) \cup M(K * \neg\phi)$. Once you follow this intuition, this is all trivial. Note that the resulting formalism is not necessarily monotone in the first argument: if $X \subseteq Y$, then not necessarily $X \mid Z \subseteq Y \mid Z$ — as would be the case if the distance would be applied individually — see also above remark on the Ramsey test.

2.2.11 Update

Theory update is about changing situations. The problem can be posed in a number of ways. In its most general form, it is probably as follows: I have information about the situation at time t_n, t_m, t_p, etc. and I would like to make an educated guess about the situation at time t_k, where t_k might be before or after t_n, t_p, etc. The general hypothesis is that of inertia: things stay the same as long as there is no reason that they change. Thus, developments of minimal change are considered more normal or likely. Consequently, we have to consider how we measure or compare the change of developments. In any case, this will be equivalent to an (abstract) distance.

As we speak about real developments, we will usually consider threads of models, and thus the choice by distance will be elementwise (individually), and not collectively (as usual for theory revision) — see Definition 2.3.5 below.

One thread of development involves normally several changes. (Otherwise we have just the situation of counterfactuals, see Section 2.2.13. This is the difference made by Katsuno/Mendelzon between theory revision and theory update, see [KM90]. According to their analysis, theory revision is done by collective distance, theory update by individual distance (in our terminology), see Definition 2.3.5 below.) Thus, to measure the global change, we have to form somehow a sum — in the roughest, least abstract form, this is just the set of changes. We can then measure or compare these sums. The choice of the best or normal cases can then be made in various ways and along the same lines as for nonmonotonic logics.

- In Section 6.3, we will examine real sums of distances for update.
- D. Lehmann has suggested a very simple criterion: proper subsequences of changes are more plausible than the longer sequence, see [BLS99] for a discussion and representation result.

Thus, we have based update on some sort of sums of distances — or a special ranking of sequences.

2.2.12 Counterfactual conditionals

Counterfactual conditionals speak about possible, but not actually true cases: if it were to rain, I would open my umbrella (but it does not rain).

The — in the author's opinion very natural — semantics of Stalnaker/Lewis tries to capture this by coding the minimal change in a distance: very unlikely worlds (where, e.g. there is always a storm of 200 km/h, and using an umbrella has no sense) are distant, and it suffices to look at those worlds which differ minimally from the actual one, but where it rains. More formally:

The intuition for counterfactual conditionals seems relatively clear: If we want to evaluate $\phi > \psi$ at world m, we look at the closest (to m) worlds m' where ϕ holds, and $m \models \phi > \psi$ iff ψ holds there.

We should note two things here:

- this is the individual variant of distance, i.e. we take the pointwise closest elements — see Definition 2.3.5 below.

- there has been some discussion whether there is one global distance over the whole universe, or whether the distances might vary for the different m — which might contradict the existence of a global distance. But, as we show in Section 4.3, this does not matter if we are prepared to accept copies: Given a multitude of distances, we can construct a logically equivalent structure with one global distance only.

So, we see again distance at work in counterfactuals.

2.3 Basic semantical concepts

We have seen in the last section a number of concepts, and their (more or less total) reduction to very few basic concepts:

"Size" of sets of worlds or other entities allows us to define "majority" and "usefulness". "Distance" between worlds or other entities allows us to define "neighborhood", "worst or marginal elements", "center" and then "prototypical" (perhaps with the additional notion of a sum), "between" (again perhaps with the additional notion of a sum). We can then define "convex sets", and this gives a notion of "simplicity". A relation of "preference" between worlds or other entities allows to define "normal", "important", perhaps again "worst or marginal elements". Somewhat apart is structural information like "simplicity", which can probably only partly be captured by the notion of "convex".

The aim of Section 2.3 is to discuss and compare the basic notions "preference", "distance", and "size" in more detail. We will define several types of preference relations (smooth, ranked) over several types of domain (injective or 1-copy, and with copies), and with several types of evaluation (minimal variant, limit variant), various types of addition (the maximum wins, etc.), and various types of filters (principal, weak) to capture size, so it will be tedious to extablish a full list of correspondences. We concentrate on some we consider important, and on the conceptual side.

In all cases, it may be impossible to say that someting is big, or that a is preferred to b, or that a is as far as b, but we can sometimes only express relativized versions of these statements, like: "if a is big, then so is b", "if a is preferred to b, then c is preferred to d", etc. Such arguments can be very useful, and a good object language which shows details, should be able to express them.

2.3.1 Preference

A preferential structure is essentially a set of classical models, with a binary relation, just like a Kripke structure, but the relation is used differently.

Historically, preferential structures were first invented by Hansson, [Han69], as a semantics for deontic logics, where the relation expresses degrees of conformity with deontic statements. They were then re-invented by Shoham, [Sho87b], and in the limit version by Bossu/Siegel, [BS85]. The latter has fallen in oblivion, despite its intuitive attractiveness, as it appeared too difficult to manipulate. We will make this impression precise in Section 5.2.3, and show on the other hand that important classes of the limit variant do not go beyond the minimal version (see Section 3.4.1 and 3.10.3). Both versions work with the same kind of structure, a set with a binary relation $\prec$, but they interpret the structure differently. The minimal version considers the minimal models wrt. the relation $\prec$, whereas the limit version considers those formulas, which "hold in the limit". This is made precise below in Definitions 2.3.1 and 2.3.2, and briefly described now.

We work as usual in propositional logic. For a fixed language $\mathcal{L}$, we take the set of (for simplicity) all classical models $M_\mathcal{L}$, with a binary relation $\prec$ on $M_\mathcal{L}$. Let $\mathcal{M} =< M_\mathcal{L}, \prec>$ be the entire structure. Given $X \subseteq M_\mathcal{L}$, $x \in X$ is called $\prec$ $-$minimal (or just minimal when the relation is fixed), in X iff there is no $x' \in X$ $x' \prec x$. Then $\mu(X)$, more precisely $\mu_\prec(X)$, is the set of all $\prec$ $-$minimal elements of X. If $x \prec y$, we sometimes say that x kills y, or that x minimizes y. Note that, with this definition, $\mu(X)$ can be empty, even if X is not, e.g. if we have infinite descending chains or cycles.

We then define $T \models_\mathcal{M} \phi$ iff $\mu(M(T)) \models \phi$ classically, i.e. iff in all minimal models of T, ϕ holds, i.e. iff $\mu(M(T)) \subseteq M(\phi)$. We write sometimes, when the context is clear $T \mid\!\sim \phi$ for $T \models_{\mathcal{M}} \phi$, and $\overline{\overline{T}}$ for the set of all $\models_{\mathcal{M}}$ $-$consequences. Thus, even if T is consistent, $\overline{\overline{T}}$ will be inconsistent if $\mu(M(T)) = \emptyset$.

This leads us to the second, more general and much more difficult to manipulate, version, which we call the limit version. Let again $\mathcal{M}$ be given, and let $X \subseteq M_{\mathcal{L}}$. We consider here sets $Y \subseteq X$ with the following property:

For all $x \in X$ there is $y \in Y$ s.t. $y \preceq x$ (i.e. $y \prec x$ or $y = x$), and if $y \in Y$, $x \in X$, $x \prec y$, then $y \in Y$. So Y is then downward closed in X, and minimizes all elements of X.

There seems to be no accepted terminology for such sets, we can call them closed minimizing sets, or minimizing initial segments. We will use the latter, and abbreviate as MISE, as any abbreviation containing "C" may cause confusion with consequence, cumulativity, etc. We will use MISE indiscriminately as adjective or noun, in plural or singular.

Note that, if $\prec$ is transitive, and X, X' are MISE of Y, then so is $X \cap X'$. If $\prec$ is not transitive, this need not be the case. In particular, Y is a MISE of Y, and $\emptyset$ is not, unless Y is empty. We then define $T \models_{\mathcal{M}} \phi$ iff there is MISE X of $M(T)$ s.t. $X \models \phi$ classically, and set $\overline{\overline{T}} := \{\phi : T \models_{\mathcal{M}} \phi\}$. As we saw, if T is consistent, $\overline{\overline{T}}$ will not contain $\bot$ either, and if $\prec$ is transitive, $\overline{\overline{T}}$ is closed under (AND), $\phi, \psi \in \overline{\overline{T}} \rightarrow \phi \wedge \psi \in \overline{\overline{T}}$.

We turn to the versions with copies. (People who like to think in terms of labelling functions, will call the version without copies injective, as the labelling function is. But these are just different words for the same thing.) Here, a classical model m might occur several times in the structure, with different positions with respect to $\prec$. First, this has a tradition in modal logic, where we usually consider classically equivalent models several times. Second, it has been argued that we might sometimes be interested in minimizing only for a subset of language, so the rest of it is "invisible" to minimization, and appears only as copies. Third, we will argue in Section 3.9 that the existence of copies makes universal models possible. (Such models make exactly the $\mid\!\sim$ $-$consequences of a theory true, and we do not need several models to see the set of all consequences, as it is the case in classical logic. We work there with strict total orders on the set of models (without copies). Total orders carry maximal preferential information — they decide all odering questions, just as classical models decide all classical formulas. In the completeness proofs, we work then with sets of such total orders, just as is done in classical logic. Such sets are very close to their disjoint union, and thus to preferential structures with copies and partial orders.)

We now define $\mu(X)$ as the set of those $x \in X$ for which there is at least one copy of x, which is minimal in X.

Whatever the justification, structures with copies can do things that structures without copies cannot do. We may need several, logically different, models to minimize one other model. Take two logically equivalent copies of m, m_1 and m_2, and set $m' \prec m_1$, $m'' \prec m_2$. Now $m \in \mu(\{m, m'\})$, $m \in \mu(\{m, m''\})$, but $m \notin \mu(\{m, m', m''\})$. This has immediate repercussions in logic, as the following example shows, which will be restated as Example 3.1.1 in Section 3.1. (The first to publish such examples seem to have been D. Lehmann and his co-authors, but we may safely attribute it to folklore, as such examples must have been re-invented many times.)

Example 2.3.1

Consider the propositional language $\mathcal{L}$ of 2 propositional variables p, q, and the preferential model $\mathcal{M}$ defined by

$m \models p \wedge q$, $m' \models p \wedge q$, $m_2 \models \neg p \wedge q$, $m_3 \models \neg p \wedge \neg q$, with $m_2 \prec m$, $m_3 \prec m'$, and let $\models_{\mathcal{M}}$ be its consequence relation.

Obviously, $Th(m) \vee \{\neg p\} \models_{\mathcal{M}} \neg p$, but there is no complete theory T' s.t. $Th(m) \vee T' \models_{\mathcal{M}} \neg p$. (If there were one, T' would correspond to m, m_2, m_3, or the missing $m_4 \models p \wedge \neg q$, but we need two models to kill all copies of m.) On the other hand, if there were just one copy of m, then one other model would suffice to kill m. More formally, if we admit at most one copy of each model in a structure $\mathcal{M}$, $m \not\models T$, and $Th(m) \vee T \models_{\mathcal{M}} \phi$ for some ϕ s.t. $m \models \neg\phi$ — i.e. m is not minimal in the models of $Th(m) \vee T$ — then there is a complete T' with $T' \vdash T$ and $Th(m) \vee T' \models_{\mathcal{M}} \phi$, i.e. there is m'' with $m'' \models T'$ and $m'' \prec m$. □

The following property is quite often overlooked: Copies give an asymmetry to preferential structures. We may need many x to kill one y, but if x kills y and y' separately, it will kill them all together, too. Thus, one element can be stronger than two others separately, but overcome to their united strength, but the analogue in the other sense is impossible. Shortly, we have some kind of addition on the left, but none on the right. We will come back to this later when we speak about addition.

In all cases, the logical side is only the more or less trivial upper structure, and what we really do (cum grano salis, we will see more fine grained problems when we speak about definability preservation) is to look at the μ function. But once we are at this level of abstraction, it is easy and natural

descending chains, etc.), so, even worse, preferential logic may introduce inconsistencies where there were none classically.

This is different from the above situation with birds and penguins: Here we have a classically inconsistent base theory, above, we had inconsistencies in the "normal" conclusions: birds (normally) fly, penguins normally do not fly, and x is a penguin and a bird, but we have no inconsistency, as we first take the full classical theory, and only then the preferential consequences, and do not mix everything together.

We now make all our definitions official. We will work in some universe U, with a binary relation $\prec$ on U or on a set of copies of elements of U. It is sometimes useful to consider $\mu(X)$ even when $X \not\subseteq U$. This codes that our structure may have "holes" in it. Note that in all definitions normality is relative, and not absolute. We are not necessarily interested in the most normal elements of the universe, it suffices to be most normal in the set considered.

The following Definition 2.3.1 is the central one for preferential models. Together with Definition 2.3.2, they are perhaps the most important definitions of the whole book. The reader who is new to the field should first read the simple versions (without copies, minimal variant), and grasp the essential idea, and then turn to the more complicated cases. He may also leave the limit case by the side until he really needs it.

Definition 2.3.1

Fix $U \neq \emptyset$, and consider arbitrary X. Note that this X has not necessarily anything to do with U, or $\mathcal{U}$ below. Thus, the functions $\mu_{\mathcal{M}}$ below are in principle functions from V to V — where V is the set theoretical universe we work in.

(A) Preferential models or structures.

(1) The version without copies:

A pair $\mathcal{M} :=< U, \prec>$ with U an arbitrary set, and $\prec$ an arbitrary binary relation is called a preferential model or structure.

(2) The version with copies:

A pair $\mathcal{M} :=< \mathcal{U}, \prec>$ with $\mathcal{U}$ an arbitrary set of pairs, and $\prec$ an arbitrary binary relation is called a preferential model or structure.

If $< x, i > \in \mathcal{U}$, then x is intended to be an element of U, and i the index of the copy.

(B) Minimal elements, the functions $\mu_{\mathcal{M}}$, and minimizing initial segments (MISE)

(1) The minimal version or variant:

(1.1) The version without copies:

Let $\mathcal{M} :=< U, \prec>$, and define

$\mu_{\mathcal{M}}(X) := \{x \in X : x \in U \wedge \neg\exists x' \in X \cap U.x' \prec x\}.$

$\mu_{\mathcal{M}}(X)$ is called the set of minimal elements of X (in $\mathcal{M}$).

(1.2) The version with copies:

Let $\mathcal{M} :=< \mathcal{U}, \prec>$ be as above. Define

$\mu_{\mathcal{M}}(X) := \{x \in X : \exists < x, i > \in \mathcal{U}.\neg\exists < x', i' > \in \mathcal{U}(x' \in X \wedge < x', i' >' \prec < x, i >)\}.$

Again, by abuse of language, we say that $\mu_{\mathcal{M}}(X)$ is the set of minimal elements of X in the structure.

Recall that we sometimes say that $< x, i >$ "kills" or "minimizes" $< y, j >$ iff $< x, i > \prec < y, j >$. By abuse of language we also say a set X kills or minimizes a set Y if for all $< y, j > \in \mathcal{U}$, $y \in Y$ there is $< x, i > \in \mathcal{U}$, $x \in X$ s.t. $< x, i > \prec < y, j >$.

(2) The limit version or variant:

(2.1) The version without copies:

Let $\mathcal{M} :=< U, \prec>$. Define

$Y \subseteq X \subseteq U$ is a minimizing initial segment, or MISE, of X iff:

(a) $\forall x \in X \exists x \in Y.y \preceq x$ — where $y \preceq x$ stands for $x \prec y$ or $x = y$

and

(b) $\forall y \in Y, \forall x \in X(x \prec y \rightarrow x \in Y)$.

(2.2) The version with copies:

Let $\mathcal{M} :=< \mathcal{U}, \prec>$ be as above. Define for $Y \subseteq X \subseteq \mathcal{U}$

Y is a minimizing initial segment, or MISE, of X iff:

(a) $\forall < x, i > \in X \exists < y, j > \in Y. < y, j > \preceq < x, i >$

and

(b) $\forall < y, j > \in Y, \forall < x, i > \in X(< x, i > \prec < y, j > \rightarrow < x, i > \in Y)$.

(C)

(1) $\mathcal{M}$ is also called injective or 1-copy, iff there is always at most one copy

$< x, i >$ for each x.

(2) We say that $\mathcal{M}$ is transitive, irreflexive, etc., iff $\prec$ is.

(3) Finally, we say that a set $\mathcal{X}$ of MISE is cofinal in another set of MISE $\mathcal{X}'$ (for the same base set X) iff for all $Y' \in \mathcal{X}'$, there is $Y \in \mathcal{X}$, $Y \subseteq Y'$.

(As we will see in the next definition, cofinal MISE sets are just as good as the original MISE sets: This is just as in analysis, where any cofinal subsequence of a converging sequence converges to the same point.)

In the case of ranked structures (see Definition 2.3.4 below), we may assume without loss of generality that the MISE sets have a particularly simple form, we will postpone the definition until Section 3.10.3, Definition 3.10.4.

Definition 2.3.2

We define the consequence relation of a preferential structure for a given propositional language $\mathcal{L}$.

(A)

(1) If m is a classical model of a language $\mathcal{L}$, we say by abuse of language

$< m, i > \models \phi$ iff $m \models \phi$,

and if X is a set of such pairs, that

$X \models \phi$ iff for all $< m, i > \in X$ $m \models \phi$.

(2) If $\mathcal{M}$ is a preferential structure, and X is a set of $\mathcal{L}-$models for a classical propositional language $\mathcal{L}$, or a set of pairs $< m, i >$, where the m are such models, we call $\mathcal{M}$ a classical preferential structure or model.

(B)

Validity in a preferential structure, or the semantical consequence relation defined by such a structure:

Let $\mathcal{M}$ be as above.

(1) The minimal variant.

We define:

$T \models_{\mathcal{M}} \phi$ iff $\mu_{\mathcal{M}}(M(T)) \models \phi$, i.e. $\mu_{\mathcal{M}}(M(T)) \subseteq M(\phi)$.

We also write sometimes $T^{\mathcal{M}} := \{\phi : T \models_{\mathcal{M}} \phi\}$.

(1.1) $\mathcal{M}$ will be called definability preserving iff for all $X \in \boldsymbol{D}_{\mathcal{L}}$ $\mu_{\mathcal{M}}(X) \in \boldsymbol{D}_{\mathcal{L}}$.

(2) The limit variant.

We define:

$T \models_{\mathcal{M}} \phi$ iff there is a MISE $Y \subseteq \mathcal{U}\lceil M(T)$ s.t. $Y \models \phi$.

($\lceil$ is defined in Definition 1.6.1: $\mathcal{U}\lceil M(T) := \{< x, i > \in \mathcal{U} : x \in M(T)\}$ — if there are no copies, we simplify in the obvious way.)

(2.1) A MISE X is called definable iff $\{x : \exists < x, i > \in X\} \in \boldsymbol{D}_{\mathcal{L}}$.

Discussion of the limit variant:

Note that the same structure can be read in the minimal and the limit variant, and both readings do not always give the same results. In particular, the limit reading is sometimes adequate if there are infinite descending chains without minimal elements.

As interesting as the limit variant seems at first sight, it reveals itself essentially as hopelessly complicated or unnecessarily complicated:

(1) The general case has only arbitrarily complex characterizations, as we will see in Section 5.2.3. The same holds for general ranked structures, and general, distance defined revision.

(2) The natural simpler classes, transitive structures (without copies) with:

  (2.1) cofinally often definable MISE

  (2.2) considering only formulas on the left in the resulting logic

  do not go beyond the much simpler minimal variant. (See Fact 3.4.6 and Fact 3.4.4, summarized in Proposition 3.4.7.)

(3) The natural simpler classes in the ranked case (without copies) with:

  (3.1) cofinally often definable MISE

  (3.2) considering only formulas on the left in the resulting logic

  again do not go beyond the much simpler minimal variant. (See Fact 3.10.18 and Proposition 3.10.19.)

This looks like a coup de grâce for the limit variant as a reasoning tool.

Of course, this does not mean that there might not still exist an interesting subclass which is neither desperate nor trivial. This is an interesting open problem. But, for the meantime, one might probably declare the demise of the limit variant as a natural reasoning tool.

On the other hand, the limit variant might find redemption as a tool to investigate differences between finitary and infinitary versions of logical rules, as can be seen in Section 3.4.1, where we show that finitary cumulativity can hold without the infinitary version, and where we also differentiate between the finitary and the infinitary version of (PR), using the limit version. In a way, this is not surprising: limits speak about infinitary approximations, so we may expect some subtleties when using them as a tool of investigation of infinite properties.

Discussion of the relation:

So far for the basics of the definition, the use of the relation and the question of copies. What about the relation itself?

The most general case is that of an arbitrary binary relation. Perhaps the most natural requirements are freedom from cycles, and transitivity.

We show in Section 3.1 below, Lemma 3.1.1, that we can always choose a cycle-free (and irreflexive) relation generating an equivalent structure, preserving transitivity, too. (We replace cycles by infinite descending chains.)

We show in Section 3.2 and 3.3 that for structures with copies the relation can be chosen transitive without any additional properties in the general and the smooth case. This is not true for structures without copies:

Example 2.3.2

Consider the structure $a \prec b \prec c$, but $a \not\prec c$. This is not equivalent to any transitive structure with one copy each, i.e. there is no transitive structure whose function μ' is equal to μ : As $\mu(\{a,b\}) = \{a\}$, and $\mu(\{b,c\}) = \{b\}$, we must have $a \prec b \prec c$, but then by transitivity $a \prec c$, so $\mu'(\{a,c\}) = \{a\}$ in the new structure, contradicting $\mu(\{a,c\}) = \{a,c\}$ in the old structure.

A strong requirement for the relation, which we find difficult to justify intuitively as a relation property, is smoothness. Essentially, it says that elements are either minimal, or there is a minimal element below them.

Definition 2.3.3

Let $\mathcal{Z} \subseteq \mathcal{P}(U)$. (In applications to logic, $\mathcal{Z}$ will be $\boldsymbol{D}_{\mathcal{L}}$.)

A preferential structure $\mathcal{M}$ is called $\mathcal{Z}-$smooth iff in every $X \in \mathcal{Z}$ every element $x \in X$ is either minimal in X or above an element minimal in X. More precisely:

(1) The version without copies:

If $x \in X \in \mathcal{Z}$, then either $x \in \mu(X)$ or there is $x' \in \mu(X).x' \prec x$.

(2) The version with copies:

If $x \in X \in \mathcal{Z}$, and $< x, i > \in \mathcal{U}$, then either there is no $< x', i' > \in \mathcal{U}$, $x' \in X$, $< x', i' > \prec < x, i >$ or there is $< x', i' > \in \mathcal{U}$, $< x', i' > \prec < x, i >$, $x' \in X$, s.t. there is no $< x'', i'' > \in \mathcal{U}$, $x'' \in X$, with $< x'', i'' > \prec < x', i' >$.

When considering the models of a language $\mathcal{L}$, $\mathcal{M}$ will be called smooth iff it is $\boldsymbol{D}_{\mathcal{L}}-$smooth; $\boldsymbol{D}_{\mathcal{L}}$ is the default.

In the finite case without copies, smoothness is a trivial consequence of transitivity and lack of cycles. But note that in the other cases infinite descending chains might still exist, even if the smoothness condition holds, they are just "short-circuited": we might have such chains, but below every element in the chain is a minimal element.

The attractiveness of smoothness comes from two sides:

First, it generates a very valuable logical property, cumulativity (CUM): If $\mathcal{M}$ is smooth, and $\overline{\overline{T}}$ is the set of $\models_{\mathcal{M}}$ −consequences, then for $T \subseteq \overline{T'} \subseteq \overline{\overline{T}}$ $\Rightarrow \overline{\overline{T}} = \overline{\overline{T'}}$. We can see this property as normal use of lemmas: We have worked hard and found some lemmas. Now we can take a rest, and come back again with our new lemmas. Adding them to the axioms will neither add new theorems, nor prevent old ones to hold.

Second, for certain approaches, it facilitates completeness proofs, as we can look directly at "ideal" elements, without having to bother about intermediate stages. See in particular the work by Lehmann and his co-authors, [KLM90], [LM92].

We will see in Section 3.6 that cumulativity can also be achieved without smoothness, by topological means.

Another property seems easier to justify intuitively, rankedness:

Definition 2.3.4

A relation $\prec_U$ on U is called ranked iff there is an order-preserving function from U to a total order O, $f : U \rightarrow O$, with $u \prec_U u'$ iff $f(u) \prec_O f(u')$, equivalently, if x and x' are $\prec_U$ −incomparable, then ($y \prec_U x$ iff $y \prec_U x'$) and ($y \succ_U x$ iff $y \succ_U x'$) for all y.

(See Fact 3.10.2.)

The property of rankedness is very strong, and the proof techniques differ substantially between the ranked and the other cases.

It is easily seen (and proved formally in Section 3.10, Lemma 3.10.4) that copies are largely unnecessary in ranked structures. The rough argument is as follows: suppose we have two copies of x, x_1 and x_2, and $y \prec x_1$, $z \prec x_2$. If, e.g. $x_1 \prec x_2$, then x_2 is superflous (at least in the transitive case, and ranked structures are transitive). If x_1 and x_2 are incomparable, then both z and y are smaller than both copies, again they are superfluous.

This might be the place to give another reading for preferential structures: the relation expresses an abstract distance from an ideal point of maximal normality — which might not exist. The lower an element, the more it is an approximation to this point. This interpretation is perhaps particularly adapted to ranked structures, where two incomparable elements have then the same distance from the ideal.

A consequence of rankedness, which seems perhaps too strong intuitively, is that one single smaller element will always suffice to kill arbitrarily many bigger ones. This very strong property reminds us of the consistency property in default reasoning: one single consistent case suffices to make the default "fire".

We can preserve the general idea of ranked structures, but still avoid this perhaps overly strong property, if we have a real distance which cooperates with the ranking: x can kill y only iff y is higher up, and if x is "between" "the ideal case" and y, or if y is behind x, seen from the ideal case. In this way, x can kill several y's, one behind the other, but not on the same level. More precisely: Ranking is seen from an ideal, fixed point I. If we have, in addition, a distance $d(a,b)$ on the domain (and the ideal point I), where a and b can both vary, we can see whether x is between the ideal point and y, or not, i.e. whether $d(I,y) = d(I,x) + d(x,y)$ or not.

We do not know whether there are still other properties of the relation $\prec$ of general interest.

2.3.2 Distance

It is not surprising that we see distance so often as a basic notion in the semantics of common-sense reasoning, which has so much to do with analogy, small changes, and similar concepts. It is more surprising that its important role in many situations has not been more emphasized earlier.

Recall that, when we fix a point x in the universe, and consider only distances from this fixed x, we are in the same situation as in preferential structures. Thus, for fixed x, we can consider partial orders, total orders, ranked orders, smooth ones, etc. So all considerations of the last section apply. In particular, there can be situations where the limit approach seems the more adequate one. This approach is again much more complicated, and sometimes subject to quite sophisticated domain closure properties. Such problems are discussed in Section 4.2.5. Usually, a ranked order, where two distances are either comparabel or equal — with values in the reals or not — will probably appear as the most natural choice.

Some more questions which we can pose already for distances from one fixed point are:

(1) Do we have addition of the values?

(2) Are we only interested in relative values (y is more distant from x than z is), or also in absolute values (like 1.5)? (In the first case, we can make the triangle inequality hold at no extra cost by choosing the values as 0 or in the interval from 0.5 to 1.0.)

Usually, however, we measure the distance from different points. It is not surprising that this makes the situation much more complicated. We have thus a number of new questions.

(1) Is the distance symmetric? Often yes, but think of the distance between Bachenthal and Ebershöhe — where Bachenthal is in the valley, and Ebershöhe on a hill, and you are cycling.

(2) Can we compare distances from different points? We will see in Section 4.2 that there are realistic situations where this is impossible, and that such situations can cause much trouble, and make finite representation sometimes impossible.

(3) Do we measure distance independently from each point, or do we respect coherences which are present with a common measure — see Section 4.3.

(4) We saw above (to be proved formally in Section 3.10) that copies have almost no meaning in ranked structures. But these are copies seen from a fixed point x. Here, we may make copies also of the points of origin (i.e. on the left). We will see again in Section 4.3 that this is an interesting question, even if the distance is ranked (and even if it is a metric).

(5) Are we only interested in the closest elements, or do the others come into consideration, too (with perhaps less influence)? In the latter case, is there a more sophisticated way of weighting involved? Dually, are we interested in the farthest elements (perhaps to be avoided in a very cautious approach)?

(6) When considering the closest elements only, are we interested in a collective approach, i.e. looking at the closest to the set of origin, or do we take an individual approach, i.e. look from each element of the set of origin to the ones closest to this element? The first approach is connected to theory revision, the second to counterfactual conditionals (Lewis/Stalnaker semantics) and update (see the work of Katsuno/Mendelsohn).

We make this formal in the following Definition 2.3.5, which is again at a strategic point of the book, in particular for Chapter 4.

Definition 2.3.5

We define the collective and the individual variant of choosing the closest elements in the second operand by:

$X \mid Y := \{y \in Y : \exists x_y \in X.\forall x' \in X, \forall y' \in Y(d(x_y, y) \leq d(x', y'))\}$

(the collective variant)

and

$X \uparrow Y := \{y \in Y : \exists x_y \in X.\forall y' \in Y(d(x_y, y) \leq d(x_y, y'))\}$

(the individual variant).

It might be worth while to stress already here that we will often encounter domain closure (or, if you prefer observability) problems, which make representation complicated, preventing finite characterizations. They occur so often in this context for the following intuitive reason: When looking for close elements, other elements may interfere, and obscure some observations. On the positive side, this allows uniform constructions (uniform metrics for counterfactuals, see Section 4.3), on the negative side, we cannot always compare distances, and paths can be partly hidden, forcing us to take infinite characterizations, see Section 4.2.4.

2.3.3 Size

There are different ways to determine the size of a subset, e.g.:

(1) by counting in the finite case,

(2) by using some measure in the sense of mathematical measure theory,

(3) by a (weak) filter.

We discuss only the concepts of (weak) filters in more detail. Counting is trivial, and measure theory is amply discussed in mathematical textbooks. Both counting and measure theory do not seem to play a very important role in common-sense reasoning, perhaps because their distinctions are usually finer than those of (weak) filters.

Recall the definition of a filter from definition 1.6.2. If $X' \subseteq X$, then the set of all Y s.t. $X' \subseteq Y \subseteq X$ is a filter over X. Such filters, which are generated by a smallest set, are called principal filters. Trivially, all filters over a finite set are principal filters. The following is an example of a nonprincipal filter: Take all subsets A of the set of natural numbers $\boldsymbol{N}$ s.t. $\boldsymbol{N} - A$ is finite. Obviously, the only filters compatible with counting in the finite case are the trivial ones: All sets are big, or none but the universe is big. If the universe has size one in some measure, all sets of size one form a filter.

A principal weak filter is automatically a filter, but we can, of course, look at minimal big sets in a weak filter, if they exist.

A (weak) filter is a mathematical abstraction, capturing the essential aspects of the intuitive abstractions "small and big subsets" — much coarser notions than those given by counting or measure theory, and thus perhaps better suited to capture rough common-sense reasoning. We should recall however, that the notion of a filter was created for other purposes than ours, so there is a priori no need to take it over without change. Of course, we may become convinced that this abstraction is also the right thing for us, but we should keep in mind that abstractions should follow the essential intuitive aspects, and not the other way round.

Real filters, by their idealistic approach, cooperate well with other notions, e.g. that of a proof: Say that all axioms suppose that the number of their exceptions is small, and all rules have the form "if $\alpha_1 \ldots \alpha_2$ hold, then so will α, and the set of exceptions to α is a subset of the union of the exceptions to the α_i", then proofs (by their finiteness) preserve smallness of the set of exceptions, so filters allow us to forget the history of deduction. But this does not mean that filters are always the adequate notion, as growing uncertainty in common-sense deduction, or the lottery paradox, show. Depending on the kind of reasoning, we might conclude that our results get ever weaker in certainty as reasoning goes on.

We turn to the question how much bigger than small is big. We thus look in more detail into the "algebra of small/big", and assume $U \neq \emptyset$ to avoid trivialities.

If the notions "small" and "big" are to have any meaning, no set should be small and big at the same time. Consequently, the degenerate filter

$\mathcal{F} = \mathcal{P}(U)$ should be excluded. By the same reasoning, and recalling that the complements of big sets are small, the union of two small sets should never be the universe: otherwise, both will be small and big.

In normal filters, no finite union of small sets will ever be big, it will not even be medium size, as finite unions of small sets stay small. By its very definition, the union of two medium sets can be big (if A is medium, so is its complement), so medium and big are not very far from each other, and "medium" is better considered just an auxiliary notion, without any more profound meaning. In weak filters, the union of two small sets can be big (and, consequently, the union of three small sets can be the universe): Consider an universe of three elements, and every subset of at least two elements is big. So, weak filters satisfy only the minimal requirements for small and big in the general case. On other hand, in a real filter, any finite union of small sets is small (but the union of countably many small sets can be big, and thus even the universe — recall above example of the filter of cofinite subsets of $\boldsymbol{N}$). Thus, weak filters formalize that small sets are smaller than big ones, real filters say that they are much much smaller, it takes infinitely many of them to grow. We can ask whether it is useful to introduce some intermediate notion, is there anything between 3 and infinity? We have some doubts, it seems too arbitrary.

In the following, we do not only consider one filter, but systems of filters over different base sets. Thus, our perspective is more abstract, and we allow for more substitution, we cannot only speak about size relative to a fixed set, but relative to several sets, which we will sometimes call base or reference sets. This will lead naturally to coherence properties, as we will see in a moment.

First, given a system of filters, there is a natural way to define "very small" or "very big": Let $\mathcal{F}$ be filter over U, $X \in \mathcal{F}$, and $\mathcal{F}'$ a filter over $U - X$, with $X' \in \mathcal{F}'$. Then it seems natural to call $X \cup X'$ very big (and thus its complement in U very small), equivalently, if $A \subseteq B$ is small, $B \subseteq C$ is small, then $A \subseteq C$ may be called very small.

We now introduce three conditions which speak about such coherence. We first define, and then discuss them. The subject will be presented and discussed in more detail in Section 7.3. Despite their simplicity, they contain important ideas of many systems of nonmonotonic logics. The reader is advised to become familiar with them, as he or she will find these same ideas under many disguises.

Definition 2.3.6

(Coh0) If A is a small subset of B, then A will also be a small subset of any superset B' of B.

(CohCUM) If A and A' are small subsets of B, then $A - A'$ will still be a small subset of $B - A'$.

(CohRM) If A is a small subset of B, and A' not a big subset of B, i.e. A' has at most medium size, then $A - A'$ will still be a small subset of $B - A'$.

We first discuss (Coh0).

(Coh0) is in parallel to and as natural as the simple filter condition, if $A \subseteq B \subseteq X$, and $A \in \mathcal{F}(X)$, then $B \in \mathcal{F}(X)$:

If "small" and "big" have any abstract meaning, (Coh0) should certainly hold.

When we have a filter on B, and one on B', we can now express the (OR) condition, when we interpret $\alpha \mid\sim \beta$ by "$\alpha \wedge \neg\beta$ is small in α", or, more precisely, "$M(\alpha \wedge \neg\beta)$ is a small subset of $M(\alpha)$": Let A be a small subset of B, A' of B', then A and A' will be small subsets of $B \cup B'$ (by (Coh0)), so by the filter or ideal property $A \cup A'$ will be small in $B \cup B'$. Thus, $\alpha \mid\sim \beta \Rightarrow M(\alpha \wedge \neg\beta) \subseteq M(\alpha)$ is small, $\alpha' \mid\sim \beta \Rightarrow M(\alpha' \wedge \neg\beta) \subseteq M(\alpha')$ is small, so $M(\alpha \wedge \neg\beta) \cup M(\alpha' \wedge \neg\beta) = M((\alpha \vee \alpha') \wedge \neg\beta) \subseteq M(\alpha \vee \alpha')$ is small, so $\alpha \vee \alpha' \mid\sim \beta$.

Condition (CohCUM) is more delicate. (Coh0) says that increasing a base set will keep small sets small. (CohCUM) says that diminishing base sets by a small amount will keep small subsets small. This goes in the wrong direction, so we have to be careful. We cannot diminish arbitrarily, e.g., if A is a small subset of B, A should not be a small subset of $B-(B-A) = A$. It still seems quite safe, if "small" is a robust notion, i.e. defined in an abstract way, and not anew for each set, and, if "small" is sufficiently far from "big", as, for example in a filter. This results now in cautious monotony: If $\beta \mid\sim \alpha$, $\beta \mid\sim \alpha'$, $M(\beta \wedge \neg\alpha')$ is small in $M(\beta)$, so by this principle in $M(\beta \wedge \alpha)$, so $\beta \wedge \alpha \mid\sim \alpha'$ (assume disjointness for simplicity).

There is, however, an important conceptual distinction to make here. Filters express "size" in an abstract way, in the context of NML, $\alpha \mid\sim \beta$ iff the set of $\alpha \wedge \neg\beta$ is small in α. But here, we were interested in "small" changes in the reference set X (or α in our example). So we have two quite different uses of "size", one for NML, abstractly expressed by a filter, the other for coherence conditions. It is possible, but not necessary, to consider both essentially the same notions. But we should not forget that we have two conceptually different uses of size here.

(CohRM) is obviously a stronger variant of (CohCUM).

Now, (in comparison to (CohCUM)) A' can be a medium size subset of B. As a matter of fact, (CohRM) is a very big strengthening of (CohCUM): Consider a principal filter $\mathcal{F} := \{X \subseteq B : B' \subseteq X\}$, $b \in B'$. Then $\{b\}$ has at least medium size, so any small set A of B is smaller than $\{b\}$ — and this is, of course, just rankedness. If we only have (CohCUM), then we need the whole generating set B' to see that A is small. This is the strong substitution property of rankedness: any b as above will show that A is small. It is easy to see that (RM) holds, if (CohRM) holds.

The more we see size as an abstract notion, and the more we see "small" different from "big" (or "medium"), the more we can go from one base set to another and find the same sizes — the more we have coherence when we reason with small and big subsets. (CohCUM) works with iterated use of "small", just as do filters, but not weak filters. So it is not surprising that weak filters and (CohCUM) do not cooperate well: Let A, B, C be small subsets of X — pairwise disjoint, and $A \cup B \cup C = X$, this is possible. By (CohCUM) B and C will be small in $X - A$, so again by (CohCUM) C will be small in $(X - A) - B = C$, but this is absurd.

If we think that filters are too strong, but we still want some coherence, i.e. abstract size, we can consider the following property: If A is a small subset of B, and A' of B', and B and B' are disjoint, then $A \cup A'$ is a small subset of $B \cup B'$. It expresses a uniform approach to size, or distributivity, if you like. It holds, e.g. for considering a small set to be one smaller than a certain fraction when counting. The important point is here that by disjointness, the big subsets do not get "used up". This property generalizes in a straightforward way to the infinite case.

The following idea and considerations are due to David Makinson:

Definition 2.3.7

Let $A \triangle B := (A - B) \cup (B - A)$, the symmetrical difference between A and B. Define $A \sim B$ iff $A \triangle B$ is a small subset of $A \cup B$.

Obviously, $\sim$ is symmetric and reflexive (leaving aside the pathological case where $A = \emptyset$).

Assuming now (Coh0), (CohCUM) and the filter property, then, as it is easy to see:

(1) $\sim$ is transitive,

(2) if $A \sim B$, then $A \sim X$ for any X s.t. $A \cap B \subseteq X \subseteq A \cup B$,

(3) $\sim$ is a congruence relation wrt. $\cup : A \sim A'$, $B \sim B' \rightarrow A \cup B \sim A' \cup B'$.

(4) Note that we do not want full congruence for meet, for even when A is almost the same as B we might intersect each of them with a set X that leaves very small meets, between which the difference remains small in absolute terms but large when compared with the join of the two meets.

(5) Nevertheless, properties (1) and (2) above give us a mitigated kind of congruence wrt. meet: $A \sim B$, i.e. $A \cap X \sim B \cap X$ for all $X \sim A$. Verification: Suppose $A \sim B$ and $X \sim A$. By (2) and symmetry from (1), $A \cap X \sim A$, and likewise using also transitivity from (1) we have $B \cap X \sim B$. So by transitivity again, $A \cap X \sim B \cap X$.

Remarks:

(1) Here transitivity does much of the work.

(2) It should be interesting (and is an exercise/open problem) to see exactly which of these constraints on $\sim$ (and thus, indirectly, on " small") are needed to validate the various postulates of preferential reasoning, under the definition $\alpha \mid\!\sim \beta$ iff $M(\alpha \wedge \beta) \sim M(\alpha)$.

Finally, it can also be reasonable to relativize comparison of size: Inside X, A and $B \subseteq X$ might have different size, but seen from higher up, in a bigger X', this may not any longer be the case, but we will not pursue this any further.

Transformations:

We look now briefly at some transformations of one notion to another:

The construction of filters from preferences:

Given a preferential structure, can we define a filter? First in the nonlimit reading: Yes, even a system of (principal) filters, in two ways. The basic idea is to take the set of minimal elements as the generating sets for principal filters. Given any subset X of the universe (this is the first reason for a system of filters), define $\mathcal{F}(X) := \{Y \subseteq X : \mu(X) \subseteq Y\}$, the filter generated by the set of minimal elements of X. We will see below that such filter systems $\{\mathcal{F}(X) : X \subseteq U\}$ have certain (coherence) properties. Second, given fixed X, we can form a nested system of filters (expressing ever increasing certainty) generated by $\mu(X)$, $\mu(X) \cup \mu(X - \mu(X))$, etc. This seems especially natural in the case of ranked structures.

Now to the limit reading: In the transitive case, the closed minimizing sets form the basis of a filter (they are closed under finite intersections), in the nontransitive case, they form the basis of a weak filter: the intersection of two cannot be empty. We will see below in Sections 3.4.1 and 3.10 that the limit reading and the minimal reading coincide in some cases (if certain closure properties of the domain hold).

We consider now the converse, from filters to preferences:

A suitable system of (principal) filters naturally generates a preferential structure — see the characterizations of the choice functions in Chapter 3.

In the case of ranked structures, we have also characterized suitable bases for filters generated in the limit case, see Section 3.10.

In the following Section 2.3.3.1, we will compare a system of size relations with a system of filters. This will be elaborated in Section 7.3.

2.3.3.1 Sums

Various sums:

From the intuitive point of view, addition is interesting if parallel is different from serial, i.e. if we can do several tasks one after the other, but not in parallel, or if resources are used up.

There are many possible different ways to define sums.

If we consider the size of sets, it is natural to consider addition, which corresponds closely to disjoint union. Assume now a fixed universe U relative to which we calculate sizes, and suppose for simplicity that all sets mentioned are disjoint. As we have already done the case with three values (small, medium, big), we assume more values here. So assume a set of possible sizes $s_1, \ldots, s_k$, which are ordered by a total or partial (transitive) order. We can also postulate that $\emptyset$ has minimal size, and U maximal size. We want to calculate the size S of $X_1 \cup \ldots \cup X_n$ from their individual sizes. Thus, we want to forget about individual sets, and retain only their abstract sizes s_i.

The first possibility to calculate S is "the winner takes it all", i.e. if (in the case of a total order) s_m is the biggest of the s_i, then the sum is just this s_m. A second possibility is to count the number of times the maximum is present, if this is r, then the sum is $r * max$ — whatever this might be. Recall that filters express a very simple addition: No matter how many (but a finite number) small sets we put togther, we will never get one big set. But two medium size sets can be big.

Note that these procedures can be justified by different assumptions about the sizes at hand and using standard addition — under these assumptions they give the same result as the usual sum. If, for instance, we have reasons to believe that the different sizes are very far from each other, for instance $s << s' << s''$, then any set of size s' will obscure anyone of size s, and no matter how many sets of size s' we have, we will never be able to come close to a set of size s'', so mentioning s' several times makes no sense. It all depends on our hypotheses about the case at hand. The closer the values are, the more we have to go into detail, so there seems to be no universal solution.

Addition and preferential structures:

We have seen that preferential structures without copies cannot add. If they have copies, they can add on the left: x and y separately may not suffice to minimize z, but x and y together may. On the other hand, there is no way to add on the right of $\prec$. If we try to capture this, we would have to code this by some additonal information, that one copy can only kill one other copy at the same time. (This is, of course, possible, but would lead beyond traditional structures. Another possibility is to work immediately with sets of elements.)

Filters from addition of individual values:

Far more interesting is the procedure to obtain in a natural way a (system of) filter(s) from a weight on the single elements in the finite case. As a matter of fact, we had used this procedure in the past (without really understanding it) in order to define revision from size — see Section 7.4 below. We present now the general idea. (D. Makinson pointed out to the author that this idea has been treated independently and in a more general form — the first to do so seems to be P. Snow — and is known under the name of "big stepped probabilities", see [Sno94], and, e.g. [BGLS02], [Luk02], [Luk04a], [Luk04b].)

In the finite case, the only filter compatible with counting is the trivial filter. If we give different weight to the elements, the situation is more complicated. Let X be finite and let x be the only element with minimal weight w. Then $X - \{x\}$ is a true subset with maximal weight, and a natural candidate for a big subset, defining the filter $\{X - \{x\}, X\}$. If x, x' have same minimal weight w, then by fairness, both $X - \{x\}$ and $X - \{x'\}$ should be treated the same way, but then their intersection has to be big, too. If x'' has now weight $\leq 2 * w$, then again by fairness wrt. $X - \{x, x'\}$, $X - \{x''\}$ should also be big. In a way, x and x' form a coalition to put x'' on their side. Consequently, $X - \{x, x', x''\}$ should be big, etc.

We have therefore the following definitions:

(1) $A \subseteq X$ is a stable subset of X iff for all $a \in A$ and all $B \subseteq X - A$ $w(a) > w(B)$, where $w(.)$ is the weight of elements, extended to sets by

addition. We can also say that the elements in A are much smaller than the elements outside A.

(2) $\mathcal{F}$ is a filter compatible with w iff $\mathcal{F}$ is generated by a stable set.

Thus, a suitable weight on elements can generate a nested system of filters over the same set — big, very big, very very big, etc. subsets.

The construction described above seems to merit further comments and discussion.

The process is highly asymmetric, as only weaker elements can form coalitions to "knock other elements out", and, by this fact, the process is absolute only on the side of the stronger elements, as the amount of possible coalitions depends on the background set — so possible coalitions are context dependent. This asymmetry reminds us of the asymmetry of preferential models: minimizing elements can form alliances to minimize all copies of another element.

Forming coalitions is not commutative: a might be too weak to win c directly, it first has to win b, and using b as an ally, it can win c. The same procedure can be seen, e.g. in default reasoning: we start with knowing x, from x we conclude y, and we need x and y to conclude z, we cannot reach z first.

We do not use up our forces to win allies, any ally once convinced will only increase our strength. In this sense, the process might differ from progressive reasoning, as progressing in the reasoning might weaken our conviction in the result — they are not independent arguments, but all are based on the start.

In our approach, coalition forming was via sums, of course, there might be other ways to form coalitions. In argumentation, coalition forming will mostly be more symmetric, each side can try to win allies, so our process seems a very special one of argumentation.

A (partially) open problem is representation:

(1) Neglecting context sensitivity, it is easy (in the finite case) to generate any ranked structure by such coalitions: Give to all elements in the top layer weight 1. Let n elements be in the top layer, then give to all elements in the second layer from top weight $n+1$, let there be n' elements, in the next layer, elements will have weight $n + n' * (n+1) + 1$, etc. This will give again the same ranking. Conversely, any such coalition forming will give a ranking.

(2) The case of context sensitive clustering is more complicated: Let $X \subseteq U$ be any subset of the universe. Then, inside X, we form layers, by considering

only the elements of X. Working in a superset $X' \supseteq X$, different layers of X might collaps, as we may have more possibilities to form coalitions. Relative to small sets, we have finer differentiation than relative to bigger sets: "from the point of view of God, dwarfs and giants are alike".

We do not know the requirements of such systems of layers to be generated by coalition forming via sums — this remains a (to our knowledge) open problem.

2.4 Coherence

Coherence conditions are not just nice abstract toys. They allow sometimes reasoning by strictly controlled analogy, and can thus speed up reasoning significantly (in the average case, if we memorize results).

On the abstract level, considering logics as coherent systems

- gives a uniform view on many logical systems, and thus allows the transfer of results from one problem to another,
- simplifies considerably at least intuitive reasoning about such systems,
- thus brings to light more clearly many questions, like
  - representation problems,
  - basic notions like certainty, size, addition, and distance
- generates new approaches to problems like revision (see (3) below).

(1) What is coherence?

The rule

(AND) $\phi \vdash \psi, \phi \vdash \psi' \to \phi \vdash \psi \wedge \psi'$

differs from the rule

(Monotony) $\phi \vdash \psi \to \phi \wedge \phi' \vdash \psi$

as the first one leaves the left hand side of $\vdash$ unchanged, and the second one changes the left hand side.

We call properties of the second type coherence properties.

We will use "coherence" as an expression, but we will not present a general theory of coherence. The notion might even be too general to create a powerful general theory about it.

Thus, coherence properties allow the transfer of conclusions from one situation to another — we do not always have to start from scratch again. This is particularly important for nonmonotonic and similar logics, as we do not necessarily have the strong coherence property of monotony, which allows powerful transfer. The more we have coherence properties, the more we have regularities, the more we can do efficient reasoning. For instance, the rule (CUM) is stronger than what we have in classical logic: If $T \subseteq \overline{T'} \subseteq \overline{\overline{T}}$, then T and T' have exactly the same consequences, as the choice function evaluates to the same results for T and T'.

For the logics we consider, rules of the second type are often more important than those of the first type. In slight exaggeration, one might say that the logics we consider are essentially characterized by properties of the second type, like (OR), (CUM), and (RM). It is not surprising that such properties are at the center of our interest, as we treat mostly "generalized modal logics", which work with suitable choice functions on model sets, e.g. $\mu(M(T))$ in the preferential case. Thus, the properties of $\mid\!\sim$ for fixed ϕ or T on the left will just be the properties of the classical $\vdash$, but usually for a different ϕ' or T', as we do not consider directly $M(\phi)$ or $M(T)$, but rather $M(\phi')$ or $M(T')$, where, e.g. $M(T') = \mu(M(T))$.

In the following technical Chapters 3–7, we will usually characterize the coherence properties of choice functions (and not directly their logics) resulting from structures like preferential models. The coherence properties bring to light the essential properties of the logics we consider in abstract terms, and thus allow meaningful classification, comparison, and parallel development. In the completeness results, coherence conditions are imposed by the underlying structure. When we consider a semantics based on size, coherence conditions will only be plausible, but not imposed. To cite the same example again, if $A \subseteq B \subseteq C$, and A is a big subset of C, then it is plausible, but not necessary, that it should be a big subset of B, too.

It is reasonable and not just a formal exercise to generalize the concept a little in two directions: First, e.g. the usual deduction theorem can be seen as a coherence property: $\phi \wedge \psi \vdash \sigma \Rightarrow \phi \vdash \psi \rightarrow \sigma$ — here, not only the left hand side changes, but the right hand side changes, too, in, of course, well-defined ways. Second, the case of defaults shows, that sometimes, not all properties will be transferred (as in the case of (CUM)), but only those which figure explicitly as a default: normally ϕ will be transferred to subsets of the universe, but not arbitrary ϕ' will be transferred.

Coherence is, of course, at the very center of analogical and related reasoning: we transfer a conclusion ϕ from a situation s to a situation s' which we consider similar. This is ubiquious in common-sense reasoning. We even express it by words like "similar", "irrelevant changes", etc. Abstraction and substitution express the same idea: we can abstract from certain properties, which are irrelevant, and we can substitute one situation for another, without major changes. Of course, what is similar and what is relevant are properties of the world, and not of logics, and we can only reason relative to some such similarity, we cannot find it in our logics itself. We can perhaps conclude that if we are permitted to transfer ϕ from s to s', then also from s to s'' for a class of situations s, s', s'', etc., but we cannot justify it in the absolute — unless we have coded it already into our picture of the world like in cumulative logics.

We can also see coherence as a hypothesis of locality — changing the situation a little will not change consequences very much. It can be found in many considerations, like the help of already existing plans as guidelines, limiting the search space (see, e.g. Bratman [Bra87]). A formal version is to consider intervals, expressing "between": $f([x, x']) \subseteq [f(x), f(x')]$ — a condition of continuity.

(2) Where do we see coherence?

As already said, coherence properties are often generated by underlying structures (preferential relations, distances ...). They are the "glue" which keeps the different bits of information together. When, for instance, we try to construct a preferential relation to represent a logic, or its choice function for model sets, the essential coherence property tells us that if $x \prec y$ holds in X, then so it will in Y, if $X \subseteq Y$. This imposes restrictions on the construction. We cannot have y minimal in Y, but not in X, if $X \subseteq Y$.

This is perhaps the place to describe this role of coherence properties in completeness proofs in somewhat more detail, though still on an abstract level.

In general, we may have the following possible cases for building the structure in completeness proofs:

(1) We have information which generates the structure, i.e. gives precise information. E.g. if $\mu(\{x, y\}) = \{x\}$, then we know that $x \prec y$ (in preferential structures).

(2) Some information excludes certain cases, but does not decide everything, e.g. $\{a, b\} \mid \{c, d\}$ for distance based theory revision does not always allow to have full information, see Example 4.2.1.

(3) Redundancy: We can have more information than needed to generate

the structure, e.g. $\mu(\{x,y\}) = \{x\}$, and $\mu(\{x,y,z\}) = \{x\}$, both tell us $x \prec y$ in the transitive case.

(4) The information has to be coherent (according to the properties of the structure generated, e.g. a ranked order imposes more regularities than a simple relation). E.g. in the example in (3), we cannot have $\mu(\{x,y\}) = \{x\}$, and $\mu(\{x,y,z\}) = \{y\}$.

(5) If we do not have enough information to construct the full structure, we need a uniform method of construction, i.e. of complementation of the missing information, and we should not have to try everything.

Whereas coherence looked at so far is absolute, i.e. expresses a complete transfer of information (any ϕ will be transferred), coherence generated by defaults is the opposite, the information to be transferred is made precise, but the target is much less so — see also below in (5).

(3) A new use of coherence (revision of NML)

Reasoning with the normal case, by its very nature, risks to be wrong. Thus, it has the revision problem built in. Fortunately, it has the solution built in, too.

Suppose $\alpha \mid\!\sim \psi$, but we find α and $\neg\psi$. The problem of theory revision is how far to go. We can go all the way and have nothing but the new information, this is usually too radical. But we have to move some way. Now, in nonmonotonic logic with, e.g. a system of filters, the system itself gives the answer: We weaken $N(X)$ to $N(X) \cup N(X - N(X))$ — the best and second best elements — and see whether this helps. This is perhaps best seen with ranked structures: we go up as far as necessary.

(In [ALS99], we discussed another approach: we revised there one order by another. The reader is referred there for more information.)

(4) Coherence as abstract conclusion transfer

When we look at coherence from a more abstract point of view, we can pose questions like the following for model choice functions f. Do such properties hold as:

$f(f(A)) = f(A)$,

$A \subseteq B \rightarrow f(A) \subseteq f(B)$,

$f(A \cup B) = f(A) \cup f(B)$,

$f(A \cap B) = f(A) \cap f(B)$,

injectivity, surjectivity,

etc.?

And if such properties seem interesting, we can ask what kind of properties they impose on some underlying structure as preferences, etc.

We can also try to find general properties like:

- If we transfer conclusion ϕ from T to T', and T'' is closer to T than T' is, then we should also be able to transfer ϕ to T''. Here, the distance would code our experience with the world. Recall our similar considerations when we shortly described a possible way of reasoning with prototypes or with marginal cases, and extrapolation.
- If we can transfer ϕ from T to T', but only modified to ϕ', and if ψ is close to ϕ, and ψ' close to ϕ', can we then transfer ψ from T to T', with result ψ'?
- If we can transfer ϕ from T to T' and to T'', both stronger than T, can we transfer also to $T' \cup T''$? (The problem of "admissible" predicates.)
- Do we want to put certainty in the process? If we transfer ϕ from T to T', and T'' is more distant from T than T' is, are we less certain to do the transfer to T''?
- How far are we prepared to transfer ϕ from T, given a distance?

In the case of defaults and inheritance, we know which kind of conclusion can be transferred. They are specified explicitly. In the general case, this is not true, and will depend on the situation. For instance, in a static situation, we cannot transfer from point a to point b the exact position they have, but in a dynamic situation, we can say that object x has stayed at the same place over time. Logic cannot tell us which properties to transfer and which not.

Usually, we do not transfer information naively, but we do so because we suspect a common mechanism behind the cases: birds are "built" such that they have wings, lay eggs, etc., cars move because they have an engine inside, etc. We think, information transfer without this background is often unnessarily naive, and this transfer will benefit from a theory of causation (or action) in the background, which may restrict possible changes to those we may reasonably expect. So the (still naive, but perhaps somewhat less so) picture we have in mind is: there is a common principle underlying several cases, this common principle may differ somewhat from one case to the other, but not too much, and, there might be disturbances which prevent the principle to "express" itself (a platonistic idea) — the match might be wet, flying might have become unnecessary in the course of evolution, etc. — see also (5) below. Of course, this is very similar to counterfactuals: in the closest ideal or normal case, something holds.

If we know nothing else, the bigger the distance is between source and destination of information transfer, the bigger the chance that some contradictory information might enter by transfer "from the side" (whatever its strength). If we transfer information from animal to blackbird, there are more possibilities that disturbing information affects blackbirds (but not animals), than if we transfer just from birds to blackbirds. This consideration is based on the criterion of specificity. Of course, there might by a very strong transfer of information over "big" distances. E.g. dead animals will never fly by their own means, and this transfers to small sets of animals in a strong way — in this case, it is even a universal quantifier.

(5) Coherence and range of application

We can see the notion of coherence as the abstract view from above, we see then the whole system.

There is another, more local view, or concept, of reasoning, of rules, or even of logics: Their range of application. This seems to be an important concept for the description and analysis of real human reasoning — and should thus be one for the logics of common-sense reasoning. This concept deserves in the author's opinion future attention and research.

Our "rules of thumb" (even in highly structured reasoning, like in physics, cf. quantum physics, and the theory of relativity, if the author has understood correctly) have the form "birds fly" — but we know that their range of application is not just limited by "bird". One kind of limit is given by specificity. It is also quite possible that one type of logic is well adapted to one situation, and another one for another situation, so even logics may have their range of application. This range of application can have the form

(a) of a restriction of the domain of application, or

(b) the kind of information we reason about can be restricted.

As said already above, there are some forms of such restrictions coded into different logics:

- Quantifiers may be "softened" from $\forall$ to ∇ — an example of restriction type (a).
- (Single) Reiter defaults speak, type (b), about just one formula to

be transferred, but, type (a), restrict the domain only by consistency. (Specificity as limit can be coded, but this is artificial.)

- Rules like cumulativity speak about arbitrary formulas, but, (a), restrict the domain precisely (everything between T and the set of its consequences).

- Inheritance networks make range and kind of information to be transferred precise: (b), the kind of information is limited by the precise and simple language, (a), the range is limited by specificity.

- Judea Pearl's networks of causation do a similar thing: they, (b), have a limited language, and, (a), treat essentially indirect causation: A has B as consequence, B has C as consequence, so C is an indirect consequence of A. In this setting, B can block A's influence on C.

- We often also have competing influences, a hammer and a feather both "feel" gravity and wind, but to different degrees. So gravity's influence is less "extended" to feathers than to hammers. Although a precise, quantitative description is possible in this case, we will often face situations which are not so clear, and where we will have to do with some rough qualitative reasoning, using reasoning with and about ranges of application. We may use some reasoning with distance, e.g. when we think that the influence decreases with distance. Clearly, indirect causation and competing influences are diffent phenomena.

- Interpolation from extreme cases as discussed in Section 2.2.3 is a case of restriction (a).

We can now perhaps describe our reasoning as follows: we have some procedures or rules of reasoning which have proved useful in the past. In different domains, different procedures or logics may prove useful. When confronted with a new problem, we may try to extend the domain of application of the various procedures we know, so to include the new problem. Often, the different candidates will give different results for the problem at hand. As we are reasoning about a not totally known world, we cannot be sure how to proceed, and use some reasoning with and about domains of application.

In the author's opinion, these phenomena deserve a systematic treatment of their own in future research, and their successful analysis can help make the logics of common-sense reasoning better suited for real applications. In short, "range of application" seems to be an important concept of common-sense reasoning, and an interesting problem for future research. This is then a special case of a more general problem: The author is not sure that all important concepts of common-sense reasoning and its formalizations have been found already. Perhaps we are still blindfolded by the habits from classical logic, which was created for other purposes.

Chapter 3

Preferences

3.1 Introduction

Recall that we gave the main concepts and definitions on preferential structures in Section 2.3.1. The reader should be familiar with them, or be prepared to consult them while reading this chapter.

3.1.1 General discussion

This chapter on preferential structures is the formally most developed part of the book. For this reason, the introduction will also be more detailed than that of the other chapters.

We discuss preferential structures, stressing

- general constructions,
- domain closure questions.

Domain closure problems will be seen in particular in Section 3.5 on a counterexample to the KLM characterization (this is the reason we repeat it here), in Section 3.7 on plausibility logic, in Section 3.4.1 and 3.10.3 on the limit version, and in our remarks on definability preservation.

Definability preservation will be assumed in this part, but we will come back to the problem later in Chapter 5, describing problems, as well as general results without definability preservation, and their techniques. The reason we postpone the discussion is, that exactly the same problems are seen in revision, they are solved in the same way, so it seems better to present them after the discussion of revision. We will also show there that a usual characterization in the not definability preserving case is not possible.

In general, one allows multiple copies (or noninjective labelling functions) for preferential structures. This has a tradition in classical modal logics, and there is some justification of this tradition for preferential structures in Section 3.9, where we work with sets of total orders instead of single partial orders. We start with this multi-copy approach, and append in each case a remark on the case with one copy each.

We first present some basic results about preferential structures.

We then treat the (probably) most general case: arbitrary, not necessarily smooth, structures, without transitivity. The technique used there is basic for much of the rest. We show that adding transitivity gives nothing new. Next, we present the general case with one copy each. We then attack the smooth case, again first with copies admitted, then without. Again, with copies, adding transitivity gives nothing new. We finish these cases with a logical characterization, and show that, in two important classes, the limit version is equivalent to the minimal version.

We then turn to ranked structures, show basic properties and characterize the minimal variant. We then turn to the limit variant, and finally show equivalence of the minimal and the limit variant in important cases again.

Recall that we use the Axiom of Choice without further mention.

In more detail:

The present Chapter 3 is one of the central technical parts of the book.

It discusses representation results and problems for various types of preferential structures. It can be read at several levels. First, more superficially, one can look at the main results, without descending to auxiliary lemmas and proofs. At a deeper level, one will read the results of the lemmas, and then go down to read and check all proofs. At an even deeper level, one will try to identify the main problems and ideas which are behind the fundamental questions and the constructions which attack them. For readers interested in this level, we have tried to elaborate the main ideas (at least as far as we did understand them ourselves) before and in the course of the proofs. Beyond these comments, there are results which in themselves reach deeper into the underlying questions.

Among these are:

- all problems and results concerning closure properties of the domain
  - the counterexample to the KLM — with "KLM" we refer indiscriminately to the article [KLM90], the authors Krauss, Lehman, Magidor, or to their approach and technique, context will tell what is meant — characterization and its analysis (see Section 3.5),
  - the negative and positive results for plausibility logic (see Section 3.7),
  - the equivalence of the limit variant or version with the minimal version in certain cases (see Section 3.4.1 and 3.10.3);
- obtaining cumulativity without smoothness (see Section 3.6), by a topological approach;
- a discussion of the limit variant of ranked structures, which uses heavily topological properties, particularly in the examples (see Section 3.10.3 again);
- more detailed investigations in the number of copies of models which might be needed to represent a logic (see Section 3.8 and in a different direction: Section 3.9).

By their very nature, these parts are addressed primarily to those readers who want to do their own research on the subject (or related ones) and progress deeper than the present text does. Even stronger, they are intended as pointers to such deeper problems and perhaps (partial) answers and their more systematic treatment than the author was able to see at the moment of writing this text. They point, we hope, towards a research project on fundamental problems of representation in the domain.

We have tried to reflect the various levels of reading in the exposition, which will not always group matter by contents. Thus, e.g. the problem of multiple copies is just briefly addressed at the beginning (see Example 3.1.1) to motivate their existence, but discussed in more detail only in Section 3.8, in order not to present a less central problem at the very beginning. We then first give the positive representation results for the general, the general transitive, the smooth, and the smooth and transitive case, for the algebraic and the logical case. This is a relatively homogenous development and a successive refinement of the basic technique, and we wanted to present it to the reader as such. The KLM-counterexample and its analysis (see Section 3.5), cumulativity by topological means (see Section 3.6), and plausibility logic (see Section 3.7) are presented after, as they lead into more subtle questions, and risk to deviate the reader from the mainstream development.

There is a major cleavage between the general and smooth case on the one side, and the ranked case on the other. The cleavage is due to the very strong rankedness condition, which groups many elements on the same level. More precisely, if a and b are incomparable (they are on the same level), and c is smaller (or bigger) than a, then it will also be smaller (or bigger) than b. Not only does the importance of multiple copies (almost) disappear (see Section 3.10.1), but we have a strong equivalence between different elements on the same level. This simplifies many things enormously and the techniques used are quite different for the case with and the case without rankedness. This is reflected in the organization of the text, as we first discuss (almost) the complete nonranked case, and turn only then to the ranked case.

The problem of not necessarily definability preserving structures is discussed only in Chapter 5, in order to better elaborate the common approach for preferential structures and distance based revision. We also show there that general, i.e. not necessarily definability preserving preferential structures, as well as the general limit version do not have a normal characterization at all, even if we admit infinitary characterizations. A similar result for ranked structures and general distance based theory revision is also shown. Thus, we choose to group by proof and description techniques, and not by subject, as the former seems to be the more important connection.

The section on total orders as basic entities of preferential reasoning (Section 3.9) stands a little apart, as its main motivation is more philosophical. We give here a justification to the use of copies, and, at the same time, move closer to the notion of a classical model. A classical model has maximal information, it decides every formula. In the context of orders, a total order has maximal information. It is therefore natural to consider total orders as the basic entities or models of preferential reasoning. Completeness proofs will then have the form known from classical logic, i.e. every formula which is not a consequence, has a counter-model, but there need not be universal structures, which make exactly those formulas true, which are consequences. We join the mainstream of preferential structures by considering disjoint unions of such total orders, and have, essentially, copies of classical models.

In general, to explain the main conceptual and proof ideas and techniques, we will try to elaborate in each case

- what has to be done,
- where the main problems are,

- what the main idea of solution is,
- which, if any, problems arise for its execution,
- which the main auxiliary results to be used are,
- the importance of the technique used for the further development,
- a general assessment of the positive and negative aspects of the idea or technique.

As we concentrate on the technical development, we will not discuss the basic concepts presented at some length in the previous Chapter 2 any more.

A suggested sequence of reading for readers interested in proofs

- the introduction (Section 2.3.1), with definitions on the fly, as they are needed (we have grouped them together for easier referencing),
- the basic construction for the not necessarily transitive case (Proposition 3.2.2),
- briefly jump to the logical representation result (Proposition 3.4.1) to get its flavor,
- the basic transitive case (Proposition 3.2.4) or the smooth case without transitivity (Proposition 3.3.4),
- the smooth case with transitivity (Proposition 3.3.8),
- the equivalence of the minimal and the limit variant in special cases (Section 3.4.1),
- the introduction to the ranked case (Section 3.10.1),
- the main results for the minimal variant of the ranked case (Section 3.10.2),
- the equivalence of the minimal and the limit variant in special cases of ranked structures (Section 3.10.3.2),
- the rest in any order.

At the risk of being too redundant, we briefly describe here the main strand of development, though it will be repeated again below, but in more dispersed manner.

We start with the algebraic characterizations. The first case is that of the general preferential structures. We show that they are characterized by the conditions

$(\mu \subseteq)$ $f(X) \subseteq X$

and

(μPR) $X \subseteq Y \rightarrow f(Y) \cap X \subseteq f(X)$.

It is easy to see that they really hold in all preferential structures: minimization is upward absolute, every element which not minimal in some set X, will also be nonminimal in any superset Y of X.

Given then a function with these properties, we have to find a structure which represents it. We take the most general approach possible, and use all elements of X to minimize x, if $x \in X - f(X)$. The use of copies of x (done by indexing with choice functions) is crucial here. The relation defined between the copies is, on the contrary, the dumbest possible, but also the adequate one, given that we should not make an arbitrary choice.

The proof that the construction does what it should is short and straightforward.

Next, we attack the transitive case. We see that indexing by trees is the natural construction, as this allows us to look ahead to all direct and indirect successors and note them as nodes in a tree. We give an intuitive argument why transitivity does not require additional properties, and pour this idea into mathematical form by considering trees which express this.

We discuss shortly the one-copy case. The crucial property is that any element, if it is minimized at all, is minimized by one single element (or several single elements separately). We consider two properties which express this, and see that the finitary version does not suffice for the general case, but the infinitary one does. We also show that transitivity imposes new conditions, if we work with one copy only.

We then turn to the smooth case, start with minimizing by all of $f(X)$, if $x \in X - f(X)$, but we have to do suitable repair operations sufficiently far from X, as we might have destroyed minimality in other sets, and have, in order to achieve smoothness, to add suitable pairs to the relation. This again might have to be repaired, and so on, in an infinite process. The proof needs in a crucial way certain domain closure properties, as will become apparent in Section 3.7 on plausibility logic.

We now combine the ideas from the general transitive case, and from the smooth case, to tackle the transitive smooth case. Indexing is again via trees, to have control over successors, and we have to work again far enough from the original set we treat. The case is complicated by the fact that we also have to stay away from other sets, collected in the inductive process.

Translation into logic is straightforward, using classical soundness and completeness, and definability preservation. Getting rid of definability preservation is a more intricate matter, and discussed in Chapter 5 — essentially, we admit small sets of exceptions, but loose normal characterization. We also show in this Section 3.4 that in important classes (transitive structures, but only formulas on the left of $\mid\sim$, or cofinally many definable closed minimizing sets), the limit version is equivalent to the minimal version.

In Section 3.9, we take a different approach to preferential structures. This section stands a little apart, as its more philosophical motivations and the technical development are closely intertwined; due to its technical character, we have put it in this chapter. First, we consider strict total orders on the set of classical models of the underlying language as the basic entity. Such structures have maximal preferential information, just as classical propositional models have maximal propositional information. Second, we will work in completeness proofs with sets of such total orders and thus again closely follow the strategy of classical logic, whereas the traditional approach for preferential structures works with one canonical structure. More precisely, in classical logic, one shows $T \vdash \phi$ iff $T \models \phi$, by proving soundness, and that for every ϕ s.t. $T \not\vdash \phi$ there is a $T-$model $m_{T,\neg\phi}$ where ϕ fails. In traditional preferential logic, one constructs a canonical structure $\mathcal{M}$, which satisfies exactly the consequences of T. We work in this Section 3.9, just as in classical logic, with sets of structures to show completeness. Third, our approach will also shed new light on the somewhat obscure question of multiple copies (equivalent to noninjective labelling functions) present in most constructions (see, e.g. the work of the author, or [KLM90], [LM92]). In our approach, it is natural to consider disjoint unions of sets of total orders over the classical models. They have (almost) the same properties as these sets have. As disjoint unions are structures with multiple copies, we have thus justified multiple copies of models or labelling functions in a natural way.

Next, we discuss ranked structures. As already said, they are very different from the nonranked case, as the rankedness condition is very strong, and many complicated situations arising in the general case do not apply any more. In particular, the need for and usefulness of copies is drastically restricted. On the other hand, there is a multitude of very similar and natural conditions, which, however, often show subtle differences. For this reason, we first show a number of positive and negative results comparing those conditions, and only then attack representation problems. We give three representation results, the first two differ slightly in their conditions, and the closure conditions imposed on the domain, the last one discusses a more general case.

The part on the limit version of ranked structures is probably the most interesting one of the material on ranked structures. We show by a number of examples, which use divergence between the order limit and the topological limit (in the standard topology of propositional logics) several quite interesting results:

- logics working with formulas or full theories on the left of $\mid\!\sim$ have quite different behavior under the limit version,
- the full theory version is not equivalent to the minimal version.

On the positive side, we show that rankedness makes the limit version amenable, we give a completeness result, and, probably most importantly, show that the limit version is equivalent to the minimal version (not necessarily with the same order, however), as long as we consider just formulas on the left of $\mid\!\sim$, or if the definable closed minimizing sets are cofinal.

In Section 3.2.4, we mention very briefly X-logics, which are in many interesting cases just an unorthodox way to note preferential logics.

Organization:

In Section 3.2, we discuss the minimal variant of general preferential structures. In Section 3.2.1, we work with arbitrarily many copies, and present the most general result about preferential structures we know. We then show in Section 3.2.2 that the transitive case gives nothing new. In Section 3.2.3, we discuss the 1-copy variant, and show that transitivity matters here. X-logics are briefly analyzed in Section 3.2.4.

In Section 3.3, we investigate the minimal variant of smooth structures, first for arbitrarily many copies, show again that transitivity gives nothing new, turn to the 1-copy case, and show again that transitivity matters there.

In Section 3.4, we present the logical characterization of the cases investigated so far. We also show equivalence of the limit and the minimal version in important classes.

In Section 3.5, we give the KLM characterization and show that it does not carry over to the case of full theories.

In Section 3.6, we show that there are also nonsmooth models of cumulativity.

In Section 3.7, we discuss the example of plausibility logic, showing the importance of closure under finite unions for smooth structures: The standard technique will fail, but we can modify it to make it work.

In Section 3.8, we present more detailed results on the role of copies, and in Section 3.9 a different approach to preferential models, based on total orders.

In Section 3.10, we turn to ranked structures. First, in Section 3.10.1, we discuss some basic properties. In Section 3.10.2, we present the minimal version of smooth and not necessarily smooth cases. In Section 3.10.3, we discuss the limit version, and show that in particular but important cases, it is equivalent to the minimal variant — again due to closure properties of the domain.

3.1.2 The basic results

We use the basic definitions from Section 2.3.1 without mentioning. Recall furthermore from Chapter 1, Definition 1.6.1:

A child (or successor) of an element x in a tree t will be a direct child in t. A child of a child, etc. will be called an indirect child. Trees will be supposed to grow downwards, so the root is the top element.

We first show that every preferential structure has an equivalent irreflexive one — perhaps by adding copies.

Lemma 3.1.1

For any preferential structure $\mathcal{Z} =< \mathcal{X}, \prec>$, there is a preferential structure $\mathcal{Z}' :=< \mathcal{X}', \prec'>$ s.t.

(1) $\mu_{\mathcal{Z}} = \mu_{\mathcal{Z}'}$,

(2) $\mathcal{Z}'$ is irreflexive,

(3) if $\mathcal{Z}$ is transitive, then so is $\mathcal{Z}'$.

Proof of Lemma 3.1.1:

Let $\mathcal{X}' := \{< x, < i, n >>: < x, i > \in \mathcal{X},\ n \in \omega\}$ and

$< x', < i', n' >> \prec' < x, < i, n >>$ iff

(i) $n' > n$ and

(ii) $< x', i' > \prec < x, i >$.

(1) Let Y be any set, we have to show $\mu_{\mathcal{Z}}(Y) = \mu_{\mathcal{Z}'}(Y)$.

"$\subseteq$": Suppose $y \in \mu_{\mathcal{Z}}(Y)$, but $y \notin \mu_{\mathcal{Z}'}(Y)$. Take $< y, i > \in \mathcal{X}$ s.t. there is no $< y', i' > \in \mathcal{X} \lceil Y$, $< y', i' > \prec < y, i >$. Consider $u := < y, < i, 0 >> \in \mathcal{X}' \lceil Y$. By $y \notin \mu_{\mathcal{Z}'}$, there is $u' := < y', < i', n' >> \in \mathcal{X}' \lceil Y$, $u' \prec' u$, but then $< y', i' > \prec < y, i >$, contradiction. "$\supseteq$": Suppose $y \in \mu_{\mathcal{Z}'}(Y)$, but $y \notin \mu_{\mathcal{Z}}(Y)$. Take $u := < y, < i, n >> \in \mathcal{X}' \lceil Y$ s.t. there is no $u' := < y', < i', n' >> \in \mathcal{X}' \lceil Y$, $u' \prec' u$. Then $< y, i > \in \mathcal{X} \lceil Y$, so there is $< y', i' > \in \mathcal{X} \lceil Y$ s.t. $< y', i' > \prec < y, i >$. But then $< y', < i', n+1 >> \prec' < y, < i, n >>$, contradiction.

(2) is trivial by the condition $n' > n$.

(3) Let $< x'', < i'', n'' >> \prec' < x', < i', n' >> \prec' < x, < i, n >>$. Then $< x'', i'' > \prec < x', i' > \prec < x, i >$, so by transitivity of $\prec$, $< x'', i'' > \prec < x, i >$. Moreover, $n'' > n' > n$, so $n'' > n$, and thus $< x'', < i'', n'' >> \prec' < x, < i, n >>$. □ (Lemma 3.1.1)

We conclude this introduction with a simple example illustrating the importance of copies. Such examples seem to have appeared for the first time in print in [KLM90], but can probably be attibuted to folklore. A more detailed discussion is in Section 3.8 below.

Example 3.1.1

Consider the propositional language $\mathcal{L}$ of two propositional variables p, q, and the classical preferential model $\mathcal{M}$ defined by

$m \models p \wedge q$, $m' \models p \wedge q$, $m_2 \models \neg p \wedge q$, $m_3 \models \neg p \wedge \neg q$, with $m_2 \prec m$, $m_3 \prec m'$, and let $\models_{\mathcal{M}}$ be its consequence relation. (m and m' are logically identical.)

Obviously, $Th(m) \vee \{\neg p\} \models_{\mathcal{M}} \neg p$, but there is no complete theory T' s.t. $Th(m) \vee T' \models_{\mathcal{M}} \neg p$. (If there were one, T' would correspond to m, m_2, m_3, or the missing $m_4 \models p \wedge \neg q$, but we need two models to kill all copies of m.) On the other hand, if there were just one copy of m, then one other model, i.e. a complete theory would suffice. More formally, if we admit at most one copy of each model in a structure $\mathcal{M}$, $m \not\models T$, and $Th(m) \vee T \models_{\mathcal{M}} \phi$ for some ϕ s.t. $m \models \neg\phi$ — i.e. m is not minimal in the models of $Th(m) \vee T$ — then there is a complete T' with $T' \vdash T$ and $Th(m) \vee T' \models_{\mathcal{M}} \phi$, i.e. there is m'' with $m'' \models T'$ and $m'' \prec m$. □

3.2 General preferential structures

We discuss first general preferential structures with arbitrarily many copies. We recall the main conditions and develop the results.

Condition 3.2.1

For a function $\mu : \mathcal{Y} \to \mathcal{Y}$, we consider the conditions:

$(\mu \subseteq)$ $\mu(X) \subseteq X$,

(μPR) $X \subseteq Y \to \mu(Y) \cap X \subseteq \mu(X)$,

(μCUM) $\mu(X) \subseteq Y \subseteq X \to \mu(X) = \mu(Y)$,

$(\mu\emptyset)$ $\mu(Y) \neq \emptyset$ if $Y \neq \emptyset$

(for all $X, Y \in \mathcal{Y}$).

Note that, if $\mathcal{Y}$ is closed under finite intersections, in the presence of $(\mu \subseteq)$, (μPR) is equivalent to $(\mu PR')$, where

$(\mu PR')$ $\mu(X) \cap Y \subseteq \mu(X \cap Y)$.

Discussion:

There are two main possibilities: an element $x \in X - \mu(X)$ is minimized by some element in X (perhaps by x itself in an infinite descending chain or a cycle), or, we need a set of elements in X to minimize x. The first possibility works directly with elements x, the second variant needs copies: we have to destroy all copies, and for this, we need a full set of elements.

In the general case with copies, transitivity does not need new conditions, the case without copies does.

We present first the algebraic constructions for the general case with copies in its transitive and not necessarily transitive variant. At first sight, it might be surprising that transitivity does not impose additional conditions. But, look: transitivity is about replacement. If $a \succ b \succ c$, then, if the relation is transitive, we can replace b by c to minimize a. This has to be read more precisely in set terms: If a is minimized by a set containing b, we can substitute b by any set minimizing b. Yet, in the most general construction, we will also have b present in any set which minimizes b (it might kill itself), as we do not know in general which c minimizes b. Thus, we replace b by a set containing b, and it is no surprise that the new set also minimizes a. If, on the other hand, we know that c as element or singleton suffices to minimize b, the situation is different, and transitivity may impose supplementary conditions. (On the other hand, as the case of smoothness shows, which is in itself a weak form of transitivity, there are more specific situations, where transitivity does not change conditions either, see the discussion of the smooth case below.)

The main problem in both the transitive and the not necessarily transitive case is to choose the elements which minimize a given element. Usually, we have not enough information to decide which, as a consequence, we need full sets of elements to minimize a given element, and this lack of information is coded into copies, which all together are minimized by such a set of elements. This is the basic construction given in Section 3.2.1 and explained in some more detail below. The construction of the copies is the most general possible, but the construction of the relation is the most brutal possible (all copies of the smaller element are made smaller, where just one would suffice), but, as we have no criterion for decision, choosing all is again the best possible choice. On the other hand, this choice prevents transitivity to hold (see Example 3.2.1). As we cannot do much about the construction of the copies, we have to restrict the relation. This is done in a look-ahead technique of bookkeeping. (As a matter of fact, we even add new copies, but they behave as the old ones do, they are just added for clarity.)

Look now at a relation and the transitivity condition. Suppose $a \succ b$, $a \succ b'$, $b \succ c$, $b \succ c'$, $b' \succ d$, $b' \succ d'$. If we write this down all at once, we have a tree (where branches may "fuse" again, of course, as, e.g. $c = d$ is possible, but this is unimportant here — it seems easier to think in terms of trees). There might be infinite branches, but this is not important, as the transitivity condition speaks about finite chains. Thus, for given a, it suffices to consider the tree of height $\leq \omega$ of direct and indirect successors of a. If we index a with this tree, it fully describes the behavior as far as transitivity is concerned for a. If, then, b is a successor of a, and we take care that its tree is the subtree of a's tree beginning at b, we have "synchronized" a's and b's behavior, and full control about successors. The tree used for bookkeeping describes immediately all direct and indirect successors of a. This idea, which, again, seems the most general and natural possible construction for transitivity, is basic for all transitive constructions in the general and the smooth case. We also conjecture that this type of construction can be used in other contexts. There is a much simpler direct proof, but with a technique we cannot re-use for the smooth case, it is given in Proposition 3.2.8 — a similar result was already shown in [Sch92].

3.2.1 General minimal preferential structures

The following construction, already used in [Sch92], is the basis for all other constructions for nonranked minimal preferential structures in Sections 3.2 and 3.3. Thus, the reader interested in proofs should understand it. For this reason, we explain it more leisurely than the other constructions. The basic idea is very simple. We first describe the problem, and then show how a suitable choice of copies solves it in a very simple manner.

We have to find suitable elements $x' \in X$, $x' \prec x$, if $x \in X - f(X)$. In the general case, we do not have enough information to choose one or more such x', and this will lead us naturally to a construction with copies. One might perhaps say that the use of copies is the key element in the construction which allows to work with less than complete knowledge, it will code the OR of incomplete knowledge.

Suppose then $x \in X - f(X)$.

To represent f, we have to find some $x' \in X$, $x' \prec x$. But we may not know which x'. This is often the main problem in completeness proofs: we have to make some choice, but we do not know which alternative to choose. If there is a (inclusion-) minimal $X' \subseteq X$ s.t. $x \in X' - f(X')$, we could choose in X'. But such a minimal X' need not exist. And even if it were to exist, this need not solve the problem: there might be $X'' \subseteq X'$ s.t. $x \in f(X'')$, and such X'' may even cover all of X' as a set, $X' = \bigcup\{X'' : X'' \subseteq X',$ $x \in f(X'')\}$. The possible existence of copies solves all these problems at the same time. We do as if we knew nothing else than $x \in X - f(X)$. For each $x' \in X$, we make a copy $< x, x' >$ which is minimized by $x' : < x, x' > \succ x'$. (Note that x might minimize itself in an infinite descending chain.) This guarantees that the full set X will minimize all copies of x — but nothing less. If $X' \subset X$, then at least one copy will not be minimized, and x will be in $\mu(X')$. It is a sort of big OR: we need all elements of X to do the minimizing. This settles our X.

Of course, there might be other X' s.t. $x \in X' - f(X')$, even X' with $X' \subset X$. Thus, we have to repeat the procedure for all such X' : every copy of x has to be minimized by by some $x' \in X'$ for any X' s.t. $x \in X' - f(X')$. This is a kind of big AND: it has to be minimized by all such X'. We now have to combine the OR and the AND. The cartesian product does exactly the choice we want to do: we consider all copies $< x, g >$, where $g \in \Pi\{X : x \in X - f(X)\}$, and make $< x, g >$ bigger than any $x' \in ran(g)$. Every such copy will be minimized in all such X, as it chooses some element in X, but nothing less does the job: there will be one copy which is not minimized by any element in any X where $x \in f(X)$. This is proven in

Claim 3.2.1.

Of course, any superset X' of some X which minimizes x, will also minimize x — this is the basic condition any f representable by preferential structures has to satisfy.

The construction is now the most general possible: we cannot do less to code minimization.

There is still one step to do: we have to repeat the construction for all other x. Thus, we will consider copies $< x, g >$ and $< x', g' >$. Now, we have some liberty: It suffices to make one copy $< x', g' >$ smaller than $< x, g >$, if $x' \in ran(g)$, but we do not know which. The best choice at this time seems to be to make ALL such $< x', g' >$ smaller than $< x, f >$, independent from g'. This is the most "brutal" construction possible — justified, of course, by our lack of knowledge —, and it is important to see that we could do with much less. We will refine the construction for the transitive case, where we want exactly to have more control over the copies $< x', g' >$ which are smaller than $< x, g >$ for given x'.

To summarize:

The construction is basic for many results on preferential structures. We code our lack of knowledge into copies, and combine the conditions on several sets by using the cartesian product. The construction is then straightforward. It is most general in one aspect (choice of copies), and most brutal in another (choice of the relation), showing where refinements are possible.

At the price of being perhaps overly redundant, we give an example: Suppose $x \in X - \mu(X)$. If μ is to be represented by a preferential structure $\mathcal{Z}$, x, (or, more precisely, all copies of x) cannot be minimal in the structure. Thus, for each copy $< x, i >$ of x, there must be some $< x', i' >$ s.t. $x' \in X$, $< x', i' > \prec < x, i >$, i.e. which "kills" $< x, i >$. An element $x \in X$ might be nonminimal in X and X'. At the same time, X might need two elements, say y and y', to kill (all copies of) x, and X' might need two elements, say z and z', to kill x. But we do not know which copy is killed by which elements. If these are all the possibilities to kill the copies of x, any Y which kills x has to contain $\{y, y'\}$ or $\{z, z'\}$. But this is equivalent to the fact that the range of any choice function in the product $\{y, y'\} \times \{z, z'\}$ has nonempty intersection with Y. This is the central idea, to be found in the proof of Claim 3.2.1 below, too.

Definition 3.2.1

For $x \in Z$, let $\mathcal{Y}_x := \{Y \in \mathcal{Y}: x \in Y - \mu(Y)\}$, $\Pi_x := \Pi\mathcal{Y}_x$.

Note that $\emptyset \notin \mathcal{Y}_x$, $\Pi_x \neq \emptyset$, and that $\Pi_x = \{\emptyset\}$ iff $\mathcal{Y}_x = \emptyset$.

The following Claim 3.2.1 is the core of the completeness proof.

Claim 3.2.1

Let $\mu : \mathcal{Y} \to \mathcal{Y}$ satisfy $(\mu \subseteq)$ and (μPR), and let $U \in \mathcal{Y}$. Then $x \in \mu(U) \leftrightarrow x \in U \wedge \exists f \in \Pi_x.ran(f) \cap U = \emptyset$.

Proof:

Case 1: $\mathcal{Y}_x = \emptyset$, thus $\Pi_x = \{\emptyset\}$. "$\to$": Take $f := \emptyset$. "$\leftarrow$": $x \in U \in \mathcal{Y}$, $\mathcal{Y}_x = \emptyset \to x \in \mu(U)$ by definition of $\mathcal{Y}_x$.

Case 2: $\mathcal{Y}_x \neq \emptyset$. "$\to$": Let $x \in \mu(U) \subseteq U$. It suffices to show $Y \in \mathcal{Y}_x \to Y - U \neq \emptyset$. But if $Y \subseteq U$ and $Y \in \mathcal{Y}_x$, then $x \in Y - \mu(Y)$, contradicting (μPR). "$\leftarrow$": If $x \in U - \mu(U)$, then $U \in \mathcal{Y}_x$, so $\forall f \in \Pi_x.ran(f) \cap U \neq \emptyset$. $\square$

Proposition 3.2.2 is the basic result of the whole chapter, with exception of Section 3.10. Most other results and techniques are variations and further developments of the same fundamental idea.

Proposition 3.2.2

An operation $\mu : \mathcal{Y} \to \mathcal{Y}$ is representable by a preferential structure iff μ satisfies $(\mu \subseteq)$ and (μPR).

Proof:

One direction is trivial. The central argument is: If $a \prec b$ in X, and $X \subseteq Y$, then $a \prec b$ in Y, too.

We turn to the other direction. The preferential structure is defined in Construction 3.2.1, Claim 3.2.3 shows representation. The construction has the same role as Proposition 3.2.2 — it is basic for much of the rest of the chapter.

Construction 3.2.1

Let $\mathcal{X} := \{< x, f >: x \in Z \wedge f \in \Pi_x\}$, and $< x', f' > \prec < x, f > :\leftrightarrow x' \in ran(f)$. Let $\mathcal{Z} :=< \mathcal{X}, \prec >$.

Claim 3.2.3

For $U \in \mathcal{Y}$, $\mu(U) = \mu_{\mathcal{Z}}(U)$.

Proof:

By Claim 3.2.1, it suffices to show that for all $U \in \mathcal{Y}$ $x \in \mu_{\mathcal{Z}}(U) \leftrightarrow x \in U$ and $\exists f \in \Pi_x.ran(f) \cap U = \emptyset$. So let $U \in \mathcal{Y}$. "$\rightarrow$": If $x \in \mu_{\mathcal{Z}}(U)$, then there is $< x, f >$ minimal in $\mathcal{X}\lceil U$ (recall from Definition 1.6.1 that $\mathcal{X}\lceil U := \{< x, i > \in \mathcal{X} : x \in U\}$), so $x \in U$, and there is no $< x', f' > \prec < x, f >$, $x' \in U$, so by $\Pi_{x'} \neq \emptyset$ there is no $x' \in ran(f)$, $x' \in U$, but then $ran(f) \cap U = \emptyset$. "$\leftarrow$": If $x \in U$, and there is $f \in \Pi_x$, $ran(f) \cap U = \emptyset$, then $< x, f >$ is minimal in $\mathcal{X}\lceil U$. □ (Claim 3.2.3 and Proposition 3.2.2)

3.2.2 Transitive minimal preferential structures

Discussion:

Construction 3.2.1 (also used in [Sch92]) cannot be made transitive as it is, this will be shown below in Example 3.2.1. The second construction in [Sch92] is a special one, which is transitive, but uses heavily lack of smoothness. (For completeness' sake, we give a similar proof in Proposition 3.2.8.) We present here a more flexibel and more adequate construction, which avoids a certain excess in the relation $\prec$ of the construction in Section 3.2.1: There, too many elements $< y, g >$ are smaller than some $< x, f >$, as the relation is independent from g. This excess prevents transitivity.

We refine now the construction of the relation, to have better control over successors.

Recall that a tree of height $\leq \omega$ seems the right way to encode the successors of an element, as far as transitivity is concerned (which speaks only about finite chains). Now, in the basic construction, different copies have different successors, chosen by different functions (elements of the cartesian product). As it suffices to make one copy of the successor smaller than the element to be minimized, we do the following: Let $< x, g >$, with $g \in \Pi\{X : x \in X - f(X)\}$ be one of the elements of the standard construction. Let $< x', g' >$ be s.t. $x' \in ran(g)$, then we make again copies $< x, g, g' >$, etc. for each such x' and g', and make only $< x', g' >$, but not some other $< x', g'' >$ smaller than $< x, g, g' >$, for some other $g'' \in \Pi\{X' : x' \in X' - f(X')\}$. Thus, we have a much more restricted relation, and much better control over it. More precisely, we make trees, where we mark all direct and indirect successors, and each time the choice is made by the appropriate choice functions of the cartesian product. An element with its tree is a successor of another element with its tree, iff the former is an initial segment of the latter — see the definition in Construction 3.2.2.

Recall also that transitivity is for free as we can use the element itself to minimize it. This is made precise by the use of the trees tf_x for a given element x and choice function f_x. But they also serve another purpose. The trees tf_x are constructed as follows: The root is x, the first branching is done according to f_x, and then we continue with constant choice. Let, e.g. $x' \in ran(f_x)$, we can now always choose x', as it will be a legal successor of x' itself, being present in all X' s.t. $x' \in X' - f(X')$. So we have a tree which branches once, directly above the root, and is then constant without branching. Obviously, this is essentially equivalent to the old construction in the not necessarily transitive case. This shows two things: first, the construction with trees gives the same μ as the old construction with simple choice functions. Second, even if we consider successors of successors, nothing changes: we are still with the old x'. Consequently, considering the transitive closure will not change matters, an element $< x, tf_x >$ will be minimized by its direct successors iff it will be minimized by direct and indirect successors. If you like, the trees tf_x are the mathematical construction expressing the intuition that we know so little about minimization that we have to consider suicide a serious possibility — the intuitive reason why transitivity imposes no new conditions.

To summarize: Trees seem the right way to encode all the information needed for full control over successors for the transitive case. The special trees tf_x show that we have not changed things substantially, i.e. the new $\mu-$functions in the simple case and for the transitive closure stay the same. We hope that this construction will show its usefulness in other contexts, its naturalness and generality seem to be a good promise.

We now give the example which shows that the old construction is too brutal for transitivity to hold.

Recall that transitivity permits substitution in the following sense: If (the two copies of) x is killed by y_1 and y_2 together, and y_1 is killed by z_1 and z_2 together, then x should be killed by z_1, z_2, and y_2 together.

But the old construction substitutes too much: In the old construction, we considered elements $< x, f >$, where $f \in \Pi_x$, with $< y, g > \prec < x, f >$ iff $y \in ran(f)$, independent of g. This construction can, in general, not be made transitive, as the following example shows:

Example 3.2.1

As we consider only one set in each case, we can index with elements, instead of with functions. So suppose $x, y_1, y_2 \in X$, $y_1, z_1, z_2 \in Y$, $x \notin \mu(X)$, $y_1 \notin \mu(Y)$, and that we need y_1 and y_2 to minimize x, so there are two copies $< x, y_1 >$, $< x, y_2 >$, likewise we need z_1 and z_2 to minimize y_1, thus we have $< x, y_1 > \succ < y_1, z_1 >$, $< x, y_1 > \succ < y_1, z_2 >$, $< x, y_2 > \succ y_2$, $< y_1, z_1 > \succ z_1$, $< y_1, z_2 > \succ z_2$ (the z_i and y_2 are not killed). If we take the transitive closure, we have $< x, y_1 > \succ z_k$ for any i, k, so for any z_k $\{z_k, y_2\}$ will minimize all of x, which is not intended. □

The new construction avoids this, as it "looks ahead", and not all elements $< y_1, t_{y_1} >$ are smaller than $< x, t_x >$, where y_1 is a child of x in t_x (or $y_1 \in ran(f)$). The new construction is basically the same as Construction 3.2.1, but avoids to make too many copies smaller than the copy to be killed.

Recall that we need no new properties of μ to achieve transitivity here, as a killed element x might (partially) "commit suicide", i.e. for some i, i' $< x, i > \prec < x, i' >$, so we cannot substitute x by any set which does not contain x : In this simple situation, if $x \in X - \mu(X)$, we cannot find out whether all copies of x are killed by some $y \neq x$, $y \in X$. We can assume without loss of generality that there is an infinite descending chain of $x-$copies, which are not killed by other elements. Thus, we cannot replace any y_i as above by any set which does not contain y_i, but then substitution becomes trivial, as any set substituting y_i has to contain y_i. Thus, we need no new properties to achieve transitivity.

Proposition 3.2.4 is the basic result for the transitive case.

Proposition 3.2.4

An operation $\mu : \mathcal{Y} \to \mathcal{Y}$ is representable by a transitive preferential structure iff μ satisfies $(\mu \subseteq)$ and (μPR).

Proof:

The trivial direction follows from the trivial direction in Proposition 3.2.2.

We turn to the other direction.

There is a trivial proof, given in Proposition 3.2.8. This proof is, however, not very instructive. In particular, it cannot be generalized to the smooth case. We therefore give a more tedious, but also much more instructive, proof, whose main idea works in the smooth case, too.

The preferential structure is defined in Construction 3.2.2, Claim 3.2.6 shows representation for the simple structure, Claim 3.2.7 representation for the transitive closure of the structure.

The main idea is to use the trees tf_x, whose elements are exactly the elements of the range of the choice function f. This makes the constructions of Sections 3.2.1 and 3.2.2 basically equivalent, and shows that the transitive case is characterized by the same conditions as the general case. These trees are defined below in Fact 3.2.5, (3), and used in the proofs of Claims 3.2.6 and 3.2.7.

Again, Construction 3.2.2 contains the basic idea for the treatment of the transitive case. It can certainly be re-used in other contexts.

Construction 3.2.2

(1) For $x \in Z$, let T_x be the set of trees t_x s.t.

(a) all nodes are elements of Z,

(b) the root of t_x is x,

(c) $height(t_x) \leq \omega$,

(d) if y is an element in t_x, then there is $f \in \Pi_y := \Pi\{Y \in \mathcal{Y}\colon y \in Y - \mu(Y)\}$ s.t. the set of children of y is $ran(f)$.

(2) For $x, y \in Z$, $t_x \in T_x$, $t_y \in T_y$, set $t_x \rhd t_y$ iff y is a (direct) child of the root x in t_x, and t_y is the subtree of t_x beginning at y.

(3) Let
$\mathcal{Z} := < \{< x, t_x >\colon x \in Z,\ t_x \in T_x\}\ ,\ < x, t_x > \succ < y, t_y > \text{ iff } t_x \rhd t_y > .$

Fact 3.2.5

(1) The construction ends at some y iff $\mathcal{Y}_y = \emptyset$, consequently $T_x = \{x\}$ iff $\mathcal{Y}_x = \emptyset$. (We identify the tree of height 1 with its root.)

(2) If $\mathcal{Y}_x \neq \emptyset$, tc_x, the totally ordered tree of height ω, branching with $card = 1$, and with all elements equal to x is an element of T_x. Thus, with (1), $T_x \neq \emptyset$ for any x.

(3) If $f \in \Pi_x$, $f \neq \emptyset$, then the tree tf_x with root x and otherwise composed of the subtrees t_y for $y \in ran(f)$, where $t_y := y$ iff $\mathcal{Y}_y = \emptyset$, and $t_y := tc_y$ iff $\mathcal{Y}_y \neq \emptyset$, is an element of T_x. (Level 0 of tf_x has x as element, the t_y's begin at level 1.)

(4) If y is an element in t_x and t_y the subtree of t_x starting at y, then $t_y \in T_y$.

(5) $< x, t_x > \succ < y, t_y >$ implies $y \in ran(f)$ for some $f \in \Pi_x$. □

Claim 3.2.6 shows basic representation.

Claim 3.2.6

$\forall U \in \mathcal{Y}.\mu(U) = \mu_{\mathcal{Z}}(U)$

Proof:

By Claim 3.2.1, it suffices to show that for all $U \in \mathcal{Y}$ $x \in \mu_{\mathcal{Z}}(U) \leftrightarrow x \in U$ $\wedge\, \exists f \in \Pi_x.ran(f) \cap U = \emptyset$. Fix $U \in \mathcal{Y}$.

"$\rightarrow$": $x \in \mu_{\mathcal{Z}}(U) \rightarrow$ there is $< x, t_x >$ minimal in $\mathcal{Z}\lceil U$, thus $x \in U$ and there is no $< y, t_y > \in \mathcal{Z}$, $< y, t_y > \prec < x, t_x >$, $y \in U$. Let f define the set of children of the root x in t_x. If $ran(f) \cap U \neq \emptyset$, if $y \in U$ is a child of x in t_x, and if t_y is the subtree of t_x starting at y, then $t_y \in T_y$ and $< y, t_y > \prec < x, t_x >$, contradicting minimality of $< x, t_x >$ in $\mathcal{Z}\lceil U$. So $ran(f) \cap U = \emptyset$.

"$\leftarrow$": Let $x \in U$. If $\mathcal{Y}_x = \emptyset$, then the tree x has no $\rhd -$successors, and $< x, x >$ is $\succ -$minimal in $\mathcal{Z}$. If $\mathcal{Y}_x \neq \emptyset$ and $f \in \Pi_x$ s.t. $ran(f) \cap U = \emptyset$, then $< x, tf_x >$ is $\succ -$minimal in $\mathcal{Z}\lceil U$. □

We consider now the transitive closure of $\mathcal{Z}$. (Recall that $\prec^*$ denotes the transitive closure of $\prec$.) Claim 3.2.7 shows that transitivity does not destroy what we have achieved. The trees tf_x will play a crucial role in the demonstration.

Claim 3.2.7

Let

$\mathcal{Z}' := < \{< x, t_x >: x \in Z,\ t_x \in T_x\}, < x, t_x > \succ < y, t_y >$ iff $t_x \rhd^* t_y >$.

Then $\mu_{\mathcal{Z}} = \mu_{\mathcal{Z}'}$.

Proof:

Suppose there is $U \in \mathcal{Y}$, $x \in U$, $x \in \mu_{\mathcal{Z}}(U)$, $x \notin \mu_{\mathcal{Z}'}(U)$. Then there must be an element $< x, t_x > \in \mathcal{Z}$ with no $< x, t_x > \succ < y, t_y >$ for any $y \in U$. Let $f \in \Pi_x$ determine the set of children of x in t_x, then $ran(f) \cap U = \emptyset$, consider tf_x. As all elements $\neq x$ of tf_x are already in $ran(f)$, no element of tf_x is in U. Thus there is no $< z, t_z > \prec^* < x, tf_x >$ in $\mathcal{Z}$ with $z \in U$, so

$< x, tf_x >$ is minimal in $\mathcal{Z}'\lceil U$, contradiction. □ (Claim 3.2.7 and Proposition 3.2.4)

We give now the direct proof, which we cannot adapt to the smooth case. Such easy results must be part of the folklore, but we give them for completeness' sake.

Proposition 3.2.8

In the general case, every preferential structure is equivalent to a transitive one — i.e. they have the same $\mu-$functions.

Proof:

If $< a, i >\succ< b, j >$, we create an infinite descending chain of new copies $< b, < j, a, i, n >>$, $n \in \omega$, where $< b, < j, a, i, n >>\succ< b, < j, a, i, n' >>$ if $n' > n$, and make $< a, i >\succ< b, < j, a, i, n >>$ for all $n \in \omega$, but cancel the pair $< a, i >\succ< b, j >$ from the relation (otherwise, we would not have achieved anything), but $< b, j >$ stays as element in the set. Now, the relation is trivially transitive, and all these $< b, < j, a, i, n >>$ just kill themselves, there is no need to minimize them by anything else. We just continued $< a, i >\succ< b, j >$ in a way it cannot bother us. For the $< b, j >$, we do of course the same thing again. So, we have full equivalence, i.e. the $\mu-$functions of both structures are identical (this is trivial to see). □

3.2.3 One copy version

The following material is very simple, and does not require further comments.

The essential property of preferential structures with at most one copy each is that we never need two or more elements to kill one other element. This is expressed by the following proerty, which we give in a finitary and an infinitary version:

Definition 3.2.2

(1-fin) Let $X = A \cup B_1 \cup B_2$ and $A \cap \mu(X) = \emptyset$. Then $A \subseteq (A \cup B_1 - \mu(A \cup B_1)) \cup (A \cup B_2 - \mu(A \cup B_2))$.

(1-infin) Let $X = A \cup \bigcup\{B_i : i \in I\}$ and $A \cap \mu(X) = \emptyset$. Then $A \subseteq \bigcup\{A \cup B_i - \mu(A \cup B_i)\}$.

It is obvious that both hold in 1-copy structures, it is equally obvious that the second guarantees the 1-copy property (consider $X = \{\{x\} : x \in X\}$, if $x \notin \mu(X)$, we find at least one $x' \in X$ s.t. $x \notin \mu(\{x, x'\})$, and this gives the construction for representation, too. It is almost as obvious that the finitary version does not suffice:

Example 3.2.2

Take an infinitary language $\{p, q_i : i \in \omega\}$, and let every $p-$model be killed by any infinite set of $\neg p-$models, and nothing else. Now, if A is minimized by $B_1 \cup B_2$, $B_1 \cup B_2$ contains an infinite number of $\neg p-$models, so either B_1 or B_2 does, so (1-fin) holds, but, obviously, the structure is not equivalent to any structure with one copy at most. □

We turn to transitivity in the 1-copy case. Consider $a \prec b \prec c$, but $a \not\prec c$. So $\mu(\{a, c\}) = \{a, c\}$. By $\mu(\{a\}) = \{a\}$, $\mu(\{b\}) = \{b\}$, $\mu(\{c\}) = \{c\}$, we see that all three elements are present, so each has to be there as one copy. By $\mu(\{a, b\}) = \{a\}$ and $\mu(\{b, c\}) = \{b\}$, we see that $a \prec b \prec c$ has to hold. But then transitivity requires $a \prec c$, thus $\mu(\{a, c\}) = \{a\}$ has to hold, so the present structure is not equivalent to any transitive structure with the 1-copy property. Thus, to have transitivity, we need a supplementary condition:

Definition 3.2.3

(T) $\mu(A \cup B) \subseteq A$, $\mu(B \cup C) \subseteq B \rightarrow \mu(A \cup C) \subseteq A$.

Taking $a \prec b \prec c$ and $A := \{a\}$, $B := \{b\}$, $C := \{c\}$, we see that (T) imposes transitivity on 1-copy structures.

Note, however, that (T) does not necessarily hold in transitive structures with more that one copy — see above example.

3.2.4 A very short remark on X-logics

Introduction:

X-logics were introduced by P. Siegel et al. (see [FRS01]) as an alternative approach to usual nonmonotonic logics. They work with formula sets in an unorthodox way. We translate these logics to model set operators, this makes them easy to understand. We give the natural definition of an extension in these terms, discuss some simple examples, show as our main small result that X-logics with one extension are just preferential systems, and indicate a short proof of the converse in the finite case. It seems easy to obtain further results — if desired — on the subject, using our translation.

Definition 3.2.4

(Taken from [FRS01].) If $\mathcal{X}$, A, B are sets of $\mathcal{L}-$formulas, then $A \mid\sim_{\mathcal{X}} B$ iff $\overline{A \cup B} \cap \mathcal{X} \subseteq \overline{A} \cap \mathcal{X}$, where $\overline{C}$ is as usual the closure of C under classical logic.

Consider the following simple example: Let $\mathcal{L}$ be given, ϕ, ψ, σ be formulas in $\mathcal{L}$, $\mathcal{X} := \{\phi\}$. Let $X := M(\phi)$, $B := M(\psi)$, $C := M(\sigma)$. By definition of X-logic, $\psi \mid\sim_{\mathcal{X}} \sigma$ iff $\overline{\psi \wedge \sigma} \cap \mathcal{X} \subseteq \overline{\psi} \cap \mathcal{X}$, i.e. if $\psi \vdash \phi$, then σ can be anything, and if $\psi \not\vdash \phi$, then $\psi \wedge \sigma \not\vdash \phi$, i.e. $Con(\psi, \sigma, \neg\phi)$ has to hold.

To understand what happens, we translate into model sets: This more or less trivializes the problem. If $B \subseteq X$, then C can be any set, including $\emptyset$. If $B \not\subseteq X$, then $B \cap C \not\subseteq X$, in particular $B \cap C \neq \emptyset$, but apart from this condition, it can again be anything. In particular, for each $x \in B - X$, C can be $\{x\}$. Intuitively, any $X \in \mathcal{X}$ is a kind of asymmetrical border: if you are (partially) outside, $B \not\subseteq X$, you cannot come totally inside, $B \cap C \not\subseteq X$.

It seems plausible to call all minimal C such that $B \cap C \not\subseteq X$ extensions. Thus, simple $\mathcal{X}$'s are plagued with an inflation of extensions. (One might consider several extensions as not one set of associated models, but as a set of models, a sort of "supermodal" logic.)

We consider now richer $\mathcal{X}$, and generalize above analysis. If $\mathcal{X} = \{\phi, \phi'\}$, with $X := M(\phi)$, $X' := M(\phi')$, then $B \not\subseteq X \rightarrow B \cap C \not\subseteq X$ and $B \not\subseteq X' \rightarrow B \cap C \not\subseteq X'$.

For A and $\mathcal{X}$, let $A_{\mathcal{X}} := \{X \in \mathcal{X} : A \not\subseteq X\}$. Thus, we need $\forall X \in A_{\mathcal{X}}$ some $a \in A - X$, i.e. we have to look at choice functions f for $A_{\mathcal{X}}$ with values in $A - X$ for $X \in A_{\mathcal{X}}$. In particular, any extension corresponds now to an f with $\subseteq -$minimal $ran(f)$, or, an extension E of A is a $\subseteq -$minimal subset of A s.t. $E - X \neq \emptyset$ for all $X \in A_{\mathcal{X}}$. Obviously, extensions always exist. If some A has only one extension E, $\overline{\overline{A}}$ is just $Th(E)$. Obviously, AND holds iff there is just one extension. In the general case, $\overline{\overline{A}} = \bigcup\{Th(E) : E$ is an $\mathcal{X}-$extension for $A\}$.

Example: If $X_x := M_{\mathcal{L}} - \{x\}$, and $\mathcal{X} := \{X_x\}$, then any B with $x \in B$ is fixed: its only extension is $\{x\}$. If $\mathcal{X} := \{X_x, X_{x'}\}$, then any B with

$x, x' \in B$ has $\{x, x'\}$ as only extension, and if only $x \in B$, then its only extension is $\{x\}$, etc.

A small side remark on Lemma 2.5 in [FRS01]: Consider $\emptyset \neq D \subseteq M_{\mathcal{L}}$ fixed, and let $\mathcal{X} := \{D \cup (M_{\mathcal{L}} - \{x\}) : x \in M_{\mathcal{L}}\}$. Then, if $B \subseteq D$, its only extension is $\emptyset$, if $B \not\subseteq D$, its only extension is $B - D$. Consequently, AND can also hold if $\mathcal{L} - \mathcal{X}$ is not closed under classical consequence.

We show our main remark:

Fact 3.2.9

If the X-logic has always only one extension, then it is preferential.

Proof:

Suppose that B has only one extension E, and $B' \subseteq B$. We show that any extension of B' has to contain $E \cap B'$. Thus the central condition for preferential structures (μPR) is satisfied, so, in this case, X-logic does not go beyond preferential logic.

Let then $B' \subseteq B$, and $x \in E \cap B'$. We have to show that x is in any extension of B'. By minimality of E for B, there is $A \in B_{\mathcal{X}}$, $E - A = \{x\}$. Let $\mathcal{Y} := \{A \in B_{\mathcal{X}} : E - A = \{x\}\}$. As $x \in B'$, any $A \in \mathcal{Y}$ is in $B'_{\mathcal{X}}$, so for all $A \in \mathcal{Y}$, we need some $y_A \in B' - A$ in any extension of B'. Suppose we can find for all $A \in \mathcal{Y}$ $y_A \in B'-$A, $y_A \neq x$, i.e. replace x by the y_A. But then $E' := (E - \{x\}) \cup \{y_A : A \in \mathcal{Y}\}$ or one of its subsets would be an extension for B, a contradiction: If $A \in A_{\mathcal{X}}$ and $E - A \neq \{x\}$, then $(E - \{x\}) \cap A \neq \emptyset$. If $E - A = \{x\}$, there is a new $y_A \in E'$. □

Conversely (in the finite case):

The receipt to construct X-logic from preferences in a set U (in rough outline):

(1) If a is globally minimal, i.e. $a \in \mu(U)$, we code this by $U - \{a\} \in \mathcal{X}$. If A contains a, then any extension of A is forced to contain a, too.

(2) If b can be minimized by a and a' (separately, no copies), i.e. $a \prec b$, $a' \prec b$, then we put $U - \{b, a, a'\}$ into $\mathcal{X}$. Thus, if A contains just b, it has to be in. If it contains, e.g. a and b, then one of a or b has to be in, but a has to be in anyway provided it is globally minimal, so a alone suffices.

(3) Suppose we need a, a', a'' to minimize the three copies of b, i.e. $a \prec b_0$,

$a' \prec b_1$, $a'' \prec b_2$. We then put $U - \{a, b\}$, $U - \{a', b\}$, $U - \{a'', b\}$ into $\mathcal{X}$. Thus, if A contains, e.g. a, a' and b, but not a'', then, by the last condition, we still have to have b in.

This is essentially the dual construction of the one presented in Theorem 3.1 of [FRS01], with the advantage that we can also cover multiple copies. We leave it to the interested reader to flesh out the details. □

A small warning to the reader interested to pursue the matter: Example 2.3 in [FRS01] seems to be wrong at many points. Consider just the case $A := \emptyset$. Then both $(\neg a \wedge b \wedge \neg f) \vee (\neg a \wedge \neg b \wedge f)$ and $(\neg a \wedge b \wedge f) \vee (\neg a \wedge \neg b \wedge \neg f)$ are candidates, which do not imply any $\phi \in X$. Moreover, any ψ s.t. $\neg a \vdash \psi$ will not entail any $\phi \in X$ either.

3.3 Smooth minimal preferential structures

3.3.1 Smooth minimal preferential structures with arbitrarily many copies

Discussion:

In the smooth case, we know that if $x \in X - f(X)$, then there must be $x' \prec x$, $x' \in f(X)$ (or, more precisely, for each copy $< x, i >$ of x, there must be such x'). Thus, the freedom of choice is smaller, and at first sight, the case seems simpler. The problem is to assure that obtaining minimization for x in X does not destroy smoothness elsewhere, or, if it does, we have to repair it. Recall that smoothness says that if some element is not minimal, then there is a minimal element below it — it does not exclude that there are nonminimal elements below it, it only imposes the existence of minimal elements below it. Thus, if, during construction, we put some nonminimal elements below some element, we can and have to repair this by putting a suitable minimal one below it. Of course, we have to take care that this repairing process does not destroy something else, or, we have to repair this again, etc., and have to assure at the same time that we do not alter the choice function.

The basic idea is thus as follows for some given x, and a copy $< x, \sigma >$ to be constructed ($< x, \sigma >$ will later be minimized by all elements in the ranges of the σ_i which constitute σ) :

- First, we minimize x, where necessary, using the same idea of cartesian product as in the not necessarily smooth case, but this time choosing in $f(Y)$ for suitable $Y : \sigma_0 \in \Pi\{f(Y) : x \in Y - f(Y)\}$.

- This might have caused trouble, if X is such that $x \in f(X)$, and $ran(\sigma_0) \cap X \neq \emptyset$, we have destroyed minimality of the copy $< x, \sigma >$ under construction in X, and have to put a new element minimal in this X below it, to preserve smoothness: $\sigma_1 \in \Pi\{f(X) : x \in f(X)$ and $ran(\sigma_0) \cap X \neq \emptyset\}$.

- Again, we might have caused trouble, as we might have destroyed minimality in some X, this time by the new $ran(\sigma_1)$, so we repeat the procedure for σ_1, and so on, infinitely often.

We then show that for each x and U with $x \in f(U)$ there is such $< x, \sigma >$, s.t. all $ran(\sigma_i)$ have empty intersection with U — this guarantees minimality of x in U for some copy. As a matter of fact, we show a stronger property, that $ran(\sigma_i) \cap H(U) = \emptyset$ for all σ_i, where $H(U)$ is a sufficiently big "hull" around U. The existence of such special $< x, \sigma >$ will also assure smoothness: Again, we make in an excess of relation all copies irrespective of the second coordinate smaller than a given copy. Thus, if an element $< y, \tau >$ for $y \in f(Y)$ is not minimal in the constructed structure, the reason is that for some i $ran(\tau_i) \cap Y \neq \emptyset$. This will be repaired in the next step $i + 1$, by putting some x minimal in Y below it, and as we do not look at the second coordinate, there will be a minimal copy of x, $< x, \sigma >$ below it.

The hull $H(U)$ is defined as $\bigcup\{X : f(X) \subseteq U\}$. The motivation for this definition is that anything inside the hull will be "sucked" into U — any element in the hull will be minimized by some element in some $f(X) \subseteq U$, and thus by U. In particular, it is easier to stay altogether out of $H(U)$ in the inductive construction of σ, than to avoid U directly — which we need for our minimal elements. Note that $H(U)$ need not be an element of the domain, which is not necessarily closed under arbitrary unions. But this does not matter, as $H(U)$ will never appear as an argument of f.

Obviously, the properties of $H(U)$ as shown in Fact 3.3.1 are crucial for the inductive construction of the σ used for minimal elements. Note that closure of the domain under finite unions is used in a crucial way in this Fact 3.3.1.

To summarize: We start in a straightforward manner with σ_0, and repair successively the damage we have done. To make it possible, we use a suitable hull $H(U)$, which we avoid in order to avoid U itself, facilitating thus the construction. Once this idea is clear, and we have found a suitable $H(U)$, the execution is quite straightforward — but the author has to admit that it took him some time and several trials to find a nice $H(U)$, and to see how to use it.

The constructions:

Recall that $\mathcal{Y}$ will be closed under finite unions and finite intersections throughout this Section 3.3. We first define $H(U)$, and show some facts about it. $H(U)$ has an important role in Sections 3.3.1 and 3.3.2, for the following reason: If $u \in \mu(U)$, but $u \in X - \mu(X)$, then there is $x \in \mu(X) - H(U)$. Consequently, to kill minimality of u in X, we can choose $x \in \mu(X) - H(U)$, $x \prec u$, without interfering with u's minimality in U. Moreover, if $x \in Y - \mu(Y)$, then, by $x \notin H(U)$, $\mu(Y) \not\subseteq H(U)$, so we can kill minimality of x in Y by choosing some $y \notin H(U)$. Thus, even in the transitive case, we can leave U to destroy minimality of u in some X, without ever having to come back into U, it suffices to choose sufficiently far from U, i.e. outside $H(U)$. $H(U)$ is the right notion of "neighborhood".

Note: Not all $z \in Z$ have to occur in our structure, therefore it is quite possible that $X \in \mathcal{Y}$, $X \neq \emptyset$, but $\mu_{\mathcal{Z}}(X) = \emptyset$. This is why we have introduced the set K in Definition 3.3.2, and such X will be subsets of $Z - K$.

Let now $\mu : \mathcal{Y} \to \mathcal{Y}$.

Definition 3.3.1

Define $H(U) := \bigcup\{X : \mu(X) \subseteq U\}$.

The following Fact 3.3.1 contains the basic properties of μ and $H(U)$ which we will need for the representation construction.

Fact 3.3.1

Let A, U, U', Y and all A_i be in $\mathcal{Y}$.

$(\mu\subseteq)$ and (μPR) entail:

(1) $A = \bigcup\{A_i : i \in I\} \to \mu(A) \subseteq \bigcup\{\mu(A_i) : i \in I\}$,

(2) $U \subseteq H(U)$, and $U \subseteq U' \to H(U) \subseteq H(U')$,

(3) $\mu(U \cup Y) - H(U) \subseteq \mu(Y)$.

$(\mu\subseteq)$, (μPR), (μCUM) entail:

(4) $U \subseteq A$, $\mu(A) \subseteq H(U) \to \mu(A) \subseteq U$,

(5) $\mu(Y) \subseteq H(U) \to Y \subseteq H(U)$ and $\mu(U \cup Y) = \mu(U)$,

(6) $x \in \mu(U)$, $x \in Y - \mu(Y) \to Y \not\subseteq H(U)$,

(7) $Y \not\subseteq H(U) \to \mu(U \cup Y) \not\subseteq H(U)$.

Proof:

(1) $\mu(A) \cap A_j \subseteq \mu(A_j) \subseteq \bigcup \mu(A_i)$, so by $\mu(A) \subseteq A = \bigcup A_i$ $\mu(A) \subseteq \bigcup \mu(A_i)$.

(2) trivial.

(3) $\mu(U \cup Y) - H(U) \subseteq_{(2)} \mu(U \cup Y) - U \subseteq_{(\mu\subseteq)} \mu(U \cup Y) \cap Y \subseteq_{(\mu PR)} \mu(Y)$.

(4) $\mu(A) = \bigcup\{\mu(A) \cap X : \mu(X) \subseteq U\} \subseteq_{(\mu PR')} \bigcup\{\mu(A \cap X) : \mu(X) \subseteq U\}$. But if $\mu(X) \subseteq U \subseteq A$, then by $\mu(X) \subseteq X$, $\mu(X) \subseteq A \cap X \subseteq X \to_{(\mu CUM)} \mu(A \cap X) = \mu(X) \subseteq U$, so $\mu(A) \subseteq U$.

(5) Let $\mu(Y) \subseteq H(U)$, then by $\mu(U) \subseteq H(U)$ and (1) $\mu(U \cup Y) \subseteq \mu(U) \cup \mu(Y) \subseteq H(U)$, so by (4) $\mu(U \cup Y) \subseteq U$ and $U \cup Y \subseteq H(U)$. Moreover, $\mu(U \cup Y) \subseteq U \subseteq U \cup Y \to_{(\mu CUM)} \mu(U \cup Y) = \mu(U)$.

(6) If not, $Y \subseteq H(U)$, so $\mu(Y) \subseteq H(U)$, so $\mu(U \cup Y) = \mu(U)$ by (5), but $x \in Y - \mu(Y) \to_{(\mu PR)} x \notin \mu(U \cup Y) = \mu(U)$, contradiction.

(7) $\mu(U \cup Y) \subseteq H(U) \to_{(5)} U \cup Y \subseteq H(U)$. □

Assume now $(\mu \subseteq)$, (μPR), (μCUM) to hold.

Definition 3.3.2

For $x \in Z$, let $\mathcal{W}_x := \{\mu(Y)\colon Y \in \mathcal{Y} \wedge x \in Y - \mu(Y)\}$, $\Gamma_x := \Pi\mathcal{W}_x$, and $K := \{x \in Z\colon \exists X \in \mathcal{Y}.x \in \mu(X)\}$.

Note that we consider here now $\mu(Y)$ in $\mathcal{W}_x$, and not Y as in $\mathcal{Y}_x$ in Definition 3.2.1.

Remark 3.3.2

(1) $x \in K \to \Gamma_x \neq \emptyset$,

(2) $g \in \Gamma_x \to ran(g) \subseteq K$.

Proof:

(1) We have to show that $Y \in \mathcal{Y}$, $x \in Y - \mu(Y) \to \mu(Y) \neq \emptyset$. By $x \in K$, there is $X \in \mathcal{Y}$ s.t. $x \in \mu(X)$. Suppose $x \in Y$, $\mu(Y) = \emptyset$. Then $x \in X \cap Y$, so by $x \in \mu(X)$ and (μPR) $x \in \mu(X \cap Y)$. But $\mu(Y) = \emptyset \subseteq X \cap Y \subseteq Y$, so by (μCUM) $\mu(X \cap Y) = \emptyset$, contradiction.

(2) By definition, $\mu(Y) \subseteq K$ for all $Y \in \mathcal{Y}$. □

The following claim is the analogue of Claim 3.2.1 above.

Claim 3.3.3

Let $U \in \mathcal{Y}$, $x \in K$. Then

(1) $x \in \mu(U) \leftrightarrow x \in U \wedge \exists f \in \Gamma_x.ran(f) \cap U = \emptyset$,

(2) $x \in \mu(U) \leftrightarrow x \in U \wedge \exists f \in \Gamma_x.ran(f) \cap H(U) = \emptyset$.

Proof:

(1) Case 1: $\mathcal{W}_x = \emptyset$, thus $\Gamma_x = \{\emptyset\}$. "$\rightarrow$": Take $f := \emptyset$. "$\leftarrow$": $x \in U \in \mathcal{Y}$, $\mathcal{W}_x = \emptyset \rightarrow x \in \mu(U)$ by definition of $\mathcal{W}_x$.

Case 2: $\mathcal{W}_x \neq \emptyset$. "$\rightarrow$": Let $x \in \mu(U) \subseteq U$. It suffices to show $Y \in \mathcal{W}_x \rightarrow \mu(Y) - H(U) \neq \emptyset$. But $Y \in \mathcal{W}_x \rightarrow x \in Y - \mu(Y) \rightarrow$ (by Fact 3.3.1, (6)) $Y \not\subseteq H(U) \rightarrow$ (by Fact 3.3.1, (5)) $\mu(Y) \not\subseteq H(U)$. "$\leftarrow$": If $x \in U - \mu(U)$, $U \in \mathcal{W}_x$, moreover $\Gamma_x \neq \emptyset$ by Remark 3.3.2, (1) and thus (or by the same argument) $\mu(U) \neq \emptyset$, so $\forall f \in \Gamma_x.ran(f) \cap U \neq \emptyset$.

(2): The proof is verbatim the same as for (1). □ (Claim 3.3.3)

Proposition 3.3.4 is the basic representation result for the smooth case.

Proposition 3.3.4

Let $\mathcal{Y}$ be closed under finite unions and finite intersections, and $\mu : \mathcal{Y} \rightarrow \mathcal{Y}$. Then there is a $\mathcal{Y}-$smooth preferential structure $\mathcal{Z}$, s.t. for all $X \in \mathcal{Y}$ $\mu(X) = \mu_{\mathcal{Z}}(X)$ iff μ satisfies $(\mu \subseteq)$, (μPR), (μCUM).

Proof:

"$\rightarrow$" is again easy and left to the reader.

Outline of "$\leftarrow$": We first define a structure $\mathcal{Z}$ (in a way very similar to Construction 3.2.1) which represents μ, but is not necessarily $\mathcal{Y}-$smooth, refine it to $\mathcal{Z}'$ and show that $\mathcal{Z}'$ represents μ too, and that $\mathcal{Z}'$ is $\mathcal{Y}-$smooth.

In the structure $\mathcal{Z}'$, all pairs destroying smoothness in $\mathcal{Z}$ are successively repaired, by adding minimal elements: If $< y, j >$ is not minimal, and has no minimal $< x, i >$ below it, we just add one such $< x, i >$. As the repair process might itself generate such "bad" pairs, the process may have to be repeated infinitely often. Of course, one has to take care that the representation property is preserved.

The proof given is close to the minimum one has to show (except that we avoid $H(U)$, instead of U — as was done in the old proof of [Sch96-1]). We could simplify further, we do not, in order to stay closer to the construction that is really needed. The reader will find the simplification as building block of the proof in Section 3.3.2. (In the simplified proof, we would consider for x, U s.t. $x \in \mu(U)$ the pairs $< x, g_U >$ with $g_U \in \Pi\{\mu(U \cup Y) : x \in Y \not\subseteq H(U)\}$, giving minimal elements. For the U s.t. $x \in U - \mu(U)$, we would choose $< x, g >$ s.t. $g \in \Pi\{\mu(Y) : x \in Y \in \mathcal{Y}\}$ with $< x', g'_U > \prec < x, g >$ for $< x', g'_U >$ as above.)

Construction 3.3.1 represents μ. The structure will not yet be smooth, we will mend it afterwards in Construction 3.3.2.

Construction 3.3.1

(Construction of $\mathcal{Z}$) Let $\mathcal{X} := \{< x, g >: x \in K,\ g \in \Gamma_x\}$, $< x', g' > \prec < x, g > :\leftrightarrow x' \in ran(g)$, $\mathcal{Z} := < \mathcal{X}, \prec >$.

Claim 3.3.5

$\forall U \in \mathcal{Y}.\mu(U) = \mu_{\mathcal{Z}}(U)$.

Proof:

Case 1: $x \notin K$. Then $x \notin \mu(U)$ and $x \notin \mu_{\mathcal{Z}}(U)$.

Case 2: $x \in K$. By Claim 3.3.3, (1) it suffices to show that for all $U \in \mathcal{Y}$ $x \in \mu_{\mathcal{Z}}(U) \leftrightarrow x \in U \wedge \exists f \in \Gamma_x.ran(f) \cap U = \emptyset$. Fix $U \in \mathcal{Y}$.

"$\rightarrow$": $x \in \mu_{\mathcal{Z}}(U) \rightarrow$ ex. $< x, f >$ minimal in $\mathcal{X}\lceil U$, thus $x \in U$ and there is no $< x', f' > \prec < x, f >$, $x' \in U$, $x' \in K$. But if $x' \in K$, then by Remark 3.3.2, (1), $\Gamma_{x'} \neq \emptyset$, so we find suitable f'. Thus, $\forall x' \in ran(f).x' \notin U$ or $x' \notin K$. But $ran(f) \subseteq K$, so $ran(f) \cap U = \emptyset$.

"$\leftarrow$": If $x \in U$, $f \in \Gamma_x$ s.t. $ran(f) \cap U = \emptyset$, then $< x, f >$ is minimal in $\mathcal{X}\lceil U$. □ (Claim 3.3.5)

We now construct the refined structure $\mathcal{Z}'$.

Construction 3.3.2

(Construction of $\mathcal{Z}'$)

σ is called x-admissible sequence iff

1. σ is a sequence of length $\leq \omega$, $\sigma = \{\sigma_i : i \in \omega\}$,
2. $\sigma_o \in \Pi\{\mu(Y)\colon Y \in \mathcal{Y} \wedge x \in Y - \mu(Y)\}$,
3. $\sigma_{i+1} \in \Pi\{\mu(X)\colon X \in \mathcal{Y} \wedge x \in \mu(X) \wedge ran(\sigma_i) \cap X \neq \emptyset\}$.

By 2., σ_0 minimizes x, and by 3., if $x \in \mu(X)$, and $ran(\sigma_i) \cap X \neq \emptyset$, i.e. we have destroyed minimality of x in X, x will be above some y minimal in X to preserve smoothness.

Let Σ_x be the set of x-admissible sequences, for $\sigma \in \Sigma_x$ let $\widehat{\sigma} := \bigcup\{ran(\sigma_i) : i \in \omega\}$. Note that by the argument in the proof of Remark 3.3.2, (1), $\Sigma_x \neq \emptyset$, if $x \in K$.

Let $\mathcal{X}' := \{< x, \sigma >\colon x \in K \wedge \sigma \in \Sigma_x\}$ and $< x', \sigma' > \prec' < x, \sigma > :\leftrightarrow x' \in \widehat{\sigma}$. Finally, let $\mathcal{Z}' :=< \mathcal{X}', \prec' >$, and $\mu' := \mu_{\mathcal{Z}'}$.

It is now easy to show that $\mathcal{Z}'$ represents μ, and that $\mathcal{Z}'$ is smooth. For $x \in \mu(U)$, we construct a special x-admissible sequence $\sigma^{x,U}$ using the properties of $H(U)$ as described at the beginning of this section.

Claim 3.3.6

For all $U \in \mathcal{Y}$ $\mu(U) = \mu_{\mathcal{Z}}(U) = \mu'(U)$.

Proof:

If $x \notin K$, then $x \notin \mu_{\mathcal{Z}}(U)$, and $x \notin \mu'(U)$ for any U. So assume $x \in K$. If $x \in U$ and $x \notin \mu_{\mathcal{Z}}(U)$, then for all $< x, f > \in \mathcal{X}$, there is $< x', f' > \in \mathcal{X}$ with $< x', f' > \prec < x, f >$ and $x' \in U$. Let now $< x, \sigma > \in \mathcal{X}'$, then $< x, \sigma_0 > \in \mathcal{X}$, and let $< x', f' > \prec < x, \sigma_0 >$ in $\mathcal{Z}$ with $x' \in U$. As $x' \in K$, $\Sigma_{x'} \neq \emptyset$, let $\sigma' \in \Sigma_{x'}$. Then $< x', \sigma' > \prec' < x, \sigma >$ in $\mathcal{Z}'$. Thus $x \notin \mu'(U)$. Thus, for all $U \in \mathcal{Y}$, $\mu'(U) \subseteq \mu_{\mathcal{Z}}(U) = \mu(U)$.

It remains to show $x \in \mu(U) \rightarrow x \in \mu'(U)$.

Assume $x \in \mu(U)$ (so $x \in K$), $U \in \mathcal{Y}$, we will construct minimal σ, i.e. show that there is $\sigma^{x,U} \in \Sigma_x$ s.t. $\widehat{\sigma^{x,U}} \cap U = \emptyset$. We construct this $\sigma^{x,U}$ inductively, with the stronger property that $ran(\sigma_i^{x,U}) \cap H(U) = \emptyset$ for all $i \in \omega$.

$\underline{\sigma_0^{x,U}}$: $x \in \mu(U)$, $x \in Y - \mu(Y) \rightarrow \mu(Y) - H(U) \neq \emptyset$ by Fact 3.3.1, (6) + (5). Let $\sigma_0^{x,U} \in \Pi\{\mu(Y) - H(U) : Y \in \mathcal{Y}, x \in Y - \mu(Y)\}$, so $ran(\sigma_0^{x,U}) \cap H(U) = \emptyset$.

$\underline{\sigma_i^{x,U} \to \sigma_{i+1}^{x,U}}$: By induction hypothesis, $ran(\sigma_i^{x,U}) \cap H(U) = \emptyset$. Let $X \in \mathcal{Y}$ be s.t. $x \in \mu(X)$, $ran(\sigma_i^{x,U}) \cap X \neq \emptyset$. Thus $X \not\subseteq H(U)$, so $\mu(U \cup X) - H(U) \neq \emptyset$ by Fact 3.3.1, (7). Let $\sigma_{i+1}^{x,U} \in \Pi\{\mu(U \cup X) - H(U) : X \in \mathcal{Y},\ x \in \mu(X),\ ran(\sigma_i^{x,U}) \cap X \neq \emptyset\}$, so $ran(\sigma_{i+1}^{x,U}) \cap H(U) = \emptyset$. As $\mu(U \cup X) - H(U) \subseteq \mu(X)$ by Fact 3.3.1, (3), the construction satisfies the $x-$admissibility condition.
□

It remains to show:

Claim 3.3.7

$\mathcal{Z}'$ is $\mathcal{Y}-$smooth.

Proof:

Let $X \in \mathcal{Y}$, $< x, \sigma > \in \mathcal{X}' \lceil X$.

Case 1, $x \in X - \mu(X)$: Then $ran(\sigma_0) \cap \mu(X) \neq \emptyset$, let $x' \in ran(\sigma_0) \cap \mu(X)$. Moreover, $\mu(X) \subseteq K$. Then for all $< x', \sigma' > \in \mathcal{X}'$ $< x', \sigma' > \prec < x, \sigma >$. But $< x', \sigma^{x',X} >$ as constructed in the proof of Claim 3.3.6 is minimal in $\mathcal{X}' \lceil X$.

Case 2, $x \in \mu(X) = \mu_{\mathcal{Z}}(X) = \mu'(X)$: If $< x, \sigma >$ is minimal in $\mathcal{X}' \lceil X$, we are done. So suppose there is $< x', \sigma' > \prec < x, \sigma >$, $x' \in X$. Thus $x' \in \widehat{\sigma}$. Let $x' \in ran(\sigma_i)$. So $x \in \mu(X)$ and $ran(\sigma_i) \cap X \neq \emptyset$. But $\sigma_{i+1} \in \Pi\{\mu(X')$: $X' \in \mathcal{Y} \wedge x \in \mu(X') \wedge ran(\sigma_i) \cap X' \neq \emptyset\}$, so X is one of the X', moreover $\mu(X) \subseteq K$, so there is $x'' \in \mu(X) \cap ran(\sigma_{i+1}) \cap K$, so for all $< x'', \sigma'' > \in \mathcal{X}'$ $< x'', \sigma'' > \prec < x, \sigma >$. But again $< x'', \sigma^{x'',X} >$ as constructed in the proof of Claim 3.3.6 is minimal in $\mathcal{X}' \lceil X$. □ (Claim 3.3.7 and Proposition 3.3.4)

3.3.2 Smooth and transitive minimal preferential structures

Discussion:

In a certain way, it is not surprising that transitivity does not impose stronger conditions in the smooth case either. Smoothness is itself a weak kind of transitivity: If an element is not minimal, then there is a minimal element below it, i.e., $x \succ y$ with y not minimal is possible, there is $z' \prec y$, but then there is z minimal with $x \succ z$. This is "almost" $x \succ z'$, transitivity.

To obtain representation, we will combine here the ideas of the smooth, but not necessarily transitive case with those of the general transitive case — as the reader will have suspected. Thus, we will index again with trees, and work with (suitably adapted) admissible sequences for the construction of the trees. In the construction of the admissible sequences, we were careful to repair all damage done in previous steps. We have to add now reparation of all damage done by using transitivity, i.e., the transitivity of the relation might destroy minimality, and we have to construct minimal elements below all elements for which we thus destroyed minimality. Both cases are combined by considering immediately all Y s.t. $x \in Y - H(U)$. Of course, the properties described in Fact 3.3.1 play again a central role.

The (somewhat complicated) construction will be commented on in more detail below.

Note that even beyond Fact 3.3.1, closure of the domain under finite unions is used in the construction of the trees. This — or something like it — is necessary, as we have to respect the hulls of all elements treated so far (the predecessors), and not only of the first element, because of transitivity. For the same reason, we need more bookkeeping, to annotate all the hulls (or the union of the respective U's) of all predecessors to be respected. One can perhaps do with a weaker operation than union — i.e. just look at the hulls of all U's separately, to obtain a transitive construction where unions are lacking, see the case of plausibility logic below — but we have not investigated this problem.

To summarize: we combine the ideas from the transitive general case and the simple smooth case, using the crucial Fact 3.3.1 to show that the construction goes through. The construction leaves still some freedom, and modifications are possible as indicated below in the course of the proof. The construction is perhaps the most complicated in the entire book, as it combines several ideas, some of which are already somewhat involved. If necessary, the proof can certainly still be elaborated, and its main points (use of a suitable $H(U)$ to avoid U, successive repair of damage done in the construction, trees as indexing) may probably be used in other contexts, too.

The construction:

Recall that $\mathcal{Y}$ will be closed under finite unions and finite intersections in this section, and let again $\mu : \mathcal{Y} \to \mathcal{Y}$.

Proposition 3.3.8 is the representation result for the smooth transitive case.

Proposition 3.3.8

Let $\mathcal{Y}$ be closed under finite unions and finite intersections, and $\mu : \mathcal{Y} \to \mathcal{Y}$. Then there is a $\mathcal{Y}-$smooth transitive preferential structure $\mathcal{Z}$, s.t. for all $X \in \mathcal{Y}$ $\mu(X) = \mu_{\mathcal{Z}}(X)$ iff μ satisfies $(\mu \subseteq)$, (μPR), (μCUM).

Proof:

The idea:

We have to adapt Construction 3.3.2 ($x-$admissible sequences) to the transitive situation, and to our construction with trees. If $< \emptyset, x >$ is the root, $\sigma_0 \in \Pi\{\mu(Y) : x \in Y - \mu(Y)\}$ determines some children of the root. To preserve smoothness, we have to compensate and add other children by the $\sigma_{i+1} : \sigma_{i+1} \in \Pi\{\mu(X) : x \in \mu(X), ran(\sigma_i) \cap X \neq \emptyset\}$. On the other hand, we have to pursue the same construction for the children so constructed. Moreover, these indirect children have to be added to those children of the root, which have to be compensated (as the first children are compensated by σ_1) to preserve smoothness. Thus, we build the tree in a simultaneous vertical and horizontal induction.

This construction can be simplified, by considering immediately all $Y \in \mathcal{Y}$ s.t. $x \in Y \not\subseteq H(U)$ — independent of whether $x \notin \mu(Y)$ (as done in σ_0), or whether $x \in \mu(Y)$, and some child y constructed before is in Y (as done in the σ_{i+1}), or whether $x \in \mu(Y)$, and some indirect child y of x is in Y (to take care of transitivity, as indicated above). We make this simplified construction.

There are two ways to proceed. First, we can take as $\lhd^*$ in the trees the transitive closure of $\lhd$. Second, we can deviate from the idea that children are chosen by selection functions f, and take nonempty subsets of elements instead, making more elements children than in the first case. We take the first alternative, as it is more in the spirit of the construction.

We will suppose for simplicity that $Z = K$ — the general case in easy to obtain by a technique similar to that in Section 3.3.1, but complicates the picture.

For each $x \in Z$, we construct trees t_x, which will be used to index different copies of x, and control the relation $\prec$.

These trees t_x will have the following form:

(a) the root of t is $< \emptyset, x >$ or $< U, x >$ with $U \in \mathcal{Y}$ and $x \in \mu(U)$,

(b) all other nodes are pairs $< Y, y >$, $Y \in \mathcal{Y}$, $y \in \mu(Y)$,

(c) $ht(t) \leq \omega$,

(d) if $< Y, y >$ is an element in t_x, then there is some $\mathcal{Y}(y) \subseteq \{W \in \mathcal{Y} : y \in W\}$, and $f \in \Pi\{\mu(W) : W \in \mathcal{Y}(y)\}$ s.t. the set of children of $< Y, y >$ is $\{< Y \cup W, f(W) >: W \in \mathcal{Y}(y)\}$.

The first coordinate is used for bookkeeping when constructing children, in particular for condition (d).

The relation $\prec$ will essentially be determined by the subtree relation.

We first construct the trees t_x for those sets U where $x \in \mu(U)$, and then take care of the others. In the construction for the minimal elements, at each level $n > 0$, we may have several ways to choose a selection function f_n, and each such choice leads to the construction of a different tree — we construct all these trees. (We could also construct only one tree, but then the choice would have to be made coherently for different x, U. It is simpler to construct more trees than necessary.)

We control the relation by indexing with trees, just as it was done in the not necessarily smooth case before.

Definition 3.3.3

If t is a tree with root $< a, b >$, then t/c will be the same tree, only with the root $< c, b >$.

Construction 3.3.3

(A) The set T_x of trees t for fixed x:

(1) Construction of the set $T\mu_x$ of trees for those sets $U \in \mathcal{Y}$, where $x \in \mu(U)$:

Let $U \in \mathcal{Y}$, $x \in \mu(U)$. The trees $t_{U,x} \in T\mu_x$ are constructed inductively, observing simultaneously:

If $< U_{n+1}, x_{n+1} >$ is a child of $< U_n, x_n >$, then

(a) $x_{n+1} \in \mu(U_{n+1}) - H(U_n)$,

and

(b) $U_n \subseteq U_{n+1}$.

Set $U_0 := U$, $x_0 := x$.

Level 0: $< U_0, x_0 >$.

Level $n \to n+1$: Let $< U_n, x_n >$ be in level n. Suppose $Y_{n+1} \in \mathcal{Y}$, $x_n \in Y_{n+1}$, and $Y_{n+1} \not\subseteq H(U_n)$. Note that $\mu(U_n \cup Y_{n+1}) - H(U_n) \neq \emptyset$ by Fact 3.3.1, (7), and $\mu(U_n \cup Y_{n+1}) - H(U_n) \subseteq \mu(Y_{n+1})$ by Fact 3.3.1, (3).

Choose $f_{n+1} \in \Pi\{\mu(U_n \cup Y_{n+1}) - H(U_n) : Y_{n+1} \in \mathcal{Y},\ x_n \in Y_{n+1} \not\subseteq H(U_n)\}$ (for the construction of this tree, at this element), and let the set of children of $< U_n, x_n >$ be $\{< U_n \cup Y_{n+1}, f_{n+1}(Y_{n+1}) >: Y_{n+1} \in \mathcal{Y},\ x_n \in Y_{n+1} \not\subseteq H(U_n)\}$. (If there is no such Y_{n+1}, $< U_n, x_n >$ has no children.) Obviously, (a) and (b) hold.

We call such trees $U, x-$trees.

(2) Construction of the set T'_x of trees for the nonminimal elements. Let $x \in Z$. Construct the tree t_x as follows (here, one tree per x suffices for all U):

Level 0: $< \emptyset, x >$

Level 1: Choose arbitrary $f \in \Pi\{\mu(U) : x \in U \in \mathcal{Y}\}$. Note that $U \neq \emptyset \rightarrow \mu(U) \neq \emptyset$ by $Z = K$ (by Remark 3.3.2, (1)). Let $\{< U, f(U) >: x \in U \in \mathcal{Y}\}$ be the set of children of $< \emptyset, x >$. This assures that the element will be nonminimal.

Level > 1: Let $< U, f(U) >$ be an element of level 1, as $f(U) \in \mu(U)$, there is a $t_{U,f(U)} \in T\mu_{f(U)}$. Graft one of these trees $t_{U,f(U)} \in T\mu_{f(U)}$ at $< U, f(U) >$ on the level 1. This assures that a minimal element will be below it to guarantee smoothness.

Finally, let $T_x := T\mu_x \cup T'_x$.

(B) The relation $\lhd$ between trees: For $x, y \in Z$, $t \in T_x$, $t' \in T_y$, set $t \rhd t'$ iff for some Y $< Y, y >$ is a child of the root $< X, x >$ in t, and t' is the subtree of t beginning at this $< Y, y >$.

(C) The structure $\mathcal{Z}$: Let
$\mathcal{Z} := < \{< x, t_x >: x \in Z,\ t_x \in T_x\},\ < x, t_x > \succ < y, t_y >$ iff $t_x \rhd^* t_y >$.

The rest of the proof are simple observations.

Fact 3.3.9

(1) If $t_{U,x}$ is an $U, x-$tree, $< U_n, x_n >$ an element of $t_{U,x}$, $< U_m, x_m >$ a direct or indirect child of $< U_n, x_n >$, then $x_m \not\in H(U_n)$.

(2) Let $< Y_n, y_n >$ be an element in $t_{U,x} \in T\mu_x$, t' the subtree starting at $< Y_n, y_n >$, then t' is a Y_n, y_n-tree.

(3) $\prec$ is free from cycles.

(4) If $t_{U,x}$ is an $U, x-$tree, then $< x, t_{U,x} >$ is $\prec -$minimal in $\mathcal{Z}\lceil U$.

(5) No $< x, t_x >$, $t_x \in T'_x$ is minimal in any $\mathcal{Z}\lceil U$, $U \in \mathcal{Y}$.

(6) Smoothness is respected for the elements of the form $< x, t_{U,x} >$.

(7) Smoothness is respected for the elements of the form $< x, t_x >$ with $t_x \in T'_x$.

(8) $\mu = \mu_{\mathcal{Z}}$.

Proof:

(1) trivial by (a) and (b).

(2) trivial by (a).

(3) Note that no $< x, t_x >$ $t_x \in T'_x$ can be smaller than any other element (smaller elements require $U \neq \emptyset$ at the root). So no cycle involves any such $< x, t_x >$. Consider now $< x, t_{U,x} >$, $t_{U,x} \in T\mu_x$. For any $< y, t_{V,y} > \prec < x, t_{U,x} >$, $y \notin H(U)$ by (1), but $x \in \mu(U) \subseteq H(U)$, so $x \neq y$.

(4) This is trivial by (1).

(5) Let $x \in U \in \mathcal{Y}$, then f as used in the construction of level 1 of t_x chooses $y \in \mu(U) \neq \emptyset$, and some $< y, t_{U,y} >$ is in $\mathcal{Z}\lceil U$ and below $< x, t_x >$.

(6) Let $x \in A \in \mathcal{Y}$, we have to show that either $< x, t_{U,x} >$ is minimal in $\mathcal{Z}\lceil A$, or that there is $< y, t_y > \prec < x, t_{U,x} >$ minimal in $\mathcal{Z}\lceil A$.

Case 1, $A \subseteq H(U)$: Then $< x, t_{U,x} >$ is minimal in $\mathcal{Z}\lceil A$, again by (1).

Case 2, $A \not\subseteq H(U)$: Then A is one of the Y_1 considered for level 1. So there is $< U \cup A, f_1(A) >$ in level 1 with $f_1(A) \in \mu(A) \subseteq A$ by Fact 3.3.1, (3). But note that by (1) all elements below $< U \cup A, f_1(A) >$ avoid $H(U \cup A)$. Let t be the subtree of $t_{U,x}$ beginning at $< U \cup A, f_1(A) >$, then by (2) t is one of the $U \cup A, f_1(A)-$trees, and $< f_1(A), t >$ is minimal in $\mathcal{Z}\lceil U \cup A$ by (4), so in $\mathcal{Z}\lceil A$, and $< f_1(A), t > \prec < x, t_{U,x} >$.

(7) Let $x \in A \in \mathcal{Y}$, $< x, t_x >$, $t_x \in T'_x$, and consider the subtree t beginning at $< A, f(A) >$, then t is one of the $A, f(A)-$trees, and $< f(A), t >$ is minimal in $\mathcal{Z}\lceil A$ by (4).

(8) Let $x \in \mu(U)$. Then any $< x, t_{U,x} >$ is $\prec -$minimal in $\mathcal{Z}\lceil U$ by (4), so $x \in \mu_{\mathcal{Z}}(U)$. Conversely, let $x \in U - \mu(U)$. By (5), no $< x, t_x >$ is minimal in U. Consider now some $< x, t_{V,x} > \in \mathcal{Z}$, so $x \in \mu(V)$. As $x \in U - \mu(U)$, $U \not\subseteq H(V)$ by Fact 3.3.1, (6). Thus U was considered in the construction of level 1 of $t_{V,x}$. Let t be the subtree of $t_{V,x}$ beginning at $< V \cup U, f_1(U) >$, by $\mu(V \cup U) - H(V) \subseteq \mu(U)$ (Fact 3.3.1, (3)), $f_1(U) \in \mu(U) \subseteq U$, and $< f_1(U), t > \prec < x, t_{V,x} >$. □ (Fact 3.3.9 and Proposition 3.3.8)

One copy again

The same techniques as in the general case will work again, details are left to the reader.

3.4 The logical characterization of general and smooth preferential models

Discussion:

The proofs and properties are straightforward. One has to pay attention, however, to the fact that we can go back and forth between model sets on the one hand side, and theories and their consequences on the other, due to the fact that we have classical soundness and completeness, and that the model set operators are assumed to be definability preserving. The reader will see in Chapter 5 that the lack of definability preservation complicates things, as we might "overlook" exceptions, when we do not separate carefully model sets from the sets of all models of a theory. The complication is very serious, as we will also show that, in this case, there is no normal characterization at all possible.

The properties (LLE) and (CCL) are, of course, essentially void, and hold for all logics defined via model sets. (SC) is trivial and expresses the subset condition, and only (PR) and (CUM) are really interesting and expressive.

We will see in Chapter 5, Example 5.1.2, (given already in [Sch92]) that condition (PR) may fail, if the structure is not definability preserving.

The problem of definability preservation occurs in other situations, too, e.g. in distance based revision, see [LMS01], and Example 4.2.3 below. A solution to the problem of definability preservation in another context (revision of defeasible databases) was examined in [ALS99]. A characterization of general, not necessarily definability preserving, preferential structures (by fundamentally nonlogical, and much uglier conditions than those presented here) is given in [Sch00-2], and, with more results, in Chapter 5 below.

In Section 3.4.1, we will show that, in an important class of examples, the limit version is equivalent to the minimal version. We consider there transitive structures in the limit interpretation, where

- either the set of definable closed minimizing sets is cofinal, or
- we consider only formulas on the left of $\mid\sim$

and show that both satisfy the laws of the minimal variant, so the generated logics can be represented by a minimal preferential structure (but, of course, perhaps with a different relation).

Proposition 3.4.1

Let $\mid\!\sim$ be a logic for $\mathcal{L}$. Recall from Definition 2.3.2 $T^{\mathcal{M}} := Th(\mu_{\mathcal{M}}(M(T)))$, where $\mathcal{M}$ is a preferential structure.

(1) Then there is a (transitive) definability preserving classical preferential model $\mathcal{M}$ s.t. $\overline{\overline{T}} = T^{\mathcal{M}}$ iff

(LLE) $\overline{T} = \overline{T'} \to \overline{\overline{T}} = \overline{\overline{T'}}$,

(CCL) $\overline{\overline{T}}$ is classically closed,

(SC) $T \subseteq \overline{\overline{T}}$,

(PR) $\overline{\overline{T \cup T'}} \subseteq \overline{\overline{\overline{\overline{T}} \cup T'}}$

for all $T, T' \subseteq \mathcal{L}$.

(2) The structure can be chosen smooth, iff, in addition

(CUM) $T \subseteq \overline{T'} \subseteq \overline{\overline{T}} \to \overline{\overline{T}} = \overline{\overline{T'}}$

holds.

The proof is an immediate consequence of Proposition 3.4.2 and the respective results of Sections 3.2 and 3.3. This proposition (or its analogue) was already shown in [Sch92] and [Sch96-1] and is repeated here for completeness' sake.

Proposition 3.4.2

Consider for a logic $\mid\!\sim$ on $\mathcal{L}$ the properties

(LLE) $\overline{T} = \overline{T'} \to \overline{\overline{T}} = \overline{\overline{T'}}$,

(CCL) $\overline{\overline{T}}$ is classically closed,

(SC) $T \subseteq \overline{\overline{T}}$,

(PR) $\overline{\overline{T \cup T'}} \subseteq \overline{\overline{\overline{\overline{T}} \cup T'}}$,

(CUM) $T \subseteq \overline{T'} \subseteq \overline{\overline{T}} \to \overline{\overline{T}} = \overline{\overline{T'}}$

for all $T, T' \subseteq \mathcal{L}$,

and for a function $\mu : \boldsymbol{D}_{\mathcal{L}} \to \mathcal{P}(M_{\mathcal{L}})$ the properties

(μdp) μ is definability preserving,

$(\mu \subseteq)$ $\mu(X) \subseteq X$,

(μPR) $X \subseteq Y \to \mu(Y) \cap X \subseteq \mu(X)$,

(μCUM) $\mu(X) \subseteq Y \subseteq X \to \mu(X) = \mu(Y)$

for all $X, Y \in \boldsymbol{D}_{\mathcal{L}}$.

It then holds:

(a) If μ satisfies (μdp), $(\mu \subseteq)$, (μPR), then $\mid\!\sim$ defined by $\overline{\overline{T}} := T^{\mu} := Th(\mu(M(T)))$ (see Definition 1.6.4) satisfies (LLE), (CCL), (SC), (PR).

If μ satisfies in addition (μCUM), then (CUM) will hold, too.

(b) If $\mid\!\sim$ satisfies (LLE), (CCL), (SC), (PR), then there is $\mu : \boldsymbol{D}_{\mathcal{L}} \to \mathcal{P}(M_{\mathcal{L}})$ s.t. $\overline{\overline{T}} = T^{\mu}$ for all $T \subseteq \mathcal{L}$ and μ satisfies (μdp), $(\mu \subseteq)$, (μPR).

If, in addition, (CUM) holds, then (μCUM) will hold, too.

Proof of Proposition 3.4.2:

We recall that, as $\boldsymbol{D}_{\mathcal{L}}$ is closed under finite intersections, in the presence of $(\mu \subseteq)$, (μPR) is equivalent to $(\mu PR')$ $\mu(X) \cap Y \subseteq \mu(X \cap Y)$, we work with $(\mu PR')$ in the proof.

(a) Suppose $\overline{\overline{T}} = T^{\mu}$ for some such μ, and all T.

(LLE): If $\overline{T} = \overline{T'}$, then $M_T = M_{T'}$, so $\mu(M_T) = \mu(M_{T'})$, and $T^{\mu} = T'^{\mu}$. (CCL) and (SC) are trivial.

We show (PR): Let now $\phi \in \overline{\overline{T \cup T'}}$, so ϕ holds in all $m \in \mu(M_{T \cup T'}) = \mu(M_T \cap M_{T'})$, so by $(\mu PR')$, ϕ holds in all $m \in \mu(M_T) \cap M_{T'}$. By (μdp), $\mu(M_T) = M_{T^{\mu}} = M_{\overline{\overline{T}}}$, so ϕ holds in all $m \in M_{\overline{\overline{T}}} \cap M_{T'} = M_{\overline{\overline{T}} \cup T'}$, so $\overline{\overline{T}} \cup T' \models \phi$, and $\phi \in \overline{\overline{\overline{T}} \cup T'}$.

We turn to (CUM): Assume $T \subseteq \overline{T'} \subseteq \overline{\overline{T}}$, so $M_{\overline{\overline{T}}} = \mu(M_T) \subseteq M_{T'} \subseteq M_T$ by (μdp). If $\phi \in \overline{\overline{T'}} = \overline{\overline{T \cup T'}}$, then by (PR) $\phi \in \overline{\overline{\overline{\overline{T}} \cup T'}} = (\overline{\overline{T}}) = \overline{\overline{T}}$. Let $\phi \in \overline{\overline{T}}$, so ϕ holds in all $m \in \mu(M_T) = \mu(M_{T'}) = M_{\overline{\overline{T'}}}$ by (μCUM) and (μdp). Thus $\overline{\overline{T'}} \vdash \phi$, but then by (CCL), $\phi \in \overline{\overline{T'}}$.

(b) Let $\mid\!\sim$ satisfy (LLE)–(CUM) for all T. We define μ and show $\overline{\overline{T}} = T^{\mu}$. (CUM) will be needed only to show (μCUM).

If $X = M_T$ for some $T \subseteq \mathcal{L}$, set $\mu(X) := M_{\overline{\overline{T}}}$. If $X = M_T = M_{T'}$, then $\overline{T} = \overline{T'}$, thus $\overline{\overline{T}} = \overline{\overline{T'}}$ by (LLE), so $M_{\overline{\overline{T}}} = M_{\overline{\overline{T'}}}$, and μ is well-defined. Moreover, μ satisfies (μdp), and by (SC), $\mu(X) \subseteq X$. We show $\overline{\overline{T}} = T^{\mu}$:

Let now $T \subseteq \mathcal{L}$ be given. Then $\phi \in T^{\mu} :\leftrightarrow \forall m \in \mu(M_T).m \models \phi \leftrightarrow \forall m \in M_{\overline{\overline{T}}}.m \models \phi \leftrightarrow \overline{\overline{T}} \vdash \phi \leftrightarrow \phi \in \overline{\overline{T}}$ (as $\overline{\overline{T}}$ is classically closed).

Next, we show that the above defined μ satisfies $(\mu PR')$. Suppose $X := M_T$, $Y := M_{T'}$. Let $m \in \mu(X) \cap Y = M_{\overline{\overline{T}}} \cap M_{T'}$, so $m \models \overline{\overline{T}} \cup T'$, and $m \models \overline{\overline{\overline{T}} \cup T'}$, so by (PR) $m \models \overline{\overline{T \cup T'}}$. As $X \cap Y = M_T \cap M_{T'} = M_{T \cup T'}$, $\mu(X \cap Y) = M_{\overline{\overline{T \cup T'}}}$ by (μdp), so $m \in \mu(X \cap Y)$.

It remains to show (μCUM). So let $X = M_T$, $Y = M_{T'}$, and $\mu(M_T) := M_{\overline{\overline{T}}} \subseteq M_{T'} \subseteq M_T \rightarrow \overline{T} \subseteq \overline{T'} \subseteq \overline{\overline{T}} = \overline{\overline{(\overline{T})}} \rightarrow \overline{\overline{T}} = \overline{\overline{(\overline{T})}} = \overline{\overline{(\overline{T'})}} = \overline{\overline{T'}} \rightarrow \mu(M_T) = M_{\overline{\overline{T}}} = M_{\overline{\overline{T'}}} = \mu(M_{T'})$, thus $\mu(X) = \mu(Y)$. □ (Proposition 3.4.2)

3.4.1 Simplifications of the general transitive limit case

We show here that the transitive limit version does not go beyond the minimal version, when we restrict ourselves to formulas on the left. This is not true for full theories on the left. We also show that the limit variant does not go beyond the minimal variant if the set of definable minimizing initial segments is cofinal.

So we work in this Section 3.4.1 with a transitive relation, and recall that we abbreviate "closed minimizing set" or "minimizing initial segment" by "MISE", used rather sloppily — see Definition 2.3.1.

Fact 3.4.3 contains some basic facts about MISE.

Fact 3.4.3

Let the relation $\prec$ be transitive.

(1) If X is MISE for A, and $X \subseteq B \subseteq A$, then X is MISE for B.

(2) If X is MISE for A, and $X \subseteq B \subseteq A$, and Y is MISE for B, then $X \cap Y$ is MISE for A.

(3) If X is MISE for A, Y MISE for B, then there is $Z \subseteq X \cup Y$ MISE for $A \cup B$.

Proof:

(1) trivial.

(2)

(2.1) $X \cap Y$ is closed in A : Let $< x, i > \in X \cap Y$, $< y, j > \prec < x, i >$, then $< y, j > \in X$. If $< y, j > \notin B$, then $< y, j > \notin X$, contradiction. So $< y, j > \in B$, but then $< y, j > \in Y$.

(2.2) $X \cap Y$ minimizes A : Let $< a, i > \in A$.

(a) If $< a, i > \in X - Y \subseteq B$, then there is $< y, j > \prec < a, i >$, $< y, j > \in Y$. By closure of X, $< y, j > \in X$.

(b) If $< a, i > \notin X$, then there is $< a', i' > \in X \subseteq B$, $< a', i' > \prec < a, i >$, continue by (a).

(3)

Let $Z := \{< x, i > \in X \colon \neg\exists < b, j > \preceq < x, i > . < b, j > \in B - Y\} \cup \{< y, j > \in Y \colon \neg\exists < a, i > \preceq < y, j > . < a, i > \in A - X\}$, where $\preceq$ stands for $\prec$ or $=$.

(3.1) Z minimizes $A \cup B$: We consider A, B is symmetrical. (a) We first show: If $< a, k > \in X - Z$, then there is $< y, i > \in Z$. $< a, k > \succ < y, i >$. Proof: If $< a, k > \in X - Z$, then there is $< b, j > \preceq < a, k >$, $< b, j > \in B - Y$. Then there is $< y, i > \prec < b, j >$, $< y, i > \in Y$. But $< y, i > \in Z$, too: If not, there would be $< a', k' > \preceq < y, i >$, $< a', k' > \in A - X$, but $< a', k' > \prec < a, k >$, contradicting closure of X. (b) If $< a'', k'' > \in A - X$, there is $< a, k > \in X$, $< a, k > \prec < a'', k'' >$. If $< a, k > \notin Z$, continue with (a).

(3.2) Z is closed in $A \cup B$: Let then $< z, i > \in Z$, $< u, k > \prec < z, i >$, $< u, k > \in A \cup B$. Suppose $< z, i > \in X$ — the case $< z, i > \in Y$ is symmetrical. (a) $< u, k > \in A - X$ cannot be, by closure of X. (b) $< u, k > \in B - Y$ cannot be, as $< z, i > \in Z$, and by definition of Z. (c) If $< u, k > \in X - Z$, then there is $< v, l > \preceq < u, k >$, $< v, l > \in B - Y$, so $< v, l > \prec < z, i >$, contradicting (b). (d) If $< u, k > \in Y - Z$, then there is $< v, l > \preceq < u, k >$, $< v, l > \in A - X$, contradicting (a).

□

In the limit variant holds now:

Fact 3.4.4

If $\prec$ is transitive, then

(1) (AND) holds,

(2) (OR) holds,

(3) $\overline{\overline{\phi \wedge \phi'}} \subseteq \overline{\overline{\overline{\phi}} \cup \{\phi'\}}$

Proof:

Let $\mathcal{Z}$ be the structure.

(1) Immediate by Fact 3.4.3, (2) — set $A = B$.

(2) Immediate by Fact 3.4.3, (3).

(3) Let $\psi \in \overline{\overline{\phi \wedge \phi'}}$, so there is X MISE for $\mathcal{Z}\lceil M(\phi \wedge \phi')$, $X \subseteq \mathcal{Z}\lceil M(\psi)$. Consider $\phi \wedge \neg\phi'$, $\mathcal{Z}\lceil M(\phi \wedge \neg\phi')$ is MISE for itself, so by Fact 3.4.3, (3) there is $Z \subseteq (\mathcal{Z}\lceil M(\phi \wedge \neg\phi')) \cup X$, MISE for $\mathcal{Z}\lceil M(\phi)$, and $Z \models \psi \vee \neg\phi' = \phi' \to \psi$, so $\phi \mid\sim \phi' \to \psi$, and $\psi = \phi' \wedge (\phi' \to \psi) \in \overline{\overline{\overline{\phi}} \cup \phi'}$. □

The following example (together with Fact 3.4.4) shows that the limit version separates the finitary and infinitary versions of (PR). We see a similar result below (Fact 3.4.5 and Example 3.4.2) for the finitary and infinitary versions of cumulativity. The limit version thus reveals itself as an interesting tool of abstract investigation.

Example 3.4.1

$\overline{\overline{T \cup T'}} \subseteq \overline{\overline{\overline{T}} \cup T'}$ can be wrong in the transitive limit version.

Any not definability preserving structure, where (PR) fails, serves as a counterexample, as minimal structures are special cases of the limit variant. Here is still another example.

Let $v(\mathcal{L}) := \{p_i : i < \omega\}$. Let $m \models p_i : i < \omega$, and $m' \models \neg p_0$, $m' \models p_i : 0 < i < \omega$, with $m \prec m'$ (this is the entire relation).

Let $T := \emptyset$, $T' := Th(\{m, m'\})$, then $T \cup T' = T'$, $\overline{\overline{T'}} = Th(\{m\})$, so $T \cup T' \mid\sim p_0$, $\overline{\overline{T}} = \overline{T} = \overline{\emptyset}$, and $\overline{\overline{\overline{T}} \cup T'} = \overline{T'} = T'$, but $p_0 \notin T'$, contradiction. □

Note:

The structure is not definability preserving, and (PR) holds neither in the minimal nor in the limit variant.

Fact 3.4.5

Finite cumulativity holds in transitive limit structures:

If $\phi \mid\sim \psi$, then $\overline{\overline{\phi}} = \overline{\overline{\phi \wedge \psi}}$.

Proof:

Suppose $\phi \mid\sim \psi$, $\phi \mid\sim \sigma$, and let X, Y be MISE for $\mathcal{Z}\lceil M(\phi)$, with $X \subseteq \mathcal{Z}\lceil M(\psi)$, $Y \subseteq \mathcal{Z}\lceil M(\sigma)$. Then $X \cap Y$ is MISE for $\mathcal{Z}\lceil M(\phi)$ by Fact 3.4.3, (2), thus for $\mathcal{Z}\lceil M(\phi \wedge \psi)$ by Fact 3.4.3, (1), so $\phi \wedge \psi \mid\sim \sigma$.

Conversely, let $\phi \mid\sim \psi$, X MISE for $\mathcal{Z}\lceil M(\phi)$, $X \subseteq \mathcal{Z}\lceil M(\psi)$, and $\phi \wedge \psi \mid\sim \sigma$, with Y MISE for $\mathcal{Z}\lceil M(\phi \wedge \psi)$, $Y \subseteq \mathcal{Z}\lceil M(\sigma)$. By Fact 3.4.3, (2), $X \cap Y$ is MISE for $\mathcal{Z}\lceil M(\phi)$, so $\phi \mid\sim \sigma$.

□

Example 3.4.2

Infinitary cumulativity may fail in transitive limit structures.

Consider the same language as in Example 3.4.1, set again $m < m'$, so $Th(\{m, m'\}) \mid\sim p_0$, but this time, we add more pairs to the relation: m and m' will now be the topmost models, and we put below all other models, making more and more p_i, $i \neq 0$, true, but alternating p_0 with $\neg p_0$, resulting in a total order (i.e. a ranked structure). Set $\phi := p_0 \vee \neg p_0$. Thus $\overline{\overline{\phi}} = Th(\{m, m'\})$, so $\mid\sim$ is not even idempotent, $\overline{\overline{\phi}} \neq \overline{\overline{(\overline{\overline{\phi}})}}$, as $\overline{\overline{Th(\{m, m'\})}} = Th(m)$. □

Consequently:

1. Our proofs show that, for the transitive case, on the left formulas only (perhaps the most important case), any limit version structure is equivalent to a minimal version structure. The proof uses closure properties (closure under set difference). Conversely, we can read any smooth minimal version as a trivial limit version, so the two are in an important class (transitive, formulas on the left) equivalent. The author was somewhat surprised by this result. This fact and the next point will be summarized in Proposition 3.4.7.

2. The KLM results show that they are equivalent to a smooth minimal structure. (We work in the other sections with the strong condition, which fails here, see Example 3.4.2.)

3. Example 3.4.2 separates the two versions of (CUM) — we can have one thing after the other in the limit version, but not all together. In a way, this is not surprising, limit was exactly about that.

We conclude with

Fact 3.4.6

Having cofinally many definable sets trivializes the problem (again in the transitive case).

Proof:

We show that under above condition, any instance of the limit version is equivalent to an instance of the minimal version.

First, we make the condition precise: We postulate that if T is any theory, and X is MISE in $M(T)$, then there is $X' \subseteq X$ X' MISE in $M(T)$ s.t. $\{x : \exists < x, i > \in X'\} = M(T')$ for some theory. We call such X' definable MISE.

Suppose the condition holds. We show that $\phi \in \overline{\overline{T \cup T'}}$ implies $\phi \in \overline{\overline{\overline{T}} \cup T'}$, so the main condition for the minimal variant (PR) is satisfied (the others are so trivially). Let $\mathcal{Z}$ be the structure considered.

Let then $\phi \in \overline{\overline{T \cup T'}}$, so there is A MISE in $M(T \cup T') = M(T) \cap M(T')$, $A \models \phi$. Consider $X := \mathcal{Z}\lceil(M(T) - M(T'))$ (as a set), let B be MISE in X, e.g. X itself. By above Fact 3.4.3 (3), there is $B' \subseteq B \cup A$ MISE for $M(T)$. Take $B'' \subseteq B'$ definable MISE for $M(T)$. We have then: ϕ holds in A, so in $B'' \cap M(T')$, $B'' \cap \mathcal{Z}\lceil M(T')$ more precisely. As B'' is definable, $B'' = M(S)$ for some $S \subseteq \overline{\overline{T}}$, so in particular, ϕ holds in all $m \in M(\overline{\overline{T}}) \cap M(T') = M(\overline{\overline{T}} \cup T')$, thus $\overline{\overline{T}} \cup T' \vdash \phi$. □

We summarize our main positive results on the limit variant of general preferential structures:

Proposition 3.4.7

Let the relation be transitive. Then

(1) Every instance of the the limit version, where the definable closed minimizing sets are cofinal in the closed minimizing sets, is equivalent to an

instance of the minimal version.

(2) If we consider only formulas on the left of $\mid\!\sim$, the resulting logic of the limit version can also be generated by the minimal version of a (perhaps different) preferential structure. Moreover, the structure can be chosen smooth.

Proof:

(1) This was shown in Fact 3.4.6. Note that there are instances of the minimal version which do not correspond to the limit version: if $X \neq \emptyset$, but $\mu(X) = \emptyset$, then $Con(T) \not\Rightarrow Con(\overline{\overline{T}})$.

(2) By Fact 3.4.4 (3), the basic law (restricted to formulas) of preferential structures in the minimal variant holds. The other laws do so trivially. So the limit variant with formulas on the left can be represented by the minimal variant by Proposition 3.4.1 (restrict the domain to formula definable model sets).

By Fact 3.4.4 (2), (OR) holds for formulas, by Fact 3.4.5 (CUM) holds for formulas, it is trivial that (LLE), (RW), (SC) hold for formulas. It was shown in [KLM90] that any such logic can be represented by a smooth preferential structure.

Conversely, it is easy to see that for a smooth structure, the minimal and the limit reading coincide.

Thus, they are equivalent.

□

3.5 A counterexample to the KLM-system

Discussion:

In [KLM90], S. Kraus, D. Lehmann, M. Magidor have shown that the finitary restrictions of all supraclassical, cumulative, and distributive inference operations are representable by preferential structures. In [Sch92], we have shown that this does not generalize to the arbitrary infinite case. Before we present the counterexample given there, we analyze the situation, and show that it is again due to a problem of domain closure. This is also the reason why we repeat this old result here: to show the ubiquity of the problem (and to give now a neater analysis in the context of domain closure questions).

First, we recall or define the notions.

Definition 3.5.1

We say that $\mid\!\sim$ satisfies

Distributivity iff $\overline{\overline{A}} \cap \overline{\overline{B}} \subseteq \overline{\overline{A \cap B}}$ for all theories A, B of $\mathcal{L}$.

Leaving aside questions of definability preservation, it translates into the following model set condition, where μ is the model choice function:

(μD) $\mu(X \cup Y) \subseteq \mu(X) \cup \mu(Y)$.

We have shown that condition

(μPR) $X \subseteq Y \rightarrow \mu(Y) \cap X \subseteq \mu(X)$

essentially characterizes preferential structures, and its validity was seen as soon as the definition of minimal preferential structures was known. In these terms, the problem is whether $(\mu \subseteq) + (\mu CUM) + (\mu D)$ entail (μPR) in the general case. Now, we see immediately:

$(\mu PR) + (\mu \subseteq)$ entail (μD) : $\mu(X \cup Y) = (\mu(X \cup Y) \cap X) \cup (\mu(X \cup Y) \cap Y)$ $\subseteq \mu(X) \cup \mu(Y)$.

Second, if the domain is closed under set difference, then $(\mu D) + (\mu \subseteq)$ entail (μPR) : Let $U \subseteq V$, $V = U \cup (V - U)$. Then $\mu(V) \cap U \subseteq (\mu(U) \cup \mu(V - U)) \cap U$ $= \mu(U)$.

The condition of closure under set difference is, of course, satisfied for formula defined model sets, but not in the general case of theory defined model sets.

To make the problem more palatable, we formulate it as follows: $\mu(X)$ is something like the "core" of X. A counterexample to (μPR), i.e. a case of $X \subseteq Y$ and $\mu(Y) \cap X \not\subseteq \mu(X)$ says that small sets do not always "protect" their core as well as big ones — the contrary to preferential structures, which "autodestruct" their "outer part", and, the bigger the set, the more elements will be destroyed. (μD) says that protection is immune to finite unions: their components protect their cores as well their union does. Now, it seems not so difficult to find a counterexample, and, in hindsight, it is surprising that it took some time to arrive there.

Intuitively, we can work from the inside, where smaller sets destroy more elements, or from the outside, where bigger sets protect their elements better. The original counterexample is of the first type: smaller sets decide more things, and we use this decision to make the logic stronger: the decision of infinitely many p_i will destroy all $\neg r-$models in the definition of the choice function. We might also work from the outside: say a point is in the core (or protected) iff there is a (nontrivial) sequence of elements converging to it in some suitable topology, e.g. the natural one in propositional logic. Then small sets are less protective, and the property will be robust under finite operations. If we take a little care, we can make the operations cumulative. (Note that we use a similar idea in Section 5.2.3, where we construct "bad" logics using the fact that big model sets protect their elements better.)

The original example is formulated directly on the logics side (and this is, we guess, the reason it took so long to find it: people, the author included, worked on the wrong (i.e. logics) side of the problem, instead of attacking it semantically). The approach taken there has some interest of its own, as it gives an easy way to construct funny logics by sufficiently long inductive constructions. For this reason, we give it here in all details.

To summarize: Distributivity says only something about well-behavior for finite unions, this leaves a large area to move about, and we can protect the core of small sets better or less well than that of big sets. Both conditions are really quite far from each other. Again, we see that a semantical description clarifies the picture a lot.

3.5.1 The formal results

S. Kraus, D. Lehmann, and M. Magidor have shown that for any logic $\mid\!\sim$ for $\mathcal{L}$, which is supraclassical, cumulative, and distributive, there is a $\boldsymbol{D}-$smooth preferential model $\mathcal{M}$, s.t. for all *finite* $T \subseteq \mathcal{L}$ $T^{\mathcal{M}} = \overline{\overline{T}}$ — where $T^{\mathcal{M}}$ is the logic defined by the structure. ([KLM90], see also [Mak94], Observation 3.4.7.) We show that the restriction to finite T is necessary, by providing a counterexample for the infinite case. We start by quoting a lemma by D. Makinson.

Both Lemma 3.5.1 and our counterexample Example 3.5.1 have appeared in [Mak94], Section 3.4 (Lemma 3.4.9, Observation 3.4.10). The reader less familiar with transfinite ordinals can find there a more algebraic proof that our counterexample satisfies the logical properties claimed. Our technique of constructing a logic inductively by a mixed iteration of suitable length has, however, proved useful in other situations as well (see [Sch91-2]), moreover, it is very fast and straightforward: once you have the necessary ingredients, the machinery will run almost by itself.

Lemma 3.5.1

Let a logic $\mid\!\sim$ on $\mathcal{L}$ be representable by a classical preferential model structure. Then, for all $A \subseteq \mathcal{L}$, $x \in \mathcal{L}$, $x \notin \overline{\overline{A}}$ there is a maximal consistent (under $\vdash$) $\Delta \subseteq \mathcal{L}$ s.t. $\overline{\overline{A}} \subseteq \Delta$, $x \notin \Delta$, and $\overline{\overline{\Delta}} \neq \mathcal{L}$.

Proof:

Let $\mathcal{M} = (\mathcal{X}, \prec)$ be a representation of $\mid\!\sim$, i.e. $\overline{\overline{A}} = A^{\mathcal{M}}$ for all $A \subseteq \mathcal{L}$. Let $A \subseteq \mathcal{L}$, $x \in \mathcal{L}$, and $x \notin \overline{\overline{A}}$. Then there is $< m, i >$ minimal in $\mathcal{X}\lceil M_A$, with $m \not\models x$. Note that by minimality, $m \models \overline{\overline{A}}$. $\Delta := \{y \in \mathcal{L}\colon\ m \models y\}$ is maximal consistent, $x \notin \Delta$, $\overline{\overline{A}} \subseteq \Delta$, and $< m, i >$ is also minimal in $\mathcal{X}\lceil M_\Delta$, by $M_\Delta \subseteq M_A$. Thus, $m \models \overline{\overline{\Delta}}$, and by classicality of the models, $\overline{\overline{\Delta}} \neq \mathcal{L}$. □

We now construct a supraclassical, cumulative, distributive logic, and show that the logic so defined fails to satisfy the condition of Lemma 3.5.1, and is thus not representable by a preferential structure.

Example 3.5.1

Let $v(\mathcal{L})$ contain the propositional variables $p_i : i \in \omega$, r. (Note that we do not require $\mathcal{L}$ to be countable, we leave plenty of room for modifications of the construction!) We shall violate compactness badly "in both directions" by adding the rules (infinitely many p_i) $\mid\!\sim r$ and (infinitely many $\neg p_i$) $\mid\!\sim r$. To account for distributivity, we shall add for all $\phi \in \mathcal{L}$ (infinitely many $p_i \vee \phi$) $\mid\!\sim r \vee \phi$ and (infinitely many $\neg p_i \vee \phi$) $\mid\!\sim r \vee \phi$. Closing under $\mid\!\sim$ and classical logic ω_1 many times to take care of the countably infinite rules will give the result.

The details:

We define the logic $\mid\!\sim$ by a mixed iteration: For $B \subseteq \mathcal{L}$ define

$I^+_{B,\phi} := \{i < \omega\colon\ p_i \vee \phi \in B\}$, $I^-_{B,\phi} := \{i < \omega\colon\ \neg p_i \vee \phi \in B\}$.

Define now inductively

$A_0 :=$ A

for successor ordinals (α a limit or 0, $i \in \omega$):

$A_{\alpha+2i+1} := \overline{A_{\alpha+2i}}$

$A_{\alpha+2i+2} := A_{\alpha+2i+1} \cup \{r \vee \phi\colon\ I^+_{A_{\alpha+2i+1},\phi}$ is infinite or $I^-_{A_{\alpha+2i+1},\phi}$ is infinite $\}$

for limit λ:

$A_\lambda := \bigcup\{A_i : i < \lambda\}$

$\overline{\overline{A}} := A_{\omega_1}$.

We show $\mid\!\sim$ is as desired. Note that the defined logic is monotone.

1) $\overline{A} \subseteq \overline{\overline{A}}$ is trivial.

2) $A \subseteq B \subseteq \overline{\overline{A}} \to \overline{\overline{A}} = \overline{\overline{B}}$:

2.1) $\overline{\overline{A}} \subseteq \overline{\overline{B}}$ by monotony

2.2) $\overline{\overline{B}} \subseteq \overline{\overline{A}}$: Let $\phi \in \overline{\overline{B}}$. In deriving ϕ in $\overline{\overline{B}}$, we have used only countably many elements from B. This is seen as follows. Let β be minimal such that $\phi \in B_\beta$. ϕ can be derived from at most countably many $\phi_i \in B_{\beta-1}$ (β has to be a successor ordinal). Arguing backwards, and using $\omega.\omega = \omega$ (cardinal multiplication), we see what we wanted. (This is, of course, the outline for an inductive proof.) As $B \subseteq \overline{\overline{A}}$, using regularity of ω_1, we see that there is some $\alpha < \omega_1$ s.t. all ϕ_j used in the derivation of ϕ from B are in A_α. But then $\phi \in A_{\alpha+\beta}$.

3) Distributivity: We show by induction on the derivation of a, b that $a \in \overline{\overline{A}}$, $b \in \overline{\overline{B}} \to a \vee b \in \overline{\overline{\overline{A} \cap \overline{B}}}$. To get started, use $A_0 \subseteq A_1 = \overline{A}$, and $a \in \overline{A}$, $b \in \overline{B} \to a \vee b \in \overline{A} \cap \overline{B}$. By symmetry, it suffices to consider the cases for a. Let $a_1, \ldots, a_n \vdash a$ by classical inference. By induction hypothesis, $a_1 \vee b, \ldots, a_n \vee b \in \overline{\overline{\overline{A} \cap \overline{B}}}$, but then $a \vee b \in \overline{\overline{\overline{A} \cap \overline{B}}}$, as the latter is closed under $\vdash$. Assume now $a = r \vee \phi \in A_\alpha$ has been derived from infinitely many $p_i \vee \phi$ $(i \in I)$ in $A_{\alpha-1}$. By induction hypothesis, $p_i \vee \phi \vee b \in \overline{\overline{\overline{A} \cap \overline{B}}}$. So $p_i \vee \phi \vee b \in (\overline{A} \cap \overline{B})_{\beta_i}$ for $\beta_i < \omega_1$. Again by regularity of ω_1, all $p_i \vee \phi \vee b \in (\overline{A} \cap \overline{B})_\beta$ $(i \in I)$ for some $\beta < \omega_1$. But then $r \vee \phi \vee b = a \vee b \in (\overline{A} \cap \overline{B})_{\beta+2}$. The case $\neg p_i \vee \phi$ is similar. □

We use the lemma to obtain the negative result, as the logic constructed above does not satisfy the lemma's condition:

Consider now $A := \emptyset$. Assume there is ϕ s.t. infinitely many $p_i \vee \phi \in \overline{A}$, thus there is ϕ s.t. infinitely many $p_i \vee \phi$ are tautologies. But then ϕ has to be a tautology (consider $(p_i \vee \phi) \leftrightarrow (\neg\phi \to p_i)$ and finiteness of ϕ!), thus ϕ and $\phi \vee r \in \overline{A}$. Likewise for $\neg p_i \vee \phi$. So, the rules (infinitely many $p_i \vee \phi$) $\mid\!\sim r \vee \phi$, etc. give nothing new, and $\overline{A} = \overline{\overline{A}}$. In particular, $r \notin \overline{\overline{A}}$.

Assume now $\Delta \subseteq \mathcal{L}$ to be maximal consistent. So Δ decides all $p_i : i \in \omega$. Thus either infinitely many p_i, or $\neg p_i$ in Δ. Thus, $r \in \overline{\overline{\Delta}}$. Hence $\mid\!\sim$ is not representable by a preferential structure. □

3.6 A nonsmooth model of cumulativity

Discussion:

The idea behind the construction in this section is extremely simple, and, again, related to closure properties of the domain.

Smoothness says that in each set of the domain, every element is either minimal, or there must be an element below it which is minimal in this set. Smoothness on the model side entails cumulativity on the logics side — but not vice versa, as the construction to be presented shows. Cumulativity can be violated in preferential structures, i.e. $\phi \mid\sim \psi$, $\phi \mid\sim \tau$, but not $\phi \wedge \psi \mid\sim \tau$, may hold, as there might be a $\phi \wedge \psi \wedge \neg\tau-$model m, which is not minimal in the set of $\phi-$models, but minimal in the set of $\phi \wedge \psi-$models. There will then be a $\phi \wedge \neg\psi-$model m' smaller than m, but no $\phi \wedge \psi-$model smaller than m, in particular no minimal $\phi-$model smaller than m, as smoothness postulates. To do without smoothness is simple. We just have to assure that in all formula definable sets (and it is here where the domain properties matter), which contain the set of $\phi-$minimal models $\mu(\phi)$, there must be an element smaller than m. But this is guaranteed if we have a sequence of models smaller than m converging in the standard topology to $\mu(\phi)$. Then any set of models for any ψ which contains $\mu(\phi)$ will contain some element of this sequence, so m will not be minimal.

Much in the argument to be presented shows that we have indeed to do without smoothness — this is imposed by the framework given by [FLM90].

To summarize: The [FLM90] framework forces us to consider nonsmooth structures. It is natural to have cumulativity without smoothness through a topological construction. The topological view demonstrates thus again its utility and naturalness. We might say it is a subtle bridge between the semantics and the logics.

3.6.1 The formal results

We need to define some further rules:

Definition 3.6.1

(NR) (Negation Rationality): $\alpha \mid\sim \beta \Rightarrow \alpha \wedge \gamma \mid\sim \beta$ or $\alpha \wedge \neg\gamma \mid\sim \beta$ (for any γ),

(WD) (Weak Determinacy): $true \mid\sim \neg\alpha \Rightarrow \alpha \mid\sim \beta$ or $\alpha \mid\sim \neg\beta$ (for any β) (we say that such α decide),

(DR) (Disjunctive Rationality): $\alpha \vee \beta \mid\sim \gamma \Rightarrow \alpha \mid\sim \gamma$ or $\beta \mid\sim \gamma$.

We also recall the rule (which is part of the system P):

(CM) (Cautious Monotony): $\alpha \mid\sim \beta$ and $\alpha \mid\sim \gamma \Rightarrow \alpha \wedge \gamma \mid\sim \beta$.

We use $\neg$(NR) as shorthand for the existence of α, β, γ such that $\alpha \mid\sim \beta$, but neither $\alpha \wedge \gamma \mid\sim \beta$, nor $\alpha \wedge \neg\gamma \mid\sim \beta$.

The main idea:

We work here in transitive structures — as this was a prerequisite in the [FLM90] paper.

If (CM) is violated, there are ϕ, ψ, τ such that $\phi \mid\sim \psi$, $\phi \mid\sim \tau$, but $\phi \wedge \psi \not\mid\sim \tau$. As all minimal models of ϕ are then minimal models of $\phi \wedge \psi$, there must be a new minimal model of $\phi \wedge \psi$, which is not a minimal model of ϕ, weakening the set of consequences of $\phi \wedge \psi$, compared to the set of consequences of ϕ. Smoothness assures that this cannot happen, as any model of $\phi \wedge \psi$ must be above some minimal model of ϕ. If smoothness cannot hold, we must prevent the existence of "dangerous" (i.e. consequence changing) new minimal $\phi \wedge \psi-$models by other means, which (by transitivity) can only be infinite descending chains of $\phi \wedge \psi-$models.

But smoothness cannot hold in an injective structure showing joint consistency of the system P, (WD), and $\neg$(NR):

Fact 3.6.1

There is no smooth injective preferential structure validating (WD) and $\neg$(NR).

Proof:

Suppose (NR) is false, so there are α, β, γ with $\alpha \mid\sim \beta$, $\alpha \wedge \gamma \not\mid\sim \beta$, $\alpha \wedge \neg\gamma \not\mid\sim \beta$. Let $\alpha \mid\sim \beta$, so $true \mid\sim \alpha \rightarrow \beta$. (If m is a minimal model of true, and if $m \models \alpha$, then m is a minimal model of α, so $m \models \beta$.) So $true \mid\sim \neg(\alpha \wedge \neg\beta)$. If $\vdash \phi \rightarrow \alpha \wedge \neg\beta$, then $true \mid\sim \neg\phi$. Thus, by (WD), if

$\vdash \phi \rightarrow \alpha \wedge \neg\beta$, ϕ decides, thus, by injectivity, ϕ has at most one minimal model in the structure.

Let now $\alpha \wedge \gamma \not\mid\!\sim \beta$, $\alpha \wedge \neg\gamma \not\mid\!\sim \beta$, thus there is a minimal model m_1 of $\alpha \wedge \gamma$, where $\neg\beta$ holds, and a minimal model m_2 of $\alpha \wedge \neg\gamma$, where $\neg\beta$ holds. Thus, m_1 is a minimal model of $\alpha \wedge \gamma \wedge \neg\beta$, m_2 a minimal model of $\alpha \wedge \neg\gamma \wedge \neg\beta$.

(a) Suppose m_1 is not a minimal model of $\alpha \wedge \neg\beta$, then by smoothness, there is $m < m_1$, m a minimal model of $\alpha \wedge \neg\beta$. $\neg\gamma$ has to hold in m, so m is a minimal model of $\alpha \wedge \neg\beta \wedge \neg\gamma$. By uniqueness, $m = m_2$, so $m_2 < m_1$, and m_2 is a minimal model of $\alpha \wedge \neg\beta$.

(b) If m_2 is not a minimal model of $\alpha \wedge \neg\beta$, then, analogously, m_1 is, and $m_1 < m_2$.

(c) m_1 and m_2 are minimal models of $\alpha \wedge \neg\beta$: Impossible, as $\alpha \wedge \neg\beta$ decides.

Suppose now, e.g. m_2 is the minimal model of $\alpha \wedge \neg\beta$, and $m_2 < m_1$. As $\alpha \mid\!\sim \beta$, m_2 cannot be a minimal model of α, so there must be $m' \models \alpha$ below m_2. $m' \models \alpha \wedge \neg\beta$ is impossible (by minimality of m_2), so $m' \models \alpha \wedge \beta$. (Note that we did not need smoothness for this argument.) But $m' \models \gamma$, or $m' \models \neg\gamma$, contradicting minimality of m_1 or of m_2. The other case is analogous. □

Thus, we have such "dangerous" $\phi \wedge \psi-$models, as we will see now. By failure of (NR), there are α, β, γ such that $\alpha \mid\!\sim \beta$, $\alpha \wedge \gamma \not\mid\!\sim \beta$, $\alpha \wedge \neg\gamma \not\mid\!\sim \beta$. Thus we have a minimal model m_1 of $\alpha \wedge \gamma \wedge \neg\beta$, and a minimal model m_2 of $\alpha \wedge \neg\gamma \wedge \neg\beta$. By (WD), there will be at most one minimal $\alpha \wedge \neg\beta-$model, so they cannot both be minimal models of $\alpha \wedge \neg\beta$. Suppose m_1 is not, the other case is analogous. A simple analysis shows that there cannot be a minimal model of $\alpha \wedge \neg\beta$ below m_1, so smoothness is indeed violated, and we must have an infinite descending chain X of $\alpha \wedge \neg\beta-$models below m_1. Let now $\phi := \alpha \wedge \neg\beta$, and m be the unique (if it exists — if not, a similar argument applies) minimal $\alpha \wedge \neg\beta-$model, and suppose $m \models \psi$, so $\phi \mid\!\sim \psi$. If there were now a minimal model m' of $\phi \wedge \psi$ in X, cumulativity would be violated: By injectivity of the structure, m' is logically different from m, and the theory determined by $\{m\}$ is stronger than the one determined by $\{m, m'\}$ (finiteness of $\{m\}$ is crucial here). Thus, in X either ψ will be infinitely often true, or not at all. We will make it infinitely often true, so "X approximates m logically".

3.6.1.1 A nonsmooth injective structure validating P, (WD), ¬(NR)

Definition 3.6.2

A sequence f of models converges to a set of models M, $f \to M$, iff $\forall\phi(M \models \phi \to \exists i \forall j \geq i. f_j \models \phi)$. If $M = \{m\}$, we will also write $f \to m$.

Fact 3.6.2

Let f be a sequence composed of n subsequences $f^1, \ldots, f^n$, e.g. $f_{n*j+0} = f^1_j$, etc., and $f^i \to M_i$. Let ϕ be a formula unboundedly often true in f. Then there is $1 \leq i \leq n$ and $m \in M_i$ s.t. $m \models \phi$.

Proof:

If for all i and all $m \in M_i$ $m \not\models \phi$, then $M_i \models \neg\phi$, so there are j_i s.t. for all $j \geq j_i$ $f^i_j \models \neg\phi$, so there is k s.t. for all $j \geq k$ $f_j \models \neg\phi$. □

Example 3.6.1

(A nonsmooth transitive injective structure validating system P, (WD), ¬(NR))

As any transitive acyclic relation over a finite structure is necessarily smooth, and an injective structure over a finite language is finite, Fact 3.6.1 shows that we need an infinite language.

Take the language defined by the propositional variables $r, s, t, p_i : i < \omega$.

Take four models m_i, $i = 1, \ldots, 4$, where for all i, j $m_i \models p_j$ (to be definite), and let $m_0 \models r, \neg s, t$, $m_1 \models r, \neg s, \neg t$, $m_2 \models r, s, t$, $m_3 \models r, s, \neg t$. It is important to make m_2 and m_3 identical except for t, the other values for the p_j are unimportant.

Let $m_2 < m_1$. (The other m_i are incomparable.)

Define two sequences of models $f^1 \to m_1$, $f^3 \to m_3$ s.t. for all i, j $f^i_j \models r, \neg t$. This is possible, as $m_1 \models r, \neg t$, $m_3 \models r, \neg t$.

All models in these sequences can be chosen different, and different from the m_i — this is no problem, as we have for all consistent ϕ uncountably many models where ϕ holds.

Let f be the mixture of f^i, e.g. $f_{2n+0} := f^1_n$, etc.

Put m_0 above f, with f in <u>descending</u> order. Arrange the rest of the 2^ω models above m_0 ordered as the ordinals, i.e. every subset has a minimum. Thus, there is one long chain C (i.e. C is totally ordered) of models, at its lower end a descending countable chain f, directly above f m_0, above m_0 all other models except $m_1 - m_3$, arranged in a well-order. The models $m_1 - m_3$ form a separate group. See Figure 3.6.1.

Figure 3.6.1

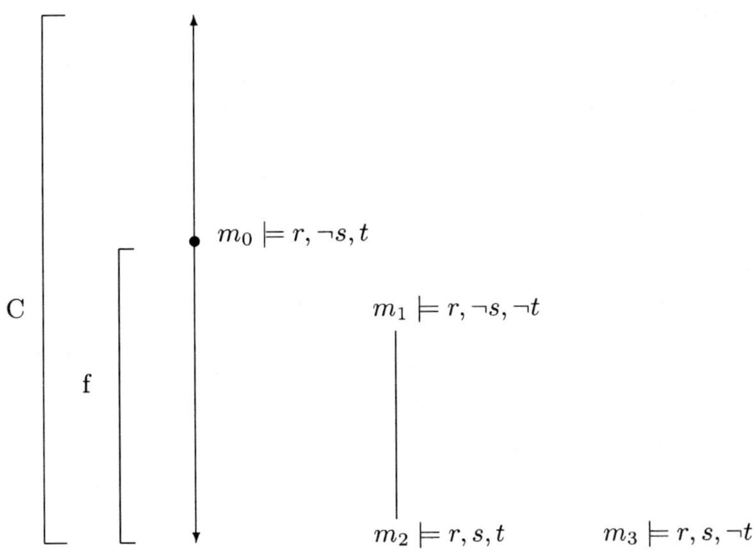

Figure 3.6.1

Note that m_0 is a minimal model of t.

Obviously, (NR) is false, as $r \mid\!\sim s$, but neither $r \wedge t \mid\!\sim s$, nor $r \wedge \neg t \mid\!\sim s$.

The usual rules of P hold, as this is a preferential structure, except perhaps for (CM), which holds in smooth structures, and our construction is not smooth. (This is the real problem.)

Note that (CM) says $\phi \mathrel{|\!\sim} \psi \rightarrow \overline{\overline{\phi}} = \overline{\overline{\phi \wedge \psi}}$, so it suffices to show for all ϕ $\phi \mathrel{|\!\sim} \psi \rightarrow \mu(\phi) = \mu(\phi \wedge \psi)$. This is the point of the construction. The infinite descending chains converge to some minimal model, so if α holds in this minimal model, then α holds infinitely often in the chain, too. Thus there are no new minimal models of α, which might weaken the consequences.

For (WD), we have to show by Fact 3.6.1 that, if $M(\phi) \cap \mu(true) = \emptyset$, then $\mu(\phi)$ contains at most one model (where $\mu(true) = \{m_2, m_3\}$).

We examine the possible cases of $\mu(\phi)$ ($\emptyset$, $\{m_1\}$, $\{m_2\}$, $\{m_3\}$, $\{m_1, m_3\}$, $\{m_2, m_3\}$, and $\mu(\phi) \cap C \neq \emptyset$).

For (CM):

Case 1: $\mu(\phi) = \{m_2, m_3\}$: Then $\phi \mathrel{|\!\sim} \psi$ iff $\{m_2, m_3\} \models \psi$. So if $\phi \mathrel{|\!\sim} \psi$, then $\phi \wedge \psi$ holds in m_3, so by f^3, $\phi \wedge \psi$ is (downward) unboundedly often true in f, so $\mu(\phi \wedge \psi) = \{m_2, m_3\}$.

Case 2: $\mu(\phi) = \{m_1\}$ and Case 3: $\mu(\phi) = \{m_3\}$: as above, by f^1 and f^3.

Case 4: $\mu(\phi) = \{m_2\}$: As $m_2 \models \phi$, and $m_3 \not\models \phi$, ϕ is of the form $\phi' \wedge t$, so none of the f_i is a model of ϕ, so ϕ has a minimal model in the chain C, so this is impossible.

Case 5: $\mu(\phi) = \{m_1, m_3\}$: Then $\phi \mathrel{|\!\sim} \psi$ iff $\{m_1, m_3\} \models \psi$. So as in Case 1, if $\phi \mathrel{|\!\sim} \psi$, $\phi \wedge \psi$ is unboundedly often true in f^1 (and in f^3), and $\mu(\phi \wedge \psi) = \{m_1, m_3\}$.

Case 6: $\mu(\phi) = \emptyset$: This is impossible by Fact 3.6.2: If ϕ is unboundedly often true in C, then it must be true in one of m_1, m_3.

Case 7: $\mu(\phi) \cap C \neq \emptyset$: Then below each $m \models \phi$, there is $m' \in \mu(\phi)$. Thus, the usual argument which shows cumulativity in smooth structures applies.

For (WD):

We only have to consider the cases where $m_2, m_3 \not\in M(\phi)$, so the only possible cases are: Case 2, Case 7.

In Case 2, there is nothing to show, $\mu(\phi)$ is a singleton.

In Case 7, (WD) is trivial, we have a unique minimum: $m_2, m_3 \not\in M(\phi)$ by prerequisite. But if $m_1 \models \phi$, then ϕ would be true unboundedly often in f, so it would not have a minimal model in C. Thus, $\mu(\phi)$ is a singleton.

□

3.7 Plausibility logic — problems without closure under finite union

3.7.1 Introduction

As plausibility logic is given as a sequent calculus, we begin with:

A short general remark on different axiom systems

We compare Hilbert style axiomatizations with sequent calculi.

In Hilbert style axiomatizations, the axioms describe the universe U, the set of all models. E.g. the axiom $\phi \rightarrow (\psi \rightarrow \phi)$ says that $M(\phi \rightarrow (\psi \rightarrow \phi)) = U$. The rules, e.g. MP, say: $X \cap Y \subseteq Z$, if $x \in M(\phi)$ and $x \in M(\phi \rightarrow \psi)$, then $x \in M(\psi)$, they describe inclusion. Thus, the system characterizes U, and how to preserve U. If the axiomatization is complete, we find all (definable) sets equivalent to the universe. To show that $T \vdash \phi$ corresponds to $T \models \phi$, we need $X \subseteq Y$ iff $\boldsymbol{C}(X) \cup Y = U$, or something equivalent.

In sequent calculi, we try to find all X, Y, s.t. $X \subseteq Y$, i.e. we characterize inclusion. We begin, e.g. with the "axiom" $X \subseteq X$, and say, if $A \subseteq B$, then $C \subseteq D$, etc. This gives directly $T \vdash \phi$, i.e. $M(T) \subseteq M(\phi)$, but no theorems. For this, we need one tautology $\boldsymbol{T}$, and this suffices: we look then at $\boldsymbol{T} \mid\sim \phi$, or whether $M(\boldsymbol{T}) \subseteq M(\phi)$. It is important to recall for the intuition that a sequence $X \mid\sim Y$ is to be interpreted as $\bigwedge X \mid\sim \bigvee Y$.

Finally, in plausibility logic, a sequent calculus, we do not describe $X \mid\sim Y$, or semantically $X \subseteq Y$, but directly $\mu(X) \subseteq Y$ — this is the way to read sequences in plausibility logic intuitively.

We turn to a

Discussion of plausibility logic

Plausibility logic was introduced by D. Lehmann [Leh92a], [Leh92b] as a sequent calculus in a propositional language without connectives. Thus, a plausibility logic language $\mathcal{L}$ is just a set, whose elements correspond to propositional variables, and a sequent has the form $X \mid\sim Y$, where X, Y are *finite* subsets of $\mathcal{L}$, thus, in the intuitive reading, $\bigwedge X \mid\sim \bigvee Y$. (We use $\mid\sim$ instead of the $\vdash$ used in [Leh92a], [Leh92b] and continue to reserve $\vdash$ for classical logic.)

For lack of space, and as our main thrust is in other directions, we will not discuss the logic itself or its motivation here in detail, but concentrate on representation questions and associated issues. The reader interested in motivation is referred to the original articles [Leh92a], [Leh92b].

A small historical remark: The representation problem held some surprise to the author. First, it was relatively easy to show that the weak version had a representation by general preferential structures, and it looked as if the full version would correspond to smooth structures. All that was missing, and which seemed quite innocent, was that the domain was not closed under finite union, as an OR is missing in the language. As the author had never understood why he needed this closure property (which, of course, is guaranteed in classical logic), he had thought that he just had not thought hard enough to find a representation result without this closure condition. After trying some time in vain, the author began to work for a counterexample. And, after some fiddling (with a small handwritten program to calculate the (finite) closure), he had one. This was the first time he realized the importance of closure conditions for the domain.

Back to contents. We show here the following:

- a proof that the weak system is complete for general preferential structures,
- a counterexample which shows that the full system is not complete for smooth preferential structures,
- we improve or modify our representation result for smooth (not necessarily transitive) preferential structures when the domain is not necessarily closed under finite unions.

The last proof follows closely the lines of the old proof (see Section 3.3.1), so one might wonder whether to unite them into one single proof. Yet, the new proof seems to need more complicated hulls which also depend on the elements, of the form $H(U,x)$, and it might not be a good idea to complicate the picture in the proof of Proposition 3.3.4 by a supplementary argument x, which will not be needed there. Consequently, we preferred a certain redundancy. As a matter of fact, we simply replace the old Fact 3.3.1 by sufficiently strong conditions on $H(U,x)$, see Definition 3.7.6 of the property (HU). Thus, as we see time and again in this book, if the domain is not nice, the conditions will not be, either.

As we are mainly interested in the basic algebraic ideas, we did not add conditions to plausibility logic to make it complete for smooth models, and left this (open problem) to the interested reader.

It is also an open problem to characterize transitive smooth structures without closure of the domain under finite unions — our proof of the transitive case uses unions.

A word of warning: Even if the domain is not closed under unions, we may well consider unions $X \cup Y$, only we cannot be sure that the function f to be represented will be defined there: $f(X \cup Y)$ need not be defined, even if $f(X)$ and $f(Y)$ are defined. A look at the proof of the transitive smooth case shows that we work there with $\mu(U_n \cup Y_{n+1})$, etc. — and this need not be defined any more. The simple union is, of course, defined, but it need not be a legal argument of the function considered any more, it need not be in its domain.

To summarize: The poor language of plausibility logic results in poor closure conditions for the domain. These can cause problems, i.e. result in more complicated conditions for representation. This was the first encounter with this phenomenon for the author. But it appears again and again in various forms (lack of finite representation, etc.), and seems to have been too much neglected so far.

The details in this Section 3.7 are somewhat involved and farther away from the mainstream, just as the logic itself.

The details:

We abuse notation, and write $X \mid\!\sim a$ for $X \vdash \{a\}$, $X, a \mid\!\sim Y$ for $X \cup \{a\} \mid\!\sim Y$, $ab \mid\!\sim Y$ for $\{a, b\} \mid\!\sim Y$, etc. When discussing plausibility logic, X, Y, etc. will denote finite subsets of $\mathcal{L}$, a, b, etc. elements of $\mathcal{L}$.

We first define the logical properties we will examine.

Definition 3.7.1

X and Y will be finite subsets of $\mathcal{L}$, a, etc. elements of $\mathcal{L}$. The base axiom and rules of plausibility logic are (we use the prefix "Pl" to differentiate them from the usual ones):

(PlI) (Inclusion): $X \mid\!\sim a$ for all $a \in X$,

(PlRM) (Right Monotony): $X \mid\!\sim Y \Rightarrow X \mid\!\sim a, Y$,

(PlCLM) (Cautious Left Monotony): $X \mid\!\sim a$, $X \mid\!\sim Y \Rightarrow X, a \mid\!\sim Y$,

(PlCC) (Cautious Cut): $X, a_1 \ldots a_n \mid\!\sim Y$, and for all $1 \leq i \leq n$ $X \mid\!\sim a_i, Y$ $\Rightarrow X \mid\!\sim Y$,

and as a special case of (PlCC):

(PlUCC) (Unit Cautious Cut): $X, a \mid\!\sim Y$, $X \mid\!\sim a, Y \Rightarrow X \mid\!\sim Y$.

and we denote by PL, for plausibility logic, the full system, i.e. (PlI)+(PlRM)+(PlCLM)+(PlCC). □

We will show in this section that:

(1) The system (PlI)+(PlRM)+(PlCC) is sound and complete for minimal preferential models (as adapted to plausibility logic).

(2) The system (PlI)+(PlRM)+(PlCC)+(PlCLM) is not complete for smooth minimal preferential models. This is somewhat surprising, and in contrast to standard propositional nonmonotonic logics, where comparable axiom systems are complete for smooth minimal preferential models. Incompleteness is essentially due to the absence of an "or" on the left hand side of $\mid\sim$, so the sets of models of plausibility formulas are not closed under finite union, violating one of the prerequisites of Proposition 3.3.4.

(3) We show how to mend the representation result for smooth structures to work without closure under finite unions.

Before we turn to the main results, we show a considerable simplification of plausibility logic (which will be used later), and introduce the main definition.

Fact 3.7.1

We note that (partially by finiteness of all sets involved)

1. In the presence of (PlRM), (PlI) is equivalent to

(PlI′): $X \mid\sim Y$ for $X \cap Y \neq \emptyset$.

2. (PlRM) is equivalent to

(PlRM′): $X \mid\sim Y \rightarrow X \mid\sim Z, Y$ for all Z.

3. (PlCC) is equivalent to

(PlCC′): $X, Z \mid\sim Y$ and $X \mid\sim z, Y$ for all $z \in Z \rightarrow X \mid\sim Y$ for all X, Y, Z s.t. $(X \cup Y) \cap Z = \emptyset$. □

Let PL' denote (PlCLM)+(PlCC').

PL and PL' are equivalent in the following sense, i.e. (PlRM) can essentially be omitted as rule:

Fact 3.7.2

Let $\mathcal{A}$ be a set of $\mathcal{L}-$sequences, $\mathcal{A}'$ be the closure of $\mathcal{A}$ under PL, and $\mathcal{A}''$ be the closure of $\{X \mid\sim Y\colon X \cap Y \neq \emptyset\} \cup \{X \mid\sim Y\colon X \mid\sim Y' \in \mathcal{A}$ for some $Y' \subseteq Y\}$ under PL'. Then $\mathcal{A}' = \mathcal{A}''$.

Proof:

"$\supseteq$" is trivial.

"$\subseteq$": Any application of (PlRM) can be pulled back through applications of (PlCLM) and (PlCC).

More formally, let, e.g. Π be a proof of $X', a' \mid\sim a, Y'$ terminating with (PlCLM), followed by (PlRM): $X' \mid\sim a'$, $X' \mid\sim Y' \Rightarrow_{\text{(PlCLM)}} X', a' \mid\sim Y'$ $\Rightarrow_{\text{(PlRM)}} X', a' \mid\sim a, Y'$. Then $X' \mid\sim Y' \Rightarrow_{\text{(PlRM)}} X' \mid\sim a, Y'$ and $X' \mid\sim a'$, $X' \mid\sim a, Y' \Rightarrow_{\text{(PlCLM)}} X', a' \mid\sim a, Y'$ also is a proof of $X', a' \mid\sim a, Y'$. (The case (PlCC) is analogous.)

Thus, any proof can be assumed to start with instances of (PlI) or elements from $\mathcal{A}$, followed by some applications of (PlRM), and only then applications of (PlCLM) or (PlCC). □

This remark allows to reduce considerably the number of possible proofs of a sequence.

We now adapt the definition of a preferential model to plausibility logic. This is the central definition on the semantic side.

Definition 3.7.2

Fix a plausibility logic language $\mathcal{L}$. A model for $\mathcal{L}$ is then just an arbitrary subset of $\mathcal{L}$.

If $\mathcal{M} :=< M, \prec>$ is a preferential model s.t. M is a set of (indexed) $\mathcal{L}-$models, then for a finite set $X \subseteq \mathcal{L}$ (to be imagined on the left hand side of $\mid\sim$!), we define

(a) $m \models X$ iff $X \subseteq m$

(b) $M(X) := \{m\colon < m, i >\in M$ for some i and $m \models X\}$

(c) $\mu(X) := \{m \in M(X)\colon \exists < m, i >\in M.\neg\exists < m', i' >\in M\ (m' \in M(X) \wedge < m', i' >\prec< m, i >)\}$

(d) $X \models_{\mathcal{M}} Y$ iff $\forall m \in \mu(X).m \cap Y \neq \emptyset$. □

(a) reflects the intuitive reading of X as $\bigwedge X$, and (d) that of Y as $\bigvee Y$ in $X \mid\sim Y$. Note that X is a set of "formulas", and $\mu(X) = \mu_{\mathcal{M}}(M(X))$.

We note as trivial consequences of the definition.

Fact 3.7.3

(a) $a \models_{\mathcal{M}} b$ iff for all $m \in \mu(a).b \in m$,

(b) $X \models_{\mathcal{M}} Y$ iff $\mu(X) \subseteq \bigcup\{M(b) : b \in Y\}$,

(c) $m \in \mu(X) \wedge X \subseteq X' \wedge m \in M(X') \rightarrow m \in \mu(X')$. □

3.7.2 Completeness and incompleteness results for plausibility logic

We have taken an old proof of ours (with a correction of Fact 3.7.6, due to D. Lehmann), and not the one published in [Sch96-3], as this (somewhat more complicated) proof fits better into the general proof strategy.

3.7.2.1 (PlI)+(PlRM)+(PlCC) is complete (and sound) for preferential models

((PlI)+(PlRM)+(PlCC) is the system of [Leh92b].)

Fact 3.7.4

If $X_i : i \in I$ are all finite, and Y is s.t. $Y \cap X_i \neq \emptyset$ for all $i \in I$, then there is $f \in \Pi X_i$ s.t.

(1) $ran(f) \subseteq Y$,

(2) $\neg\exists g \in \Pi X_i.ran(g) \subset ran(f)$.

Proof:

(a) There is $Y' \subseteq Y$ minimal with $Y' \cap X_i \neq \emptyset$ for all i: If not, there is an infinite descending sequence $Y_\alpha : \alpha < \kappa$, $lim(\kappa)$, $Y_0 = Y$, and for all $\alpha < \kappa$

$Y_\alpha \cap X_i \neq \emptyset$ for all i, but $\bigcap Y_\alpha \cap X_i = \emptyset$ for some fixed i, contradicting finiteness of X_i.

(b) Let $f \in \Pi(Y' \cap X_i)$, any $g \in \Pi X_i$, $ran(g) \subset ran(f) \subseteq Y'$ would show Y' not minimal. □

Definition 3.7.3

For $T \subseteq \mathcal{L}$ let $\overline{\overline{T}} := \{Y : T \mid\!\sim Y\}$.

Let $\mathcal{P}_f(X)$ denote the set of finite subsets of X.

Let $\Pi'\overline{\overline{T}} := \{f \in \Pi\overline{\overline{T}}$: there is no $g \in \Pi\overline{\overline{T}}$ s.t. $ran(g) \subset ran(f)\}$.

Given a function $\mu : \mathcal{P}_f(\mathcal{L}) \to \mathcal{PP}(\mathcal{L})$, define for $T, Y \in \mathcal{P}_f(\mathcal{L})$

$T \models_\mu Y :\leftrightarrow \forall m \in \mu(T).m \cap Y \neq \emptyset$.

Lemma 3.7.5

Let $\mid\!\sim$ satisfy (PlRM), and μ be s.t.

1. If $T \mid\!\sim \emptyset$, then $\mu(T) = \emptyset$.

2. If $T \not\mid\!\sim \emptyset$, then $\mu(T)$ has the properties

(a) $\{ran(f) : f \in \Pi'\overline{\overline{T}}\} \subseteq \mu(T)$,

(b) if $m \in \mu(T)$, then there is $f \in \Pi'\overline{\overline{T}}$ s.t. $ran(f) \subseteq m$.

Then $\models_\mu = \mid\!\sim$.

Proof:

We have to show $T \mid\!\sim Y \leftrightarrow T \models_\mu Y$ for all $T, Y \in \mathcal{P}_f(\mathcal{L})$.

Note that if $T \not\mid\!\sim \emptyset$, then $\mu(T) \neq \emptyset$ by (a) and Fact 3.7.4.

Case 1: $T \mid\!\sim \emptyset : T \mid\!\sim \emptyset \leftrightarrow \mu(T) = \emptyset \leftrightarrow \forall m \in \mu(T).m \cap \emptyset \neq \emptyset \leftrightarrow T \models_\mu \emptyset$.

Case 2: $T \not\mid\!\sim \emptyset$: "$\rightarrow$": Let $T \mid\!\sim Y$, so for all $f \in \Pi\overline{\overline{T}}$ $ran(f) \cap Y \neq \emptyset \rightarrow$ (by (b)) $\forall m \in \mu(T).m \cap Y \neq \emptyset \rightarrow T \models_\mu Y$ "$\leftarrow$": Suppose $T \not\mid\!\sim Y$, then by (PlRM), for no $X \subseteq Y$ $T \mid\!\sim X$. Thus, for all $X \in \overline{\overline{T}}$ $X - Y \neq \emptyset$, let $f \in \Pi\{X - Y : X \in \overline{\overline{T}}\}$, so $f \in \Pi\overline{\overline{T}}$, and $ran(f) \cap Y = \emptyset$. Let $f' \in \Pi'\overline{\overline{T}}$ s.t. $ran(f') \subseteq ran(f)$ (this exists by Fact 3.7.4), so $ran(f') \in \mu(T)$, and $T \not\models_\mu Y$. □

Thus, if (PlRM) holds for $\mid\sim$, then for μ defined by

$$\mu(T) := \begin{cases} \emptyset & \text{iff} \quad T \mid\sim \emptyset \\ \{ran(f) \colon f \in \Pi' \overline{\overline{T}}\} & \text{otherwise} \end{cases}$$

$\models_\mu = \mid\sim$.

Definition 3.7.4

For $\mathcal{M} := < M, \mu >$ with $M \subseteq \mathcal{P}(\mathcal{L})$, $\mu : \mathcal{P}_f(\mathcal{L}) \rightarrow \mathcal{P}(M)$, let $\models_\mathcal{M} := \models_\mu$. Let $\mid\sim$ be given, define $\mathcal{M} := < M, \mu >$ by:

1.) $M := \mathcal{P}(\mathcal{L})$,

2.)

$$\mu(T) := \begin{cases} \emptyset & \text{iff} \quad T \mid\sim \emptyset \\ \{ m \subseteq \mathcal{L} \colon & \text{otherwise} \\ 1.\ m \cap Y \neq \emptyset \text{ for all } Y \in \overline{\overline{T}}, & \\ 2.\ \text{there is } T' \subseteq T \text{ s.t. } m = ran(f) & \\ \text{for some } f \in \Pi' \overline{\overline{T'}}\} & \end{cases}$$

By Fact 3.7.4, μ satisfies the prerequisites of Lemma 3.7.5, so if $\mid\sim$ satisfies (PlRM), then $\models_\mathcal{M} = \mid\sim$.

For $T \subseteq \mathcal{L}$, let $M(T) := \{X \subseteq \mathcal{L} : T \subseteq X\}$ and $\mathcal{Y} := \{M(T) : T \subseteq \mathcal{L}\}$, thus $\mathcal{Y} \subseteq \mathcal{P}(M)$. Define $F : \mathcal{Y} \rightarrow \mathcal{P}(M)$ by $F(M(T)) := \mu(T)$, this is well-defined, as $T = \bigcap M(T)$.

We work now with μ and F as just defined, and first note two auxiliary facts:

Fact 3.7.6

Let $\mid\sim$ satisfy (PlRM), $f \in \Pi' \overline{\overline{S}}$, $R \subseteq ran(f)$ be finite, then there is $Y_{S,f,R}$ s.t.

1. $S \mid\sim r, Y_{S,f,R}$ for all $r \in R$,
2. $Y_{S,f,R} \cap ran(f) = \emptyset$.

Proof:

By minimality of $ran(f)$, for each $q \in ran(f)$ exists $Z_q \in \overline{\overline{S}}$ with $Z_q \cap ran(f) = \{q\}$. Let $Y_q := Z_q - \{q\}$, so $Y_q \cap ran(f) = \emptyset$ and $S \mid\sim q, Y_q$. Let $Y_{S,f,R} := \bigcup \{Y_r : r \in R\}$, so $Y_{S,f,R} \cap ran(f) = \emptyset$ and for $r \in R$ $S \mid\sim r, Y_{S,f,R}$

(by (PlRM)). □

Fact 3.7.7

Let $\mid\!\sim$ satisfy (PlRM)+(PlCC), $f \in \Pi'\overline{\overline{S}}$, $S \subseteq S' \subseteq ran(f)$, $S' \mid\!\sim Y$, then $ran(f) \cap Y \neq \emptyset$.

Proof:

If $S = S'$, we are done. Let $R := \{r_1 \ldots r_n\}$, $S' := S \cup R$, and $Y_{S,f,R}$ as in Fact 3.7.6. so $S \mid\!\sim r_i, Y_{S,f,R}$, and by (PlRM) and $S' \mid\!\sim Y$ $S \cup \{r_1 \ldots r_n\} \mid\!\sim Y, Y_{S,f,R}$ and $S \mid\!\sim r_i, Y, Y_{S,f,R}$. So by (PlCC), $S \mid\!\sim Y, Y_{S,f,R}$. By $f \in \Pi\overline{\overline{S}}$, $ran(f) \cap (Y \cup Y_{S,f,R}) \neq \emptyset$, but $ran(f) \cap Y_{S,f,R} = \emptyset$. □

Lemma 3.7.8

(a) If $\mid\!\sim$ satisfies (PlI), then $F(M(T)) \subseteq M(T)$.

(b) If $\mid\!\sim$ satisfies (PlRM)+(PlCC), then $M(T') \subseteq M(T) \rightarrow F(M(T)) \cap M(T') \subseteq F(M(T'))$.

Proof:

(a) If $T \mid\!\sim \emptyset$, we are done. Otherwise, let $m \in \mu(T)$, but by (PlI), $T \mid\!\sim x$ for all $x \in T$, so $T \subseteq m$.

(b) By $M(T') \subseteq M(T)$, $T = \bigcap M(T)$, and $T' = \bigcap M(T')$, $T \subseteq T'$ holds. We have to show $\mu(T) \cap M(T') \subseteq \mu(T')$. Let $m \in \mu(T) \cap M(T')$, so $T \subseteq T' \subseteq m$. Moreover, there is $T'' \subseteq T \subseteq T'$, $m = ran(f)$ for some $f \in \Pi'\overline{\overline{T''}}$. It thus remains to show $m \cap Y \neq \emptyset$ for all $Y \in \overline{\overline{T'}}$. But this follows from Fact 3.7.7. □

We have shown in Section 3.2.2 above, Proposition 3.2.4, that such F can be represented by a (transitive, irreflexive) preferential structure.

□ (Completeness)

3.7.2.2 Incompleteness of full plausibility logic for smooth structures

We work in PL and construct a counterexample, a set of formulas which satisfies the axiom and rules of plausibility logic, but violates Fact 3.7.9 below, and thus cannot be represented by a smooth preferential model.

We note the following fact for smooth preferential models:

Fact 3.7.9

Let U, X, Y be any sets, $\mathcal{M}$ be smooth for at least $\{Y, X\}$ and let $\mu(Y) \subseteq U \cup X$, $\mu(X) \subseteq U$, then $X \cap Y \cap \mu(U) \subseteq \mu(Y)$.

Proof:

We show $m \in X \cap Y$, $m \notin \mu(Y) \rightarrow m \notin \mu(U)$. As $m \in Y - \mu(Y)$, by smoothness, there is $m' \prec m$, $m' \in \mu(Y)$. Case 1: $m' \in U$: we are done. Case 2 : $m' \in X$. Thus $m \in X - \mu(X)$, so by smoothness, there is $m'' \prec m$, $m'' \in \mu(X)$, but $\mu(X) \subseteq U$, so we are done again. (An analogous proof holds when we work with copies.) □

Example 3.7.1

Let $\mathcal{L} := \{a, b, c, d, e, f\}$, and $\mathcal{X} := \{a \mid\sim b,\ b \mid\sim a,\ a \mid\sim c,\ a \mid\sim fd,\ dc \mid\sim ba,\ dc \mid\sim e,\ fcba \mid\sim e\}$. We show that $\mathcal{X}$ does not have a smooth representation.

Fact 3.7.10

$\mathcal{X}$ does not entail $a \mid\sim e$.

Proof:

Let $\mathcal{A} := \{a \mid\sim b,\ a \mid\sim c,\ a \mid\sim ed,\ a \mid\sim fd,\ b \mid\sim a,\ b \mid\sim c,\ b \mid\sim ed,\ b \mid\sim fd,\ ba \mid\sim c,\ ba \mid\sim ed,\ ba \mid\sim fd,\ ca \mid\sim b,\ ca \mid\sim ed,\ ca \mid\sim fd,\ cb \mid\sim a,\ cb \mid\sim ed,\ cb \mid\sim fd,\ cba \mid\sim ed,\ cba \mid\sim fd,\ dc \mid\sim ba,\ dc \mid\sim e,\ edc \mid\sim ba,\ fcba \mid\sim e\}$.

Set $\mathcal{A}_0 := \{X \mid\sim Y\colon\ X \cap Y \neq \emptyset\}$, $\mathcal{A}_1 := \{X \mid\sim Y\colon$ there is $Y' \subseteq Y$ s.t. $X \mid\sim Y' \in \mathcal{A}\}$, and $\mathcal{A}'' := \mathcal{A}_0 \cup \mathcal{A}_1$.

As $\mathcal{A}''$ contains $\mathcal{X}$, but not $a \mid\sim e$, it suffices to show that $\mathcal{A}''$ is a plausibility logic, i.e. is closed under PL. By Fact 3.7.2, this is equivalent to showing that $\mathcal{A}''$ is closed under (PlCLM) + (PlCC'). We note

Remark 3.7.11

(a) For $X \in \{a, b, ba, ca, cb, cba\}$ and $Y \in \{a, b, c, ed, fd\}$ $X \mid\sim Y \in \mathcal{A}''$,

(b) for $X \in \{dc, edc, fcba, fecba\}$, $Y \in \{e, ba\}$ $X \mid\sim Y \in \mathcal{A}''$

□ (Remark 3.7.11)

Note also that all cases of $\mathcal{A}$ occur as cases of (a) or (b).

We first show closure of $\mathcal{A}''$ under (PlCLM): $X' \mid\sim a'$, $X' \mid\sim Y' \rightarrow X', a' \mid\sim Y'$ $(a' \notin X')$:

Thus, $X' \mid\sim a' \in \mathcal{A}$, and $X' = a$ and $a' = b$ or $a' = c$, $X' = b$ and $a' = a$ or $a' = c$, $X' = ba$ and $a' = c$, $X' = ca$ and $a' = b$, $X' = cb$ and $a' = a$, $X' = dc$ and $a' = e$, $X' = fcba$ and $a' = e$.

The case $X' \mid\sim Y' \in \mathcal{A}_0$ is trivial. Suppose $X' \mid\sim Y' \in \mathcal{A}_1$, so there is $Y \subseteq Y'$ and $X' \mid\sim Y \in \mathcal{A}$. It suffices to show that then $X', a' \mid\sim Y \in \mathcal{A}''$, as $\mathcal{A}''$ is obviously closed under (PlRM). But all cases are handled by Remark 3.7.11 (a) or (b): If $X' \mid\sim a' \in \mathcal{A}$ and $X' \mid\sim Y \in \mathcal{A}$, then X' is one of the X and Y is one of the Y in (a) or (b). But then X', a' is also one of the X in (a) or (b).

We turn to closure under (PlCC'). We have to show for all X', Y', Z' with $Z' \cap (X' \cup Y') = \emptyset$, $Z' \neq \emptyset$: $X', Z' \mid\sim Y'$, $X' \mid\sim z', Y'$ for all $z' \in Z'$ $\rightarrow$ $X' \mid\sim Y'$. As for no Y $\emptyset \mid\sim Y \in \mathcal{A}''$, $X' \neq \emptyset$.

The case $X', Z' \mid\sim Y' \in \mathcal{A}_0$ is again trivial, as $Z' \cap Y' = \emptyset$, likewise the case $X' \mid\sim z', Y' \in \mathcal{A}_0$ for some $z' \in Z'$, as $Z' \cap X' = \emptyset$.

So assume without loss of generality $X' \neq \emptyset$, $Z' \neq \emptyset$, $X' \cap Z' = \emptyset$, $Y' \cap Z' = \emptyset$, $X', Z' \mid\sim Y' \in \mathcal{A}_1$, and for all $z' \in Z'$ $X' \mid\sim z', Y' \in \mathcal{A}_1$. We have to show $X' \mid\sim Y' \in \mathcal{A}''$. Note that by definition of $\mathcal{A}_1$, and $X' \mid\sim z', Y' \in \mathcal{A}_1$, X' has to occur on the left hand side in $\mathcal{A}$, so $X' \in \{a, b, ba, ca, cb, cba, dc, edc, fcba\}$. As $X', Z' \mid\sim Y' \in \mathcal{A}_1$, there is some $Y \subseteq Y'$ with $X', Z' \mid\sim Y \in \mathcal{A}$. Moreover, $X' \cup Z'$ has at least two elements.

Case 1: $X' \cup Z' \in \{ba, ca, cb, cba\}$. X' is a proper, nonempty subset of $X' \cup Z'$. As $X' \neq c$, Remark 3.7.11 (a) shows that $X' \mid\sim Y$ too, and thus $X' \mid\sim Y'$.

Case 2: $X' \cup Z' \in \{dc, edc, fcba\}$. The possible cases are:

(1) $X' \cup Z' = edc$, $X' = dc$ and $Z' = e$,

(2) $X' \cup Z' = fcba$, $X' \in \{a, b, ba, ca, cb, cba\}$, $Z' = fcba - X'$, so $f \in Z'$.

In (1), we are done by Remark 3.7.11 (b).

(2): As $X' \cup Z' \mid\sim Y' \in \mathcal{A}_1$, Y' has to contain e. Moreover, by $f \in Z'$, $X' \mid\sim f, Y' \in \mathcal{A}_1$, so there must be some $Y'' \subseteq f, Y'$ with $X' \mid\sim Y'' \in \mathcal{A}$. If $Y'' \subseteq Y'$, we are done, as then $X' \mid\sim Y' \in \mathcal{A}_1 \subseteq \mathcal{A}''$. But if $f \in Y''$, then $Y'' = fd$, so $d \in Y'$. Thus, $d, e \in Y'$. But by Remark 3.7.11 (a), $X' \mid\sim ed \in \mathcal{A}''$, so $X' \mid\sim Y' \in \mathcal{A}''$.

□ ($\mathcal{X}$ does not entail $a \mid\sim e$, Fact 3.7.10)

Suppose now that there is a smooth preferential model $\mathcal{M} =< M, \prec>$ for plausibility logic which represents $\mid\sim$, i.e. for all X, Y finite subsets of $\mathcal{L}$ $X \mid\sim Y$ iff $X \models_{\mathcal{M}} Y$. (See Definition 3.7.2 and Fact 3.7.3.)

$a \mid\sim a$, $a \mid\sim b$, $a \mid\sim c$ implies for $m \in \mu(a)$ $a, b, c \in m$. Moreover, as $a \mid\sim df$, then also $d \in m$ or $f \in m$. As $a \not\mid\sim e$, there must be $m \in \mu(a)$ s.t. $e \notin m$. Suppose now $m \in \mu(a)$ with $f \in m$. So $a, b, c, f \in m$, thus by $m \in \mu(a)$ and Fact 3.7.3, $m \in \mu(a, b, c, f)$. But $fcba \mid\sim e$, so $e \in m$. We thus have shown that $m \in \mu(a)$ and $f \in m$ implies $e \in m$. Consequently, there must be $m \in \mu(a)$ s.t. $d \in m$, $e \notin m$. Thus, in particular, as $cd \mid\sim e$, there is $m \in \mu(a)$, $a, b, c, d \in m$, $m \notin \mu(cd)$. But by $cd \mid\sim ab$, and $b \mid\sim a$, $\mu(cd) \subseteq M(a) \cup M(b)$ and $\mu(b) \subseteq M(a)$ by Fact 3.7.3. Let now $T := M(cd)$, $R := M(a)$, $S := M(b)$, and $\mu_{\mathcal{M}}$ be the choice function of the minimal elements in the structure $\mathcal{M}$, we then have by $\mu(S) = \mu_{\mathcal{M}}(M(S))$:

1. $\mu_{\mathcal{M}}(T) \subseteq R \cup S$,
2. $\mu_{\mathcal{M}}(S) \subseteq R$,
3. there is $m \in S \cap T \cap \mu_{\mathcal{M}}(R)$, but $m \notin \mu_{\mathcal{M}}(T)$,

but this contradicts above Fact 3.7.9. □ (Counterexample 3.7.1)

3.7.2.3 Discussion and remedy

We show here only how to do the semantical construction of Section 3.3.1 without closure of the domain under finite unions. The syntactical side is left open. The presentation is rather succinct, as this Section 3.7.2.3 is more for the specialists.

The important point in the counterexample in Section 3.7.2.2 is that the condition

$\mu(T) \subseteq R \cup S$ and $\mu(S) \subseteq R$ imply $S \cap T \cap \mu(R) \subseteq \mu(T)$

holds in all smooth models, but not in the example.

Thus, we need new conditions, which take care of the "semi-transitivity" of smoothness, coding it directly and not by a simple condition, which uses finite union. For this purpose, we modify the definition of $H(U)$, and replace it by $H(U,x)$:

Definition 3.7.5

Definition of $H(U,x)$:

$H(U,x)_0 := U$

$H(U,x)_{i+1} := H(U,x)_i \cup \bigcup\{U' : x \in \mu(U'),\ \mu(U') \subseteq H(U,x)_i\}$

We take unions at limits.

$H(U,x) := \bigcup\{H(U,x)_i : i < \kappa\}$ for κ sufficiently big.

Definition 3.7.6

(HU) is the property:

$x \in \mu(U),\ x \in Y - \mu(Y) \rightarrow \mu(Y) \not\subseteq H(U,x)$.

We then have:

Fact 3.7.12

(1) $x \in \mu(Y),\ \mu(Y) \subseteq H(U,x) \rightarrow Y \subseteq H(U,x)$,

(2) (HU) holds in all smooth models.

Proof:

(1) Trivial by definition.

(2) Suppose not. So let $x \in \mu(U)$, $x \in Y - \mu(Y)$, $\mu(Y) \subseteq H(U,x)$. By smoothness, there is $x_1 \in \mu(Y)$, $x \succ x_1$, and let κ_1 be the least κ s.t. $x_1 \in H(U,x)_{\kappa_1}$. κ_1 is not a limit, and $x_1 \in U'_{x_1} - \mu(U'_{x_1})$ with $x \in \mu(U'_{x_1})$ for some U'_{x_1}, so as $x_1 \notin \mu(U'_{x_1})$, there must be (by smoothness) some other $x_2 \in \mu(U'_{x_1}) \subseteq H(U,x)_{\kappa_1 - 1}$ with $x \succ x_2$. Continue with x_2, we thus construct a descending chain of ordinals, which cannot be infinite, so there must be $x_n \in \mu(U'_{x_n}) \subseteq U$, $x \succ x_n$, contradicting minimality of x in U. (More precisely, this works for all copies of x.) □

This suffices for the construction:

We patch the proof of the smooth case in Section 3.3.1 a little. $H(U)$ is replaced by $H(U,x)$, Fact 3.3.1 is replaced by above Fact 3.7.12, and we avoid unions.

For intersections: We can either stipulate closure under intersections, and work with Remark 3.7.13 below, or demand that $X \neq \emptyset \rightarrow \mu(X) \neq \emptyset$, or, perhaps the simplest thing, just demand that if $x \in K$, $Y \in \mathcal{Y}$, $x \in Y - \mu(Y)$, then $\mu(Y) \neq \emptyset$, as shown in 3.7.13.

Suppose now $(\mu \subseteq)$, (μPR), (μCUM) to hold, as well as property (HU), as defined in Definition 3.7.6.

We first show two basic facts and then turn to the main result, Proposition 3.7.15.

Definition 3.7.7

For $x \in Z$, let $\mathcal{W}_x := \{\mu(Y)\colon Y \in \mathcal{Y} \wedge x \in Y - \mu(Y)\}$, $\Gamma_x := \Pi\mathcal{W}_x$, and $K := \{x \in Z\colon \exists X \in \mathcal{Y}.x \in \mu(X)\}$.

Remark 3.7.13

(1) $x \in K \rightarrow \Gamma_x \neq \emptyset$,

(2) $g \in \Gamma_x \rightarrow ran(g) \subseteq K$.

Proof:

(1) See above remark on closure under intersections. We have to show that $Y \in \mathcal{Y}$, $x \in Y - \mu(Y) \rightarrow \mu(Y) \neq \emptyset$. By $x \in K$, there is $X \in \mathcal{Y}$ s.t. $x \in \mu(X)$. Suppose $x \in Y$, $\mu(Y) = \emptyset$. Then $x \in X \cap Y$, so by $x \in \mu(X)$ and (μPR) $x \in \mu(X \cap Y)$. But $\mu(Y) = \emptyset \subseteq X \cap Y \subseteq Y$, so by (μCUM) $\mu(X \cap Y) = \emptyset$, contradiction.

(2) By definition, $\mu(Y) \subseteq K$ for all $Y \in \mathcal{Y}$. □

Claim 3.7.14

Let $U \in \mathcal{Y}$, $x \in K$. Then

(1) $x \in \mu(U) \leftrightarrow x \in U \wedge \exists f \in \Gamma_x.ran(f) \cap U = \emptyset$,

(2) $x \in \mu(U) \leftrightarrow x \in U \wedge \exists f \in \Gamma_x.ran(f) \cap H(U,x) = \emptyset$.

Proof:

(1) Case 1: $\mathcal{W}_x = \emptyset$, thus $\Gamma_x = \{\emptyset\}$. "$\rightarrow$": Take $f := \emptyset$. "$\leftarrow$": $x \in U \in \mathcal{Y}$, $\mathcal{W}_x = \emptyset \rightarrow x \in \mu(U)$ by definition of $\mathcal{W}_x$.

Case 2: $\mathcal{W}_x \neq \emptyset$. "$\rightarrow$": Let $x \in \mu(U) \subseteq U$. By (HU), if $Y \in \mathcal{W}_x$, then $\mu(Y) - H(U,x) \neq \emptyset$. "$\leftarrow$": If $x \in U - \mu(U)$, $\mu(U) \in \mathcal{W}_x$, moreover $\Gamma_x \neq \emptyset$ by Remark 3.7.13, (1) and thus (or by the same argument) $\mu(U) \neq \emptyset$, so $\forall f \in \Gamma_x.ran(f) \cap U \neq \emptyset$.

(2): The proof is verbatim the same as for (1). □ (Claim 3.7.14)

The following Proposition 3.7.15 is the main result of Section 3.7.2.3 and shows how to characterize smooth structures in the absence of closure under finite unions. The strategy of the proof follows closely the proof of Proposition 3.3.4.

Proposition 3.7.15

Let $\mathcal{Y}$ be closed under finite intersections (or some other condition, see above), and $\mu : \mathcal{Y} \rightarrow \mathcal{Y}$. Then there is a $\mathcal{Y}-$smooth preferential structure $\mathcal{Z}$, s.t. for all $X \in \mathcal{Y}$ $\mu(X) = \mu_{\mathcal{Z}}(X)$ iff μ satisfies $(\mu \subseteq)$, (μPR), (μCUM), and (HU) above.

Proof:

"$\rightarrow$" is again easy and left to the reader. (HU) was shown in Fact 3.7.12.

The strategy is the same as in Section 3.3.1, we recall it:

Outline of "$\leftarrow$": We first define a structure $\mathcal{Z}$ which represents μ, but is not necessarily $\mathcal{Y}-$smooth, refine it to $\mathcal{Z}'$ and show that $\mathcal{Z}'$ represents μ too, and that $\mathcal{Z}'$ is $\mathcal{Y}-$smooth.

In the structure $\mathcal{Z}'$, all pairs destroying smoothness in $\mathcal{Z}$ are successively repaired, by adding minimal elements: If $< y, j >$ is not minimal, and has no minimal $< x, i >$ below it, we just add one such $< x, i >$. As the repair process might itself generate such "bad" pairs, the process may have to be repeated infinitely often. Of course, one has to take care that the representation property is preserved.

Construction 3.7.1

(Construction of $\mathcal{Z}$)

Let $\mathcal{X} := \{< x, g >: x \in K,\ g \in \Gamma_x\}$, $< x', g' > \prec < x, g > :\leftrightarrow x' \in ran(g)$,

$\mathcal{Z} :=< \mathcal{X}, \prec >$.

Claim 3.7.16

$\forall U \in \mathcal{Y}.\mu(U) = \mu_{\mathcal{Z}}(U)$

Proof:

Case 1: $x \notin K$. Then $x \notin \mu(U)$ and $x \notin \mu_{\mathcal{Z}}(U)$.

Case 2: $x \in K$. By Claim 3.7.14, (1) it suffices to show that for all $U \in \mathcal{Y}$ $x \in \mu_{\mathcal{Z}}(U) \leftrightarrow x \in U \wedge \exists f \in \Gamma_x.ran(f) \cap U = \emptyset$. Fix $U \in \mathcal{Y}$. "$\rightarrow$": $x \in \mu_{\mathcal{Z}}(U) \rightarrow$ ex. $< x, f >$ minimal in $\mathcal{X}\lceil U$, thus $x \in U$ and there is no $< x', f' > \prec < x, f >$, $x' \in U$, $x' \in K$. But if $x' \in K$, then by Remark 3.7.13, (1), $\Gamma_{x'} \neq \emptyset$, so we find suitable f'. Thus, $\forall x' \in ran(f).x' \notin U$ or $x' \notin K$. But $ran(f) \subseteq K$, so $ran(f) \cap U = \emptyset$. "$\leftarrow$": If $x \in U$, $f \in \Gamma_x$ s.t. $ran(f) \cap U = \emptyset$, then $< x, f >$ is minimal in $\mathcal{X}\lceil U$. □ (Claim 3.7.16)

We now construct the refined structure $\mathcal{Z}'$.

Construction 3.7.2

(Construction of $\mathcal{Z}'$)

σ is called $x-$admissible sequence iff

1. σ is a sequence of length $\leq \omega$, $\sigma = \{\sigma_i : i \in \omega\}$,
2. $\sigma_o \in \Pi\{\mu(Y)$: $Y \in \mathcal{Y} \wedge x \in Y - \mu(Y)\}$,
3. $\sigma_{i+1} \in \Pi\{\mu(X)$: $X \in \mathcal{Y} \wedge x \in \mu(X) \wedge ran(\sigma_i) \cap X \neq \emptyset\}$.

By 2., σ_0 minimizes x, and by 3., if $x \in \mu(X)$, and $ran(\sigma_i) \cap X \neq \emptyset$, i.e. we have destroyed minimality of x in X, x will be above some y minimal in X to preserve smoothness.

Let Σ_x be the set of $x-$admissible sequences, for $\sigma \in \Sigma_x$ let $\widehat{\sigma} := \bigcup\{ran(\sigma_i) : i \in \omega\}$. Note that by the argument in the proof of Remark 3.7.13, (1), $\Sigma_x \neq \emptyset$, if $x \in K$.

Let $\mathcal{X}' := \{< x, \sigma >: x \in K \wedge \sigma \in \Sigma_x\}$ and $< x', \sigma' > \prec' < x, \sigma > :\leftrightarrow x' \in \widehat{\sigma}$.

Finally, let $\mathcal{Z}' :=< \mathcal{X}', \prec' >$, and $\mu' := \mu_{\mathcal{Z}'}$.

It is now easy to show that $\mathcal{Z}'$ represents μ, and that $\mathcal{Z}'$ is smooth. For $x \in \mu(U)$, we construct a special $x-$admissible sequence $\sigma^{x,U}$ using the properties of $H(U,x)$ as described at the beginning of this section.

Claim 3.7.17

For all $U \in \mathcal{Y}$ $\mu(U) = \mu_{\mathcal{Z}}(U) = \mu'(U)$.

Proof:

If $x \notin K$, then $x \notin \mu_{\mathcal{Z}}(U)$, and $x \notin \mu'(U)$ for any U. So assume $x \in K$. If $x \in U$ and $x \notin \mu_{\mathcal{Z}}(U)$, then for all $<x,f> \in \mathcal{X}$, there is $<x',f'> \in \mathcal{X}$ with $<x',f'> \prec <x,f>$ and $x' \in U$. Let now $<x,\sigma> \in \mathcal{X}'$, then $<x,\sigma_0> \in \mathcal{X}$, and let $<x',f'> \prec <x,\sigma_0>$ in $\mathcal{Z}$ with $x' \in U$. As $x' \in K$, $\Sigma_{x'} \neq \emptyset$, let $\sigma' \in \Sigma_{x'}$. Then $<x',\sigma'> \prec' <x,\sigma>$ in $\mathcal{Z}'$. Thus $x \notin \mu'(U)$. Thus, for all $U \in \mathcal{Y}$, $\mu'(U) \subseteq \mu_{\mathcal{Z}}(U) = \mu(U)$.

It remains to show $x \in \mu(U) \to x \in \mu'(U)$.

Assume $x \in \mu(U)$ (so $x \in K$), $U \in \mathcal{Y}$, we will construct minimal σ, i.e. show that there is $\sigma^{x,U} \in \Sigma_x$ s.t. $\overbrace{\sigma^{x,U}} \cap U = \emptyset$. We construct this $\sigma^{x,U}$ inductively, with the stronger property that $ran(\sigma_i^{x,U}) \cap H(U,x) = \emptyset$ for all $i \in \omega$.

$\underline{\sigma_0^{x,U}}$: $x \in \mu(U)$, $x \in Y - \mu(Y) \to \mu(Y) - H(U,x) \neq \emptyset$ by (HU). Let $\sigma_0^{x,U} \in \Pi\{\mu(Y) - H(U,x) : Y \in \mathcal{Y}, x \in Y - \mu(Y)\}$, so $ran(\sigma_0^{x,U}) \cap H(U,x) = \emptyset$.

$\underline{\sigma_i^{x,U} \to \sigma_{i+1}^{x,U}}$: By the induction hypothesis, $ran(\sigma_i^{x,U}) \cap H(U,x) = \emptyset$. Let $X \in \mathcal{Y}$ be s.t. $x \in \mu(X)$, $ran(\sigma_i^{x,U}) \cap X \neq \emptyset$. Thus $X \not\subseteq H(U,x)$, so $\mu(X) - H(U,x) \neq \emptyset$ by Fact 3.7.12, (1). Let $\sigma_{i+1}^{x,U} \in \Pi\{\mu(X) - H(U,x) : X \in \mathcal{Y}, x \in \mu(X), ran(\sigma_i^{x,U}) \cap X \neq \emptyset\}$, so $ran(\sigma_{i+1}^{x,U}) \cap H(U,x) = \emptyset$. The construction satisfies the $x-$admissibility condition. □

It remains to show:

Claim 3.7.18

$\mathcal{Z}'$ is $\mathcal{Y}-$smooth.

Proof:

Let $X \in \mathcal{Y}$, $<x,\sigma> \in \mathcal{X}' \lceil X$.

Case 1, $x \in X - \mu(X)$: Then $ran(\sigma_0) \cap \mu(X) \neq \emptyset$, let $x' \in ran(\sigma_0) \cap \mu(X)$. Moreover, $\mu(X) \subseteq K$. Then for all $<x',\sigma'> \in \mathcal{X}'$ $<x',\sigma'> \prec <x,\sigma>$.

But $< x', \sigma^{x',X} >$ as constructed in the proof of Claim 3.7.17 is minimal in $\mathcal{X}' \lceil X$.

Case 2, $x \in \mu(X) = \mu_{\mathcal{Z}}(X) = \mu'(X)$: If $< x, \sigma >$ is minimal in $\mathcal{X}' \lceil X$, we are done. So suppose there is $< x', \sigma' > \prec < x, \sigma >$, $x' \in X$. Thus $x' \in \widehat{\sigma}$. Let $x' \in ran(\sigma_i)$. So $x \in \mu(X)$ and $ran(\sigma_i) \cap X \neq \emptyset$. But $\sigma_{i+1} \in \Pi\{\mu(X'):$ $X' \in \mathcal{Y} \wedge x \in \mu(X') \wedge ran(\sigma_i) \cap X' \neq \emptyset\}$, so X is one of the X', moreover $\mu(X) \subseteq K$, so there is $x'' \in \mu(X) \cap ran(\sigma_{i+1}) \cap K$, so for all $< x'', \sigma'' > \in \mathcal{X}'$ $< x'', \sigma'' > \prec < x, \sigma >$. But again $< x'', \sigma^{x'',X} >$ as constructed in the proof of Claim 3.7.17 is minimal in $\mathcal{X}' \lceil X$.

□ (Claim 3.7.18 and Proposition 3.7.15)

Comments:

(1) We have not looked into the transitive case, my first (weak) conjecture is that we need new conditions for the transitive case.

(2) We have again the situation where weaker closure conditions impose more complicated representation conditions.

(3) Note that the construction of $H(U, x)$ is "vertically" essentially finite, as the argument when using it, shows. Smoothness is like transitivity, where it suffices to show closure under finite chains. The horizontal part is arbitrary, but a look at the argument in the old construction shows that we do not need unions there. These, we think, are the basic reasons why closure under finite unions can replace the infinite construction above.

3.8 The role of copies in preferential structures

We discuss the importance of copies in preferential structures now in more detail. The material in this Section 3.8 was published in [Sch96-1]. These results are not important for the rest of the book, but they round off the picture.

Most representation results for preferential structures use in their constructions several copies of logically identical models (see, e.g. [KLM90], or Sections 3.2 and 3.3 of this book). Thus, we may have in those constructions m and m' with the same logical properties, but with different "neighborhoods" in the preferential structure, for example, there may be some m''

with $m'' \prec m$, but $m'' \not\prec m'$. D. Makinson and H. Kamp had asked the author whether such repetitions of models are sometimes necessary to represent a logic, we now give a (positive) answer. For the connection of the question to ranked structures see Section 3.10, in particular Lemma 3.10.4.

We have already given a simple example (see Example 3.1.1 above) illustrating the importance in the finite case. We discuss now more subtle situations.

The infinite case

Let κ, λ be infinite cardinals. Let $\mathcal{L}$ have κ propositional variables, p_i, $i < \kappa$. Consider any $\vdash$ $-$consistent $\mathcal{L}-$theory T, a model m s.t. $m \not\models T$, and the following structure $\mathcal{M}$: $\mathcal{X} := \{< m, n >: n \models T\} \cup \{< n, 0 >: n \models T\}$, with $< n, 0 > \prec < m, n >$. Let $\phi \in T$ be s.t. $m \not\models \phi$. Obviously, $T \vee Th(m) \models_{\mathcal{M}} \phi$, as all copies of m are destroyed by the full set of models of T, but no T' truly stronger than T will do, as some copy of m will not be destroyed.

In general, however, the same logic as defined by $\mathcal{M}$ can be represented by structures with considerably less copies. It suffices to find a set of models $M \subset M_T$, where exactly the formulas of the classical closure of T hold — i.e. $M \models \phi$ iff $T \vdash \phi$, we shall then call M dense in M_T - and to take as $\mathcal{M}'$ the structure $\mathcal{X}' := \{< m, n >: n \in M\} \cup \{< n, 0 >: n \in M\}$, again with $< n, 0 > \prec < m, n >$. So we can rephrase the question to: What is the minimal size of M dense in M_T?

(a) A nice case:

Take for m the model that makes all p_i true, and $T := \{\neg p_0\}$, so $card(M_T) = 2^\kappa$, and the first construction of $\mathcal{M}$ as above will need 2^κ copies of m. As $\mathcal{L}$ has only κ formulas, and any subset of M_T makes all formulas of T true, we see that there is a dense subset $M \subseteq M_T$ of size κ: For any ϕ s.t. $T \not\vdash \phi$ take some $m_\phi \in M_T$ s.t. $m_\phi \not\models \phi$. But, in our nice case, considerably less than κ models might do: Assume there is $\lambda < \kappa$ s.t. $2^\lambda \geq \kappa$, so there is an injection $h : \{p_i : 0 < i < \kappa\} \to \mathcal{P}(\lambda)$. Let now $0 < i \neq j < \kappa$. For $\alpha < \lambda$, define the model m_α by $m_\alpha \models \neg p_0$ and $m_\alpha \models p_i :\leftrightarrow \alpha \in h(p_i)$. By $h(p_i) \neq h(p_j)$, there is $\alpha < \lambda$ s.t. $\alpha \in h(p_i) - h(p_j)$ or $\alpha \in h(p_j) - h(p_i)$, so $m_\alpha \models p_i \wedge \neg p_j$ or $m_\alpha \models \neg p_i \wedge p_j$, i.e. there is some m_α which discerns p_i, p_j. This is essentially enough: Let M be the closure of $\{m_\alpha : \alpha < \lambda\}$ under the finite operations $-, +, *$ defined by

$(-m) \models p_i :\leftrightarrow m \models \neg p_i$

$(m + m') \models p_i :\leftrightarrow m \models p_i$ or $m' \models p_i$

$(m * m') \models p_i :\leftrightarrow m \models p_i$ and $m' \models p_i$.

M still has cardinality λ, and $M \subseteq M_T$.

Let ϕ be s.t. $\neg p_0 \not\vdash \phi$, we have to find $m \in M$ s.t. $m \models \neg\phi$. Let $\neg\phi \equiv \phi_0 \vee \ldots \vee \phi_n$, where each $\phi_k = \pm p_{i_0} \wedge \ldots \wedge \pm p_{i_r}$ for some $i_0 \ldots i_r$. By $\neg p_0 \not\vdash \phi$, $Con(\neg p_0, \neg\phi)$ ($\vdash$ $-$consistency), so $Con(\neg p_0, \phi_k)$ for some $0 \leq k \leq n$. Fix such $\phi_k = \pm p_{i_0} \wedge \ldots \wedge \pm p_{i_r}$, say $\phi_k = p_{j_0} \wedge \ldots \wedge p_{j_s} \wedge \neg p_{g_0} \wedge \ldots \wedge \neg p_{g_t}$. By $Con(\neg p_0, \phi_k)$, p_0 is none of the p_{j_x}. (If one of the $\neg p_{g_y}$ is $\neg p_0$, it can be neglected, it will come out true anyway.) Fix $0 \leq x \leq s$, let $0 \leq y \leq t$. Then there is m_α s.t. $m_\alpha \models p_{j_x} \wedge \neg p_{g_y}$ or $-m_\alpha \models p_{j_x} \wedge \neg p_{g_y}$. Let $m_{x,y}$ be the m_α or $-m_\alpha$, and set $m_x := m_{x,0} * \ldots * m_{x,t}$. Then $m_x \models p_{j_x} \wedge \neg p_{g_0} \wedge \ldots \wedge \neg p_{g_t}$. For $m := m_0 + \ldots + m_s$, $m \models \phi_k$, so $m \models \neg\phi$, and $m \in M$.

On the other hand, in our example, λ many models with $2^\lambda < \kappa$ will not do: Assume that for each $0 < i \neq j < \kappa$ there is $\alpha < \lambda$ and $m_\alpha \in M_T$ with $m_\alpha \models p_i \wedge \neg p_j$. Then there is a function $f : 2^\lambda \to \kappa - \{0\}$ onto: For $A \subseteq \lambda$, let $f(A) := \bigcup\{j : 0 < j < \kappa \wedge \forall\alpha \in A.m_\alpha \models p_j\}$. But, for $0 < i < \kappa$, and $A_i := \{\alpha < \lambda : m_\alpha \models p_i\}$ $f(A_i) = i$: Obviously, for $\alpha \in A_i$, $m_\alpha \models p_i$. But, if $i \neq j$, then there is $\alpha \in A_i$ with $m_\alpha \models p_i \wedge \neg p_j$.

(b) There are, however, examples where we need the full size κ:

Let $\mathcal{L}$ be as above, consider $m^- \models \{\neg p_j : j < \kappa\}$, $T := \{p_i \vee p_j : i \neq j < \kappa\}$, and let $m^+ \models \{p_j : j < \kappa\}$ and $m_i^- \models \{\neg p_i\} \cup \{p_j : i \neq j < \kappa\}$ for $i < \kappa$.

Let the structure $\mathcal{M}$ be defined by $\mathcal{X} := \{< m^-, m^+ >\} \cup \{< m^-, m_i^- >: i < \kappa\} \cup \{< m^+, 0 >\} \cup \{< m_i^-, 0 >: i < \kappa\}$ and $< n, 0 > \prec < m^-, n >$ for $n = m^+$ or $n = m_i^-$, some $i < \kappa$. Then $Th(m^-) \vee T \models_{\mathcal{M}} T$. But there is no $M \subseteq M_T$ dense with $card(M) < \kappa$. Obviously, $M_T = \{m^+\} \cup \{m_i^- : i < \kappa\}$, and $\{m_i^- : i < \kappa\} \subseteq M_T$ is dense (see [Sch92]), but taking away any m_i^- will change T: p_i becomes true.

We turn now to a different approach to copies.

3.9 A new approach to preferential structures

3.9.1 Introduction

This section deals with some fundamental concepts and questions of preferential structures. A model for preferential reasoning will, in this section, be a total order on the models of the underlying classical language. Instead of working in completeness proofs with a canonical preferential structure as done traditionally, we work with sets of such total orders. We thus stay close to the way completeness proofs are done in classical logic. Our new approach will also justify multiple copies (or noninjective labelling functions) present in most work on preferential structures. A representation result for the finite case is given.

3.9.1.1 Main concepts and results

We address in this Section 3.9 some fundamental questions of preferential structures. Our guiding principle will be classical propositional (or first order) logic.

First, we reconsider the concept of a model for preferential reasoning. Traditionally, such a model is a strict partial order on the set of classical models of the underlying language. Instead, we will work here with strict <u>*total*</u> orders on the set of classical models of the underlying language. Such structures have maximal preferential information, just as classical propositional models have maximal propositional information.

Second, we will work in completeness proofs with sets of such total orders and thus closely follow the strategy for classical logic, whereas the traditional approach for preferential models works with one canonical structure. More precisely, in classical logic, one shows $T \vdash \phi$ iff $T \models \phi$, by proving soundness and that for every ϕ s.t. $T \not\vdash \phi$ there is a $T-$model $m_{T,\neg\phi}$, where ϕ fails. In traditional preferential logic, one constructs a canonical structure $\mathcal{M}$, which satisfies exactly the consequences of T, i.e. $T \models_{\mathcal{M}} \phi$ iff $T \mid\sim \phi$, simultaneously for all T and ϕ (where $T \models_{\mathcal{M}} \phi$ iff $\mu(T) \subseteq M(\phi)$, i.e. iff in all minimal models of T in $\mathcal{M}$ ϕ holds).

Third, our approach will also shed new light on the somewhat obscure question of multiple copies (equivalent to noninjective labelling functions) present in most constructions (see, e.g. the work of the author, or [KLM90], [LM92]). In our approach, it is natural to consider disjoint unions of sets of total orders over the classical models. They have (almost) the same properties as these sets have. As disjoint unions are structures with multiple copies, we have justified multiple copies of models or noninjective labelling functions in a natural way.

3.9.1.2 Motivation and overview

Our work started as an analysis of the different ways completeness proofs are made in classical logic and traditional preferential structures — the first is folklore, for the second see, e.g. [KLM90]. The idea to consider total orders as models can essentially also be found in [ALS99], where we revised nonmonotonic databases. The justification of multiple copies arose naturally with the definition of a disjoint union of total orders.

The concept of a disjoint union of preferential structures raises the question whether a property which holds in all individual structures will also hold in the disjoint union of the structures. This is (trivially) true for entailment relations, but not necessarily for inference rules. A counterexample for the latter using definability problems is given in Section 3.9.3, Example 3.9.1.

In the rest of this Section 3.9.1 we will first argue in more detail that total orders can be considered as the models of preferential reasoning. We then introduce some notation and basic facts. Finally, we will describe the type of our representation result.

In Section 3.9.2, we examine the kind of properties which describe preferential structures and their logics, and their interpretation in our new approach.

In Section 3.9.3, we introduce disjoint unions of preferential structures, and present our results on preservation of properties from sets of structures to their disjoint union. We also show that not all preferential structures are equivalent to a disjoint union of total orders, and justify in more detail the existence of multiple copies of models or labelling functions in traditional preferential structures.

In Section 3.9.4, we formulate and prove our representation result for the finite case, taking our usual algebraic detour which has proved useful in so many cases. We first characterize the choice functions of sets of total orders (or, equivalently, their disjoint unions), and translate this characterization by a standard argument to logic.

Strict total orders are the models of preferential reasoning

A classical propositional or first order model has maximal propositional or first order information: every formula is decided, either the formula or its negation holds. A set of models (corresponding to an incomplete formula, i.e. to a formula ϕ s.t. there is a formula ψ with neither $\phi \vdash \psi$, nor $\phi \vdash \neg\psi$) has less information. Preferential reasoning reasons about preferences between the classical models of a given language $\mathcal{L}$. Maximal preferential information is given by a strict total order between these classical models. A strict partial order can also be considered as the set of total orders which extend it (as a set of pairs). Thus, strict total orders on the set of classical models <u>*are*</u>, in this sense, the models of preferential reasoning, just as classical propositional models are the models of propositional reasoning.

3.9.1.3 Basic definitions and facts

Recall from Definition 1.6.1 that by a child (or successor) of an element x in a tree t we mean a direct child in t. A child of a child, etc. will be called an indirect child. Trees will be supposed to grow downwards, so the root is the top element.

Definition 3.9.1

For a given language $\mathcal{L}$, TO, etc. will stand for a strict total order on $M_{\mathcal{L}}$. Considering TO as a preferential model, we will slightly abuse notation here: as there will only be one copy per model, we will omit the indices i.

$\mathcal{O}$, etc. will stand for sets of such structures.

If $\mathcal{O}$ is such a set, we set $\mu_{\mathcal{O}}(X) := \bigcup\{\mu_{\mathcal{M}}(X) : \mathcal{M} \in \mathcal{O}\}$, and define $T \models_{\mathcal{O}} \phi$ iff $T \models_{\mathcal{M}} \phi$ for all $\mathcal{M} \in \mathcal{O}$.

Note that for all T and all strictly totally ordered structures TO, $\mu_{TO}(T)$ is either a singleton, or empty, so TO is definability preserving.

Definition 3.9.2

Let $\mathcal{Z} :=< \mathcal{X}, \prec>$ be a preferential structure. For $< x,i >\in \mathcal{X}$, let

$< x,i >_{\mathcal{Z}}^{-}:= \{< y,j >\in \mathcal{X} :< y,j >\prec< x,i >\}$, and

$< x,i >_{\mathcal{Z}}^{*}:= \{y : \exists < y,j >\in \mathcal{X}. < y,j >\prec< x,i >\}$.

When the context is clear, we omit the index $\mathcal{Z}$.

Fact 3.9.1

Let $\mathcal{Z} :=< \mathcal{X}, \prec>$, $\mathcal{Z}' :=< \mathcal{X}', \prec'>$ be two preferential structures.

(1) Let $x \in X$. Then $x \in \mu_{\mathcal{Z}}(X)$ iff $\exists < x,i >\in \mathcal{X}. X\cap < x,i >_{\mathcal{Z}}^{*}= \emptyset$.

(2) If $\forall < x,i >\in \mathcal{X} \exists < x,i' >\in \mathcal{X}'. < x,i' >_{\mathcal{Z}'}^{*}\subseteq< x,i >_{\mathcal{Z}}^{*}$ and $\forall < x,i' >\in \mathcal{X}' \exists < x,i >\in \mathcal{X}. < x,i >_{\mathcal{Z}}^{*}\subseteq< x,i' >_{\mathcal{Z}'}^{*}$, then $\mu_{\mathcal{Z}} = \mu_{\mathcal{Z}'}$.

Proof:

(1) $x \in \mu_{\mathcal{Z}} \Leftrightarrow \exists < x,i >\in \mathcal{X}. \neg\exists < y,j >\in \mathcal{X}. < y,j >\prec< x,i > \wedge y \in X$ $\Leftrightarrow \exists < x,i >\in \mathcal{X}. < x,i >_{\mathcal{Z}}^{*} \cap X = \emptyset$.

(2) Let $x \in \mu_{\mathcal{Z}}(X)$, then by (1) $\exists < x,i >\in \mathcal{X}. X\cap < x,i >_{\mathcal{Z}}^{*}= \emptyset$. By prerequisite, $\exists < x,i' >\in \mathcal{X}'. < x,i' >_{\mathcal{Z}'}^{*}\subseteq< x,i >_{\mathcal{Z}}^{*}$, so $x \in \mu_{\mathcal{Z}'}(X)$ by (1). The other direction is symmetrical. □

Fact 3.9.2

If $\mathcal{O}$ is a set of preferential structures, then $T \models_{\mathcal{O}} \phi$ iff $\mu_{\mathcal{O}}(M_T) \models \phi$.

Proof:

$\mu_{\mathcal{O}}(M_T) \models \phi \Leftrightarrow \mu_{\mathcal{O}}(M_T) := \bigcup\{\mu_{\mathcal{Z}}(M_T) : \mathcal{Z} \in \mathcal{O}\} \subseteq M(\phi) \Leftrightarrow \forall \mathcal{Z} \in \mathcal{O}.\mu_{\mathcal{Z}}(M_T) \subseteq M(\phi) \Leftrightarrow$
$\forall \mathcal{Z} \in \mathcal{O}.T \models_{\mathcal{Z}} \phi \Leftrightarrow T \models_{\mathcal{O}} \phi.$ □

3.9.1.4 Outline of our representation results and technique

We describe here the kind of representation result we will show in Section 3.9.4.

We have characterized in Sections 3.3 and 3.4 usual smooth preferential structures first algebraically by conditions on their choice functions, and only then logically by corresponding conditions. More precisely, given a function μ satisfying certain conditions, we have shown that there is a preferential structure $\mathcal{Z}$, whose choice function $\mu_{\mathcal{Z}}$ is exactly μ. The choice functions correspond to the logics by the equation $\mu(M(T)) = M(\overline{\overline{T}})$.

We will take a similar approach here, but will first analyze the form a representation theorem will have in our context.

Our starting point was that classical completeness proofs have the following form: For each ϕ s.t. $T \not\vdash \phi$, find $m_{T,\neg\phi}$ s.t. $m_{T,\neg\phi} \models T, \neg\phi$, or, equivalently, find a set of models $\boldsymbol{M}_T$ s.t. for each such ϕ there is suitable $m_{T,\neg\phi}$ in $\boldsymbol{M}_T$. Then, by soundness, $Th(\boldsymbol{M}_T) = \overline{T}$. Our construction will have a similar form.

First, given any strict total order TO (or any set $\mathcal{O}$ of strict total orders) over $M_{\mathcal{L}}$, the logic defined by $T \mid\!\sim \phi :\Leftrightarrow T \models_{TO} \phi$ (or $:\Leftrightarrow T \models_{\mathcal{O}} \phi$) satisfies our conditions (LLE), (CCL), (SC), (PR), (CUM) (see Proposition 3.9.5). Second, given a logic $\mid\!\sim$ satisfying (LLE), (CCL), (SC), (PR), (CUM), there is a set $\mathcal{O}$ of strict total orders over $M_{\mathcal{L}}$ s.t. $T \mid\!\sim \phi \Leftrightarrow T \models_{\mathcal{O}} \phi$. Thus, the <u>set</u> $\mathcal{O}$ represents exactly $\mid\!\sim$, contrary to usual preferential structures, where a single structure represents exactly $\mid\!\sim$.

We work again first via an algebraic characterization, and show the following: Given any strict total order TO (or any set $\mathcal{O}$ of strict total orders), the choice function μ_{TO} (the choice function $\mu_{\mathcal{O}}$) satisfies our algebraic conditions $(\mu \subseteq)$, (μPR), (μCUM) (see Proposition 3.9.6). Conversely, given a choice function μ satisfying $(\mu \subseteq)$, (μPR), (μCUM), there is a set $\mathcal{O}$ of strict total orders s.t. $\mu = \mu_{\mathcal{O}}$.

The logical part will then follow easily via a standard argument.

To summarize:

We present here a new analysis of preferential reasoning. We have modified the notion of a model, and, consequently, the way completeness results are proved. Both modifications keep us close to the concepts and techniques of classical propositional and first order logic. We have also given a new justification of preferential structures with multiple copies (or labelling functions) by introducing disjoint unions of structures. We have shown that some, but not all, properties are preserved by going from sets of structures to their disjoint unions. Finally, we have given a representation result for the finite case.

The main open problem seems to be a characterization of the infinite case, or at least the infinite smooth case.

3.9.2 Validity in traditional and in our preferential structures

We distinguish here validity of type 1 and type 2, where type 1 validity is validity of entailment like $T \mid\sim \phi$, and type 2 validity is validity of rules like $\phi \mid\sim \psi \wedge \sigma \Rightarrow \phi \mid\sim \psi$.

(The set $\mathcal{O}$ used in this section is motivated by Example 3.9.1, where we do not consider all totally ordered sets, but only those satisfying a certain property.)

Definition 3.9.3

(1) Validity of type 1:

This is validity of expressions like $\phi \mid\sim \psi$ (or $T \mid\sim \psi$), and is defined for a given preferential structure $\mathcal{M}$ in the usual sense by $\phi \models_{\mathcal{M}} \psi$ (or $T \models_{\mathcal{M}} \psi$). In our new interpretation we read this as: $\phi \models_{\mathcal{O}} \psi$ (or $T \models_{\mathcal{O}} \psi$).

(2) Validity of type 2:

This is validity of rules of, e.g. the type

(2.1) $\phi \mid\sim \psi, \phi \mid\sim \psi' \Rightarrow \phi \mid\sim \psi \wedge \psi'$,

(2.2) $\phi \mid\sim \psi \Rightarrow (\phi \mid\sim \neg\phi'$ or $\phi \wedge \phi' \mid\sim \psi)$,

(2.3) $\overline{\overline{T \cup T'}} \subseteq \overline{\overline{\overline{T}} \cup T'}$.

As strict total orders are definability preserving, we can argue semantically when dealing with them. More precisely, there is a 1-1 correspondence between theories (and formulas) and sets of models: If $\mathcal{M}$ is a definability preserving preferential model, and T a theory, then $M(\{\phi : T \models_{\mathcal{M}} \phi\}) = \mu_{\mathcal{M}}(M(T))$, so setting $\overline{\overline{T}} := \{\phi : T \models_{\mathcal{M}} \phi\}$, we have for instance $T' \vdash \overline{\overline{T}}$ iff $M(T') \subseteq M(\overline{\overline{T}})$.

Discussion of (2.1):

In usual preferential structures, we read (2.1) as: If in a fixed structure $\mathcal{M}$ $\phi \models_{\mathcal{M}} \psi$ and $\phi \models_{\mathcal{M}} \psi'$ hold, then so will $\phi \models_{\mathcal{M}} \psi \wedge \psi'$.

In our new approach, we read (2.1) now as: If $\phi \models_{\mathcal{O}} \psi$ and $\phi \models_{\mathcal{O}} \psi'$ hold, then $\phi \models_{\mathcal{O}} \psi \wedge \psi'$ will also hold. In semantical terms: If $\mu_{\mathcal{O}}(\phi) \subseteq M(\psi)$ and $\mu_{\mathcal{O}}(\phi) \subseteq M(\psi')$, then $\mu_{\mathcal{O}}(\phi) \subseteq M(\psi) \cap M(\psi')$.

This is the exact analogue of the classical definition: $\alpha \models \beta$ iff in all classical models where α (and perhaps some other property, too) holds, β will also hold. Our α is here of the form $\phi \mid\sim \psi$ (or $\phi \models_{TO} \psi$), etc.

Discussion of (2.2):

The usual approach is similar to the one for rule (2.1).

For the new approach, we have to be careful with distributivity. A comparison with classical logic helps. In all classical models it is true that if $\alpha \vee \beta$ holds, then α holds, or β holds (by definition of validity of "OR"). But we do not say that $\alpha \vee \beta \models \alpha$ or $\alpha \vee \beta \models \beta$ holds, as this would imply either that in all models where $\alpha \vee \beta$ holds, α holds, or that in all models where $\alpha \vee \beta$ holds, β holds, which is usually false.

So a rule of type (2.2) holds iff $\phi \models_{\mathcal{O}} \psi$ implies $\phi \models_{\mathcal{O}} \neg\phi'$ or $\phi \wedge \phi' \models_{\mathcal{O}} \psi$. In semantical terms: A rule of type (2.2) holds iff $\mu_{\mathcal{O}}(\phi) \subseteq M(\psi)$ implies $\mu_{\mathcal{O}}(\phi) \subseteq M(\neg\phi')$ or $\mu_{\mathcal{O}}(\phi \wedge \phi') \subseteq M(\psi)$.

Note that (2.2) holds in all strict total orders on $M_{\mathcal{L}}$, as such structures are ranked. But in a set of such structures, it is usually wrong, as it is usually not true that either in all these structures $\phi \mid\sim \neg\phi'$ holds, or that in all these structures $\phi \wedge \phi' \mid\sim \psi$ holds.

Discussion of (2.3):

(2.3) stands for: If $T \cup T' \mid\sim \phi$, then there are $\phi_1, \ldots, \phi_n$ and $\phi'_1, \ldots, \phi'_m$ s.t. $T \mid\sim \phi_i$ and $\phi'_i \in T'$, and $\{\phi_1, \ldots, \phi_n, \phi'_1, \ldots, \phi'_m\} \vdash \phi$.

So, in usual preferential structures, (2.3) holds in structure $\mathcal{M}$, iff: If $T \cup T' \models_{\mathcal{M}} \phi$, then there are $\phi_1, \ldots, \phi_n$ and $\phi'_1, \ldots, \phi'_m$ s.t. $T \models_{\mathcal{M}} \phi_i$ and $\phi'_i \in T'$, and $\{\phi_1, \ldots, \phi_n, \phi'_1, \ldots, \phi'_m\} \vdash \phi$.

In our new approach, (2.3) holds iff in all strict total orders $TO \in \mathcal{O}$ $T \cup T' \models_{TO} \phi$, there are $\phi_1, \ldots, \phi_n$ and $\phi'_1, \ldots, \phi'_m$ s.t. $T \models_{TO} \phi_i$ and $\phi'_i \in T'$, and $\{\phi_1, \ldots, \phi_n, \phi'_1, \ldots, \phi'_m\} \vdash \phi$.

The discussion in semantical terms clarifies the role of the existential quantifiers (which are "ORS" — see the discussion of (2.2)): Condition (2.3) reads now: in all $TO \in \mathcal{O}$ $\mu_{TO}(T) \cap M(T') \subseteq \mu_{TO}(T \cup T')$ holds (and thus also $\mu_{\mathcal{O}}(T) \cap M(T') \subseteq \mu_{\mathcal{O}}(T \cup T')$).

3.9.3 The disjoint union of models and the problem of multiple copies

The central concepts in this Section 3.9.3 are disjoint unions of models and the preservation of validity in such unions.

3.9.3.1 Disjoint unions and preservation of validity in disjoint unions

We introduce the disjoint union of preferential structures and examine the question whether a property Φ which holds in all $\mathcal{M}_i$, $i \in I$, will also hold in their disjoint union $\biguplus\{\mathcal{M}_i : i \in I\}$. This is true for type 1 validity, but not for type 2 validity in the general infinite case.

Disjoint unions and preservation of type 1 validity

Definition 3.9.4

Let $\mathcal{M}_i :=< M_i, \prec_i>$ be a family of preferential structures. Let then $\biguplus\{\mathcal{M}_i : i \in I\} :=< M, \prec>$, where $M := \{< x, < k, i >>: i \in I, < x, k >\in M_i\}$, and $< x, < k, i >>\prec< x', < k', i' >>$ iff $i = i'$ and $< x, k >\prec_i< x', k' >$. Thus, $\biguplus\{\mathcal{M}_i : i \in I\}$ is the disjoint union of the sets and the relations, and we will call it so.

Fact 3.9.3

Let μ_i be the choice functions of the $\mathcal{M}_i$.

Then $\mu_{\biguplus\{\mathcal{M}_i : i \in I\}}(X) = \bigcup\{\mu_i(X) : i \in I\}$, so $\mu_{\biguplus\{\mathcal{M}_i : i \in I\}} = \mu_{\{\mathcal{M}_i : i \in I\}}$.

Proof:

(Trivial.) Let $\mu := \mu_{\biguplus\{\mathcal{M}_i:i\in I\}}$. Let $<x,<k,i>>\in \mu(X)$, then $x \in X$, and there is no $x' \in X$ with $<x',<k',i>>\prec<x,<k,i>>$ for some $<x',k'>\in M_i$, so $x \in \mu_i(X)$. The converse holds by a similar argument. □

Fact 3.9.4

$T \models_{\biguplus\{\mathcal{M}_i:i\in I\}} \phi$ iff for all $i \in I$ $T \models_{\mathcal{M}_i} \phi$. Thus $T \models_{\biguplus\{\mathcal{M}_i:i\in I\}} \phi$ iff $T \models_{\{\mathcal{M}_i:i\in I\}} \phi$, and disjoint unions preserve type 1 validity.

Proof:

(Trivial.) Let again $\mu := \mu_{\biguplus\{\mathcal{M}_i:i\in I\}}$. $T \models_{\biguplus\{\mathcal{M}_i:i\in I\}} \phi$ iff in all $m \in \mu(T)$ ϕ holds. If for all $i \in I$ in all $m \in \mu_i(T)$ ϕ holds, then ϕ holds in all $m \in \mu(T)$ by Fact 3.9.3. But if there is some $i \in I$ and $m \in \mu_i(T)$ s.t. ϕ fails in m, then ϕ will fail in some $m \in \mu(T)$, too, again by Fact 3.9.3. □

Preservation of type 2 validity

Rules of type (2.1) are preserved: This is a direct consequence of Fact 3.9.4, the argument is similar to the following one for type (2.2) rules.

Rules of type (2.2) are preserved: We show that if in all strict total orders TO where $\phi \mid\sim \psi$ (and perhaps some other property) holds, $\phi \mid\sim \neg\phi'$ holds, then $\phi \mid\sim \neg\phi'$ holds in the disjoint union $\mathcal{M}$ of these structures, and, if in all strict total orders TO where $\phi \mid\sim \psi$ (and perhaps some other property) holds, $\phi \wedge \phi' \mid\sim \psi$ holds, then $\phi \wedge \phi' \mid\sim \psi$ holds in the disjoint union $\mathcal{M}$ of these structures. But, it is a direct consequence of Fact 3.9.4 that in the first case $\phi \models_{\mathcal{M}} \neg\phi'$, and in the second case $\phi \wedge \phi' \models_{\mathcal{M}} \psi$.

Rules of type (2.3) are not necessarily preserved — at least not in the general infinite case:

Example 3.9.1

(This is the — slightly adapted — Example 5.1.2, which shows failure of infinite conditionalization in a case where definability preservation fails.)

Consider the language $\mathcal{L}$ defined by the propositional variables p_i, $i \in \omega$. Let $T_0^+ := \{p_0\} \cup \{p_i : 0 < i < \omega\}$, $T_0^- := \{\neg p_0\} \cup \{p_i : 0 < i < \omega\}$, set $T' := T_0^+ \vee T_0^-$, and $T := \emptyset$. Let the classical model m_0^+ (m_0^-) be the unique model satisfying T_0^+ (T_0^-), so $M(T') = \{m_0^+, m_0^-\}$. Consider the set $\mathcal{O}$ of all strict total orders TO on $M_{\mathcal{L}}$ satisfying $T' \models_{TO} T_0^-$. Obviously $T' \models_{TO} T_0^-$ iff $m_0^- \prec_{TO} m_0^+$. If $TO \in \mathcal{O}$ has no (global) minimum, then $T \models_{TO} \bot$, so $\neg p_0 \in \overline{\overline{\overline{T}} \cup T'}$ — where $\overline{\overline{T}} := \{\phi : T \models_{TO} \phi\}$. If TO has a minimum, which is neither m_0^+ nor m_0^-, then $\overline{\overline{T}} \cup T'$ is inconsistent, and again $\neg p_0 \in \overline{\overline{\overline{T}} \cup T'}$. The minimum cannot be m_0^+, so in all cases $\neg p_0 \in \overline{\overline{\overline{T}} \cup T'}$. But now every model except m_0^+ can be minimal, so in the disjoint union $\mathcal{M} := \biguplus \mathcal{O}$ of these structures, $\mu_{\mathcal{M}}(T) = M_{\mathcal{L}} - \{m_0^+\}$. Thus $\overline{\overline{T}} = \overline{T}$ (in $\mathcal{M}$), and $\overline{\overline{\overline{T}} \cup T'} = \overline{T'}$, but $\neg p_0 \notin \overline{T'}$. In particular, the example shows that rule (2.3) of Section 3.9.2 might hold in all components of a disjoint union, but fail in the union: As any total order TO is definability preserving, (2.3) holds in TO, by the results of Section 3.4. On the other hand, $\neg p_0 \in \overline{\overline{T \cup T'}}$ (in $\mathcal{M}$), so (2.3) fails in $\mathcal{M}$. □

Remarks:

(1) Failure of definability preservation in $\mathcal{M}$ is crucial for our example. More generally, definability preserving disjoint unions preserve rule (2.3). We know this already from Section 3.4, but give a direct argument to illustrate which kinds of rules of type 2 will be preserved in definability preserving disjoint unions. Let $\mathcal{X}$ be some set of strict total orders and $\mathcal{M} = \biguplus \mathcal{X}$. We have to show $M(\overline{\overline{\overline{T}} \cup T'}) \subseteq M(\overline{\overline{T \cup T'}})$ (in $\mathcal{M}$). If $\mathcal{M}$ is definability preserving, then $M(\overline{\overline{T}}) = \mu_{\mathcal{M}}(T)$, so $M(\overline{\overline{T}} \cup T') = M(\overline{\overline{T}}) \cap M(T') = \mu_{\mathcal{M}}(T) \cap M(T') = \bigcup\{\mu_{TO}(T) : TO \in \mathcal{X}\} \cap M(T') = \bigcup\{\mu_{TO}(T) \cap M(T') : TO \in \mathcal{X}\} \subseteq \bigcup\{\mu_{TO}(T \cup T') : TO \in \mathcal{X}\} = \mu_{\mathcal{M}}(T \cup T') = M(\overline{\overline{T \cup T'}})$. (In the inclusion, we have used property $(\mu PR')$, which holds in all preferential structures.) Thus $\overline{\overline{T \cup T'}} \subseteq \overline{\overline{\overline{T}} \cup T'}$, and as $\neg p_0 \in \overline{\overline{T \cup T'}}$ in our Example 3.9.1, the example would not work.

(2) The general argument showing preservation of a rule in a definability preserving structure will argue semantically as above, i.e. that the rule is preserved under union: $\Phi(\mu_i(X), \mu_i(Y), \ldots)$ implies $\Phi(\bigcup \mu_i(X), \bigcup \mu_i(Y), \ldots)$. The semantical argument is possible by $M(\overline{\overline{T}}) = \mu_{\mathcal{M}}(T)$.

Equivalence of general preferential structures with sets of total orders

Ideally, one would like every preferential structure to be (or at least, to be equivalent for type 1 validity to) a disjoint union of strictly totally ordered structures. This is not the case.

Example 3.9.2

Consider the language defined by one variable, p. Let $m \models p$, $m' \models \neg p$, and consider the structure $< m, 0 > \succ < m', 0 > \succ < m', 1 > \succ < m', 2 > \succ \ldots$. Then $\mu(true) = \emptyset$, but $\mu(p) = \{m\}$. There are only two possible total orders: $m \prec m'$, $m' \prec m$. $m \prec m'$ gives $\mu(\emptyset) = \{m\}$, $m' \prec m$ gives $\mu(\emptyset) = \{m'\}$, $(m \prec m') \uplus (m' \prec m)$ gives $\mu(\emptyset) = \{m, m'\}$. (Omitting some models totally will not help, either.)

Thus, traditional preferential structures are more expressive than strict total orders (or their disjoint union).

In Section 3.9.4, we will construct an equivalent structure in the finite cumulative case.

3.9.3.2 Multiple copies

The usual constructions with multiple copies (the author's notation) or noninjective labelling functions (notation, e.g. of Kraus, Lehmann, Magidor) have always intrigued the author for their intuitive justification, which seemed somewhat weak (e.g. different languages of description and reasoning, as discussed in [Imi87]). We give here a purely formal one. Recall that we have discussed in Section 3.8 the expressive strength of structures with multiple copies in more detail.

Fact 3.9.4 shows that we can construct a usual structure with multiple copies out of a set of strictly totally ordered sets of classical models (without multiple copies), preserving validity of type 1. Example 3.9.1 shows that validity of type 2 is usually not preserved. For its failure, we needed a not definability preserving structure, which exists only for infinite languages. We thus conjecture that validity of type 2 is also preserved in the case of finite languages.

Thus, considering sets of strict total orders of models leads us naturally to consider their disjoint unions — at least largely equivalent structures - which are constructions with multiple copies.

3.9.4 Representation in the finite case

We show in this Section 3.9.4 our main result, Proposition 3.9.5, a representation theorem for the finite cumulative case. The infinite case stays an open problem.

As done before, we first show an algebraic representation result, Proposition 3.9.6, whose proof is the main work, and translate this result by routine methods to the logical representation problem.

It is easily seen that the consequence relations of the structures examined will be cumulative: First, it is well known (see, e.g. [KLM90], or Section 3.4) that smooth structures define cumulative consequence relations. Second, transitive relations over finite sets are smooth, and, third, we will see that our structures will be finite (see the modifications in the proof of Proposition 3.9.6).

Let us explain why the result of Proposition 3.9.5 is precisely the result to be expected. Classical logic defines exactly one consequence relation, $\vdash$. The conditions for preferential structures (system P of [KLM90], or our conditions of Proposition 3.4.1) do not describe _one_ consequence relation, but a whole class, which have to obey certain principles. The representation theorem of classical logic states $T \vdash \phi$ iff in _all_ models, if T holds, then so will ϕ. This unrestricted universal quantifier fixes _one_ consequence relation, $\vdash$. This cannot be expected in our case. In our case, each preferential consequence relation $\mid\sim$, i.e. each relation $\mid\sim$ satisfying our conditions, will have to correspond to one particular set $\mathcal{O}_{\mid\sim}$ of total orders, in the sense that $T \mid\sim \phi$ iff in all $TO \in \mathcal{O}_{\mid\sim}$ $T \models_{TO} \phi$. The quantifier is restricted to $\mathcal{O}_{\mid\sim}$. This is the completeness part of Proposition 3.9.5. The soundness part shows that _any_ set $\mathcal{O}$ of total orders satisfies the conditions, thus a fortiori any total order will do so. Looking back at traditional preferential structures, and, e.g. the classical paper [KLM90], we see the exact correspondence to our result. There, it was shown in the soundness part that _every_ preferential structure satisfies the system P. The completeness part there shows that there is _one_ preferential structure $\mathcal{M}$ s.t. $T \mid\sim \phi$ iff $T \models_{\mathcal{M}} \phi$, if $\mid\sim$ satisfies system P. As preferential structures in the usual sense correspond to sets of total orders, we see that our result is the exact analogue of, e.g. the KLM result. To summarize, we show the exact analogue to usual preferential structures, and the closest analogue possible to classical logic.

We state now our main result, logical characterization.

Proposition 3.9.5

Let $\mathcal{L}$ be a propositional language defined by a finite set of variables.

(A) (Soundness) Let $\mathcal{O}$ be a set of strict total orders over $M_{\mathcal{L}}$, defining a logic $\mid\sim$ by $T \mid\sim \phi :\Leftrightarrow T \models_{\mathcal{O}} \phi$. Then $\mid\sim$ satisfies (LLE), (CCL), (SC), (PR), (CUM).

(B) (Completeness) If a logic $\mid\sim$ for $\mathcal{L}$ satisfies (LLE), (CCL), (SC), (PR), (CUM), then there is a set $\mathcal{O}$ of strict total orders over $M_{\mathcal{L}}$ s.t. $T \mid\sim \phi \Leftrightarrow T \models_{\mathcal{O}} \phi$.

For the algebraic representation result, we will consider some $\mathcal{Y} \subseteq \mathcal{P}(Z)$, closed under finite unions and finite intersections, and a function $\mu : \mathcal{Y} \to \mathcal{Y}$. $\mathcal{Y}$ is intended to be $\boldsymbol{D}_{\mathcal{L}}$ for some propositional language $\mathcal{L}$.

The proof uses the following algebraic characterization, and is given after the proof of the latter.

Proposition 3.9.6

Let Z be a finite set, let $\mathcal{Y} \subseteq \mathcal{P}(Z)$ be closed under finite unions and finite intersections, and $\mu : \mathcal{Y} \to \mathcal{Y}$.

(A) (Soundness) Let $\mathcal{O}$ be a set of strict total orders over Z, then $\mu_{\mathcal{O}}$ satisfies $(\mu \subseteq)$, (μPR), (μCUM).

(B) (Completeness) Let μ satisfy $(\mu \subseteq)$, (μPR), (μCUM), then there is a set $\mathcal{O}$ of strict total orders over Z s.t. $\mu = \mu_{\mathcal{O}}$.

The proof of Proposition 3.9.6 will be a modification of a proof for traditional preferential structures as shown in Section 3.3.

Proof of Proposition 3.9.6:

By Fact 3.9.3, $\mu_{\mathcal{O}} = \mu_{\biguplus \mathcal{O}}$, so we can work with the set or its disjoint union.

(A) Soundness:

Conditions $(\mu \subseteq)$ and (μPR) hold for arbitrary preferential structures, and (μCUM) holds for smooth preferential structures (see Sections 3.2 and 3.3). Strict total orders over finite sets are smooth, so is their disjoint union.

(B) Completeness:

We will modify the construction in the proof of Proposition 3.3.8. We have constructed there for a function μ satisfying $(\mu \subseteq)$, (μPR), (μCUM) a transitive smooth preferential structure $\mathcal{Z} =< \mathcal{X}, \prec>$ representing μ. We first show in (a) that the construction is finite for finite languages. We then eliminate in (b) unnecessary copies, and construct in (c) for each remaining $< x, i >$ a total order $TO_{<x,i>}$ such that the set of all these $TO_{<x,i>}$ represents μ.

(a) Finiteness of the construction:

First, if the language $\mathcal{L}$ is finite, the constructed structure is finite, too: As $v(\mathcal{L})$ is finite, $Z = M_{\mathcal{L}}$ is finite. For each nonminimal element $x \in Z$, there is one tree in T'_x, so this is easy. Now, for the set T_x. T_x consists of trees $t_{U,x}$ where the elements of $t_{U,x}$ are pairs $< U', x' >$ with $U' \in \mathcal{Y} \subseteq \mathcal{P}(Z)$ and $x' \in Z$, so there are finitely many such pairs. Each element in the tree has at most $card(\mathcal{P}(Z))$ successors, and by Fact 3.3.9, (1), if $< U_m, x_m >$ is a direct or indirect successor in the tree of $< U_n, x_n >$, then $x_m \notin H(U_n)$, but $x_n \in U_n \subseteq H(U_n)$, so $x_n \neq x_m$, so branches have length at most $card(Z)$. So there is a uniform upper bound on the size of the trees, so there are only finitely many of such trees.

(b) Elimination of unnecessary copies:

Next, if, for each $x \in Z$, there is a finite number of copies, then "best" copies $< x, i >$ in the sense that there is no $< x, i' > \prec < x, i >$ in $\mathcal{Z}$ exist, so we can eliminate the "not so good" copies $< x, i >$ for which there is $< x, i' > \prec < x, i >$, without changing representation. (Note that, instead of arguing with finiteness, we can argue here with smoothness, too, as singletons are definable.)

Representation is not changed, as the following easy argument shows: Let $\mathcal{Z}' =< \mathcal{X}', \prec >$ be the new structure, we have to show that $\mu_{\mathcal{Z}} = \mu_{\mathcal{Z}'}$. Suppose $X \in \mathcal{Y}$, and $x \in \mu_{\mathcal{Z}}(X)$. Then there is $< x, i >$ minimal in $\mathcal{Z}\lceil X$. But then $< x, i > \in \mathcal{X}'$ too, and, as we have not introduced new smaller elements, $x \in \mu_{\mathcal{Z}'}(X)$. Suppose now $x \in \mu_{\mathcal{Z}'}$, then there is some $< x, i >$ minimal in $\mathcal{Z}'\lceil X$. If there were $< y, j >$ smaller than $< x, i >$ in $\mathcal{Z}$, $y \in X$, then $< y, j >$ would have been eliminated, as there is minimal $< y, k >$ below $< y, j >$, but then, by transitivity, $< y, k >$ is smaller than $< x, i >$, too, but $< y, k >$ is kept in $\mathcal{Z}'$, so $< x, i >$ would not be minimal in $\mathcal{Z}'$, either. Thus, $\mu_{\mathcal{Z}} = \mu_{\mathcal{Z}'}$.

(c) Construction of the total orders:

We take now the modified construction $\mathcal{Z}'$ to construct a set of total orders. $< x, i >^-$, etc. will now be relative to $\mathcal{Z}'$.

We construct for each $x \in Z$ a set $\mathcal{O}_x = \{TO_{<x,i>} :< x, i > \in \mathcal{X}'\}$ of total orders. $\biguplus \mathcal{O} := \biguplus \{TO : TO \in \mathcal{O}_x, x \in Z\}$ will be the final structure, equivalent to $\mathcal{Z}$. $TO_{<x,i>}$ is constructed as follows: We first put all elements $y \in < x, i >^*$ below x, and all $y \neq x$, $y \notin < x, i >^*$ above x. We then order $< x, i >^*$ totally, staying sufficiently close to the order of $\mathcal{Z}'$, and finally do the same with the remaining elements.

Fix now $< x, i >$, and let $< := <_{TO_{<x,i>}}$ be the strict total order on Z to be constructed.

First, set $y < x$ iff $y \in < x, i >^*$, and set $x < y$ iff $y \neq x$ and $y \notin < x, i >^*$.

We construct in (α) the part of the total order below x, and then in (β) the part above x.

(α) Work now inside $< x, i >^*$, and construct a total order $<$ on $< x, i >^*$ in three steps.

(1) Extend the partial order $\prec$ on $< x, i >^-$ to a total order $\lhd$.

(2) If $< y, j > \lhd < y, j' >$, eliminate $< y, j' >$. By finiteness, one copy of y survives.

(3) For $y, z \in < x, i >^*$, let $y < z$ iff there are $< y, j >$, $< z, k >$ with $< y, j > \lhd < z, k >$ left in step (2).

By step (2), $<$ in $< x, i >^*$ is free from cycles, by elimination of unnecessary elements in the construction of $\mathcal{Z}'$ x does not occur in $< x, i >^*$, so the entire relation constructed so far is free from cycles.

Note that for $y \in < x, i >^*$, there is some $< y, j >$ s.t. $< y, j >^* \subseteq \{z : z < y\}$: Let $< y, j >$ be the $\lhd-$least copy of y, i.e. the one which survives step (2). Then by (1), all $< z, k > \in < y, j >^-$ are $\lhd-$below $< y, j >$. But if some such $< z, k >$ is eliminated in (2), there is an even smaller $< z, k' > \lhd < y, j >$ which survives, so $z < y$ in step (3).

(β) Work now on $\mathcal{X}' - (\{< x, i >\} \cup < x, i >^-)$.

(1) Extend the order $\prec$ on $\mathcal{X}' - (\{< x, i >\} \cup < x, i >^-)$ to a total order $\lhd$.

(2) Eliminate again $< y, j' >$, if $< y, j > \lhd < y, j' >$, but eliminate also all $< y, j' >$ s.t. $y = x$ or $y \in < x, i >^*$.

(3) Let $y < z$ iff there are $< y, j >$, $< z, k >$ with $< y, j > \lhd < z, k >$ left in step (2).

By the same argument as above, we see that for any y $<-$above x, there is some $< y, j >$ s.t. $< y, j >^* \subseteq \{z : z < y\}$.

Let finally $\mathcal{O} = \{TO_{<x,i>} : x \in Z, < x, i > \in \mathcal{X}'\}$ and consider $\biguplus \mathcal{O}$. Let $< x, i > \in \mathcal{X}'$. By construction, $< x, TO_{<x,i>} >^*_{\mathcal{O}} = x^*_{TO_{<x,i>}} = < x, i >^*_{\mathcal{Z}'}$. Consider now arbitrary $TO_{<x,i>}$, and $y \in Z$. It was shown in the construction that there is $< y, j > \in \mathcal{X}'$ s.t. $< y, j >^*_{\mathcal{Z}'} \subseteq y^*_{TO_{<x,i>}} = < y, TO_{<x,i>} >^*_{\mathcal{O}}$. So by Fact 3.9.1, (2) $\mu_{\mathcal{Z}'} = \mu_{\biguplus \mathcal{O}} = \mu_{\mathcal{O}}$.

□ (Proposition 3.9.6)

Proposition 3.4.2 completes the proof of Proposition 3.9.5. (Note that by finiteness of the language, μ is definability preserving.)

Proof of Proposition 3.9.5:

Let $\mathcal{O}$ be any set of strict total orders over $M_{\mathcal{L}}$, then $\mu_{\mathcal{O}} = \mu_{\biguplus \mathcal{O}}$ satisfies $(\mu \subseteq)$, (μPR), (μCUM) by Proposition 3.9.6, so the logic defined by $T \mid\sim \phi :\Leftrightarrow \mu_{\biguplus \mathcal{O}}(M_T) \models \phi \Leftrightarrow T \models_{\biguplus \mathcal{O}} \phi$ ($\Leftrightarrow T \models_{\mathcal{O}} \phi$) satisfies (LLE), (CCL), (SC), (PR), (CUM) by Proposition 3.4.2. Conversely, given a logic $\mid\sim$ which satisfies (LLE), (CCL), (SC), (PR), (CUM), then the model choice function μ defined by $\mu(M_T) := M_{\overline{\overline{T}}}$ satisfies $(\mu \subseteq)$, (μPR), (μCUM) and $T \mid\sim \phi \Leftrightarrow \mu(M_T) \models \phi$ by Proposition 3.4.2, so by Proposition 3.9.6, there is a set $\mathcal{O}$ of strict total orders over $M_{\mathcal{L}}$ s.t. $\mu = \mu_{\mathcal{O}} = \mu_{\biguplus \mathcal{O}}$, so $T \mid\sim \phi$ iff $\mu(M_T) = \mu_{\mathcal{O}}(M_T) = \mu_{\biguplus \mathcal{O}}(M_T) \models \phi$ iff $T \models_{\mathcal{O}} \phi$, the latter by Fact 3.9.2.
□ (Proposition 3.9.5)

3.10 Ranked preferential structures

3.10.1 Introduction

First, we will present some general, basic results and discuss various properties of ranked preferential structures, which apply mostly to the minimal variant only.

We then discuss in a more systematic fashion the following versions, where (1) – (3) treat the minimal variant:

(1) Ranked structures which preserve nonemptiness (property $(\mu\emptyset)$) $X \neq \emptyset \rightarrow \mu(X) \neq \emptyset$, they are almost equivalent to smooth ranked structures.

(2) The more general case, but without copies of elements, which is very similar to case (1), as the decisive property, $(\mu\emptyset)$, still holds for finite sets. The order itself may, however, now be nonwellfounded.

(3) The general case with copies.

(4) The limit variant without copies.

(5) Equivalence of the minimal and the limit variant for an important subclass (model sets definable by formulas). This result is in parallel to those presented in Section 3.4.1 for the general case.

As the proofs are relatively straightforward, we present here a global discussion of this section on ranked structures, but will not give much more detailed comments directly before the individual results.

3.10.1.1 Detailed discussion of this section

The minimal variant

The ranked case holds some paradoxa. On the one hand, the relation is much simpler in aspect and properties, and much easier to manipulate, on the other hand, there is a multitude of conditions which are very close to each other, but still differ in more or less exotic situations.

This bio-diversity leads to the confusing list of conditions in Definition 3.10.3, of positive interrelations in Fact 3.10.9, of negative results in Fact 3.10.10 and Fact 3.10.13. They largely came as a surprise to the author, too. In particular, this teaches us to be careful with intuitions about ranked structures, what normally holds need not hold in exceptional cases, but, after all, this is a text about nonmonotonic logics, and the subject may be allowed some nonmonotonicity, too.

Such confusion is, of course, a sign that we did not delve deep enough into the subject to find deeper order where disorder reigns on the surface. So, this is an obvious area of more research.

On the other hand, this list of in(ter)dependences may be useful to study logical systems of various strengths and in many subtleties.

Now, first a word about simplicity. The crucial property is that incomparable elements have the same behavior: they are on one level, so anything which is smaller than one of them, is smaller than all of them, likewise for bigger. Formally: $a \bot b$ (i.e. neither $a \prec b$ nor $b \prec a$) and $c \prec a$ ($c \succ a$) imply $c \prec b$ ($c \succ b$). Obviously, this makes the relation much simpler. We can see the relation now as a quite realistic distance from some outside point, without information about the absolute values, but only about the relative distance.

As an immediate result, copies are (mostly) redundant: if $f(\{x\}) = \emptyset$, then either x is not at all present, or we have a cycle $x \prec x$ (this is excluded in ranked structures), or an infinite descending chain of $x-$copies. Presence (or absence) of x can be felt: If, for instance, $f(\{x,y\}) = \emptyset$, but $f(\{y\}) = \{y\}$, then x must minimize y, so it has to be present, it sucks y in like a black hole. On the other hand, if we have for instance two copies of x, x' and x'', if $x' \prec x''$, then only x' is interesting, if $x' \bot x''$, then they have the same behavior, so one copy suffices. So, we need either infinitely many copies (ω

many suffice), or just one. A formal argument is in Lemma 3.10.4.

The main positive results for minimal ranked structures are Propositions 3.10.11 and 3.10.12 for structures without copies, 3.10.14 for the general case. Proposition 3.10.14 is the most general result we show in this context. The first two are very similar, in both the condition

$(\mu =)$ $X \subseteq Y$, $\mu(Y) \cap X \neq \emptyset \rightarrow \mu(Y) \cap X = \mu(X)$

plays a central role. This is (essentially) a strengthening of the basic condition (μPR) $X \subseteq Y \rightarrow \mu(Y) \cap X \subseteq \mu(X)$, and is a very strong property. Its validity in ranked structures is obvious.

As can be expected by above remarks, the condition $(\mu\emptyset)$, $\mu(X) \neq \emptyset$ if $X \neq \emptyset$, facilitates representation. The proof of the second variant, Proposition 3.10.12 is perhaps more intuitive, as it works with pairs of elements — which are required to be in the domain. Proposition 3.10.11 will in particular be used later (in Proposition 3.10.19) to demonstrate equivalence of the limit variant with the minimal variant in a special, but important, case. Again, closure conditions for the domain (existence of pairs, closure under unions, etc.) play an important role, e.g. Proposition 3.10.12 cannot be used to show Proposition 3.10.19, the prerequisites are not satisfied.

A word on the proofs of Propositions 3.10.11, 3.10.12, 3.10.14.

Propsition 3.10.11 uses $(\mu\emptyset)$ for all sets, but does not require finite sets to be in the domain. The relation to be constructed is defined by aRb iff for some A $a \in \mu(A)$ and $b \in A$. This relation R is then extended by a higher abstract nonsense result to a total relation S, and we finally set $a \prec b$ iff aSb but not bSa. It is easy to show that this $\prec$ represents μ and is ranked.

Proposition 3.10.12 uses only $(\mu\emptyset fin)$, i.e. $(\mu\emptyset)$ for finite sets, but the finite sets are required to be in the domain. We define then directly $a \prec b$ by: $a \prec b$ iff $\mu(\{a,b\}) = \{a\}$. It is straightforward to show that $\prec$ does what it should.

In the proof of Proposition 3.10.14, we split the set into two parts, $A := \{a : \mu(\{a\}) \neq \emptyset\}$, and B the rest. We define the relation $\prec$ first on A, exactly as for Proposition 3.10.12 (finite sets are assumed to be in the domain). For $b \in B$, we consider $A_b := \{a \in A : a \notin \mu(\{a,b\})\}$, and put all $a \in A_b$ above all copies of b — for $b \in B$, we need infinite descending chains of $b-$copies.

The limit variant

Perhaps the most important results in this section concern the limit variant. We consider here the (transitive) limit variant without copies, to simplify the picture.

We will

- Show that minimizing initial segments have a particularly nice form in the ranked case. In particular, they are totally ordered by inclusion.
- Use this description to obtain an algebraic characterization of the ranked limit version.
- Show that the logic defined by the ranked limit version is equivalent to that defined by the the minimal version in two important classes:
  - if the definable MISE are cofinal,
  - if we consider only formulas on the left.

  Recall that we have shown analogous results in Section 3.4.1 for the general transitive case.
- Finally give a very rough sketch of a logical characterization of the ranked limit case — which will necessarily be infinite, see Section 5.2.3.

We first discuss the algebraic representation result just mentioned.

Suppose we are given a system of initial segments, and want to construct a representing relation. As we need not work with copies, it suffices to consider pairs to construct the relation. But, for finite sets, the minimal and the limit variant coincide. Thus, we can construct the relation for the limit variant by looking at pairs and the much simpler minimal variant. This allows then a general technique: Special finite cases allow to construct the relation in a unique way, using representation results for the minimal variant, and we only have to make sure that we have sufficient conditions to assure that the relation so constructed coincides with the properties of the whole function.

The perhaps central property — called $(\Lambda 5)$ — is that for any suitable initial segment A of X, which has nonempty intersection with $Y \subseteq X$, $A \cap Y$ is also a suitable initial segment for Y. This is a very strong condition and the analogue of $(\mu =) : X \subseteq Y,\ \mu(Y) \cap X \neq \emptyset \rightarrow \mu(Y) \cap X = \mu(X)$ above.

Apart from this condition, we add conditions which express that the system of intial segments is sufficiently big to allow (essential) reconstruction of the system by the relation. (The reconstruction will not be total, but sufficiently close: we will reconstruct a system in which the old one is a cofinal subsystem, and this is sufficient.)

Condition ($\Lambda 5$) serves also to reflect the situation down to the finite case (choose Y finite), where the minimal and the limit variant coincide, and allows a straightforward definition of the relation. More precisely, we define $\sigma(X) := \bigcap \Lambda(X)$, where $\Lambda(X)$ is the set of suitable initial segments. For finite X, $\sigma(X)$ is not empty, and for finite X, $\sigma(X)$ has the properties needed to apply the same strategy as used in the proof of Proposition 3.10.12, to show that the relation $\prec$ defined by $a \prec b$ iff $\sigma(\{a,b\}) = \{a\}$ is ranked. We define now $\Lambda_{\prec}(X)$, and show that for every $A \in \Lambda_{\prec}(X)$ there is $B \in \Lambda(X)$, $B \subseteq A$, and, conversely, $\Lambda(X) \subseteq \Lambda_{\prec}(X)$.

We turn now more to logics and consider problems — and solutions — inspired by the systems of sets defined by formulas and theories. We will see that, again, closure conditions play a crucial role for what can be done.

In preparation of our main result — equivalence of the limit and the minimal variant for formulas on the left of $\mid\!\sim$ — we make the problem precise, and show a number of simple results and give some examples and counterexamples.

First, we define $T \models_{\Lambda} \phi$ for a given system of initial segments, by $T \models_{\Lambda} \phi$ iff there is $A \in \Lambda(M(T))$ s.t. $A \models \phi$ (classically). Thus, $T \models_{\Lambda} \phi$ means that ϕ holds in the limit of $T-$models, or, finally, ϕ becomes true.

Example 3.10.1 prepares for the more complicated and crucial Example 3.10.2. The examples always let the logical (i.e. of the standard topology of classical logic) and the order (of the relation $\prec$) limit diverge. We use infinite descending sequences of models, which converge logically to some model m, but we put the model m not at its natural place, but elsewhere. Then, Λ converges to m, but m is not there.

Fact 3.10.17 shows some easy results about the logics defined by Λ.

One of the problems to handle is that the initial segments need not be definable by any theory. A natural idea for a simplification is an overkill: If we assume that the systems $\Lambda(M(T))$ contains a cofinal subset of definable sets, we trivialize the problem, it becomes equivalent to the minimal variant (see Fact 3.10.18).

The important Example 3.10.2 shows that formulas and full theories on the left of $\mid\!\sim$ (or $\models_{\Lambda}$) have a very different behavior. We can reorder structures without any influence on $\phi \models_{\Lambda} \psi$, but with drastic influences on $T \models_{\Lambda} \psi$. In particular, it is possible to find logics where the formula-fragment can be represented by the ranked limit variant, but not the full logic with theories on the left of $\mid\!\sim$.

The main result, Proposition 3.10.19, shows the former: Given a logic defined for the formula part by a ranked structure, interpreted in the limit variant, we can find a logically equivalent ranked structure interpreted in the minimal variant — with, perhaps a different ordering of the models. Thus, we have reduced the much more complicated limit variant to the minimal variant for formulas, but also shown in Example 3.10.2 that the same is impossible for full theories.

The strategy, as well as its execution, of the proof of Proposition 3.10.19 are simple. We show that for $f(M(\phi)) := M(\{\psi : \phi \models_\Lambda \psi\})$, f has the properties of the prerequisites of Proposition 3.10.11, and can thus be represented by a minimal ranked structure. The crucial property is again (Λ5).

We finish by a sketch of a characterization for the general (full theory) case.

To summarize:

Rankedness is a very strong condition, which simplifies many problems, and necessitates new (and simpler) techniques for representation. There are a number of closely related, but nonetheless differing conditions of ranked structures, we show implications and independences. Three representation results are given, for various structures and various closure conditions of the domain. We also examine the limit version, which is considerably simpler than the limit version of general preferential structures, obtain a representation result, and can show that the formula part, but not the general case, can be reduced to the minimal variant.

3.10.1.2 Introductory facts and definitions

We give here some definitions, and show elementary facts about ranked structures. We also prove a general abstract nonsense fact about extending relations, to be used here and again in Section 4.2 on theory revision.

The crucial fact will be Lemma 3.10.4, it shows that we can do with either one or infinitely many copies of each model. The reason behind it is the following: Suppose we have exactly two copies of one model, m, m', where m and m' have the same logical properties. If, e.g., $m \prec m'$, then, as we consider only minimal elements, m' will be "invisible". If m and m' are incomparable, then, by rankedness (modularity), they will have the same elements above (and below) themselves: they have the same behavior in the preferential structure. An immediate consequence is the "singleton property" of Fact 3.10.6: One element suffices to destroy minimality, and it suffices to look at pairs (and singletons).

The material of Fact 3.10.2 – Fact 3.10.6 is taken from [Sch96-1], and is mostly folklore.

We first note the following trivial

Fact 3.10.1

In a ranked structure, smoothness and the condition

$(\mu\emptyset)$ $X \neq \emptyset \rightarrow \mu(X) \neq \emptyset$

are (almost) equivalent.

Proof:

Suppose $(\mu\emptyset)$ holds, and let $x \in X - \mu(X)$, $x' \in \mu(X)$. Then $x' \prec x$ by rankedness. Conversely, if the structure is smooth and there is an element $x \in X$ in the structure (recall that structures may have "gaps", but this condition is a minor point, which we shall neglect here — this is the precise meaning of "almost"), then either $x \in \mu(X)$ or there is $x' \prec x$, $x' \in \mu(X)$, so $\mu(X) \neq \emptyset$. □

Note further that if we have no copies (and there is some $x \in X$ in the structure), $(\mu\emptyset)$ holds for all finite sets, and this will be sufficient to construct the relation for representation results, as we shall see.

We introduce some more notation:

Notation 3.10.1

(1) $A = B \parallel C$ stands for: $A = B$ or $A = C$ or $A = B \cup C$.

(2) Recall from Definition 1.6.1: Given $\prec$, $a \bot b$ means: neither $a \prec b$ nor $b \prec a$.

Fact 3.10.2

Let $\prec$ be an irreflexive, binary relation on X, then the following two conditions are equivalent:

(1) There is Ω and an irreflexive, total, binary relation $\prec'$ on Ω and a function $f : X \rightarrow \Omega$ s.t. $x \prec y \leftrightarrow fx \prec' fy$ for all $x, y \in X$ (we sometimes write fx for $f(x)$, etc.).

(2) Let $x, y, z \in X$ and $x \bot y$ wrt. $\prec$ (i.e. neither $x \prec y$ nor $y \prec x$), then $z \prec x \rightarrow z \prec y$ and $x \prec z \rightarrow y \prec z$.

Proof:

(1) $\rightarrow$ (2): Let $x \bot y$, thus neither $fx \prec' fy$ nor $fy \prec' fx$, but then $fx = fy$.

Let now $z \prec x$, so $fz \prec' fx = fy$, so $z \prec y$. $x \prec z \rightarrow y \prec z$ is similar.

(2) $\rightarrow$ (1): For $x \in X$ let $[x] := \{x' \in X : x \bot x'\}$, and $\Omega := \{[x] : x \in X\}$. For $[x], [y] \in \Omega$ let $[x] \prec' [y] :\leftrightarrow x \prec y$. This is well-defined: Let $x \bot x'$, $y \bot y'$ and $x \prec y$, then $x \prec y'$ and $x' \prec y'$. Obviously, $\prec'$ is an irreflexive, total binary relation. Define $f : X \rightarrow \Omega$ by $fx := [x]$, then $x \prec y \leftrightarrow [x] \prec' [y] \leftrightarrow fx \prec' fy$. $\Box$

We repeat from Section 2.3.1

Definition 3.10.1

Call an irreflexive, binary relation $\prec$ on X, which satisfies (1) (equivalently (2)) of Fact 3.10.2, ranked. By abuse of language, we also call the structure $< X, \prec >$ ranked.

Fact 3.10.3

If $\prec$ on X is ranked, and free of cycles, then $\prec$ is transitive.

Proof:

Let $x \prec y \prec z$. If $x \bot z$, then $y \succ z$, resulting in a cycle of length 2. If $z \prec x$, then we have a cycle of length 3. So $x \prec z$. $\Box$

Definition 3.10.2

Let $\mathcal{Z} =< \mathcal{X}, \prec >$ be a preferential structure. Call $\mathcal{Z}$ $1 - \infty$ over Z, iff for all $x \in Z$ there are exactly one or infinitely many copies of x, i.e. for all $x \in Z$ $\{u \in \mathcal{X} : u =< x, i > \text{ for some } i\}$ has cardinality 1 or $\geq \omega$.

Lemma 3.10.4

Let $\mathcal{Z} =< \mathcal{X}, \prec >$ be a preferential structure and $f : \mathcal{Y} \rightarrow \mathcal{P}(Z)$ with $\mathcal{Y} \subseteq \mathcal{P}(Z)$ be represented by $\mathcal{Z}$, i.e. for $X \in \mathcal{Y}$ $f(X) = \mu_{\mathcal{Z}}(X)$, and $\mathcal{Z}$ be ranked and free of cycles. Then there is a structure $\mathcal{Z}'$, $1 - \infty$ over Z, ranked and free of cycles, which also represents f.

Proof:

We construct $\mathcal{Z}' =< \mathcal{X}', \prec' >$.

Let $A := \{x \in Z$: there is some $< x, i > \in \mathcal{X}$, but for all $< x, i > \in \mathcal{X}$ there is $< x, j > \in \mathcal{X}$ with $< x, j > \prec < x, i >\}$,

let $B := \{x \in Z$: there is some $< x, i > \in \mathcal{X}$, s.t. for no $< x, j > \in \mathcal{X}$ $< x, j > \prec < x, i >\}$,

let $C := \{x \in Z$: there is no $< x, i > \in \mathcal{X}\}$.

Let $c_i : i < \kappa$ be an enumeration of C. We introduce for each such c_i ω many copies $< c_i, n >: n < \omega$ into $\mathcal{X}'$, put all $< c_i, n >$ above all elements in $\mathcal{X}$, and order the $< c_i, n >$ by $< c_i, n > \prec' < c_{i'}, n' > :\leftrightarrow (i = i'$ and $n > n')$ or $i > i'$. Thus, all $< c_i, n >$ are comparable.

If $a \in A$, then there are infinitely many copies of a in $\mathcal{X}$, as $\mathcal{X}$ was cycle-free, we put them all into $\mathcal{X}'$. If $b \in B$, we choose exactly one such minimal element $< b, m >$ (i.e. there is no $< b, n > \prec < b, m >$) into $\mathcal{X}'$, and omit all other elements. (For definiteness, assume in all applications $m = 0$.) For all elements from A and B, we take the restriction of the order $\prec$ of $\mathcal{X}$. This is the new structure $\mathcal{Z}'$.

Obviously, adding the $< c_i, n >$ does not introduce cycles, irreflexivity and rankedness are preserved. Moreover, any substructure of a cycle-free, irreflexive, ranked structure also has these properties, so $\mathcal{Z}'$ is $1 - \infty$ over Z, ranked and free of cycles.

We show that $\mathcal{Z}$ and $\mathcal{Z}'$ are equivalent. Let then $X \subseteq Z$, we have to prove $\mu(X) = \mu'(X)$ $(\mu := \mu_{\mathcal{Z}}$, $\mu' := \mu_{\mathcal{Z}'})$.

Let $z \in X - \mu(X)$. If $z \in C$ or $z \in A$, then $z \notin \mu'(X)$. If $z \in B$, let $< z, m >$ be the chosen element. As $z \notin \mu(X)$, there is $x \in X$ s.t. some $< x, j > \prec$ $< z, m >$. x cannot be in C. If $x \in A$, then also $< x, j > \prec' < z, m >$. If $x \in B$, then there is some $< x, k >$ also in $\mathcal{X}'$. $< x, j > \prec < x, k >$ is impossible. If $< x, k > \prec < x, j >$, then $< z, m > \succ < x, k >$ by transitivity. If $< x, k > \perp < x, j >$, then also $< z, m > \succ < x, k >$ by rankedness. In any case, $< z, m > \succ' < x, k >$, and thus $z \notin \mu'(X)$.

Let $z \in X - \mu'(X)$. If $z \in C$ or $z \in A$, then $z \notin \mu(X)$. Let $z \in B$, and some $< x, j > \prec' < z, m >$. x cannot be in C, as they were sorted on top, so $< x, j >$ exists in $\mathcal{X}$ too and $< x, j > \prec < z, m >$. But if any other $< z, i >$ is also minimal in $\mathcal{Z}$ among the $< z, k >$, then by rankedness also $< x, j > \prec < z, i >$, as $< z, i > \perp < z, m >$, so $z \notin \mu(X)$. □

Assume in the sequel that $\mathcal{Y}$ contains all singletons and pairs, and fix $f : \mathcal{Y} \rightarrow \mathcal{P}(Z)$. We also fix the following notation: $A := \{x \in Z : f(x) = \emptyset\}$ and $B := Z - A$ (here and in future we sometimes write $f(x)$ for $f(\{x\})$, likewise

$f(x, x') = x$ for $f(\{x, x'\}) = \{x\}$, etc., when the meaning is obvious).

Corollary 3.10.5

If f can be represented by a ranked $\mathcal{Z}$ free of cycles, then there is $\mathcal{Z}'$, which is also ranked and cycle-free, all $b \in B$ occur in one copy, all $a \in A$ ∞ often.

Fact 3.10.6

(1) If $\mathcal{Z}'$ is as in Corollary 3.10.5, $b \in B$, $a \in A$, $f(a, b) = b$, then for all $< a, i > \in \mathcal{X}'$ $< a, i > \succ' < b, 0 >$.

(2) If f can be represented by a cycle-free ranked $\mathcal{Z}$, then it has the "singleton property": If $x \in X$, then $x \notin f(X) \leftrightarrow \exists x' \in X.x \notin f(x, x')$.

(3) If f is as in (2), $b, b' \in B$, then $f(b, b') \neq \emptyset$.

Proof:

(1) For no $< a, i > < b, 0 > \succ' < a, i >$, since otherwise $f(a, b) = \emptyset$. If $< b, 0 > \perp < a, i >$, then as there is $< a, j > \prec < a, i >$, $< a, j > \prec' < b, 0 >$ by rankedness, contradiction.

(2) "$\leftarrow$" holds for all preferential structures. "$\rightarrow$": If $x \in A$, then $x \notin f(x, x)$. Let $x \in B$, $\mathcal{Z}$ a $1 - \infty$ over Z structure representing f as above. So there is just one copy of x in $\mathcal{X}$, $< x, 0 >$, and there is some $< y, j > \prec$ $< x, 0 >$, $y \in X$, thus $x \notin f(x, y)$.

(3) In any $1 - \infty$ over Z representation of f, $< b, 0 > \perp < b', 0 >$, or $< b, 0 > \prec < b', 0 >$, or $< b', 0 > \prec < b, 0 >$. $< b, 0 > \prec < b', 0 > \prec < b, 0 >$ cannot be, as this is a cycle. □

We conclude this introduction by a generalized abstract nonsense result, taken from [LMS01], which must also be part of the folklore:

Lemma 3.10.7

Given a set X and a binary relation R on X, there exists a total preorder (i.e. a total, reflexive, transitive relation) S on X that extends R such that

$\forall x, y \in X(xSy, ySx \Rightarrow xR^*y)$

where R^* is the reflexive and transitive closure of R.

Proof:

Define $x \equiv y$ iff xR^*y and yR^*x. The relation $\equiv$ is an equivalence relation.

Let $[x]$ be the equivalence class of x under $\equiv$. Define $[x] \preceq [y]$ iff xR^*y. The definition of $\preceq$ does not depend on the representatives x and y chosen. The relation $\preceq$ on equivalence classes is a partial order. Let $\leq$ be any total order on these equivalence classes that extends $\preceq$. Define xSy iff $[x] \leq [y]$. The relation S is total (since $\leq$ is total) and transitive (since $\leq$ is transitive) and is therefore a total preorder. It extends R by the definition of $\preceq$ and the fact that $\leq$ extends $\preceq$. Suppose now xSy and ySx. We have $[x] \leq [y]$ and $[y] \leq [x]$ and therefore $[x] = [y]$ by antisymmetry. Therefore $x \equiv y$ and xR^*y. □

3.10.2 The minimal variant

3.10.2.1 Introductory definitions and results

We work now on a (nonempty) set U, and consider functions $\mu : \mathcal{Y} \to \mathcal{P}(U)$, where $\mathcal{Y} \subseteq \mathcal{P}(U)$. We first enumerate some conditions, which we will consider in the sequel. The differences between them are sometimes quite subtle, as will be seen below, e.g. in Fact 3.10.10. Facts 3.10.8 and 3.10.9 collect some positive results, Fact 3.10.10 some negative ones.

Definition 3.10.3

The conditions for the minimal case — where we recall also some of the standard conditions for easier reading — are:

$(\mu \subseteq)$ $\mu(X) \subseteq X$,

$(\mu\emptyset)$ $X \neq \emptyset \to \mu(X) \neq \emptyset$,

$(\mu\emptyset fin)$ $X \neq \emptyset \to \mu(X) \neq \emptyset$ for finite X,

(μPR) $X \subseteq Y \to \mu(Y) \cap X \subseteq \mu(X)$,

$(\mu =)$ $X \subseteq Y$, $\mu(Y) \cap X \neq \emptyset \to \mu(Y) \cap X = \mu(X)$,

$(\mu =')$ $\mu(Y) \cap X \neq \emptyset \to \mu(Y \cap X) = \mu(Y) \cap X$,

$(\mu \parallel)$ $\mu(X \cup Y) = \mu(X) \parallel \mu(Y)$ ($\parallel$ is defined in Notation 3.10.1),

$(\mu\cup)$ $\mu(Y) \cap (X - \mu(X)) \neq \emptyset \to \mu(X \cup Y) \cap Y = \emptyset$,

$(\mu\cup')$ $\mu(Y) \cap (X - \mu(X)) \neq \emptyset \to \mu(X \cup Y) = \mu(X)$,

$(\mu \in)$ $a \in X - \mu(X) \to \exists b \in X. a \notin \mu(\{a, b\})$,

(μCUM) $\mu(Y) \subseteq X \subseteq Y \to \mu(X) = \mu(Y)$.

Note that $(\mu =')$ is very close to Rational Monotony: Rational Monotony says: $\alpha \mid\sim \beta, \alpha \mid\not\sim \neg\gamma \rightarrow \alpha \wedge \gamma \mid\sim \beta$. Or, $\mu(A) \subseteq B, \mu(A) \cap C \neq \emptyset \rightarrow \mu(A \cap C) \subseteq B$ for all A, B, C. This is not quite, but almost: $\mu(A \cap C) \subseteq \mu(A) \cap C$ (it depends how many B there are, if $\mu(A)$ is some such B, the fit is perfect).

Fact 3.10.8

In all ranked structures, $(\mu \subseteq)$, $(\mu =)$, (μPR), $(\mu =')$, $(\mu \parallel)$, $(\mu\cup)$, $(\mu\cup')$, $(\mu \in)$ will hold, if the corresponding closure conditions are satisfied.

Proof:

$(\mu \subseteq)$ and (μPR) hold in all preferential structures. $(\mu =)$ and $(\mu =')$ are trivial. $(\mu\cup)$ and $(\mu\cup')$: All minimal copies of elements in $\mu(Y)$ have the same rank. If some $y \in \mu(Y)$ has all its minimal copies killed by an element $x \in X$, by rankedness, x kills the rest, too. $(\mu \in)$: If $\mu(\{a\}) = \emptyset$, we are done. Take the minimal copies of a in $\{a\}$, they are all killed by one element in X. $(\mu \parallel)$: Case $\mu(X) = \emptyset$: If below every copy of $y \in Y$ there is a copy of some $x \in X$, then $\mu(X \cup Y) = \emptyset$. Otherwise $\mu(X \cup Y) = \mu(Y)$. Suppose now $\mu(X) \neq \emptyset$, $\mu(Y) \neq \emptyset$, then the minimal ranks decide: if they are equal, $\mu(X \cup Y) = \mu(X) \cup \mu(Y)$, etc. □

Fact 3.10.9

The following properties (2)–(9) hold, provided corresponding closure conditions for the domain $\mathcal{Y}$ are satisfied. We first enumerate these conditions.

For (3), (4), (8): closure under finite unions.

For (2): closure under finite intersections.

For (6) and (7): closure under finite unions, and $\mathcal{Y}$ contains all singletons.

For (5): closure under set difference.

For (9): suffienctly strong conditions — which are satisfied for the set of models definable by propositional theories.

Note that the closure conditions for (5), (6), (9) are quite different, for this reason, (5) alone is not enough.

(1) $(\mu =)$ entails (μPR),

(2) in the presence of $(\mu \subseteq)$, $(\mu =)$ is equivalent to $(\mu =')$,

(3) $(\mu \subseteq)$, $(\mu =) \rightarrow (\mu\cup)$,

(4) $(\mu\subseteq)$, $(\mu\emptyset)$, $(\mu =)$ entail:

(4.1) $(\mu \parallel)$,

(4.2) $(\mu\cup')$,

(4.3) (μCUM),

(5) $(\mu \subseteq) + (\mu \parallel) \rightarrow (\mu =)$,

(6) $(\mu \parallel) + (\mu \in) + (\mu PR) + (\mu \subseteq) \rightarrow (\mu =)$,

(7) $(\mu CUM) + (\mu =) \rightarrow (\mu \in)$,

(8) $(\mu CUM) + (\mu =) + (\mu \subseteq) \rightarrow (\mu \parallel)$,

(9) $(\mu PR) + (\mu CUM) + (\mu \parallel) \rightarrow (\mu =)$.

Proof:

(1) trivial.

(2) $(\mu =) \rightarrow (\mu =')$: Let $\mu(Y) \cap X \neq \emptyset$, we have to show $\mu(X \cap Y) = \mu(Y) \cap X$. By $(\mu \subseteq)$ $\mu(Y) \subseteq Y$, so $\mu(Y) \cap X = \mu(Y) \cap (X \cap Y)$, so by $(\mu =)$ $\mu(Y) \cap X = \mu(Y) \cap (X \cap Y) = \mu(X \cap Y)$. $(\mu =') \rightarrow (\mu =)$: Let $X \subseteq Y$, $\mu(Y) \cap X \neq \emptyset$, then $\mu(X) = \mu(Y \cap X) = \mu(Y) \cap X$.

(3) If not, $\mu(X \cup Y) \cap Y \neq \emptyset$, but $\mu(Y) \cap (X - \mu(X)) \neq \emptyset$. By (1), (μPR) holds, so $\mu(X \cup Y) \cap X \subseteq \mu(X)$, so $\emptyset \neq \mu(Y) \cap (X - \mu(X)) \subseteq \mu(Y) \cap (X - \mu(X \cup Y))$, so $\mu(Y) - \mu(X \cup Y) \neq \emptyset$, so by $(\mu \subseteq)$ $\mu(Y) \subseteq Y$ and $\mu(Y) \neq \mu(X \cup Y) \cap Y$. But by $(\mu =)$ $\mu(Y) = \mu(X \cup Y) \cap Y$, a contradiction.

(4.1) If X or Y or both are empty, then this is trivial. Assume then $X \cup Y \neq \emptyset$, so by $(\mu\emptyset)$ $\mu(X \cup Y) \neq \emptyset$. By $(\mu \subseteq)$ $\mu(X \cup Y) \subseteq X \cup Y$, so $\mu(X \cup Y) \cap X = \emptyset$ and $\mu(X \cup Y) \cap Y = \emptyset$ together are impossible. Case 1, $\mu(X \cup Y) \cap X \neq \emptyset$ and $\mu(X \cup Y) \cap Y \neq \emptyset$: By $(\mu =)$ $\mu(X \cup Y) \cap X = \mu(X)$ and $\mu(X \cup Y) \cap Y = \mu(Y)$, so by $(\mu \subseteq)$ $\mu(X \cup Y) = \mu(X) \cup \mu(Y)$. Case 2, $\mu(X \cup Y) \cap X \neq \emptyset$ and $\mu(X \cup Y) \cap Y = \emptyset$: So by $(\mu =)$ $\mu(X \cup Y) = \mu(X \cup Y) \cap X = \mu(X)$. Case 3, $\mu(X \cup Y) \cap X = \emptyset$ and $\mu(X \cup Y) \cap Y \neq \emptyset$: Symmetrical.

(4.2) If $X \cup Y = \emptyset$, then $\mu(X \cup Y) = \mu(X) = \emptyset$ by $(\mu \subseteq)$. So suppose $X \cup Y \neq \emptyset$. By (3), $\mu(X \cup Y) \cap Y = \emptyset$, so $\mu(X \cup Y) \subseteq X$ by $(\mu \subseteq)$. By $(\mu\emptyset)$, $\mu(X \cup Y) \neq \emptyset$, so $\mu(X \cup Y) \cap X \neq \emptyset$, and $\mu(X \cup Y) = \mu(X)$ by $(\mu =)$.

(4.3) If $Y = \emptyset$, this is trivial by $(\mu \subseteq)$. If $Y \neq \emptyset$, then by $(\mu\emptyset)$ — which is crucial here — $\mu(Y) \neq \emptyset$, so by $\mu(Y) \subseteq X$ $\mu(Y) \cap X \neq \emptyset$, so by $(\mu =)$ $\mu(Y) = \mu(Y) \cap X = \mu(X)$.

(5) Let $X \subseteq Y$, and consider $Y = X \cup (Y - X)$. Then $\mu(Y) = \mu(X) \parallel \mu(Y - X)$. As $\mu(Y - X) \cap X = \emptyset$, $\mu(Y) \cap X \subseteq \mu(X)$. If $\mu(Y) \cap X \neq \emptyset$, then by the same argument $\mu(X)$ is involved, so $\mu(X) \subseteq \mu(Y)$.

(6) Suppose $X \subseteq Y$, $x \in \mu(Y) \cap X$, we have to show $\mu(Y) \cap X = \mu(X)$. "$\subseteq$" is trivial by (μPR). "$\supseteq$": Assume $a \notin \mu(Y)$ (by $(\mu \subseteq)$), but $a \in \mu(X)$. By $(\mu \in)$ $\exists b \in Y . a \notin \mu(\{a,b\})$. As $a \in \mu(X)$, by (μPR), $a \in \mu(\{a,x\})$. By $(\mu \parallel)$, $\mu(\{a,b,x\}) = \mu(\{a,x\}) \parallel \mu(\{b\})$. As $a \notin \mu(\{a,b,x\})$, $\mu(\{a,b,x\}) = \mu(\{b\})$, so $x \notin \mu(\{a,b,x\})$, contradicting (μPR), as $a,b,x \in Y$.

(7) Let $a \in X - \mu(X)$. If $\mu(X) = \emptyset$, then $\mu(\{a\}) = \emptyset$ by (μCUM). If not: Let $b \in \mu(X)$, then $a \notin \mu(\{a,b\})$ by $(\mu =)$.

(8) By (μCUM), $\mu(X \cup Y) \subseteq X \subseteq X \cup Y \rightarrow \mu(X) = \mu(X \cup Y)$, and $\mu(X \cup Y) \subseteq Y \subseteq X \cup Y \rightarrow \mu(Y) = \mu(X \cup Y)$. Thus, if $(\mu \parallel)$ were to fail, $\mu(X \cup Y) \not\subseteq X$, $\mu(X \cup Y) \not\subseteq Y$, but then by $(\mu \subseteq)$ $\mu(X \cup Y) \cap X \neq \emptyset$, so $\mu(X) = \mu(X \cup Y) \cap X$, and $\mu(X \cup Y) \cap Y \neq \emptyset$, so $\mu(Y) = \mu(X \cup Y) \cap Y$ by $(\mu =)$. Thus, $\mu(X \cup Y) = (\mu(X \cup Y) \cap X) \cup (\mu(X \cup Y) \cap Y) = \mu(X) \cup \mu(Y)$.

(9) Suppose $(\mu =)$ does not hold. So, by (μPR), there are X, Y, y s.t. $X \subseteq Y$, $X \cap \mu(Y) \neq \emptyset$, $y \in Y - \mu(Y)$, $y \in \mu(X)$. Let $a \in X \cap \mu(Y)$. If $\mu(Y) = \{a\}$, then by (μCUM) $\mu(Y) = \mu(X)$, so there must be $b \in \mu(Y)$, $b \neq a$. Take now Y', Y'' s.t. $Y = Y' \cup Y''$, $a \in Y'$, $a \notin Y''$, $b \in Y''$, $b \notin Y'$, $y \in Y' \cap Y''$. Assume now $(\mu \parallel)$ to hold, we show a contradiction. If $y \notin \mu(Y'')$, then by (μPR) $y \notin \mu(Y'' \cup \{a\})$. But $\mu(Y'' \cup \{a\}) = \mu(Y'') \parallel \mu(\{a,y\})$, so $\mu(Y'' \cup \{a\}) = \mu(Y'')$, contradicting $a \in \mu(Y)$. If $y \in \mu(Y'')$, then by $\mu(Y) = \mu(Y') \parallel \mu(Y'')$, $\mu(Y) = \mu(Y')$, contradiction as $b \notin \mu(Y')$.

□

Fact 3.10.10

(1) $(\mu \subseteq) + (\mu PR) + (\mu =) \not\rightarrow (\mu \parallel)$,

(2) $(\mu \subseteq) + (\mu PR) + (\mu \parallel) \not\rightarrow (\mu =)$ (without closure under set difference),

(3) $(\mu \subseteq) + (\mu PR) + (\mu \parallel) + (\mu =) + (\mu \cup) \not\rightarrow (\mu \in)$ (and thus $(\mu \subseteq) + (\mu PR) + (\mu \parallel) + (\mu =) + (\mu \cup)$ do not guarantee representability by ranked structures by Fact 3.10.8).

Proof:

(1) Consider the following structure *without transitivity* $U := \{a,b,c,d\}$, c and d have ω many copies in descending order $c_1 \succeq c_2 \ldots$, etc. a, b have one single copy each. $a \succeq b$, $a \succeq d_1$, $b \succeq a$, $b \succeq c_1$. $(\mu \parallel)$ does not hold: $\mu(U) = \emptyset$, but $\mu(\{a,c\}) = \{a\}$, $\mu(\{b,d\}) = \{b\}$. (μPR) holds as in all preferential structures.
$(\mu =)$ holds: If it were to fail, then for some $A \subseteq B$, $\mu(B) \cap A \neq \emptyset$, so

$\mu(B) \neq \emptyset$. But the only possible cases for B are now: $(a \in B,\ b, d \notin B)$ or $(b \in B,\ a, c \notin B)$. Thus, B can be $\{a\}$, $\{a,c\}$, $\{b\}$, $\{b,d\}$ with $\mu(B) = \{a\}$, $\{a\}$, $\{b\}$, $\{b\}$. If $A = B$, then the result will hold trivially. Moreover, A has to be $\neq \emptyset$. So the remaining cases of B where it might fail are $B = \{a, c\}$ and $\{b, d\}$, and by $\mu(B) \cap A \neq \emptyset$, the only cases of A where it might fail, are $A = \{a\}$ or $\{b\}$ respectively. So the only cases remaining are: $B = \{a, c\}$, $A = \{a\}$ and $B = \{b, d\}$, $A = \{b\}$. In the first case, $\mu(A) = \mu(B) = \{a\}$, in the second $\mu(A) = \mu(B) = \{b\}$, but $(\mu =)$ holds in both.

(2) Work in the set of theory definable model sets of an infinite propositional language. Note that this is not closed under set difference, and closure properties will play a crucial role in the argumentation. Let $U := \{y, a, x_{i<\omega}\}$, where $x_i \to a$ in the standard topology. For the order, arrange s.t. y is minimized by any set iff this set contains a cofinal subsequence of the x_i, this can be done by the standard construction. Moreover, let the x_i all kill themselves, i.e. with ω many copies $x_i^1 \succeq x_i^2 \succeq \ldots$. There are no other elements in the relation. Note that if $a \notin \mu(X)$, then $a \notin X$, and X cannot contain a cofinal subsequence of the x_i, as X is closed in the standard topology. (A short argument: suppose X contains such a subsequence, but $a \notin X$. Then the theory of a $Th(a)$ is inconsistent with $Th(X)$, so already a finite subset of $Th(a)$ is inconsistent with $Th(X)$, but such a finite subset will finally hold in a cofinal sequence converging to a.) Likewise, if $y \in \mu(X)$, then X cannot contain a cofinal subsequence of the x_i.

Obviously, $(\mu \subseteq)$ and (μPR) hold, but $(\mu =)$ does not hold: Set $B := U$, $A := \{a, y\}$. Then $\mu(B) = \{a\}$, $\mu(A) = \{a, y\}$, contradicting $(\mu =)$.

It remains to show that $(\mu \parallel)$ holds.

$\mu(X)$ can only be $\emptyset$, $\{a\}$, $\{y\}$, $\{a, y\}$. As $\mu(A \cup B) \subseteq \mu(A) \cup \mu(B)$ by (μPR), Case 1, $\mu(A \cup B) = \{a, y\}$ is settled.

Note that if $y \in X - \mu(X)$, then X will contain a cofinal subsequence, and thus $a \in \mu(X)$.

Case 2: $\mu(A \cup B) = \{a\}$.

Case 2.1: $\mu(A) = \{a\}$ — we are done.

Case 2.2: $\mu(A) = \{y\}$: A does not contain a, nor a cofinal subsequence. If $\mu(B) = \emptyset$, then $a \notin B$, so $a \notin A \cup B$, a contradiction. If $\mu(B) = \{a\}$, we are done. If $y \in \mu(B)$, then $y \in B$, but B does not contain a cofinal subsequence, so $A \cup B$ does not either, so $y \in \mu(A \cup B)$, contradiction.

Case 2.3: $\mu(A) = \emptyset$: A cannot contain a cofinal subsequence. If $\mu(B) = \{a\}$, we are done. $a \in \mu(B)$ does have to hold, so $\mu(B) = \{a, y\}$ is the only remaining possibility. But then B does not contain a cofinal subsequence, and neither does $A \cup B$, so $y \in \mu(A \cup B)$, contradiction.

Case 2.4: $\mu(A) = \{a, y\}$: A does not contain a cofinal subsequence. If $\mu(B) = \{a\}$, we are done. If $\mu(B) = \emptyset$, B does not contain a cofinal subsequence (as $a \notin B$), so neither does $A \cup B$, so $y \in \mu(A \cup B)$, contradiction. If $y \in \mu(B)$, B does not contain a cofinal subsequence, and we are done again.

Case 3: $\mu(A \cup B) = \{y\}$: To obtain a contradiction, we need $a \in \mu(A)$ or $a \in \mu(B)$. But in both cases $a \in \mu(A \cup B)$.

Case 4: $\mu(A \cup B) = \emptyset$: Thus, $A \cup B$ contains no cofinal subsequence. If, e.g. $y \in \mu(A)$, then $y \in \mu(A \cup B)$, if $a \in \mu(A)$, then $a \in \mu(A \cup B)$, so $\mu(A) = \emptyset$.

(3) Let $U := \{y, x_{i<\omega}\}$, x_i a sequence, each x_i kills itself, $x_i^1 \succeq x_i^2 \succeq \ldots$ and y is killed by all cofinal subsequences of the x_i. Then for any $X \subseteq U$ $\mu(X) = \emptyset$ or $\mu(X) = \{y\}$.

$(\mu \subseteq)$ and (μPR) hold obviously.

$(\mu \parallel)$: Let $A \cup B$ be given. If $y \notin X$, then for all $Y \subseteq X$ $\mu(Y) = \emptyset$. So, if $y \notin A \cup B$, we are done. If $y \in A \cup B$, if $\mu(A \cup B) = \emptyset$, one of A, B must contain a cofinal sequence, it will have $\mu = \emptyset$. If not, then $\mu(A \cup B) = \{y\}$, and this will also hold for the one y is in.

$(\mu =)$: Let $A \subseteq B$, $\mu(B) \cap A \neq \emptyset$, show $\mu(A) = \mu(B) \cap A$. But now $\mu(B) = \{y\}$, $y \in A$, so B does not contain a cofinal subsequence, neither does A, so $\mu(A) = \{y\}$.

$(\mu \cup)$: $(A - \mu(A)) \cap \mu(A') \neq \emptyset$, so $\mu(A') = \{y\}$, so $\mu(A \cup A') = \emptyset$, as $y \in A - \mu(A)$.

But $(\mu \in)$ does not hold: $y \in U - \mu(U)$, but there is no x s.t. $y \notin \mu(\{x, y\})$.

□

We turn to characterizations.

3.10.2.2 Characterizations

We show first two results for the case without copies (Propositions 3.10.11 and 3.10.12), then again some negative results for the general case (Proposition 3.10.13), and conclude with a characterization of the general case (Proposition 3.10.14).

We will show now that $(\mu \subseteq)$, $(\mu\emptyset)$, and $(\mu =)$ provide a complete characterization for the case where $(\mu\emptyset)$ holds.

We give two variants. The first imposes $(\mu\emptyset)$ globally, but does not require the finite subsets to be in the domain, the second needs $(\mu\emptyset)$ only for finite sets, i.e. $(\mu\emptyset fin)$, but finite sets have to be in the domain. Thus, the first is useful for rules of the form $\phi \mid\sim \psi$, the second for rules of the form $T \mid\sim \phi$. The second fully characterizes ranked structures without copies. The first will be needed again below in the limit variant. The proof of the second variant is somewhat more constructive than the proof for the first variant.

We show the first characterization, $(\mu\emptyset)$ is assumed to hold for this result.

Proposition 3.10.11

Let $\mathcal{Y} \subseteq \mathcal{P}(U)$ be closed under finite unions. Then $(\mu \subseteq)$, $(\mu\emptyset)$, $(\mu =)$ characterize ranked structures for which for all $X \in \mathcal{Y}$ $X \neq \emptyset \to \mu_<(X) \neq \emptyset$ hold, i.e. $(\mu \subseteq)$, $(\mu\emptyset)$, $(\mu =)$ hold in such structures for $\mu_<$, and if they hold for some μ, we can find a ranked relation $<$ on U s.t. $\mu = \mu_<$. Moreover, the structure can be choosen $\mathcal{Y}-$smooth.

Proof:

(1) Soundness: For $(\mu \subseteq)$ and $(\mu =)$ see Fact 3.10.8, $(\mu\emptyset)$ is trivial.

(2) Completeness:

Note that by Fact 3.10.9 (3) + (4) $(\mu \parallel)$, $(\mu\cup)$, $(\mu\cup')$ hold.

Define aRb iff $\exists A \in \mathcal{Y}(a \in \mu(A), b \in A)$ or $a = b$. R is reflexive and transitive: Suppose aRb, bRc, let $a \in \mu(A)$, $b \in A$, $b \in \mu(B)$, $c \in B$. We show $a \in \mu(A \cup B)$. By $(\mu \parallel)$ $a \in \mu(A \cup B)$ or $b \in \mu(A \cup B)$. Suppose $b \in \mu(A \cup B)$, then $\mu(A \cup B) \cap A \neq \emptyset$, so by $(\mu =)$ $\mu(A \cup B) \cap A = \mu(A)$, so $a \in \mu(A \cup B)$.

Moreover, $a \in \mu(A)$, $b \in A - \mu(A) \to \neg(bRa)$: Suppose there is B s.t. $b \in \mu(B)$, $a \in B$. Then by $(\mu\cup)$ $\mu(A \cup B) \cap B = \emptyset$, and by $(\mu\cup')$ $\mu(A \cup B) = \mu(A)$, but $a \in \mu(A) \cap B$, contradiction.

Let by Lemma 3.10.7 S be a total, transitive, reflexive relation on U which extends R s.t. $xSy, ySx \to xRy$ (recall that R is transitive and reflexive). Define $a < b$ iff aSb, but not bSa. If $a \perp b$ (i.e. neither $a < b$ nor $b < a$), then, by totality of S, aSb and bSa. $<$ is ranked: If $c < a \perp b$, then by transitivity of S cSb, but if bSc, then again by transitivity of S aSc. Similarly for $c > a \perp b$.

$<$ represents μ and is $\mathcal{Y}-$smooth: Let $a \in A - \mu(A)$. By $(\mu\emptyset)$, $\exists b \in \mu(A)$, so bRa, but (by above argument) not aRb, so bSa, but not aSb, so $b < a$, so $a \in A - \mu_<(A)$, and, as b will then be $<-$minimal (see the next sentence), $<$ is $\mathcal{Y}-$smooth. Let $a \in \mu(A)$, then for all $a' \in A$ aRa', so aSa', so there is no $a' \in A$ $a' < a$, so $a \in \mu_<(A)$.

$\square$

For the following representation result, we assume only $(\mu\emptyset fin)$, but the domain has to contain singletons.

Proposition 3.10.12

Let $\mathcal{Y} \subseteq \mathcal{P}(U)$ be closed under finite unions, and contain singletons. Then $(\mu\subseteq)$, $(\mu\emptyset fin)$, $(\mu =)$, $(\mu \in)$ characterize ranked structures for which for all finite $X \in \mathcal{Y}$ $X \neq \emptyset \rightarrow \mu_<(X) \neq \emptyset$ hold, i.e. $(\mu\subseteq)$, $(\mu\emptyset fin)$, $(\mu =)$, $(\mu \in)$ hold in such structures for $\mu_<$, and if they hold for some μ, we can find a ranked relation $<$ on U s.t. $\mu = \mu_<$.

Proof:

(1) Soundness: See Fact 3.10.8.

(2) Completeness:

Note that by Fact 3.10.9 (3) + (4) $(\mu \parallel)$, $(\mu\cup)$, $(\mu\cup')$ hold for finite sets.

Let $a < b$ iff $a \neq b$ and $\mu(\{a,b\}) = \{a\}$. (Thus, by $(\mu\emptyset fin)$ $a \bot b$ iff $\mu(\{a,b\}) = \{a,b\}$.)

We show:

(a) $<$ is irreflexive,

(b) $<$ is transitive (and thus free from loops),

(c) $<$ is ranked,

(d) $a < b$ iff there is A s.t. $a \in \mu(A)$, $b \in A - \mu(A)$,

(e) $\mu = \mu_<$.

(a) is trivial.

(b) Let $\mu(\{a,b\}) = \{a\}$, $\mu(\{b,c\}) = \{b\}$, then by $(\mu\cup')$ $\mu(\{a,b,c\}) = \{a\}$. By $(\mu \parallel)$, $\mu(\{a,b,c\}) = \mu(\{a,c\}) \parallel \mu(\{b,c\})$, so $\mu(\{a,c\}) = \{a\}$. Thus, $<$ contains no loops.

(c) $a \bot b < c \rightarrow a < c$: By $(\mu\emptyset fin)$ $\mu(\{c\}) = \{c\}$, by $(\mu =)$ $c \notin \mu(\{a,b,c\}) = \mu(\{a,b\}) \parallel \mu(\{c\})$, so $\mu(\{a,b,c\}) = \mu(\{a,b\}) = \{a,b\}$. But $\mu(\{a,b,c\}) = \mu(\{a,c\}) \parallel \mu(\{b\})$, so $\mu(\{a,b,c\}) = \mu(\{a,c\}) \cup \mu(\{b\})$, so $\mu(\{a,c\}) = \{a\}$.

$a \bot b > c \rightarrow a > c$: $b \notin \mu(\{a,b,c\}) = \mu(\{a,b\}) \parallel \mu(\{b,c\})$, so $\mu(\{a,b,c\}) =$

$\mu(\{b, c\}) = \{c\}$. So $\{c\} = \mu(\{a, b, c\}) = \mu(\{a, c\}) \parallel \mu(\{b\})$, so $\mu(\{a, b, c\}) = \mu(\{a, c\}) = \{c\}$.

(d) Suppose there is $A \in \mathcal{Y}$ s.t. $a \in \mu(A)$, $b \in A - \mu(A)$. Then $\{a, b\} \cap \mu(A) \neq \emptyset$, so by $(\mu =)$ $\mu(\{a, b\}) = \{a\}$.

(e) Let $a \in \mu(A)$ and suppose $a \in A - \mu_<(A)$, then there is $b \in A$ s.t. $b < a$, so $\mu(\{a, b\}) = \{b\}$, contradicting $(\mu =)$. Suppose $a \in A - \mu(A)$. Case 1: There is $b \in \mu(A)$. Then by (d) $b < a$, so $a \notin \mu_<(A)$. Case 2: $\mu(A) = \emptyset$. By $(\mu \in)$, there is $b \in A$ $a \notin \mu(\{a, b\})$, so $b < a$, and $a \notin \mu_<(A)$. (This is the only place where we used $(\mu \in)$.)

□

Note that the prerequisites of Proposition 3.10.12 hold in particular in the case of ranked structures without copies, where all elements of U are present in the structure — we need infinite descending chains to have $\mu(X) = \emptyset$ for $X \neq \emptyset$.

We turn now to the general case, where every element may occur in several copies.

Fact 3.10.13

(1) $(\mu \subseteq) + (\mu PR) + (\mu =) + (\mu \cup) + (\mu \in)$ do not imply representation by a ranked structure.

(2) The infinitary version of $(\mu \parallel)$:

$(\mu \parallel \infty)$ $\mu(\bigcup\{A_i : i \in I\}) = \bigcup\{\mu(A_i) : i \in I'\}$ for some $I' \subseteq I$.

will not always hold in ranked structures.

Proof:

(1) Counterexample: Consider $\mu(\{a, b\}) = \emptyset$, $\mu(\{a\}) = \{a\}$, $\mu(\{b\}) = \{b\}$. The conditions hold trivially. This is representable, e.g. by $a_1 \succeq b_1 \succeq a_2 \succeq b_2 \ldots$ without transitivity. (Note that rankedness implies transitivity, $a \preceq b \preceq c$, but not for $a = c$.) But this cannot be represented by a ranked structure: As $\mu(\{a\}) \neq \emptyset$, there must be a copy a_i of minimal rank, likewise for b and some b_i. If they have the same rank, $\mu(\{a, b\}) = \{a, b\}$, otherwise it will be $\{a\}$ or $\{b\}$.

(2) Take as counterexample an infinite descending chain of x_i, $A_i := \{x_i\}$, then $\mu(A_i) = \{x_i\}$, but $\mu(\bigcup\{A_i : i \in I\}) = \emptyset$, so this will not hold.

□

We assume again the existence of singletons for the following representation result.

Proposition 3.10.14

Let $\mathcal{Y}$ be closed under finite unions and contain singletons. Then $(\mu\subseteq)+(\mu PR)+(\mu\parallel)+(\mu\cup)+(\mu\in)$ characterize ranked structures.

Proof:

Note that the construction is similar to the proof of Proposition 3.8 in [Sch96-1].

(1) Soundness: See Fact 3.10.8.

(2) Completeness: By Fact 3.10.9 (5), $(\mu =)$ holds for finite sets, as they are closed under set difference. Set $A := \{a \in U : \mu(\{a\}) \neq \emptyset\}$ and $B := \{a \in U : \mu(\{a\}) = \emptyset\}$.

(A) We work first only with the elements of A.

Define for $a, a' \in A$ $a < a'$ iff $\mu(\{a, a'\}) = \{a\}$.

Note that by $(\mu\parallel)$ for any finite subset $A' \subseteq A$ $\mu(A') \neq \emptyset$.

We show that so far the relation is ranked (and thus also transitive, and free from cycles). But when we look at the proof of Proposition 3.10.12, we see that we work there with finite sets of elements, as possible cycles have finitely many elements, transitivity, etc. are shown by looking at three elements. But for finite subsets of A, we can use Fact 3.10.9 (5), so $(\mu =)$ will hold, and the proof of Proposition 3.10.12 goes through for parts (a)–(c), and (d) will hold by (μPR) and $(\mu\parallel)$ and by definition of A above.

(B) We treat now the $b \in B$.

For $b \in B$, let $A_b := \{a \in A : a \notin \mu(\{a, b\})\}$. (Note that by $(\mu\parallel)$ $a \notin \mu(\{a, b\}) \leftrightarrow \mu(\{a, b\}) = \emptyset$.)

First, we show that

(1) $a \in A_b$, $a' \in A$, $a \bot a' \rightarrow a' \in A_b$

(2) $a \in A_b$, $a' \in A$, $a < a' \rightarrow a' \in A_b$

Proof:

(1) By $a \notin \mu(\{a, b\})$, $a \in \mu(\{a, a'\})$, and $(\mu\cup)$ $\mu(\{a, a', b\}) \subseteq \{b\}$. But by $(\mu\parallel)$ $\mu(\{a, a', b\}) = \mu(\{a', b\}) \parallel \mu(\{a, a'\})$, so $\mu(\{a', b\}) \subseteq \{b\}$.

(2) By $a' \notin \mu(\{a, a'\})$, $a \notin \mu(\{b, a\})$ and (μPR) $\mu(\{a, a', b\}) \subseteq \{b\}$. But by

$(\mu \parallel)$ $\mu(\{a,a',b\}) = \mu(\{a',b\}) \parallel \mu(\{a,a'\})$, so $\mu(\{a',b\}) \subseteq \{b\}$.

Corollary: $A_b \subseteq A_{b'}$ or $A_{b'} \subseteq A_b$ or $A_b = A_{b'}$. (If not, let $a \in A_b - A_{b'}$, $a' \in A_{b'} - A_b$, and consider the three possible cases $a < a'$, $a' < a$, $a \bot a'$.)

Define now $b \sim b'$ iff $A_b = A_{b'}$, and $b \prec b'$ iff $A_b \supset A_{b'}$. Inside each class $[b]_\sim$ order the b_i in some arbitrary total order $\lhd$.

We make infinite descending chains for each b, $b^1 \geq b^2 \geq \ldots$.

Finally, for each $b^i, b'^j, a (b, b' \in B,\ a \in A)$ we set: $b^i < a$ iff $a \in A_b$, $a < b^i$ iff $a \in A - A_b$, $b^i < b'^j$ iff $b \prec b'$ or $b \lhd b'$ (and close the latter two cases among each other under transitivity), to complement the relation defined so far just between elements from A.

Note that all copies of b's have the same behavior wrt. other elements.

We have to show that the new relation is ranked, too, and, finally, that it represents μ.

Rankedness is easily seen by inspection of the cases.

Representation is now trivial: $x \in X - \mu(X) \rightarrow$ by $(\mu \in)$ $\exists x' \in X.x \notin \mu(\{x,x'\}) \rightarrow x' < x$ (as a singleton, or in all copies). If $x \in \mu(X)$, then $x \in A$ and there cannot be any $x' \in X$ s.t. $x \notin \mu(\{x,x'\})$, by (μPR).

□

3.10.3 The limit variant without copies

Introduction:

In this section, we consider structures of the type $(U, \prec)$, where $\prec$ is a ranked relation, without copies. The condition $\emptyset \neq X \subseteq U \rightarrow \mu_\prec(X) \neq \emptyset$ will not necessarily hold (but it will hold for finite X as we have no copies). Instead of considering now minimal elements, we will generalize, and consider initial segments, with the aim of defining in some way $T \mid\sim \phi$, iff ϕ "finally holds" in the T-models, i.e. if we "go down sufficiently far" then ϕ will hold — see the discussion in Section 2.3.1, in particular the definition of MISE in Definition 2.3.1. Obviously, this is a generalization of the minimal case. We will see that rankedness makes this definition amenable. Again, we will first take an algebraic approach, but see later that, as long as we just characterize $\mid\sim$ for formulas, i.e. $\phi \mid\sim \psi$, or, if we have cofinally many definable closed minimizing sets (MISE), we will find an equivalent ranked structure where the minimal variant defines $\mid\sim$. This is in general not true for full theories, as we will also see. Recall that we have shown analogous results for the general case in Section 3.4.1.

3.10.3.1 Representation

We first present the main definition, show some easy but important results, and turn then to an algebraic representation result Proposition 3.10.16 and the discussion of its general technique. A similar technique will be used for revision in the limit case.

We now make the idea of closed minimizing set (MISE) precise for ranked structures, as we can simplify the concept considerably.

In a ranked structure, an initial segment consists of some elements in a first layer, and all the complete layers below. If the first and the minimal layer are the same, this first layer has to be complete. Consequently, as only initial segments interest us, it suffices to consider full layers. To summarize: wlog., we can assume that closed minimizing sets (MISE) are downward closed sets of full layers, or, in other words, just (full) initial segments, as we will call them here for simplicity — this simplifies the task considerably.

Thus, the basic definition is:

Definition 3.10.4

Given a ranked structure, let for $X \subseteq U$

$\Lambda(X) := \{A \subseteq X : \forall x \in X \exists a \in A(a \prec x \text{ or } a = x) \wedge \forall a \in A \forall x \in X(x \prec a \vee x \bot a \rightarrow x \in A)\}$ (A minimizes X and is downward and horizontally closed.)

$\Lambda(X)$ ist thus wlog. the set of MISE for X. Strictly speaking, we have to index Λ by $\prec$, but when the context is clear, we omit it.

We first note some elementary facts.

Remark 3.10.15

In ranked structures, the following hold:

(1) If $\emptyset \neq A \subseteq X$, and $\forall a \in A \forall x \in X(x \prec a \vee x \bot a \rightarrow x \in A)$, then A minimizes X.

(2) Thus, for $X \neq \emptyset$ $\Lambda(X) = \{\emptyset \neq A \subseteq X$: $\forall a \in A \forall x \in X(x \prec a \vee x \bot a \rightarrow x \in A)\}$, $\Lambda(X)$ consists of all nonempty, and downward and horizontally closed subsets of X.

(3) If $\bigcap \Lambda(X) \neq \emptyset$, then $\bigcap \Lambda(X) = \mu(X)$ (where $\mu = \mu_{\prec}$, of course).

(4) If X is finite, $\bigcap \Lambda(X) = \mu(X)$.

(5) If $x, y \in \bigcap \Lambda(X)$, then $x \perp y$.

(6) As the order is fully determined by considering pairs, we can recover all information about Λ by considering $\Lambda(X)$, or, alternatively, $\mu(X)$ for pairs $X = \{a, b\}$ — whenever $\mathcal{Y}$ contains all pairs.

Proof:

(1) Let $x \in X$, then for any $a \in A$ $x \prec a$ or $x \perp a$ or $x \succ a$. In the first two cases $x \in A$.

(3) If $\bigcap \Lambda(X) \neq \emptyset$, then this must be the bottom layer of X, which consists of its minimal elements.

(4) By finiteness, as there are no copies, $\bigcap \Lambda(X) \neq \emptyset$.

(5) By (3).

□

We consider now the following conditions for Λ, which we will use for representation:

Definition 3.10.5

($\Lambda 1$) $\Lambda(X) \subseteq \mathcal{P}(X)$,

($\Lambda 2$) $X \in \Lambda(X)$,

($\Lambda 3$) $X \neq \emptyset \rightarrow \emptyset \notin \Lambda(X)$,

($\Lambda 4$) $A, B \in \Lambda(X) \rightarrow A \subseteq B$ or $B \subseteq A$,

($\Lambda 5$) $A \in \Lambda(X)$, $Y \subseteq X$, $Y \cap A \neq \emptyset \rightarrow Y \cap A \in \Lambda(Y)$,

($\Lambda 6$) if there are X and A s.t. $A \in \Lambda(X)$, $a \in A$, $b \in X - A$, then: $a, b \in Y$ $\rightarrow \exists B \in \Lambda(Y)(a \in B, b \notin B)$,

($\Lambda 7$) $\Lambda' \subseteq \Lambda(X)$, $\bigcap \Lambda' \neq \emptyset \rightarrow \bigcap \Lambda' \in \Lambda(X)$.

A general technique for representing the limit case

We first discuss in abstract terms a general technique to do the limit case, which we will use again for the limit case of revision.

By Remark 3.10.15 (6), we can recover the full information needed to construct the ranked relation $\prec$ by considering μ for pairs. We then construct

$\prec$ using the results for μ. Finally, we have to check that the rest of the information for Λ does not interfere.

It is thus a three step process:

(1) Reflect the limit case down to the finite case.

(2) Use representation for the finite case — we do not need to use the proof, just the result.

(3) Beam the result up to the limit situation (infinite case).

This procedure presupposes, of course, that we can code the limit case by (an infinite amout of) finite information. In the general preferential case, this is impossible — we may need arbitrarily many copies. In the ranked preferential case (where we have essentially only one copy — the case with ω copies is trivial), this is possible.

This is also possible in the revision case (symmetric or not necessarily so). In the revision case, we will have to work with finite sets on the left, too, it is thus more complicated.

The details:

The conditions for the limit case (see Definition 3.10.5) can be separated into four groups, the first two are essentially independent of the particular case:

(a) Trivial conditions like $X \in \Lambda(X)$, conditions $(\Lambda 1)$–$(\Lambda 4)$ in the ranked preferential case.

(b) Conditions which express that the systems are sufficiently rich, conditions $(\Lambda 6)$–$(\Lambda 7)$ in the ranked preferential case.

(c) Conditions which reflect the limit case to the finite one, condition $(\Lambda 5)$ in the ranked preferential case.

(d) Conditions which express the specificities of the finite case — they can either be general ones, which hold for the infinite case, too, or conditions which directly treat the finite case, again condition $(\Lambda 5)$ in the preferential case.

Note that condition $(\Lambda 5)$ thus serves in the ranked preferential case a double purpose. On the one hand, Y can be chosen finite, which permits to go down, on the other hand, it expresses the basic coherence property of ranked preferential models.

We show now that conditions $(\Lambda 1)$–$(\Lambda 7)$ are sound and complete for the limit variant of ranked structures without copies, where the domain is closed under finite unions and contains all finite sets.

Completeness means here the following: If $\prec$ is the relation constructed, Λ the original set of systems satisfying $(\Lambda 1)$–$(\Lambda 7)$, $\Lambda_{\prec}$ the set of $\prec$ –initial segments, then for all $X \in \mathcal{Y}$ $\Lambda(X) \subseteq \Lambda_{\prec}(X)$, and for $A \in \Lambda_{\prec}(X)$ there is $A' \subseteq A$ $A' \in \Lambda(X)$. This is sufficient, as we are only interested in what finally holds.

The main representation result for the general limit case is now the following algebraic characterization, Proposition 3.10.16.

Proposition 3.10.16

$(\Lambda 1)$–$(\Lambda 7)$ are sound and complete for the limit variant of ranked structures without copies, where the domain is closed under finite unions and contains all finite sets.

Proof:

(a) Soundness:

$(\Lambda 1)$–$(\Lambda 7)$ hold for ranked structures and Λ as defined above.

Proof:

$(\Lambda 1)$–$(\Lambda 3)$ are trivial.

$(\Lambda 4)$: Suppose not, so there are $a \in A$–B, $b \in B$–A. But if $a \perp b$, $a \in B$ and $b \in A$, similarly if $a \prec b$ or $b \prec a$.

$(\Lambda 5)$: As $A \in \Lambda(X)$ and $Y \subseteq X$, $Y \cap A$ is downward and horizontally closed. As $Y \cap A \neq \emptyset$, $Y \cap A$ minimizes Y.

$(\Lambda 6)$: Consider $B := \{y \in Y : y \prec b\}$.

$(\Lambda 7)$: $\bigcap \Lambda'$ is downward and horizontally closed, as all $A \in \Lambda'$ are. As $\bigcap \Lambda' \neq \emptyset$, $\bigcap \Lambda'$ minimizes X.

(b) Completeness:

Given Λ, define $\sigma(X) := \bigcap \Lambda(X)$. We first show the following

Fact:

(1) $\sigma(X) \subseteq X$.

Suppose for the following that X, Y are finite.

(2) $\sigma(X) \in \Lambda(X)$,

(3) $X \neq \emptyset \rightarrow \sigma(X) \neq \emptyset$,

(4) $Y \subseteq X$, $\sigma(X) \cap Y \neq \emptyset \rightarrow \sigma(X) \cap Y = \sigma(Y)$.

Proof of this Fact:

(1) trivial.

(2) trivial by finiteness of $\Lambda(X)$ and $(\Lambda 4)$.

(3) trivial by $(\Lambda 3)$ and (2).

(4) By (2) and $(\Lambda 5)$ $\sigma(Y) \subseteq \sigma(X) \cap Y$. By $\sigma(Y) \in \Lambda(Y)$ and $(\Lambda 6)$ there is for each $x \in Y - \sigma(Y)$ some $B_x \in \Lambda(X)$ s.t. $x \notin B_x$. By definition, $\sigma(X) \subseteq \bigcap\{B_x : x \in Y - \sigma(Y)\}$, but $\bigcap\{B_x : x \in Y - \sigma(Y)\} \cap Y \subseteq \sigma(Y)$, so $\sigma(X) \cap Y \subseteq \sigma(Y)$.

Thus, $(\mu \subseteq)$, $(\mu\emptyset fin)$, $(\mu =)$ hold for σ and finite X, Y.

We work now as in the proof of Proposition 3.10.12.

Consider σ for finite sets. By Fact 3.10.9 (3) + (4), $(\mu \parallel)$, $(\mu\cup)$, $(\mu\cup')$ hold for σ and finite sets.

Define $a \prec b$ iff $\sigma(\{a, b\}) = \{a\}$, thus by $(\mu\emptyset fin)$ $a \perp b$ iff $\sigma(\{a, b\}) = \{a, b\}$. As in the proof of Proposition 3.10.12, we show that $\prec$ is irreflexive, transitive, thus free from loops, and ranked.

Moreover, we have $a \prec b \leftrightarrow \exists X, A \in \Lambda(X)(a \in A,\ b \in X - A)$: "$\rightarrow$" is trivial. "$\leftarrow$": Suppose there are X, $A \in \Lambda(X)$ s.t. $a \in A$, $b \in X - A$. By $(\Lambda 5)$, $\{a\} \in \Lambda(\{a, b\})$. As $\Lambda(X)$ is totally ordered by $\subseteq$, $\{b\} \notin \Lambda(\{a, b\})$, and as $\emptyset \notin \Lambda(\{a, b\})$, $\sigma(\{a, b\}) = \{a\}$.

It remains to show that Λ is "almost" $\Lambda_\prec$ for the relation $\prec$, more precisely for all $X \in \mathcal{Y}$:

(5) $\Lambda(X) \subseteq \Lambda_\prec(X)$,

(6) if $A \in \Lambda_\prec(X)$, then there is $A' \subseteq A$, $A' \in \Lambda(X)$.

Proof of (5) and (6):

(5) It suffices to show that each $A \in \Lambda(X)$ is downward and horizontally closed wrt. $\prec$ (it is $\neq \emptyset$ by $(\Lambda 3)$, and thus minimizing). Suppose $a \in A$, $x \in X - A$, $x \prec a$ (or $x \perp a$). Then $\{x\} \in \Lambda(\{a, x\})$ ($\{a, x\} \in \Lambda(\{a, x\})$), contradicting $(\Lambda 5)$.

(6) Let $A \in \Lambda_\prec(X)$. Fix $a \in A$. For each $x \in X - A$, there is by $a \prec x$ and $(\Lambda 6)$ $B_x \in \Lambda(X)(a \in B_x,\ x \notin B_x)$. Thus $A' := \bigcap\{B_x : x \in X - A\} \neq \emptyset$, so $A' \in \Lambda(X)$ by $(\Lambda 7)$, and $A' \subseteq A$.

□

3.10.3.2 Partial equivalence of limit and minimal ranked structures

We define the consequence relation for the limit version of ranked structures, give an introductory Example 3.10.1, show an easy Fact 3.10.17 and a trivialization result Fact 3.10.18. We then turn to the essential difference between theories and formulas on the left of $\mid\sim$, and show that the first is not equivalent to the minimal variant (see, e.g. Example 3.10.2). We then prove our main result in this Section 3.10.3 that the situation with formulas on the left of $\mid\sim$ is equivalent to a minimal ranked structure (Proposition 3.10.19).

We make the central definition for the logical part now precise:

Definition 3.10.6

$T \models_\Lambda \phi$ iff there is $A \in \Lambda(M(T))$ s.t. $A \models \phi$. We shall also write $T \mid\sim \phi$ for $\models_\Lambda$, and $\overline{\overline{T}} := \{\phi : T \models_\Lambda \phi\}$.

Comment:

The problem with the logical variant is that we do not "see" directly the closed sets. We see only the ϕ, but not the A — moreover, A need not be the model set of any theory.

Example 3.10.1

Take an infinite propositional language $p_i : i \in \omega$. We have ω_1 models. (Assume for simplicity CH.)

(1) Take the model m_0 which makes all p_i true, and put it on top. Next, going down, take all models which make p_0 false, and then all models which make p_0 true, but p_1 false, etc. in a ranked construction. So, successively more p_i will become (and stay) true. Consequently, $\emptyset \models_\Lambda p_i$ for all i. But the structure has no minimum, and the logical limit m_0 is not in the set wise limit — but, of course, a model of the theory. (Recall compactness, so $\overline{\overline{T}} = T \cup \{\phi : T \models_\Lambda \phi\}$ is consistent by inclusion, so it has a model, which must be in the set of all $T-$models.)

(2) Take exactly the same set structure, but enumerate the models differently: each consistent formula is made unboundedly often true (this is possible, as each consistent formula has ω_1 many models), so $\emptyset \models_\Lambda \phi$ iff ϕ is a tautology.

The behavior is as different as possible (under consistency — from the empty theory to a consistent complete one).

The first example shows in particular that $M(T \cup \{\phi\})$ need not be closed in $M(T)$, if $T \models_\Lambda \phi$ — the topmost model satisfies ϕ.

□

Note that the situation is quite asymmetric: If $T \models_\Lambda \phi$, then we know that all $\neg\phi$ models are minimized, from some level onward, there will be no more $\neg\phi$ models, but we do not know whether any $\overline{\overline{T}}-$model is very low, as we saw, it might be in the worst position. The best guess we had for a minimal model was the worst one.

Fact 3.10.17

The following laws hold in ranked structures interpreted as in Definition 3.10.6:

(1) $\overline{\overline{T}}$ is consistent, if T is,

(2) $\overline{T} \subseteq \overline{\overline{T}}$,

(3) $\overline{\overline{T}}$ is classically closed,

(4) $T \mid\sim \phi,\ T' \mid\sim \phi \rightarrow T \vee T' \mid\sim \phi$,

(5) If $T \mid\sim \phi$, then $T \mid\sim \phi' \leftrightarrow T \cup \{\phi\} \mid\sim \phi'$.

Proof:

Trivial.

(1) This follows from the fact that $\emptyset \not\in \Lambda(X)$, if $X \neq \emptyset$, nestedness of Λ, and compactness of classical logic.

(2) Trivial by $A \in \Lambda(X) \rightarrow A \subseteq X$.

(3) See the proof of (1).

(4) Let κ be some rank below which all $T-$models satisfy ϕ, κ' likewise for T'. Then $min\{\kappa, \kappa'\}$ will do the job for $M(T) \cup M(T')$.

(5) Suppose ϕ holds in all $T-$models below κ, likewise κ' for ϕ' and T. Then ϕ' will hold in all $T \cup \{\phi\}$ models below $min\{\kappa, \kappa'\}$. Conversely, if ϕ' holds below κ' in all $T \cup \{\phi\}$ models, then, as a subset of $M(T \cup \{\phi\})$ forms an initial segment of the $T-$ models, ϕ' holds finally in all $T-$ models. □

We have a first trivialization result:

Fact 3.10.18

Having cofinally many definable sets in the $\Lambda's$ trivializes the problem — it becomes equivalent to the minimal variant.

Proof:

Suppose each $\Lambda(X)$ contains cofinally many definable sets, let $\Lambda'(X)$ be this subset. Then $\bigcap \Lambda(X) = \bigcap \Lambda'(X)$. As $\Lambda(X)$ is totally ordered by $\subseteq$, by compactness of the standard topology, and $\emptyset \notin \Lambda$, $\bigcap \Lambda'(X) \neq \emptyset$, but then $\emptyset \neq \bigcap \Lambda'(X) = \mu(X)$, so we are in the simple $\mu-$case. □

The following example shows that there is an important difference between considering full theories and considering just formulas (on the left of $\mid\!\sim$). If we consider full theories, we can "grab" single models, and thus determine the full order. As long as we restrict ourselves to formulas, we are much more shortsighted, and see only a blurred picture. In particular, we can make sequences of models to converge to some model, but put this model elsewhere. Suitable such manipulations will pass unobserved by formulas. The example also shows that there are structures whose limit version for theories is unequal to any minimal structure.

Example 3.10.2

Let $\mathcal{L}$ be given by the propositional variables p_i, $i < \omega$. Order the atomic formulas by $p_i \prec \neg p_i$, and then order all sequences $s = < +/-p_0, +/-p_1, \ldots. >$, $i < n \leq \omega$ lexicographically, identify models with such sequences of length ω. So, in this order, the biggest model is the one making all p_i false, the smallest the one making all p_i true. Any finite sequence (an initial segment) $s = < +/-p_0, +/-p_1, \ldots +/-p_n >$ has a smallest model $< +/-p_0, +/-p_1, \ldots +/-p_n, p_{n+1}, p_{n+2}, \ldots >$, which continues all positive, call it m_s. As there are only countably many such finite sequences, the number of m_s is countable, too (and $m_s = m_{s'}$ for different s, s' can happen). Take now any formula ϕ, it can be written as a finite disjunction of sequences s of fixed length $n < +/-p_0, +/-p_1, \ldots +/-p_n >$, choose wlog. n minimal, and denote s_ϕ the smallest (in our order) of these s. E.g., if $\phi = (p_0 \wedge p_1) \vee (p_1 \wedge \neg p_2) = (p_0 \wedge p_1 \wedge p_2) \vee (p_0 \wedge p_1 \wedge \neg p_2) \vee (p_0 \wedge p_1 \wedge \neg p_2) \vee (\neg p_0 \wedge p_1 \wedge \neg p_2)$, $s_\phi =< p_0, p_1, p_2 >$.

(1) Consider now the initial segments defined by this order. In this order, the initial segments of the models of ϕ are fully determined by the smallest (in our order) s of ϕ, moreover, they are trivial, as they all contain the

minimal model $m_s = s_\phi + < p_{n+1}, p_{n+2}, \ldots >$ — where $+$ is concatenation. It is important to note that even when we take away m_s, the initial segments will still converge to m_s — but it is not there any more. Thus, in both cases, m_s there or not, $\phi \models_\Lambda s_\phi + < p_{n+1}, p_{n+2}, \ldots >$ — written a little sloppily. (A more formal argument: If $\phi \models_\Lambda \psi$, with the m_s present, then ψ holds in m_s, but ψ has finite length, so beyond some p_k the values do not matter, and we can make them negative — but such sequences did not change their rank, they stay there.)

(2) Modify the order now. Put all m_s on top of the construction. As there are only countably many, all consistent ϕ will have most of their models in the part left untouched — the m_s are not important for formulas and their initial segments.

To summarize: $\phi \models_\Lambda \psi$ is the same in both structures, as long as we consider just formulas ϕ. Of course, when considering full theories, we will see the difference — it suffices to take theories of exactly two models. Thus, just considering formulas does not suffice to fully describe the underlying structure.

Note that we can add to the information about formulas information about full theories, which will contradict rankedness (e.g., in the second variant, take three models, and make $m \bot m' \prec m''$, but not $m \prec m''$) — but this information will not touch the formula part, as far as formulas are concerned, it stays consistent, as we never miss those models m_s.

Moreover, the reordered structure (in (2)) is not equivalent to any minimal structure when considering full theories: Suppose it were. We have $\emptyset \mid\!\sim +p_i$ for all i, so the whole structure has to have exactly one minimal model, but this model is minimized by other models, a contradiction.

□

We have, however, the following result, which shows equivalence between the limit and the minimal variant (not necessarily with the same relation) for formulas. It is perhaps the most important result of Section 3.10.3. Essentially, this is again a question of domain closure (under set difference). In the proof, we define the choice function on the model set, and show that it has the properties required for representation by the minimal variant. Formulas or full theories makes really a big difference, as the contrast with the strong negative result Proposition 5.2.16 shows.

Proposition 3.10.19

When considering just formulas, in the ranked case without copies, Λ is equivalent to μ — so Λ is trivialized in this case. More precisely:

Let a logic $\phi \mid\sim \psi$ be given by the limit variant without copies, i.e. by Definition 3.10.6. Then there is a ranked structure, which gives exactly the same logic, but interpreted in the minimal variant.

(As Example 3.10.2 has shown, this is NOT necessarily true if we consider full theories T and $T \mid\sim \psi$.)

Proof:

Assume $\mid\sim$ is given by initial segments Λ, i.e. $\phi \mid\sim \psi$ iff ψ finally holds in all initial segments of the $\phi-$models.

We show that, if we define $f(M(\phi)) := M(\overline{\overline{\phi}})$, f has the properties:

$(\mu \subseteq)$ $f(X) \subseteq X$,

$(\mu\emptyset)$ $X \neq \emptyset \rightarrow f(X) \neq \emptyset$,

$(\mu =)$ $X \subseteq Y$, $f(Y) \cap X \neq \emptyset \rightarrow f(X) = f(Y) \cap X$.

Obviously, the set of $M(\phi)$'s is closed under finite unions.

The result is then a consequence of the representation result Proposition 3.10.11.

$(\mu \subseteq)$ and $(\mu\emptyset)$ are trivial.

$(\mu =)$ Assume $M(\psi) \subseteq M(\phi)$ and $M(\overline{\overline{\phi}}) \cap M(\psi) \neq \emptyset$, so $\vdash \psi \rightarrow \phi$ and $Con(\overline{\overline{\phi}}, \psi)$. We show $\overline{\overline{\psi}} = \overline{\overline{\phi}} \cup \{\psi\}$, thus $f(M(\psi)) = M(\overline{\overline{\psi}}) = M(\overline{\overline{\phi}} \cup \{\psi\}) = M(\overline{\overline{\phi}}) \cap M(\psi) = f(M(\phi)) \cap M(\psi)$.

$Con(\overline{\overline{\phi}}, \psi)$ implies $\neg\psi \notin \overline{\overline{\phi}}$, so any initial segment A of $M(\phi)$ contains a $\psi-$model. Thus, $M(\psi) \cap A \neq \emptyset$, and $M(\psi) \cap A$ is an initial segment of $M(\psi)$ by $(\Lambda 5)$. Thus, if $\phi' \in \overline{\overline{\phi}}$, ϕ' will finally hold in $M(\phi)$, so $\phi' \wedge \psi$ will finally hold in $M(\psi)$. Thus, if $\sigma \in \overline{\overline{\phi}} \cup \{\psi\}$, then $\overline{\overline{\phi}} \cup \{\psi\} \vdash \sigma$, so $\overline{\overline{\phi}} \vdash \psi \rightarrow \sigma$, so $\psi \rightarrow \sigma \in \overline{\overline{\phi}}$, so $\psi \wedge (\psi \rightarrow \sigma) \in \overline{\overline{\psi}}$, and $\sigma \in \overline{\overline{\psi}}$. Conversely, if ϕ' holds finally in $M(\psi)$, as any initial segment A' of $M(\psi)$ can be completed to an initial segment A of $M(\phi)$ (complete all levels of A') s.t. $A \cap M(\psi) = A'$, in ϕ, finally $\phi' \vee \neg\psi$ holds. (This is the only place where the fact that ψ is a formula is important.) So $\psi \mid\sim \phi'$ implies $\phi \mid\sim \phi' \vee \neg\psi$, so $\phi' \in \overline{\overline{\phi}} \cup \{\psi\}$.

(The important fact is here the closure of the domain under complements.)

□

Sketch of a characterization:

Finally, we sketch a (not very nice) logical characterization. It will make heavy use of finite sets, as they are the only ones we really have control about. The reader should bear in mind that we need many finite (model) sets to describe arbitrary theories.

(CP) $\perp \notin \overline{T} \rightarrow \perp \notin \overline{\overline{T}}$

(SC)+(CCL) $T \subseteq \overline{\overline{T}} = \overline{(\overline{\overline{T}})}$

(LLE) $\overline{\overline{T}} = \overline{\overline{(\overline{T})}}$

Properties (1)–(2) allow us to construct the relation by considering just theories with two models each, as for such theories T, $\overline{\overline{T}}$ can have one or two models. Property (3) just expresses robustness to syntactic reformulation.

We can now formulate what an initial segment is, and how intial segments behave. This is left to the reader.

(The perhaps main observation is that if $T \models_{\Lambda} \phi$, then there must be some $m \models T \cup \{\phi\}$ which is smaller than all $\neg\phi-$models in T, i.e. $Th(m) \vee (T \cup \{\neg\phi\}) \models \phi$. The converse is also true.)

Chapter 4

Distances

4.1 Introduction

We discuss in this chapter distance based revision and counterfactual conditionals. Again, there are different levels of reading for this chapter. The main results are given in Propositions 4.2.2, 4.2.5, 4.2.9, 4.2.10, 4.2.11, 4.2.12 for revision, Proposition 4.3.1 for counterfactual conditionals. The result showing equivalence between the minimal and the limit version (Proposition 4.2.12) is, of course, very close to Propositions 3.4.7 and 3.10.19 for preferential structures. Again, it reflects the importance of closure conditions, as things which are equivalent for formulas (and thus sets of subsets closed under complementation) are not equivalent any more in the more general case of theories. This result is not really surprising once we have seen the corresponding result for preferential structures. The lack of finite characterization, however, came as quite a surprise to the author, and it is again a very nice illustration of the importance of closure conditions of the domain. Given a rich enough domain, we can reflect a local hidden result (comparison of distances) on a more distant one, and thus compare. Obviously, this result and its proof and analysis are destined for the more advanced reader, who will prove his/her own results. The same applies to the abovementioned equivalence result.

The main result on counterfactual conditionals has again two levels of reading. On the one side, it is a perhaps somewhat surprising result about independent distances. On the other and perhaps deeper level, it illustrates the fact, that as long as we are only interested in the things closest to us — which hide everything else from sight — we have more liberty than if we look behind the screens. Seen like this, it is not surprising: if we blind ourselves, we see less clearly, or: shortsighted people do not see far enough.

The more advanced results and considerations are, in this chapter, perhaps less well separated from the basic ones, than in Chapter 3 on preferential structures.

We recall that the main cleavage in this chapter is between the collective approach to distance (theory revision) and the individual one (counterfactuals), see Definition 2.3.5. (The results on theory update, also the individual distance variant, are to be found in the chapter on sums.)

The lack of finite representability will reappear in several other occasions in this book. We have not grouped these negative results together, as the techniques used for their demonstration are quite different in each case (apart from the basic idea: construct arbitrarily big arbitrarily similar negative and positive examples). Their natural place is with the corresponding positive results.

Recall that the case of not necessarily definability preserving structures is discussed in Chapter 5, where the common approach for preferential structures and distance based revision will be elaborated. We also show there that general, not necessarily definability preserving, distance based revision has no "normal" characterization of any size.

First, the basic definitions for distance based revision and distance based counterfactuals.

Given some set U, and a distance d on U, and $A, B \subseteq U$, let

(1) $A \mid B := \{b \in B : \exists a_b \in A(\forall a' \in A, b' \in B.d(a_b, b) \leq d(a', b'))\}$. Thus, $A \mid B$ is the set of $B-$elements collectively closest to A, and

(2) $A \uparrow B := \{b \in B : \exists a_b \in A(\forall b' \in B.d(a_b, b) \leq d(a_b, b'))\}$. Thus, $A \uparrow B$ is the set of $B-$elements individually closest to A.

Given these operations, we define revision (collectively) by

$T * \phi := Th(M(T) \mid M(\phi))$, and

the counterfactual $\phi \Rrightarrow \psi$ by

$m \models \phi \Rrightarrow \psi$ iff $\{m\} \mid M(\phi) \subseteq M(\psi)$, where we distribute for model sets M as variant (2) dictates.

Note that in both cases, we are only interested in the closest elements, and this is all we see. This shortsightedness has the positive consequence in the case of counterfactuals that we can manipulate the distances as done in Section 4.3, and the negative consequence in the case of revision that we cannot have a finite characterization (unless we have sufficiently many "mirrors"), resulting in lack of finite characterization (Proposition 4.2.11). Now to details, and first to theory revision.

4.1.1 Theory revision

Recall that we gave the main concepts and definitions of theory revision in Section 2.2.10. We assume familiarity with the basics of the theory.

Theory revision speaks about minimal change, so, it is natural to look for a distance based semantics, and in hindsight, it is surprising that this had not been done earlier. The distance measures the change, and we look for the closest situations measured by that distance, i.e. for those, which are minimally different from the original situation. Perhaps not all changes are comparable, perhaps there is no "best" situation by this measure (only an infinite chain of ever better situations), these are possible complications.

The original AGM approach does not consider such "limit" situations, this is encoded by the consistency postulate, a limit condition, (if T, ϕ are consistent, so will be $T * \phi$), comparability by the relatively strong axioms (K7) and (K8), which together are rankedness, and rankedness says that all things are equal or comparable.

The consistency condition is a limit condition also in the technical sense, as its generalization leads naturally to consider the limit approach, this time not in (ranked) preferential structures, but in the more complex structures with distances. They are more complex, as the left point from which we measure can change now. We can then hope to find a similar result as for ranked preferential structures, which shows that as long as we work just with formulas, the limit and the minimal variant coincide — and, as a matter of fact, this is true, there is such a result.

Distance based revision goes beyond the AGM approach in a trivial but conceptually (and for applications) very important point: it also imposes conditions on iterated revision, whereas the AGM revision conditions never change the K, and so cannot say anything about iterated revision. We give an example for this extra expressivity: For instance, if A' is the set of $A-$elements closest to B, then the sets of elements in B closest to A and to A' are the same — provided the distance is symmetrical.

This increased power is especially important if we want to put theory revision into the object language, to have nestedness, boolean connectives, cooperation with other operators, etc.

When we fix the left hand side, AGM revision corresponds exactly to rankedness, or, a distance from a fixed point. Thus, the author had naively hoped that it might be possible to have separate sets of conditions for the right hand side, for the left hand side, and one or two which fuse both sides together, to make a (finite) set of conditions for variation on the left and right, i.e. a full distance based revision. The negative result (no finite characterization is possible, Proposition 4.2.11) shows that this hope was indeed very naive, the enterprise is impossible.

We discuss now the results to be presented.

We first give the basic definitions, in particular what a distance based revision is to be:

Given a distance d on the set of models, we define the revision based on this distance by

$T * \phi := Th(\{m \in M(\phi) : d(M(T), m) \leq d(M(T), m')$ for all $m' \in M(\phi)\})$
$= Th(M(T) \mid M(\phi))$

where $d(X, m) := min\{d(x, m) : x \in X\}$.

Given some caveats, it is easy to see that theory revision so defined satisfies the AGM postulates. This is shown in Fact 4.2.8 for completeness' sake.

The formal results to be shown differentiate between various distances — symmetrical, not necessarily symmetrical, respecting 0, etc. In the most general case, a distance is just a set totally ordered by some relation $<$ (i.e. the set of equivalence classes defined by a ranked set).

As usual, we first give an algebraic characterization of the associated choice functions, and turn to the logics part only later.

The central definition, given a distance d, is thus in algebraic terms

$X \mid Y := \{y \in Y : d(X, y) \leq d(X, y')$ for all $y' \in Y\}$

the set of $y \in Y$, (collectively) closest to X.

The symmetric case:

Before we start the completeness results, we give an example (Example 4.2.1) which shows that distances cannot always be compared, provided they have no common starting (or end) point. This example first just seemed somewhat annoying, as we have to "fill gaps" in the construction of the relation in a completeness proof by sufficiently judicious guessing (which is done by the abstract nonsense result 3.10.7 in Section 3.10.1). As a matter of fact, its consequences are far more serious, they prevent a finite characterization, as will be shown in Proposition 4.2.11.

We then attack the completeness proof for symmetric distances. Its main condition is a very elegant loop condition (due to M. Magidor):

$(\mid S1)$ (Loop):

$(X_1 \mid (X_0 \cup X_2)) \cap X_0 \neq \emptyset,$

$(X_2 \mid (X_1 \cup X_3)) \cap X_1 \neq \emptyset,$

$(X_3 \mid (X_2 \cup X_4)) \cap X_2 \neq \emptyset,$

....

$(X_k \mid (X_{k-1} \cup X_0)) \cap X_{k-1} \neq \emptyset$

imply

$(X_0 \mid (X_k \cup X_1)) \cap X_1 \neq \emptyset.$

Its elegance is also its defect: in a certain way, it encodes too many things at the same time: It codes symmetry by changing the starting point from left to right, it codes freedom from loops (or transitivity) in the order, and it works with the main diagnostic for determining distances. Its other defect lies in its unlimited size. As we know now, there is no real alternative: finite conditions will not suffice.

We turn to the construction of the distance.

As the distance will be symmetrical, we define (in Definition 4.2.6):

Set $\| A, B \| \leq \| A, B' \|$ iff $(A \mid B \cup B') \cap B \neq \emptyset,$

set $\| A, B \| < \| A, B' \|$ iff $\| A, B \| \leq \| A, B' \|$, but not $\| A, B \| \geq \| A, B' \|$.

We show a number of important properties in Fact 4.2.3, and define in Definition 4.2.7 an extension S of $\leq$ on the $\| A, B \|$'s, the distances between the different A, B, using the abstract nonsense result 3.10.7, and show that the restriction to singletons is a symmetric distance in Fact 4.2.4, which generates the original operation $\mid$. The proofs are quite straightforward.

The not necessarily symmetric case:

The not necessarily symmetric case is somwehat more complicated, and we present a characterization of the finite case only. (The infinite case seems to need an additional limit condition on the left.)

We define a relation R or $R_\mid$ from a binary set operator $\mid$ (essentially) by:

$(A, B)R(A', B')$ iff one of the following two cases obtains:

(1) $A = A'$ and $(A \mid B \cup B') \cap B \neq \emptyset,$

(2) $B = B'$ and $(A \cup A' \mid B) \neq (A' \mid B),$

(see Definition 4.2.8).

We consider then the following (complete) set of conditions in Proposition 4.2.5

$(\mid 1)$ $(A \mid B) \subseteq B$,

$(\mid A1)$ $(A \cup A' \mid B) \subseteq (A \mid B) \cup (A' \mid B)$,

$(\mid A2)$ If $(A, B)R^*(A, B')$, then $(A \mid B) \subseteq (A \mid B \cup B')$,

$(\mid A3)$ If $(A, B)R^*(A', B)$, then $(A \mid B) \subseteq (A \cup A' \mid B)$,

$(\mid 2)$ If $A \cap B \neq \emptyset$, then $A \mid B = A \cap B$,

$(\mid A4)$ If $(A, B)R^*(A', B')$ and $A' \cap B' \neq \emptyset$, then $A \cap B \neq \emptyset$. (R^* is the transitive closure of R.)

The fact that the conditions are on R, and not directly on $\mid$, as might be expected (and as it would certainly be nicer) reflects again the difficulty to observe pertinent results, therefore we treat R and its compositions R^* as a black box, whose operations we cannot observe directly on the domain.

The proof of completeness is a little involved. We first extend R as usual by our abstract nonsense result 3.10.7 to a total preorder S, consider the $S-$equivalence classes, and set $d(A, B) :=$ the $S-$equivalence class of (A, B).

We then show in Lemma 4.2.7: For any A, B $d(A, B) = min\{d(A, b) : b \in B\}$, $A \mid B = \{b \in B : d(A, b) = d(A, B)\}$, and $A \mid B = A \mid_d B$.

The logical characterization:

We first show the importance of definability preservation, which means, in this context, of course, that $M(T) \mid M(\phi)$ has again to be $M(\psi)$ for some ψ. Example 4.2.3 shows that the translation of the main (loop) condition need not hold when the structure is not definability preserving. It also shows that the AGM properties need not all hold in this case. This example is very similar to the one for preferential structures, and illustrates the "coarseness" of logic — we might not detect small sets which are missing. The remedy, to be shown in Chapter 5 follows the same strategy in both cases, i.e. to admit small sets of exceptions. On the other hand, we show in Section 5.2.3 that no "normal" characterization is possible.

Apart from this problem, the logical characterization is quite straightforward, and more detailed comments are not needed.

We now turn to two natural questions:

First, do we really need such complicated conditions as the loop condition to characterize distance based revision when we can modify the right and left hand of the operator?

Second, is there a way to avoid the somewhat unnatural limit condition that $X \mid Y \neq \emptyset$ if $X, Y \neq \emptyset$? After all, infinite approching chains of models might very well happen in the infinite case.

The answer to the first question is negative, and came as somewhat a surprise to the author, after having tried in vain to find finite conditions.

The answer to the second question has two parts: Yes, it is possible, there is a natural extension of the definition — which amounts to a limit version, just as in the preferential structure case — and we give a characterization of the general case, and, perhaps more importantly, if we just look at revisions of the type $\phi * \psi$, i.e. with a formula on the left of $*$ (even if the language is infinite), there is an equivalent structure in the minimal version. The strategy we follow is in close parallel to the one for ranked preferential structures, and we recommend to the reader to read first the part on the limit variant of ranked preferential structures, as the situation is simpler there.

We present now these two problems in more detail.

First to the impossibility to find finite conditions (Section 4.2.4).

We can read the loop condition (roughly) as a transitivity condition: if $x_1 \leq x_2 \leq \ldots \leq x_n$, then $x_1 \leq x_n$. Unfortunately, we cannot replace this with a condition of length two: if $x \leq y \leq z$, then $x \leq z$. The reason is that we may well be able to see $x \leq y$, and $y \leq z$, but not be able to determine $x \leq z$, as $x \leq z$ is "hidden" behind closer elements — see the Example 4.2.1 for an illustration. Thus, we cannot take "shortcuts", but have to write down arbitrarily long chains of conditions. The idea is thus very simple, but we have to take a little care in writing the details. To make a formal proof, we have to find examples of arbitrarily big size, which can only be described by arbitrarily big conditions, i.e. we need arbitrarily much information to determine whether the structures are distance representable or not. In particular, we have to exclude that "good", i.e. representable, structures can be differetiated from "bad" structures by some other "bit" of information than those long chains of $x_1 \leq x_2 \leq \ldots \leq x_n$.

For this purpose, we construct arbitrarily big "hamster wheels", bad examples, which, by just one single change can be transformed into good examples, which are otherwise totally identical to the bad case. So we need to know all of the information to determine that the examples are bad ones. (The situation is slightly more complicated, as we have to work with sets of elements, but the only interesting sets are small, i.e. have at most two elements.) The main work to do is to show that the good and the bad examples really differ by just one bit of information about the revision.

Note that the construction depends heavily on the fact that we might be unable to determine (relative) distances. If the domain is sufficiently rich, we can mirror distances away from interfering elements, and determine them indirectly — and finite characterization becomes possible. So, we have again a case where closure conditions of the domain are crucial — this time not algebraic closure conditions, but richness and homogeneity of the domain.

We turn to the second problem, the limit condition discussed in Section 4.2.5. We give a short introduction to the question, and then turn immediately to the equivalence problem. We show here that the logical version of the completeness conditions (see Proposition 4.2.9) are satisfied for the limit approach to distance defined revision, as long as we consider only revisions of the type $\phi * \psi$, i.e. with formulas on the left of $*$. The main property to show is the loop condition. For this, we show $Con(\phi, \phi' * (\phi \vee \phi''))$ iff $\forall A \in \Lambda(M(\phi'), M(\phi \vee \phi'')).M(\phi) \cap A \neq \emptyset$. ($\Lambda$ is the MISE analogue, and the reader is referred to Section 4.2.5 for details.) But, if Λ is distance defined, this is equivalent to: $\forall x' \in M(\phi')\forall y' \in M(\phi \vee \phi'')\exists x \in M(\phi')\exists y \in M(\phi).d(x,y) \leq d(x',y')$. The rest is easy.

4.1.2 Counterfactuals

The original paper by Lewis [Lew73] allowed that the metrics determining the distance from each point to other points may be totally different metrics for each point of origin. This may very well be justified philosophically, mathematically it is a very complicated structure. Of course, one single metric imposes conditions which need not hold if the metrics are independent: For example, a single metric forces $a \prec_b c$ whenever $b \prec_a c$ and $a \prec_c b$, since $d(a,b) < d(a,c) = d(c,a) < d(c,b)$ implies by the transitivity of $<$ and symmetry of a metric that $d(b,a) = d(a,b) < d(c,b) = d(b,c)$. ($a \prec_b c$ stands for: seen from b, a is closer than c is.) But such coherences need not exist if we have independent metrics. Thus, at first sight, the resulting structures are quite different. We show that, indeed, they are not.

The idea is the following: As we are interested only in the closest worlds — or, more precisely, the counterfactual semantics is defined only by what holds in the closest worlds — everything behind the closest worlds is hidden from view. We "recreate" the world now as seen from each world in the original structure, by making as many copies as necessary, and taking care to choose distances in a way that the situation as seen from other worlds is hidden from view.

For instance, we begin arbitrarily with world w. We arrange the rest of the points around w just as w's metric dictates. Take now w'. Now, we arrange all worlds (new copies, more precisely) different from w', just as w' dictates, but take care to make the distances smaller, so we do not see the old w and what we did around w. As we are only interested in closest worlds, what we did for w is invisible. If w'' is one of the new worlds (perhaps a new copy of w), we repeat the same procedure: arrange everything around w'' as w'' likes to see it, with new copies, and again smaller distances so we see neither the old w, nor the old w' — they are too far away. Thus, the idea is very simple, and we just have to take a little care that everything works as it is intended, i.e. there is no interference between the different "galaxies" we created.

In the end, we have to show that the same counterfactuals hold in the new and the old structure — but this is then trivial.

4.1.3 Summary

We first show a number of results on distance based theory revision. More precisely, to revise T by ϕ, we consider the $\phi-$models closest to the T-models as the models of $T * \phi$. We start by algebraic characterizations of the symmetric and not necessarily symmetric case, for the latter we treat only the finite case, as it seems to necessitate a second limit condition on the left in the infinite case. The symmetric case is described by a simple, elegant loop condition, which, however, can be arbitrarily long. Attempts to find simpler conditions have failed, and for a reason: it can be shown that we need infinite conditions, see below. The translation to logic is again straightforward, provided we treat definability preserving structures.

We then show by a class of examples that we really need arbitrarily complex conditions. These examples show that the difference between distance representable and not distance representable cases can be "infinitely" small, more precisely, we can find bigger and bigger "bad" examples, which differ from "good" ones just by one result of revision, so the difference is in comparison arbitrarily small. Taking some care about the language results in the proof of what we wanted to show.

We introduce again a limit version, very close to the limit version of ranked structures, and show that, for formulas on the left, this version which gets rid of the somewhat unnatural limit condition (there must be closest elements), is again equivalent to the minimal variant presented above.

Finally, we show the positive counterpart of the negative result about finitary characterizations: for counterfactual conditionals, we can fuse many independent metrics defining a semantics to one single metric. The price we have to pay is a (large) number of copies. The reason it works is that anything which could cause trouble is hidden behind closer elements.

Recommended reading:

The reader might start either with revision (Section 4.2) or counterfactuals (Section 4.3), they are independent. The section on counterfactuals should be read as a whole (or not at all), there is no natural first part. In the section on revision, the reader should first read the algebraic part (Section 4.2.2), and there the introduction and the symmetric variant. Next, he should perhaps read the logical characterization (Section 4.2.3), and only then come back to the not necessarily symmetric part — but this can wait, the "asymmetric" part is not needed for any other results. The reader who has already read the limit variant of ranked structures, might now read the limit variant of revision (Section 4.2.5). An, in the author's opinion, interesting result is the one on lack of finite characterizations (Section 4.2.4). The reader can read this part as soon as he or she has read the algebraic part on revision, and as long as his or her memory on this part is still fresh.

4.2 Revision by distance

4.2.1 Introduction

We characterize here distance based revision. We first take the purely algebraic approach, and only then turn to logic. Again, we begin by the minimal variant, and discuss the limit variant only later. Again, both approaches are equivalent for formulas, and diverge only for full theories. Problems of definability preservation pose themselves again, and are treated together with those of preferential structures later (see Chapter 5). There is, however, an important difference to preferential structures: By the very definition (we look at closest elements), some information might hide other information, unless the domain is sufficiently rich, see Example 4.2.1. Consequently, transitivity is not always directly observable, resulting in complicated conditions, and a negative result: there is no finite characterization (in a reasonable sense) of distance based revision possible.

The reader should be familiar with the basic concepts and definitions of theory revision, as summarized in Section 2.3.2.

4.2.2 The algebraic results

We give now first an abstract definition of a distance, and say what we mean by distance representation. We will then give an algebraic representation result, and transform this result in a more or less straigtforward manner only later to its logical counterpart. We turn to the symmetric case, give an — in hindsight — important example (Example 4.2.1) which shows that we cannot know everything, formulate the algebraic conditions for the binary choice function $|$, construct a first relation comparing distances, show its main properties, and extend it to a distance. The proof that this distance represents the function $|$ is then easy. We then treat the not necessarily symmetric case. We first define a relation R from the binary choice function $|$, formulate appropriate conditions for this relation, and show that we can recover $|$. Again, we extend R to construct the distance d, and show that d does what it should.

4.2.2.1 Introduction and pseudo-distances

We give here the basic definition. For a first reading, the reader may assume that d is a normal distance, but perhaps not symmetric.

We will base our semantics for revision on pseudo-distances between models. Pseudo-distances differ from distances in that their values are not necessarily reals, no addition of values has to be defined, and symmetry need not hold. All we need is a totally ordered set of values. If there is a minimal element 0 such that $d(x, y) = 0$ iff $x = y$, we say that d respects identity. Pseudo-distances which do not respect identity have their interest in situations where staying the same requires effort.

We first recollect:

Definition 4.2.1

A binary relation $\leq$ on X is a preorder, iff $\leq$ is reflexive and transitive. If $\leq$ is in addition total, i.e. iff $\forall x, y \in X$ $x \leq y$ or $y \leq x$, then $\leq$ is a total preorder.

A binary relation $<$ on X is a total order, iff $<$ is transitive, irreflexive, i.e. $x \not< x$ for all $x \in X$, and for all $x, y \in X$ $x < y$ or $y < x$ or $x = y$.

Remark 4.2.1

If $\leq$ is a total preorder on X, $\approx$ the corresponding equivalence relation defined by $x \approx y$ iff $x \leq y$ and $y \leq x$, $[x]$ the $\approx$ $-$equivalence class of x, and we define $[x] < [y]$ iff $x \leq y$, but not $y \leq x$, then $<$ is a total order on $\{[x] : x \in X\}$.

Definition 4.2.2

$d : U \times U \to Z$ is called a pseudo-distance on U iff (d1) holds:

(d1) Z is totally ordered by a relation $<$.

If, in addition, Z has a $<-$smallest element 0, and (d2) holds, we say that d respects identity:

(d2) $d(a, b) = 0$ iff $a = b$.

If, in addition, (d3) holds, then d is called symmetric:

(d3) $d(a, b) = d(b, a)$.

(For any $a, b \in U$.)

Let $\leq$ stand for $<$ or $=$.

Note that we can force the triangle inequality to hold trivially (if we can choose the values in the real numbers): It suffices to choose the values in the set $\{0\} \cup [0.5, 1]$, i.e. in the interval from 0.5 to 1, or as 0.

Definition 4.2.3

Given a pseudo-distance $d : U \times U \to Z$, let for $A, B \subseteq U$

$A \mid_d B := \{b \in B : \exists a_b \in A \forall a' \in A \forall b' \in B.d(a_b, b) \leq d(a', b')\}$

Thus, $A \mid_d B$ is the subset of B consisting of all $b \in B$ that are closest to A. Note that, if A or B is infinite, $A \mid_d B$ may be empty, even if A and B are not empty. A condition assuring nonemptiness will be imposed when necessary. (The limit version gets rid of such nonemptiness conditions.)

The aim of this Section 4.2.2 is to characterize those operators $\mid : \mathcal{P}(U) \times \mathcal{P}(U) \to \mathcal{P}(U)$, for which there is a pseudo-distance d, such that $A \mid B = A \mid_d B$. We call such $\mid$ representable:

Definition 4.2.4

An operation $\mid$ is representable iff there is a pseudo-distance $d : U \times U \to Z$ such that

(1) $A \mid B = A \mid_d B := \{b \in B : \exists a_b \in A \forall a' \in A \forall b' \in B(d(a_b, b) \leq d(a', b'))\}$.

The following is the central definition, it describes the way a revision $*_d$ is attached to a pseudo-distance d on the set of models.

Definition 4.2.5

$T *_d T' := Th(M(T) \mid_d M(T'))$.

$*$ is called representable iff there is a pseudo-distance d on the set of models s.t. $T * T' = Th(M(T) \mid_d M(T'))$.

In the following two Sections, 4.2.2.2 and 4.2.2.3, we first present a result (Example 4.2.1) which shows that some distances may be hidden by closer elements, and thus invisible. This fact has much deeper importance later on, when we show that in the general case, there is no finite characterization for distance based revision. We then formulate in Proposition 4.2.2 the conditions for the binary choice function $\mid$, where the loop property is the central one. This property is both elegant and nasty. Elegant, because it codes many things at the same time, and nasty, because it is arbitrarily big. When trying to find smaller conditions, the author noted that this is impossible, see Section 4.2.4. The main, and obvious step in the construction is in Definition 4.2.6, where we define (already essentially) the distance relation. The auxiliary Fact 4.2.3 shows its main properties, and the relation is extended in Definition 4.2.7 to the final distance. The proof that this distance represents $\mid$ is now easy (Fact 4.2.4).

Note that the algebraic representation results we are going to demonstrate in this Section 4.2.2 are independent of logic, and work for arbitrary sets U, not only for sets of models. On the other hand, if the (propositional) language $\mathcal{L}$ is defined from infinitely many propositional variables, not all sets of models are definable by a theory: There are $X \subseteq M_{\mathcal{L}}$ s.t. there is no T with $X = M(T)$, see Fact 1.6.2 for an example. Moreover, we will consider only consistent theories. This now motivates the following framework:

Let $U \neq \emptyset$, and let $\mathcal{Y} \subseteq \mathcal{P}(U)$ contain all singletons, be closed under finite $\cap$ and finite $\cup$, $\emptyset \notin \mathcal{Y}$ and consider an operation $\mid : \mathcal{Y} \times \mathcal{Y} \to \mathcal{Y}$. (For our representation results, finite $\cap$ suffices.)

We first characterize those operations $\mid$ which can be represented by a symmetric pseudo-distance, and then those representable by a not necessarily symmetric pseudo-distance.

Notation 4.2.1

For $a \in U$, $X \in \mathcal{Y}$ $a \mid X$ will stand for $\{a\} \mid X$, etc.

4.2.2.2 The representation results for the symmetric case

We work here with possibly infinite, but nonempty U.

Note that even when the pseudo-distance is a real distance, the resulting revision operator $|_d$ does not always permit to reconstruct the relations of the distances: revision is a coarse instrument to investigate distances.

Distances with common start (or end, by symmetry) can always be compared by looking at the result of revision:

$a \mid_d \{b, b'\} = b$ iff $d(a,b) < d(a,b')$,

$a \mid_d \{b, b'\} = b'$ iff $d(a,b) > d(a,b')$,

$a \mid_d \{b, b'\} = \{b, b'\}$ iff $d(a,b) = d(a,b')$.

This is not the case with arbitrary distances $d(x,y)$ and $d(a,b)$, as the following example will show.

Example 4.2.1

We work in the real plane, with the standard distance, the angles have 120 degrees. a' is closer to y than x is to y, a is closer to b than x is to y, but a' is farther away from b' than x is from y. Similarly for b, b'. But we cannot distinguish the situation $\{a, b, x, y\}$ and the situation $\{a', b', x, y\}$ through $|_d$. (See Figure 4.2.1.)

Proof:

Seen from a, the distances are in that order: y, b, x.

Seen from a', the distances are in that order: y, b', x.

Seen from b, the distances are in that order: y, a, x.

Seen from b', the distances are in that order: y, a', x.

Seen from y, the distances are in that order: $a/b, x$.

Seen from y, the distances are in that order: $a'/b', x$.

Seen from x, the distances are in that order: $y, a/b$.

Seen from x, the distances are in that order: $y, a'/b'$.

Thus, any $c \mid_d C$ will be the same in both situations (with a interchanged with a', b with b'). The same holds for any $X \mid_d C$ where X has two elements.

Thus, any $C \mid_d D$ will be the same in both situations, when we interchange a with a', and b with b'. So we cannot determine by $|_d$ whether $d(x,y) > d(a,b)$ or not. □

The following Proposition 4.2.2 is the main result for the symmetric case. It

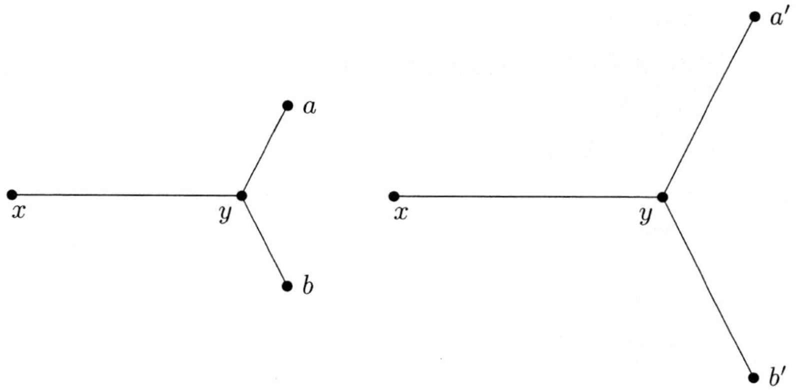

Figure 4.2.1

gives a characterization of distance definable choice functions $\mid$. The central (and almost only) condition is the loop condition $(\mid S1)$.

Proposition 4.2.2

Let $U \neq \emptyset$, $\mathcal{Y} \subseteq \mathcal{P}(U)$ be closed under finite $\cap$ and finite $\cup$, $\emptyset \notin \mathcal{Y}$.

Let $A, B, X_i \in \mathcal{Y}$.

Let $\mid : \mathcal{Y} \times \mathcal{Y} \to \mathcal{Y}$, and consider the conditions

$(\mid 1)$ $A \mid B \subseteq B$

$(\mid 2)$ $A \cap B \neq \emptyset \to A \mid B = A \cap B$

$(\mid S1)$ (Loop):

$(X_1 \mid (X_0 \cup X_2)) \cap X_0 \neq \emptyset$,

$(X_2 \mid (X_1 \cup X_3)) \cap X_1 \neq \emptyset$,

$(X_3 \mid (X_2 \cup X_4)) \cap X_2 \neq \emptyset$,

...

$(X_k \mid (X_{k-1} \cup X_0)) \cap X_{k-1} \neq \emptyset$

imply

$(X_0 \mid (X_k \cup X_1)) \cap X_1 \neq \emptyset.$

(a) $\mid$ is representable by a symmetric pseudo-distance $d : U \times U \to Z$ iff $\mid$ satisfies $(\mid 1)$ and $(\mid S1)$.

(b) $\mid$ is representable by an identity respecting symmetric pseudo-distance $d : U \times U \to Z$ iff $\mid$ satisfies $(\mid 1)$, $(\mid 2)$, and $(\mid S1)$.

Note that $(\mid 1)$ corresponds to $(*2)$, $(\mid 2)$ to $(*3)$, $(*0)$ will hold trivially, $(*1)$ holds by definition of $\mathcal{Y}$ and $\mid$, $(*4)$ will be a consequence of representation. $(\mid S1)$ corresponds to:

$d(X_1, X_0) \leq d(X_1, X_2),$

$d(X_2, X_1) \leq d(X_2, X_3),$

$d(X_3, X_2) \leq d(X_3, X_4) \leq$

...

$\leq d(X_k, X_{k-1}) \leq d(X_k, X_0)$

$\to$

$d(X_0, X_1) \leq d(X_0, X_k),$

and, by symmetry,

$d(X_0, X_1) \leq d(X_1, X_2) \leq$

...

$\leq d(X_0, X_k)$

$\to$

$d(X_0, X_1) \leq d(X_0, X_k),$

i.e. transitivity, or to absence of loops involving $<$.

Note:

The loop condition is simple (always fixed on one side) for the following reason: In the classical, observable (!), 2x2 case, $A = \{a, a'\}$, $B = \{b, b'\}$, $d(a', b') < d(a, b)$ decomposes into: $d(b', a) > d(b', a') = d(\{a, a'\}, b') < d(\{a, a'\}, b) = d(b, a)$.

We first show the hard direction via a number of auxiliary definitions and lemmas (up to Fact 4.2.4). We assume all A, B, etc. to be in $\mathcal{Y}$, and $(\mid 1)$, $(\mid S1)$ to hold from now on.

We define first:

Definition 4.2.6

Set $\| A, B \| \leq \| A, B' \|$ iff $(A \mid B \cup B') \cap B \neq \emptyset$,

set $\| A, B \| < \| A, B' \|$ iff $\| A, B \| \leq \| A, B' \|$, but not $\| A, B \| \geq \| A, B' \|$.

$\| A, B \|$ is to be read as the pseudo-distance between A and B or between B and A. Recall that the pseudo-distance will be symmetric, so $\| \ldots \|$ operates on the unordered pair $\{A, B\}$. Note that $A \mid B \neq \emptyset$, by definition of the function $\mid$.

Let $\leq^*$ be the transitive closure of $\leq$, we write also $<^*$ if it involves $<$. Write $\| a, B \|$ for $\| \{a\}, B \|$, etc.

The loop condition reads in the $\|$ $-$notation as follows:

$\| X_0, X_1 \| \leq \| X_2, X_1 \| \leq \| X_2, X_3 \| \leq \| X_4, X_3 \| \leq \ldots \leq \| X_k, X_{k-1} \| \leq$
$\| X_k, X_0 \| \rightarrow \| X_0, X_1 \| \leq \| X_0, X_k \|$

The following Fact 4.2.3 describes the main properties of $<$ and $\leq$, it is the key to the representation result.

Fact 4.2.3

(1) $\| A, B \| \not\leq \| A, B' \|$ iff $\| A, B' \| < \| A, B \|$.

(2) $B' \subseteq B \rightarrow \| A, B \| \leq \| A, B' \|$.

(3) There are no cycles of the forms

$\| A, B \| \leq \| A, B' \| \leq \ldots \leq \| A, B'' \| \leq \| A, B \|$ or

$\| A, B \| \leq \| A, B' \| \leq \ldots \leq \| A'', B \| \leq \| A, B \|$

involving $<$.

(The difference between the two cycles is that the first contains possibly only variations on one side, of the form $\| A, B'' \| \leq \| A, B \| \leq \| A, B' \|$, the second one possibly only alternating variations, of the form $\| A'', B \| \leq$ $\| A, B \| \leq \| A, B' \|$.)

(4) $b \in A \mid B \rightarrow \| A, b \| \leq \| A, B \|$.

(5) $b \notin A \mid B$, $b \in B \rightarrow \| A, B \| < \| A, b \|$.

(6) $\| A, b \| \leq^* \| A, B \|$, $b \in B \rightarrow b \in A \mid B \ldots$

(7) $b \in A \mid B$, $a_b \in b \mid A$, $a_b \in A' \subseteq A$ implies

(a) $b \in A' \mid B$,

(b) $A' \mid B \subseteq A \mid B \ldots$

(8) $b \in A \mid B$, $a_b \in b \mid A$, $a' \in A$, $b' \in B \rightarrow \parallel a_b, b \parallel \leq^* \parallel a', b' \parallel$.

(9) $b \in B$, $b \notin A \mid B$, $b' \in A \mid B$, $a_{b'} \in b' \mid A$, $a \in A$.

Then $\parallel a_{b'}, b' \parallel <^* \parallel a, b \parallel$.

If ($\mid$ 2) holds, then

(10) $A \cap B \neq \emptyset \rightarrow \parallel A, B \parallel \leq^* \parallel A', B' \parallel$.

(11) $A \cap B \neq \emptyset$, $A' \cap B' = \emptyset \rightarrow \parallel A, B \parallel <^* \parallel A', B' \parallel$.

Proof:

(1) and (2) are trivial.

(3) We prove both variants simultaneously. Case 1, length of cycle=1: $\parallel A, B \parallel < \parallel A, B \parallel$, so $(A \mid B) \cap B = \emptyset$, contradiction. Case 2: length > 1 : Let, e.g. $\parallel A_0, B_0 \parallel \leq \parallel A_0, B_1 \parallel \leq \ldots \leq \parallel A_0, B_k \parallel < \parallel A_0, B_0 \parallel$ be such a cycle. If the cycle is not yet in the form of the loop condition, we can build a loop as in the loop condition by repeating elements, if necessary, e.g.: $\parallel A_0, B_0 \parallel \leq \parallel A_0, B_1 \parallel \leq \parallel A_0, B_2 \parallel$ can be transformed to $\parallel A_0, B_0 \parallel \leq \parallel A_0, B_1 \parallel \leq_{\text{by } (2)} \parallel A_0, B_1 \parallel \leq \parallel A_0, B_2 \parallel$. By Loop, we conclude $\parallel A_0, B_0 \parallel \leq \parallel A_0, B_k \parallel$, contradicting (1).

(4) and (5) are trivial.

(6) $b \notin A \mid B \rightarrow_{\text{by } (5)} \parallel A, B \parallel < \parallel A, b \parallel$, contradicting $\parallel A, b \parallel \leq^* \parallel A, B \parallel$ by (3).

(7) (a) By (6), it suffices to show that $\parallel A', b \parallel \leq^* \parallel A', B \parallel$. But $\parallel A', b \parallel \leq_{\text{by } (2)} \parallel a_b, b \parallel \leq^*_{(4) \text{ twice}} \parallel A, B \parallel \leq_{\text{by } (2)} \parallel A', B \parallel$.

(b) Let $b' \in A' \mid B$, we show $b' \in A \mid B$.

By (6), it suffices to show $\parallel A, b' \parallel \leq^* \parallel A, B \parallel$: $\parallel A, b' \parallel \leq_{(2)} \parallel A', b' \parallel \leq_{(4)} \parallel A', B \parallel \leq^*_{(2) \text{ twice}} \parallel a_b, b \parallel \leq^*_{(4) \text{ twice}} \parallel A, B \parallel$.

(8) $\parallel a_b, b \parallel \leq^* \parallel A, B \parallel \leq^* \parallel a', b' \parallel$.

(9) $\parallel a_{b'}, b' \parallel \leq^*_{(4) \text{ twice}} \parallel A, B \parallel <_{(5)} \parallel A, b \parallel \leq_{(2)} \parallel a, b \parallel$.

(10) $\parallel A, B \parallel \leq \parallel A, B \cup B' \parallel$, as $(A \mid B \cup B') \cap B \neq \emptyset$, by $A \cap B \subseteq A \mid B \cup B'$. Likewise $\parallel A, B \cup B' \parallel \leq \parallel A \cup A', B \cup B' \parallel$. Moreover, $\parallel A \cup A', B \cup B' \parallel \leq \parallel A', B' \parallel$ by (2).

(11) We show first that $A \cap B \neq \emptyset$, $A \cap B' = \emptyset$ implies $\parallel A, B \parallel < \parallel A, B' \parallel$:

$A \mid B \cup B' = A \cap (B \cup B') = A \cap B \subseteq A$, so $(A \mid B \cup B') \cap B' = \emptyset$. Thus, $\| A, B \| \leq^*_{\text{by } (10)} \| A', A' \| < \| A', B' \|$.

□

We define:

Definition 4.2.7

Let S, by Lemma 3.10.7, be a total preorder on $\{\| A, B \|: A, B \in \mathcal{Y}\}$ extending $\leq$ s.t. $\| A, B \| \; S \; \| A', B' \|$ and $\| A', B' \| \; S \; \| A, B \|$ imply $\| A, B \| \leq^* \| A', B' \|$.

Let $\| A, B \| \approx \| A', B' \|$ iff $\| A, B \| \; S \; \| A', B' \|$ and $\| A', B' \| \; S \; \| A, B \|$, and $[\| A, B \|]$ be the set of $\approx$ −equivalence classes and define $[\| A, B \|] < [\| A', B' \|]$ iff $\| A, B \| \; S \; \| A', B' \|$ but not $\| A', B' \| \; S \; \| A, B \|$. This is a total order on $\{[\| A, B \|] : A, B \in \mathcal{Y}\}$. Define $d(A, B) := [\| A, B \|]$ for $A, B \in \mathcal{Y}$.

If $(\mid 2)$ holds, let $0 := [\| A, A \|]$ for any A. This is then well-defined by Fact 4.2.3, (10).

Note that by abuse of notation, we use $\leq$ also between equivalence classes.

Fact 4.2.4

(1) The restriction of d as just defined to singletons is a symmetric pseudo-distance; if $(\mid 2)$ holds, then d respects identity.

(2) $A \mid B = A \mid_d B$.

Proof:

(1)

(d1) Trivial. If $[\| b, c \|] < [\| a, a \|]$, then $\| b, c \| \leq^* \| a, a \|$, but not $\| a, a \| \leq^* \| b, c \|$, contradicting Fact 4.2.3, (10).

(d2) $d(a, b) = d(a, a)$ iff $\| a, b \| \leq^* \| a, a \|$ iff $a = b$ by Fact 4.2.3, (10) and (11).

(d3) $[\| a, b \|] \leq [\| b, a \|]$ is trivial.

(2)

"$\subseteq$": Let $b \in A \mid B$. Then there is $a_b \in b \mid A$. By Fact 4.2.3, (8), $\| a_b, b \| \leq^* \| a', b' \|$ for all $a' \in A$, $b' \in B$. So $d(a_b, b) \leq d(a', b')$ for all $a' \in A$, $b' \in B$ and $b \in A \mid_d B$.

"$\supseteq$": Let $b \in B$, $b \notin A \mid B$. Take $b' \in A \mid B$, $a_{b'} \in b' \mid A$, $a \in A$. Then by Fact 4.2.3, (9) $\| a_{b'}, b' \| <^* \| a, b \|$, so $b \notin A \mid_d B$.

□

It remains to show the easy direction of Proposition 4.2.2.

All conditions but $(\mid S1)$ are trivial. Define for two sets $A, B \neq \emptyset$ $d(A,B) := d(a_b, b)$, where $b \in A \mid_d B$, and $a_b \in b \mid_d A$. Then $d(A,B) = d(B,A)$ by $d(a,b) = d(b,a)$ for all a, b. Loop amounts thus to $d(X_1, X_0) \leq \ldots \leq d(X_k, X_0) \rightarrow d(X_0, X_1) \leq d(X_0, X_k)$, which is now obvious.

□ (Proposition 4.2.2)

4.2.2.3 The representation result for the finite not necessarily symmetric case

We first give an example which illustrates the expressive weakness of a not necessarily symmetric distance. We then turn to the representation problem. Given an operation $\mid$, we define again a relation, R (Definition 4.2.8), and formulate the representation conditions (mostly for R, not for $\mid$) in Proposition 4.2.5. The representation proof proceeds then via two auxiliary Lemmas (4.2.6 and 4.2.7), extending again R to define d.

Note that we work here with finite U only, $\mathcal{Y}$ will be $\mathcal{P}(U) - \{\emptyset\}$.

Example 4.2.2

This example, illustrated in Figure 4.2.2, shows that we cannot find out, in the non symmetric case, which of the elements a, a' is closest to the the set $\{b, b'\}$ (we look from a/a' to $\{b, b'\}$). In the first case, it is a', in the second case a. Yet all results about revision stay the same.

In the first case, we can take the "road" in both directions, in the second case, we have to follow the arrows. (For simplicity, the vertical parts have length 0.) Otherwise, distances are as indicated by the numbers, so, e.g. in the second case, from a' to a it is 1, from a to a' 1.2. For any $X, Y \subseteq \{a, a', b, b'\}$ $X \mid Y$ will be the same in both cases, but, seen from a or a', the distance to $\{b, b'\}$ is closer from a' in the first case, closer from a in the second.

The characterization of the not necessarily symmetric case presented in the

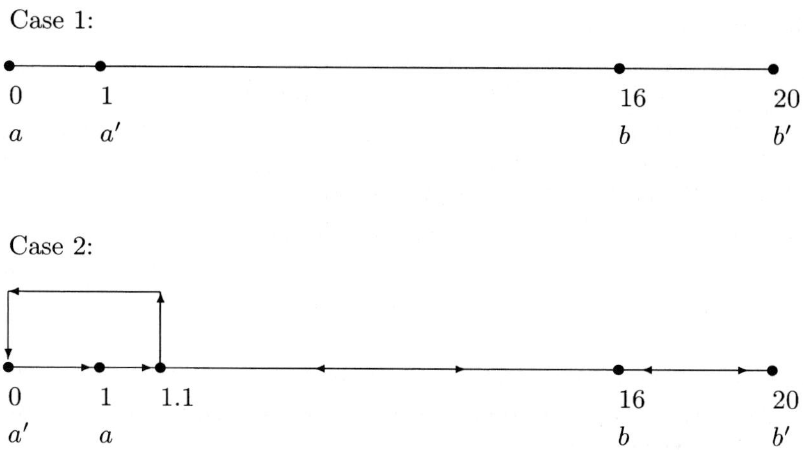

Figure 4.2.2

following does not seem perhaps very elegant at first sight, but it is straightforward and very useful in the search for more elegant characterizations of similar operations. For our characterization a definition is necessary. It associates a binary relation between pairs of nonempty subsets of U : intuitively, $(A, B)R_{|}(A', B')$ may be understood as meaning that the pseudo-distance between A and B is provably smaller than or equal to that between A' and B'. The main idea of the representation theorem is to define a relation (the relation $R_{|}$ of Definition 4.2.8) that describes all inequalities we know must hold between pseudo-distances, and require that the consequences of those inequalities are upheld (conditions $(|\ A2)$ and $(|\ A3)$ of Proposition 4.2.5). The proof of the theorem shows that Definition 4.2.8 was comprehensive enough.

Definition 4.2.8

Given an operation $|$, one defines a relation $R_{|}$ on pairs of nonempty subsets of U by: $(A, B)R_{|}(A', B')$ iff one of the following two cases obtains:

(1) $A = A'$ and $(A \mid B \cup B') \cap B \neq \emptyset$,

(2) $B = B'$ and $(A \cup A' \mid B) \neq (A' \mid B)$.

If the pseudo-distance is to respect identity, we also consider a third case:
(3) $A \cap B \neq \emptyset$.

Definition 4.2.8 can be written as:

(1) $(A \mid B \cup B') \cap B \neq \emptyset \Rightarrow (A, B) R_{|} (A, B')$,

(2) $(A \cup A' \mid B) \neq (A' \mid B) \Rightarrow (A, B) R_{|} (A', B)$,

(3) $A \cap B \neq \emptyset \Rightarrow (A, B) R_{|} (A', B')$.

In the sequel we shall write R instead of $R_{|}$.

As usual, we shall denote by R^* the reflexive and transitive closure of R.

Notice also that we do not require that the pseudo-distance between A and B be less or equal than that between A' and B' if $A' \subseteq A$ and $B' \subseteq B$, as one could expect. In fact, a theorem similar to Proposition 4.2.5 below may be proved with a definition of R that includes a fourth case: $(A, B) R (A', B')$ if $A' \subseteq A$ and $B' \subseteq B$, and its proof is slightly easier, but we prefer to prove the stronger theorem. Notice also that, in order to avoid the fourth case just mentioned, the conclusion of case (2) is $(A, B) R_{|} (A', B)$, and not the seemingly stronger but in fact weaker in the absence of the fourth case mentioned above: $(A, B) R_{|} (A \cup A', B)$.

We may now formulate our main technical result. Condition $(\mid A1)$ expresses a property of Disjunctive Rationality (see [KLM90], [LM92], [Fre93]) for the left-hand-side argument of the operation $\mid$.

Proposition 4.2.5

Consider the following conditions:

$(\mid 1)$ $(A \mid B) \subseteq B$,

$(\mid A1)$ $(A \cup A' \mid B) \subseteq (A \mid B) \cup (A' \mid B)$,

$(\mid A2)$ If $(A, B) R^* (A, B')$, then $(A \mid B) \subseteq (A \mid B \cup B')$,

$(\mid A3)$ If $(A, B) R^* (A', B)$, then $(A \mid B) \subseteq (A \cup A' \mid B)$,

$(\mid 2)$ If $A \cap B \neq \emptyset$, then $A \mid B = A \cap B$,

$(\mid A4)$ If $(A, B) R^* (A', B')$ and $A' \cap B' \neq \emptyset$, then $A \cap B \neq \emptyset$.

(a) An operation $\mid : \mathcal{Y} \times \mathcal{Y} \to \mathcal{Y}$ is representable by a pseudo-distance iff it satisfies the conditions $(\mid 1)$, $(\mid A1)$–$(\mid A3)$ for any nonempty sets $A, B \subseteq U$, where the relation R is generated by cases (1) and (2) of Definition 4.2.8.

(b) An operation $\mid : \mathcal{Y} \times \mathcal{Y} \to \mathcal{Y}$ is representable by an identity respecting pseudo-distance iff it satisfies the conditions $(\mid 1)$, $(\mid 2)$, $(\mid A1)$–$(\mid A4)$ for any nonempty sets $A, B \subseteq U$, where the relation R is generated by cases (1) – (3) of Definition 4.2.8.

Proof:

First, we shall deal with the soundness part of the theorem, and then with the more challenging completeness part. We prove (a) and (b) together.

Suppose, then, that $|$ is representable by a pseudo-distance. The function d acts on pairs of elements of U, and it may be extended to a function on pairs of nonempty subsets of U in the usual way: $d(A, B) = min\{d(a, b) : a \in A, b \in B\}$.

Then Equation (1) in Definition 4.2.4, defining representability, may be written as:

(3) $A \mid B = \{b \in B : d(A, b) = d(A, B)\}$.

We must now show that the conditions of Proposition 4.2.5 hold.

Condition ($\mid 1$) is obvious.

Condition ($\mid A1$) holds since $d(A \cup A', B) = min\{d(A, B), d(A', B)\}$.

Considering Definition 4.2.4 and the different cases of Definition 4.2.8, we shall see that $(A, B)R(A', B')$ implies $d(A, B) \leq d(A', B')$. Case 1 is obvious. Let us treat case (2). Clearly $d(A \cup A', B) = min\{d(A', B), d(A, B)\}$. We shall show that if $d(A', B) < d(A, B)$, then $A \cup A' \mid B = A' \mid B$. Suppose $d(A', B) < d(A, B)$. Then, $d(A \cup A', B) = d(A', B) < d(A, B)$. Therefore $A \cup A' \mid B = A' \mid B$. Case 2 has been taken care of. If d respects identity, Case 3 is obvious. We conclude that $(A, B)R^*(A', B')$ implies that $d(A, B) \leq d(A', B')$. Condition ($\mid A2$) holds because $d(A, B) \leq d(A, B')$ implies $d(A, B \cup B') = d(A, B)$. Condition ($\mid A3$) holds because $d(A, B) \leq d(A', B)$ implies $d(A \cup A', B) = d(A, B)$.

It remains to show that ($\mid 2$) and ($\mid A4$) follow from respect of identity:

Condition ($\mid 2$) holds because $A \mid B = \{b \in B : d(A, b) = d(A, B) = 0\}$ if $A \cap B \neq \emptyset$. Condition ($\mid A4$) holds because $d(A, B) \leq d(A', B') = 0$ implies $d(A, B) = 0$.

For the other direction, we work, unless stated otherwise, in the base situation, i.e. where at least conditions ($\mid 1$), ($\mid A1$)–($\mid A3$) hold, and the relation R is generated by at least cases (1) and (2) of Definition 4.2.8.

In our proof, a number of lemmas will be needed. These lemmas will be presented when needed, and their proof inserted in the midst of the proof of Proposition 4.2.5.

First, a simple result, analogous to the (OR) rule.

Lemma 4.2.6

For any sets A, A', B, $(A \mid B) \cap (A' \mid B) \subseteq A \cup A' \mid B$.

Proof:

Without loss of generality we may assume that $A \mid B \neq A \cup A' \mid B$. Then $(A', B)R(A, B)$ by case (2) of Definition 4.2.8, and $A' \mid B \subseteq A \cup A' \mid B$ by condition $(\mid A3)$ of Proposition 4.2.5. □

We consider the set $\mathcal{Y} \times \mathcal{Y}$ and the binary relation R on this set defined from $\mid$ by Definition 4.2.8. By Lemma 3.10.7, R may be extended to a total preorder S satisfying:

(4) $xSy, ySx \Rightarrow xR^*y$.

Let Z be the totally ordered set of equivalence classes of $\mathcal{Y} \times \mathcal{Y}$ defined by the total preorder S. The function d sends a pair of subsets A, B to its equivalence class under S.

We shall define $d(a, b)$ as $d(\{a\}, \{b\})$. Notice that we have first defined a pseudo-distance between subsets of U, and then a pseudo-distance between elements of U. It is only the pseudo-distance between elements that is required by the definition of representability. The pseudo-distance between subsets just defined must be used with caution because it does not satisfy the property: $d(A, B) = min\{d(a, b) : a \in A, b \in B\}$. It satisfies half of it, as stated in Lemma 4.2.7 below.

Clearly, $(A, B)R(A', B')$ implies $d(A, B) \leq d(A', B')$. Equation (4) also implies that if $d(A, B) = d(A', B')$, then $(A, B)R^*(A', B')$.

The following argument prepares respect of identity. Suppose that $\mid$ satisfies $(\mid 2)$ and $(\mid A4)$ too, and that R was defined including case (3) of Definition 4.2.8. Defining $0 := d(A, A)$ for any $A \in \mathcal{Y}$, we see that

(a) 0 is well-defined: By definition, $(A, A)R(B, B)$ for any $A, B \in \mathcal{Y}$.

(b) there is no $d(B, C) < 0$: By definition again, $(A, A)R(B, C)$.

(c) $d(A, B) = 0$ iff $A \cap B \neq \emptyset$: $A \cap B \neq \emptyset$ implies $(A, B)R(A, A)$, so $d(A, B) = 0$. $d(A, B) = 0$ implies $(A, B)S(A, A)S(A, B)$, so $(A, B)R^*(A, A)$, so $A \cap B \neq \emptyset$ by $(\mid A4)$.

The next lemma shows that our pseudo-distance d behaves nicely as far as its second argument is concerned.

Lemma 4.2.7

For any A,B $d(A, B) = min\{d(A, b) : b \in B\}$

and

(5) $A \mid B = \{b \in B : d(A, b) = d(A, B)\}$.

Proof:

(Remember the elements of $\mathcal{Y}$ are nonempty.) Suppose $b \in B$.

Since $(A \mid B \cup \{b\}) \cap B \neq \emptyset$ by condition $(\mid 1)$ of Proposition 4.2.5, $(A, B)R(A, b)$ by case (1) of Definition 4.2.8, and therefore $d(A, B) \leq min\{d(A, b) : b \in B\}$. If $b \in A \mid B$, then $(A \mid B) \cap \{b\} \neq \emptyset$ and, by Definition 4.2.8, case (1), $(A, b)R(A, B)$ and therefore $d(A, b) = d(A, B)$. We have shown that the left hand side of Equation (5) is a subset of the right hand side. Since $A \mid B$ is not empty there is a $b \in A \mid B$ and, by the previous remark, $d(A, B) = d(A, b)$ and therefore we conclude that $d(A, B) = min\{d(A, b) : b \in B\}$.

To see that the right hand side of Equation (5) is a subset of the left hand side, notice that $d(A, B) = d(A, b)$ implies $(A, b)R^*(A, B)$ and therefore, by condition $(\mid A2)$ of Proposition 4.2.5, $A \mid b \subseteq A \mid B$ and $b \in A \mid B$. □

To conclude the proof of (a), we must show that Equation (1) of Definition 4.2.4 holds. Suppose, first, that $b \in B$, $a \in A$ and $d(a, b) \leq d(a', b')$ for any $a' \in A$, $b' \in B$. By Lemma 4.2.7, $b \in a \mid B$ and $d(a, B) \leq d(a', B)$, for any $a' \in A$.

We want to show now that $b \in A \mid B$. We will show that, for any $a' \in A$, $b \in \{a, a'\} \mid B$. One, then, concludes that $b \in A \mid B$ by Lemma 4.2.6, remembering that U is finite. Since $b \in a \mid B$, we may, without loss of generality, assume that $a \mid B \neq \{a, a'\} \mid B$. By case (2) of Definition 4.2.8, $d(a', B) \leq d(a, B)$. But we already noticed that $d(a, B) \leq d(a', B)$. We can therefore conclude that $d(a, B) = d(a', B)$, so $(a, B)R^*(a', B)$, $a \mid B \subseteq \{a, a'\} \mid B$ and finally that $b \in \{a, a'\} \mid B$. We have shown that the right hand side of Equation (1) is a subset of the left hand side.

We proceed to show that the left hand side of Equation (1) is a subset of its right hand side.

Suppose that $b \in A \mid B$. By condition $(\mid 1)$ of Proposition 4.2.5, $b \in B$. We want to show that there exists an $a \in A$ such that $d(a, b) \leq d(a', b')$ for any $a' \in A$, $b' \in B$. Since the set U is finite, it is enough to prove that, changing the order of the quantifiers:

(6) $\forall a' \in A, b' \in B, \exists a \in A$ such that $d(a,b) \leq d(a',b')$.

Indeed, if Equation (6) holds, we get some $a \in A$ for every pair a', b', and we may take the a for which $d(a,b)$ is minimal: it satisfies the required condition. Since $A = \bigcup\{\{a',x\} : x \in A\}$ (the right-hand side is a finite union) and $b \in A \mid B$, by condition $(\mid A1)$ of Proposition 4.2.5, there is some $x \in A$ such that $b \in \{a',x\} \mid B$. We distinguish two cases. First, if $b \in a' \mid B$, by Lemma 4.2.7, $d(a',b) \leq d(a',b')$ and we may take $a = a'$. Second, suppose that $b \notin a' \mid B$. We notice that, since $b \in \{a',x\} \mid B$, condition $(\mid A1)$ of Proposition 4.2.5 implies that $b \in x \mid B$. But $b \notin a' \mid B$ also implies that $\{a',x\} \mid B \neq a' \mid B$. By Definition 4.2.8, case (2), $(x,B)R(a',B)$ and $d(x,B) \leq d(a',B)$. But, by Lemma 4.2.7, we have $d(x,b) \leq d(x,B)$ (since $b \in x \mid B$) and $d(a',B) \leq d(a',b')$. We conclude that $d(x,b) \leq d(a',b')$, and we can take $a = x$. This concludes the proof of (a).

It remains to show the rest of (b), respect of identity, i.e. that $A \cap B \neq \emptyset$ implies $A \mid B = A \cap B$, under the stronger prerequisites. Let $A \cap B \neq \emptyset$. Then for $b \in B$ $d(A,B) = 0 = d(A,b)$ iff $b \in A$. So by Equation (5) $A \mid B = A \cap B$.

□ (Proposition 4.2.5)

4.2.3 The logical results

4.2.3.1 Introduction

The translation of the algebraic results is, as usual, straigthforward if we assume definability preservation (see Definition 4.2.9). The situation without definability preservation is treated below in Chapter 5.

[ALS99] discusse a problem similar to that of definability preservation in the revision of preferential databases, and its solution. The solution adopted there necessitates much more complicated conditions, for this reason, we have not adopted it here.

In the proofs, we make heavy and tacit use of classical completeness. We work now in (as usual propositional) logic.

Definition 4.2.9

By abuse of language, a pseudo-distance d is called definability preserving iff $\mid_d$ is.

d is called consistency preserving iff $M(T) \mid_d M(T') \neq \emptyset$ for consistent T, T'.

Note that $\models T \leftrightarrow Th(M(T))$, and $T = Th(M(T))$ if T is deductively closed. Moreover, $X = M(Th(X))$ if there is some T s.t. $X = M(T)$, so if the operation $\mid$ is definability preserving, and $T * T' = Th(M(T) \mid M(T'))$, then $M(T * T') = M(T) \mid M(T')$.

The trivial Fact 4.2.8 shows that, given definability preservation and some additional easy caveats, the AGM postulates will hold in distance defined theory revision.

Fact 4.2.8

A distance based revision satisfies the AGM postulates provided:

(1) it respects identity, i.e. $d(a,a) < d(a,b)$ for all $a \neq b$,

(2) it satisfies a limit condition: minima exist,

(3) it is definability preserving.

(It is trivial to see that the first two are necessary, and Example 4.2.3 (2) below shows the necessity of (3). In particular, (2) and (3) will hold for finite languages.)

Proof:

We use $\mid$ to abbreviate $\mid_d$. As a matter of fact, we show slightly more, as we admit also full theories on the right of $*$.

$(K*1)$, $(K*2)$, $(K*6)$ hold by definition, $(K*3)$ and $(K*4)$ as d respects identity, $(K*5)$ by existence of minima.

It remains to show $(K*7)$ and $(K*8)$, we do them together, and show: If $T * T'$ is consistent with T'', then $T * (T' \cup T'') = \overline{(T * T') \cup T''}$.

Note that $M(S \cup S') = M(S) \cap M(S')$, and that $M(S * S') = M(S) \mid M(S')$. (The latter is only true if $\mid$ is definability preserving.) By prerequisite, $M(T * T') \cap M(T'') \neq \emptyset$, so $(M(T) \mid M(T')) \cap M(T'') \neq \emptyset$. Let $A := M(T)$, $B := M(T')$, $C := M(T'')$. "$\subseteq$": Let $b \in A \mid (B \cap C)$. By prerequisite, there is $b' \in (A \mid B) \cap C$. Thus $d(A, b') \geq d(A, B \cap C) = d(A, b)$. As $b \in B$, $b \in A \mid B$, but $b \in C$, too. "$\supseteq$": Let $b' \in (A \mid B) \cap C$. Thus $d(A, b') = d(A, B) \leq d(A, B \cap C)$, so by $b' \in B \cap C$ $b' \in A \mid (B \cap C)$. We conclude $M(T) \mid (M(T') \cap M(T'')) = (M(T) \mid M(T')) \cap M(T'')$, thus that $T * (T' \cup T'') = \overline{(T * T') \cup T''}$.

□

4.2.3.2 The symmetric case

We consider the following conditions for a revision function $*$ defined for arbitrary consistent theories on both sides. This is thus a slight extension of the AGM framework, as AGM work with formulas only on the right of $*$.

$(*0)$ If $\models T \leftrightarrow S$, $\models T' \leftrightarrow S'$, then $T * T' = S * S'$,

$(*1)$ $T * T'$ is a consistent, deductively closed theory,

$(*2)$ $T' \subseteq T * T'$,

$(*3)$ If $T \cup T'$ is consistent, then $T * T' = \overline{T \cup T'}$,

$(*S1)$ $Con(T_0, T_1 * (T_0 \vee T_2))$,

$Con(T_1, T_2 * (T_1 \vee T_3))$,

$Con(T_2, T_3 * (T_2 \vee T_4)) \ldots$

$Con(T_{k-1}, T_k * (T_{k-1} \vee T_0))$

imply

$Con(T_1, T_0 * (T_k \vee T_1))$.

The following Example 4.2.3 (1) shows that, in general, a revision operation defined on models via a pseudo-distance by $T * T' := Th(M(T) \mid_d M(T'))$ will not satisfy $(*S1)$, unless we require $\mid_d$ to preserve definability. But this is not proper to our new condition $(*S1)$, the same happens to the original AGM postulates, as essentially the same Example 4.2.3 (2) shows.

To see this, we summarize the AGM postulates $(K * 7)$ and $(K * 8)$ in $(*4)$:

$(*4)$ If $T * T'$ is consistent with T'', then $T * (T' \cup T'') = \overline{(T * T') \cup T''}$.

$(*4)$ may fail in the general infinite case without definability preservation, as we will see now.

Example 4.2.3

Consider an infinite propositional language $\mathcal{L}$.

Let X be an infinite set of models, m, m_1, m_2 be models for $\mathcal{L}$. Arrange the models of $\mathcal{L}$ in the real plane s.t. all $x \in X$ have the same distance < 2 (in the real plane) from m, m_2 has distance 2 from m, and m_1 has distance 3 from m.

Let T, T_1, T_2 be complete (consistent) theories, T' a theory with infinitely many models, $M(T) = \{m\}$, $M(T_1) = \{m_1\}$, $M(T_2) = \{m_2\}$. The two variants diverge now slightly:

(1) $M(T') = X \cup \{m_1\}$. T, T', T_2 will be pairwise inconsistent.

(2) $M(T') = X \cup \{m_1, m_2\}$, $M(T'') = \{m_1, m_2\}$.

Assume in both cases $Th(X) = T'$, so X will not be definable by a theory.

Now for the results:

Then $M(T) \mid M(T') = X$, but $T * T' = Th(X) = T'$.

(1) We easily verify $Con(T, T_2 * (T \vee T))$, $Con(T_2, T * (T_2 \vee T_1))$, $Con(T, T_1 * (T \vee T))$, $Con(T_1, T * (T_1 \vee T'))$, $Con(T, T' * (T \vee T))$, and conclude by Loop (i.e. $(*S1)$) $Con(T_2, T * (T' \vee T_2))$, which is wrong.

(2) So $T * T'$ is consistent with T'', and $\overline{(T * T') \cup T''} = T''$. But $T' \cup T'' = T''$, and $T * (T' \cup T'') = T_2 \neq T''$, contradicting $(*4)$.

□

We finally have the following representation result for the symmetric case:

Proposition 4.2.9

Let $\mathcal{L}$ be a propositional language.

(a) A revision operation $*$ is representable by a symmetric consistency and definability preserving pseudo-distance iff $*$ satisfies $(*0)$–$(*2)$, $(*S1)$.

(b) A revision operation $*$ is representable by a symmetric consistency and definability preserving, identity respecting pseudo-distance iff $*$ satisfies $(*0)$–$(*3)$, $(*S1)$.

Proof:

We prove (a) and (b) together.

For the first direction, let $\mathcal{Y} := \{M(T) : T$ a consistent $\mathcal{L}$–theory $\}$, and define $M(T) \mid M(T') := M(T * T')$.

By $(*0)$, this is well-defined, $\mid$ is obviously definability preserving, and by $(*1)$, $M(T) \mid M(T') \in \mathcal{Y}$.

We show the properties of Proposition 4.2.2.

$(\mid 1)$ holds by $(*2)$, if $(*3)$ holds, so will $(\mid 2)$. $(\mid S1)$ holds by $(*S1)$: E.g. $(M(T_1) \mid (M(T_0) \cup M(T_2))) \cap M(T_0) \neq \emptyset$ iff $(M(T_1) \mid M(T_0 \vee T_2)) \cap M(T_0) \neq \emptyset$ iff (by definition of $\mid$) $M(T_1 * (T_0 \vee T_2)) \cap M(T_0) \neq \emptyset$ iff $Con(T_1 * (T_0 \vee$

$T_2), T_0)$. By Proposition 4.2.2, $\mid$ can be represented by an — if $(\mid 2)$ holds, identity respecting — symmetric pseudo-distance d, so $M(T*T') = M(T) \mid M(T') = M(T) \mid_d M(T')$, and $Th(M(T*T')) = Th(M(T) \mid_d M(T'))$. As $T*T'$ is deductively closed, $T*T' = Th(M(T*T'))$.

Conversely, define $T*T' := Th(M(T) \mid_d M(T'))$. We use Proposition 4.2.2. $(*0)$ and $(*1)$ will trivially hold. By $(\mid 1)$, $(*2)$ holds, if $(\mid 2)$ holds, so will $(*3)$. As above, we see that $(*S1)$ holds by $(\mid S1)$, where now $(M(T_1) \mid_d M(T_0 \vee T_2)) \cap M(T_0) \neq \emptyset$ iff $M(T_1 * (T_0 \vee T_2)) \cap M(T_0) \neq \emptyset$ by definability preservation. □

4.2.3.3 The finite not necessarily symmetric case

Recall that we work here with a language defined by finitely many propositional variables.

For the not necessarily symmetric case, we consider the following conditions for a revision function $*$ defined for arbitrary consistent theories on both sides.

$(*0)$ If $\models T \leftrightarrow S$, $\models T' \leftrightarrow S'$, then $T*T' = S*S'$,

$(*1)$ $T*T'$ is a consistent, deductively closed theory,

$(*2)$ $T' \subseteq T*T'$,

$(*3)$ If $T \cup T'$ is consistent, then $T*T' = \overline{T \cup T'}$,

$(*A1)$ $(S \vee S')*T \vdash (S*T) \vee (S'*T)$,

$(*A2)$ If $(S,T)R^*(S,T')$, then $S*T \vdash S*(T \vee T')$,

$(*A3)$ If $(S,T)R^*(S',T)$, then $S*T \vdash (S \vee S')*T$,

$(*A4)$ If $(S,T)R^*(S',T')$ and $Con(S',T')$, then $Con(S,T)$.

Where the relation R is defined by

(1) If $Con(S*(T \vee T'), T)$, then $(S,T)R(S,T')$,

(2) If $(S \vee S')*T \neq S'*T$, then $(S,T)R(S',T)$,

and, in the identity respecting case, in addition by

(3) If $Con(S,T)$, then $(S,T)R(S',T')$.

Note that by finiteness, any pseudo-distance is automatically definability preserving. We have the representation result

Proposition 4.2.10

Let $\mathcal{L}$ be a finite propositional language.

(a) A revision operation $*$ is representable by a consistency preserving pseudo-distance iff $*$ satisfies $(*0)$–$(*2)$, $(*A1)$–$(*A3)$, where the relation R is defined from the first two cases.

(b) A revision operation $*$ is representable by a consistency preserving, identity respecting pseudo-distance iff $*$ satisfies $(*0)$–$(*3)$, $(*A1)$–$(*A4)$, where the relation R is defined from all three cases.

Proof:

We show (a) and (b) together.

We first note: If $T * T' = Th(M(T) \mid M(T'))$, then by definability preservation in the finite case $M(T * T') = M(T) \mid M(T')$, so $(M(S) \mid M(T) \cup M(T')) \cap M(T) \neq \emptyset \Leftrightarrow Con(S * (T \vee T'), T)$ and $(M(S) \cup M(S')) \mid M(T) \neq M(S') \mid M(T) \Leftrightarrow (S \vee S') * T \neq S' * T$. Thus, the relation R defined in Definition 4.2.8 between sets of models, and the relation R as just defined between theories correspond.

For the first direction, let $\mathcal{Y} := \{M(T) : T$ a consistent $\mathcal{L}-$theory $\}$, and define $M(T) \mid M(T') := M(T * T')$.

By $(*0)$, this is well-defined, and by $(*1)$, $M(T) \mid M(T') \in \mathcal{Y}$.

We show the properties of Proposition 4.2.5.

$(\mid 1)$ holds by $(*2)$.

$(\mid A1)$: We show $(M(S) \cup M(S')) \mid M(T) \subseteq (M(S) \mid M(T)) \cup (M(S') \mid M(T))$. By $(*A1)$, $(S \vee S') * T \vdash (S * T) \vee (S' * T)$, so $(M(S) \cup M(S')) \mid M(T) = M(S \vee S') \mid M(T) = M((S \vee S') * T) \subseteq M(S * T) \cup M(S' * T) = (M(S) \mid M(T)) \cup (M(S') \mid M(T))$.

For $(\mid A2)$: Let $(M(S), M(T)) R^* (M(S), M(T'))$, so by the correspondence between the relation R between sets of models, and the relation R between theories, $(S, T) R^* (S, T')$, so by $(*A2)$ $S * T \vdash S * (T \vee T')$, so $M(S) \mid M(T) \subseteq M(S) \mid (M(T) \cup M(T'))$.

$(\mid A3)$: Similar, using $(*A3)$. If $*$ satisfies $(*3)$ and $(*A4)$ and R is also generated by case (3), then $(\mid 2)$ and $(\mid A4)$ will hold by similar arguments.

Thus, by Proposition 4.2.5, there is an (identity respecting) pseudo-distance d representing $\mid$, $M(T * T') = M(T) \mid_d M(T')$ holds, so by deductive closure of $T * T'$ $T * T' = Th(M(T) \mid_d M(T'))$.

Conversely, define $T * T' := Th(M(T) \mid_d M(T'))$. We use again Proposition

4.2.5. (∗0) and (∗1) will trivially hold. The proof of the other properties closely follows the proof in the first direction. □

4.2.4 There is no finite characterization

We show here that no finite normal characterization of distance defined revision is possible.

We work on the algebraic side. The crucial example (Example 4.2.4), can be chosen arbitrarily big. We take care that the important revision results are isolated, i.e. that they have no repercussion on other results. For this purpose, we use the property that closer elements hide those farther away, as was shown already in Example 4.2.1. As a result, we obtain structures which are trivially not distance definable, but changing just one "bit" of information makes them distance definable. Consequently, in the limit, the amount of information distinguishing the representable and the not representable case becomes arbitrarily small, and we need arbitrarily much information to describe the situation. This is made formal in Proposition 4.2.11.

We will have to modify the general framework described in Section 1.6.2.1 a little, but the main idea is the same. First, we recall the positive result.

We have characterized revision representable by distance. The crucial condition was a loop condition, of the type: if $d(a_1, b_1) \leq d(a_2, b_2) \leq \ldots \leq d(a_n, b_n)$, then $d(a_1, b_1) \leq d(a_n, b_n)$.

We always thought that this condition has, despite its elegance, an ugly aspect, as it more or less directly expresses — and with arbitrary length — what we want, a loop-free order representing revision. We just did not think hard enough to find better conditions, the author felt, and tried to find better ones — without success, but with a reason:

There are no better conditions, and we can prove it.

We construct a class of examples, which provides for all $n \in \omega$ a $Y(n)$ which is not representable by distances, and for all $a_1, \ldots, a_n$ in $Y(n)$ a structure X which is representable by distances and agrees with $Y(n)$ on $a_1, \ldots, a_n$. We call the distance representable examples "legal", and the other ones "illegal".

For didactic reasons, we develop the construction from the end. The problem is to construct revision formalisms which can be transformed by a very minor change from an illegal to a legal case. Thus, the problem is to construct legal examples sufficiently close to illegal ones, and, for this, we have to define a distance. We will construct legal structures where the crucial property (freedom from loops) can be isolated from the rest of the information by a suitable choice of distances. It will then suffice to change just one bit of information to obtain an illegal example, which can be transformed back to a legal one by changing again one bit of the important information (not necessarily the same one).

Example 4.2.4

(Hamster wheels)

Recall the remark that any distance with values in $\{0\} \cup [0.5, 1]$ automatically respects the triangle inequality — $[x, y]$ is the usual (closed) interval. (As J. Arcamone remarked, multiplying with a factor > 0 preserves this property.)

Fix n sufficiently big (> 4 or s.t. the like should do). d will be the (symmetrical) distance to be defined now.

Take $\{a_i : 1 \leq i \leq n\} \cup \{b_i : 1 \leq i \leq n\}$ and join them (separately) in two "wheels", s.t. a_1 is between a_n and a_2, etc.

Let $d(a_i, a_j) := d(b_i, b_j) := 1$ for any $i \neq j$, and $d(x, x) = 0$ for all x.

Call b_i the opposite of a_i, b_{i-1} and b_{i+1} (all modulo n) the 1-opposite of a_i, and b_{i-2} and b_{i+2} the 2-opposite of a_i, etc.

Let $d(a_i, b_j) := 1.9$ if b_j is the 1-opposite of a_i, and $d(a_i, b_j) := 1.1$ if b_j is the k-opposite of a_i for $k > 1$.

Choose $d(a_i, b_i) \in [1.2, 1.8]$ arbitrarily. We call this the "choices".

Look now at $A \mid B$ (the set of closest elements in B, seen from A). We show that almost all A, B give the same results, independent of the choices of $d(a_i, b_i)$.

The case $A \cap B \neq \emptyset$ is trivial.

If there is some $a_i \in A$, and some $a_j \in B$, then $A \mid B$ will contain all $a_j \in B$. Likewise for b_i, b_j. In these cases, the distance 1 makes all other cases invisible.

Let now $A = \{a_i : i \in I\}$, and $B = \{b_j : j \in J\}$. ($A = \{b_i \ldots\}$, etc. is symmetrical.)

Case 1: $A = \{a_i\}$. Then in all choices the k-opposites, $k > 1$, have precedence over the opposites over the 1-opposites, the result does not depend on the choices.

Case 2: A contains at least three a_i. Assume that B contains at least two

b_j. If not, we are in Case 1. In this case, one b_j is k-opposite, $k > 1$, and this decides, independent from the choices of the $d(a_i, b_i)$.

Case 3: $A = \{a_i, a_j\}$. By Cases 1 and 2 and symmetry, the only interesting case is where $B = \{b_l, b_m\}$. If $j \neq i+1$, then b_l or b_m are k-opposites, $k > 1$, and the outcome is the same for all choices.

So, finally, all revision information which allows to differentiate between the different choices is of the type $\{a_i, a_{i+1}\} \mid \{b_i, b_{i+1}\}$ — and they do it, e.g. $\{a_i, a_{i+1}\} \mid \{b_i, b_{i+1}\} = \{b_i\}$ iff $d(a_i, b_i) < d(a_{i+1}, b_{i+1})$.

But, to see whether we have a legal situation, e.g. of the type $d(a_i, b_i) = d(a_j, b_j)$ for all i, j, or an illegal one of the type $d(a_1, b_1) < d(a_2, b_2) < \ldots < d(a_n, b_n) < d(a_1, b_1)$, which cannot be represented by a distance, we need the whole chain of n pieces of information. This is easy, just construct a legal case for any smaller set of information.

More precisely, define a revision operator $\mid$ as above for all but the crucial sets. The construction indicates how to define a distance which generates these results. For the illegal case, add now a loop, by working with the crucial cases. This operator cannot be generated by a distance. But omitting one step of the loop results in a structure which is distance definable. As we took care to isolate the crucial cases from the rest, the other results stay unchanged. Consequently, all sufficiently small formulas (below the upper bound) are valid (or not) in both cases.

We make this formal in the following Proposition 4.2.11.

□

Proposition 4.2.11

No small (finite) characterization of distance representable revision is possible.

Discussion and proof:

This case is slightly complicated by the fact that we do not only speak about elements, but also about sets, but those that count are small, they have exactly two elements.

Let ϕ now be a characterization of distance representable revision structures with a purely universal formula $\phi = \forall x_1, \ldots, x_k \phi'(x_1, \ldots, x_k)$, where ϕ contains set expressions (involving $\subseteq$, $\cap$, $\cup$, $-$, and $\in$), as well as the function $\mid$, as well as perhaps the constants U, $\emptyset$, and ϕ' is quantifier free.

We consider now the number of "relevant" elements involved in deciding validity of the formula in legal and illegal examples. Take first the simple case where ϕ contains no nested $\mid$ operators, say n such operators. Consider the hamster wheel W of size $m := 4*n+1$. By hypothesis, for some parameters c_i (sets or elements) $W \models \neg\phi'[c_1, \ldots, c_k]$. Let $a_1, \ldots, a_{m-1}$ be the elements of crucial (i.e. two-element) arguments of $\mid$, appearing in $\phi'[c_1, \ldots, c_k]$ (where some a_i, a_j may be equal), and take the legal structure L, which agrees with W on all $a_1, \ldots, a_{m-1}$, even for two-element arguments of $\mid$, (but differs on a_m). As all set operators evaluate the same way in W and L, as well as all $\mid$ operators where at least one argument has cardinality other than 2, and L agrees with W also on the $a_1, \ldots, a_{m-1}$ elements, $L \models \neg\phi[c_1, \ldots, c_k]$, a contradiction. Consider now the case where there are nested $\mid' s$, up to depth s, and assume each level has at most $r \mid$ operators. Consider now, e.g. level 2 $\mid$ operators. Their arguments may depend on the result of level 1 $\mid$ operators, e.g. in the form $A \mid (B - (C \mid D))$. Thus, again $4*r$ new elements may be involved, but, if the level 1 $\mid$ evaluate in both structures the same way, there are no more involved. (Otherwise, this might depend on the $(4*r)^2$ possible outcomes of the level 1 $\mid$ operators, we would have to consider $4*r*(4*r)^2$ cases.) Thus, just as in the flat case, we have to consider $4*r*s$ elements, which might take a crucial role in the evaluation of $\phi'[c_1, \ldots, c_k]$. Let then $m := 4*r*s+1$, and continue as in the flat case. By induction, we see that the outcome of the level 1 $\mid$ operators is the same in W and L, and so on to the highest level.

Note that the upper bound m depends on ϕ, and not on W or the c_i — though the number of elements really involved may be smaller and depend on W and the c_i. Thus, given ϕ, we find uniform m, and then choose W, and find c_i, and then choose L (the actual L depends on the c_i, but it is one of the $L(W)$).

Now, we are finished: $W \models \neg\phi'[c_1, \ldots, c_k]$, so $L \models \neg\phi'[c_1, \ldots, c_k]$, so $L \models \neg\phi$, i.e. $L \models \neg\forall x_1, \ldots, x_k\phi'(x_1, \ldots, x_k)$, a contradiction. □

Analysis of the example:

The problem is that we cannot always see directly by revision that, if $d(a,b) < d(c,e) < d(f,g)$, then $d(a,b) < d(f,g)$. (Other distances might get in our way.) If we could, we could have done with a rule of length 3, expressing transitivity, grouping long paths together. Revision is too coarse, too blind, if you like.

But also a sufficiently rich domain could have helped: If we had enough points, sufficiently far from each other, we could, e.g. see directly by revision that $d(a,b) = d(a',b')$, and $d(a',b') < d(f,g)$ for suitable a', b'. This amounts to having "mirrors" which reflect the situation, and which are far enough not to intervene with close elements.

We may also say that the domain lacks coherence, so changes in one place — from a counterexample to an example — do not propagate to other places, and we have arbitrarily many degrees of freedom. Having enough elements, sufficiently distant, provides the "glue" of coherence.

Real distances, the structure we want to represent with, however, being more abstract, allow more substitution, and force thus more coherence. This difference in coherence is perhaps the deeper reason why we need complicated conditions for representation.

4.2.5 The limit case

4.2.5.1 Introduction

Analogous to the case of preferential, and in particular ranked structures, we show how to characterize the limit variant of distance based revision, and, perhaps most importantly, show that, as long as we consider revisions of the form $\phi * \psi$, the limit version is equivalent to the minimal version. The technique is the same as for ranked preferential structures. We first indicate how to reflect the limit case down to the minimal case using pairs of elements, and then to beam the resulting structure up to the limit and infinite case. We then show that the limit version for formulas has the logical properties of the minimal case, thus a limit distance structure is equivalent to a minimal distance structure — with, perhaps, a different distance. Essential are, here again, closure properties of the domain.

Recall that the limit version frees us from the necessity of the existence of closest elements, i.e. essentially from the absence of infinite chains of ever closer elements. This problem can be felt when $A, B \neq \emptyset$, but $A \mid B = \emptyset$.

We remind the reader of the general remarks on the proof strategy for the limit case of ranked preferential structures in Section 3.10.3.1.

In the revision case, we have to reflect down to the finite case on the left, too. This is not necessary in the ranked preferential case, as the "point of reference" is an "imaginary" point outside the structure (which, in addition, does not change). The essential point is now: Given two sets X and Y, we are interested in systems of points in Y, which are closer and closer to X.

So, on the right, we compare $d(X, y)$ with $d(X, y')$, but, X may itself be infinite and getting closer and closer to Y without a minimum. Now, if $d(X, y) < d(X, y')$, then there is a "witness" $x \in X$ which shows this, i.e. $\exists x \in X$ s.t. $\forall x' \in X \; d(x, y) < d(x', y')$:

$d(X, y) < d(X, y')$ iff there is $x \in X$ s.t. $\forall x' \in X \; d(x, y) < d(x', y')$ — such x will be called a witness for $d(X, y) < d(X, y')$

So, we are again down to four elements, x, y, x', y'. Thus, considering sets of the type $\{x, x'\}$ on the left, $\{y, y'\}$ on the right, we can determine the distance — if there is one, and there will be one, given the necessary properties for the finite case — which will do the job in the limit (infinite) case, too. But, as long as we consider finite sets, we can use the hypothesis $A \mid B \neq \emptyset$, if $A, B \neq \emptyset$, and thus the results in Sections 4.2.2 and 4.2.3. So, it suffices to express the conditions for the finite case, and to make sure that they cooperate with the infinite case. This is the central condition which allows reflection down to the finite case. The reader should, however, note that there may be many points involved, and this is the reason why a "normal" characterization will not be possible — see Section 5.2.3. Moreover, we look here only at the algebraic side, and Section 5.2.3 discusses the logical problems, which, in this case, are more difficult.

Thus, we will consider systems $\Lambda(X, Y)$, where

$\Lambda(X, Y) \subseteq \mathcal{P}(Y)$

Given a distance d, such $\Lambda(X, Y)$ will be $\emptyset \neq \{y \in Y : d(X, y) \leq r\}$ for some r (alternatively: $d(X, y) < r$), or, more generally, for X which get themselves ever closer, $\emptyset \neq \{y \in Y : \exists x \in X.d(x, y) \leq r\}$ ($< r$ respectively). Note that for $X, Y \neq \emptyset$ any $A \in \Lambda(X, Y)$ is nonempty, too, as we do not choose r too small, and that for $A, A' \in \Lambda(X, Y)$ $A \subseteq A'$ or $A' \subseteq A$. The logical side is then defined by: $\phi \in T * T'$ iff there is $A \in \Lambda(M(T), M(T'))$ s.t. $A \models \phi$. By compactness and inclusion, $T * T'$ is consistent (if T and T' are) and deductively closed.

We leave it to the reader to fill in the details, and, after a short remark on the logics involved, concentrate on the fact that the new definition reduces to the minimal version, as long as we limit ourselves to formula defined sets.

4.2.5.2 Remarks on the logics of the revision limit case

This case is particularly complicated, and we just give a rough sketch how to treat it. We combine the ideas about the algebraic situation of the limit revision case, and those from the logics of the ranked limit case. More precisely, we first restrict our attention to theories with two models on the right (as above), then look at the witnesses on the left (as in the algebraic limit case of revision), and thus construct the distance. Sufficiently strong, but straightforward, conditions on both sides will assure that the constructed distance is compatible with the results for theories with infinitely many models.

4.2.5.3 Equivalence of the minimal and the limit case for formulas

We show here that the logical properties of the limit variant for formulas on the left of $*$ satisfy the conditions for representability in the much simpler minimal variant. The following proposition is thus a trivialization result.

Proposition 4.2.12

The limit variant of a symmetrical distance defined revision is equivalent to the minimal variant, as long as we consider formulas (and not full theories) on the left.

Proof:

We show that the conditions of Proposition 4.2.9 are satisfied for the distance defined limit version. Consequently, the limit version is equivalent to the minimal version (with, perhaps, a different distance).

The nontrivial condition to show is Loop:

$Con(\phi_0, \phi_1 * (\phi_0 \vee \phi_2))$,

$Con(\phi_1, \phi_2 * (\phi_1 \vee \phi_3))$,

$Con(\phi_2, \phi_3 * (\phi_2 \vee \phi_4)), \ldots,$

$Con(\phi_{k-2}, \phi_{k-1} * (\phi_{k-2} \vee \phi_k))$,

$Con(\phi_{k-1}, \phi_k * (\phi_{k-1} \vee \phi_0))$

imply

$Con(\phi_1, \phi_0 * (\phi_k \vee \phi_1))$.

First, we have $M(\phi) \cap A \neq \emptyset$ iff $Con(\phi, Th(A))$: $\neg Con(T, T') \to M(T) \cap M(T') = \emptyset$, thus $\neg Con(\phi, Th(A)) \to M(\phi) \cap M(Th(A)) = \emptyset \to M(\phi) \cap A = \emptyset$. Conversely, $A \cap M(\phi) = \emptyset \to \forall a \in A.a \models \neg\phi \to \neg\phi \in Th(A) \to \neg Con(Th(A), \phi)$. (Usually, $A \cap B = \emptyset \to \neg Con(Th(A), Th(B))$ is false, we use here the fact that ϕ is a formula.)

Second, if a logic is defined with a nested system $\mathcal{X}$ of model sets, i.e.

$T \mid\sim \phi$ iff there is $X \in \mathcal{X}$ s.t. $X \models \phi$, then $Con(T', \overline{\overline{T}})$ iff for all $X \in \mathcal{X}$ $Con(T', Th(X))$.

Consequently, $Con(\phi, \phi' * (\phi \vee \phi''))$ iff $\forall A \in \Lambda(M(\phi'), M(\phi \vee \phi'')).Con(\phi, Th(A))$ iff $\forall A \in \Lambda(M(\phi'), M(\phi \vee \phi'')).M(\phi) \cap A \neq \emptyset$.

But, if Λ is distance defined, this is equivalent to: $\forall x' \in M(\phi')\forall y' \in M(\phi \vee \phi'')\exists x \in M(\phi')\exists y \in M(\phi).d(x, y) \leq d(x', y')$.

We now show the loop condition using the reformulated properties.

Let then $x_k \in M(\phi_k) \cup M(\phi_1)$, $x_0 \in M(\phi_0)$. We have to find $x'_1 \in M(\phi_1)$, $x'_0 \in M(\phi_0)$ s.t. $d(x'_0, x'_1) \leq d(x_0, x_k)$. If $x_k \in M(\phi_1)$, we are done. So suppose $x_k \in M(\phi_k)$. By $Con(\phi_{k-1}, \phi_k * (\phi_{k-1} \vee \phi_0))$, there are $x'_k \in M(\phi_k)$ and $x_{k-1} \in M(\phi_{k-1})$ s.t. $d(x_{k-1}, x'_k) \leq d(x_0, x_k)$. By $Con(\phi_{k-2}, \phi_{k-1} * (\phi_{k-2} \vee \phi_k))$, there are $x'_{k-1} \in M(\phi_{k-1})$ and $x_{k-2} \in M(\phi_{k-2})$ s.t. $d(x_{k-2}, x'_{k-1}) \leq d(x_{k-1}, x'_k)$, etc., until, finally, by $Con(\phi_0, \phi_1 * (\phi_0 \vee \phi_2))$, there are $x'_0 \in M(\phi_0)$, $x'_1 \in M(\phi_1)$ s.t. $d(x'_0, x'_1) \leq d(x_1, x'_2)$, and we have $d(x'_0, x'_1) \leq d(x_1, x'_2) \leq \ldots \leq d(x_{k-2}, x'_{k-1}) \leq d(x_{k-1}, x'_k) \leq d(x_0, x_k)$. □

4.3 Local and global metrics for the semantics of counterfactuals

4.3.1 Introduction

Recall that we will work in this Section 4.3 with the individual variant of distance — see Definition 2.3.5.

We have seen above that the "trees can hide the forest", seeing only closest elements can have the consequence that there is no finite characterization, provided the universe is not sufficiently rich to allow seeing hidden elements in a mirror. We shall see now that such properties can also have positive effects, in the sense that they enable a uniform metric for counterfactual conditionals — provided we accept copies of models. The trick is to hide everything which might disturb the picture behind closer elements, and as we see only the closest ones, they disappear from sight.

Overview:

After introductory definitions, we formulate the main result (Proposition 4.3.1), and give an outline of its proof. We then give the formal proof via a number of auxiliary lemmata (Lemma 4.3.2, 4.3.3, 4.3.4). We append a small new result (Proposition 4.3.5), as corollary, combining Proposition 4.3.1 with Proposition 4.2.12, to show that the limit approach for counterfactuals is equivalent with the minimal approach.

General perspective:

Traditional counterfactual semantics allow a different metric for each point of origin. It may be argued from a philosophical point of view that for at least some models the relations should be left totally independent of each other, for "closeness" as seen from the standpoint of one possible world need have little in common with closeness as seen from another world. Now, in such a model, there will not exist any common (or global) metric d that determines it in the sense that for all worlds x, y, a in the model, $x \prec_a y$ iff $d(a,x) < d(a,y)$. For whereas the relations of such a model are quite independent of each other, the existence of such a global d creates connections. For example, it forces $a \prec_b c$ whenever $b \prec_a c$ and $a \prec_c b$, since $d(a,b) < d(a,c) = d(c,a) < d(c,b)$ implies by the transitivity of $<$ over the reals and symmetry of a metric that $d(b,a) = d(a,b) < d(c,b) = d(b,c)$.

Our main result (Proposition 4.3.1) shows that the language of counterfactual conditionals cannot distinguish between models in which the distance (or closeness) of worlds is defined by several metrics — in the extreme case by a distinct metric for each world — and models where distance is defined by a single global metric. More precisely, we show that for each model of the first kind, one can construct a model of the second kind in which exactly the same formulas of the language of counterfactual conditionals hold. The proof makes essential use of the fact that, by the very definition of the truth of the conditional $\phi \Rrightarrow \psi$ in a single world in terms of "closest worlds where ϕ holds", the closest $\phi-$worlds "fence off" all more distant worlds where ϕ may also hold — all we "see" is the closest, inner layer of $\phi-$worlds surrounding the given world.

4.3.1.1 Basic definitions

Definition 4.3.1

Let $\mathcal{L}$ be a propositional language for counterfactual conditionals, with primitive connectives $\neg$, $\wedge$, $\Rrightarrow$, and let W be a fixed set of "possible worlds", such that each $a \in W$ is associated with a classical model m_a for $\mathcal{L}$ (not all models for $\mathcal{L}$ need occur, and some may occur more than once).

For each $a \in W$ let a relation $\prec_a$ over W be given.

Then, $\mathcal{W} :=< W, \{\prec_a : a \in W\} >$ defines a model for counterfactual conditionals as follows (by simultaneous induction for $a \models_{\mathcal{W}} \phi$, $[[\phi]]$, $U_\phi(a)$ on the complexity of ϕ). For all $a \in W$, $\phi, \psi \in \mathcal{L}$:

(a) When ϕ is a propositional variable, $a \models_{\mathcal{W}} \phi :\leftrightarrow m_a \models \phi$ (remember: m_a is a classical model)

$a \models_{\mathcal{W}} \neg\phi :\leftrightarrow a \not\models_{\mathcal{W}} \phi$

$a \models_{\mathcal{W}} \phi \wedge \psi :\leftrightarrow a \models_{\mathcal{W}} \phi$ and $a \models_{\mathcal{W}} \psi$

$a \models_{\mathcal{W}} \phi \Rrightarrow \psi :\leftrightarrow U_\phi(a) \subseteq [[\psi]]$

(b) $[[\phi]] := \{a \in W : a \models_{\mathcal{W}} \phi\}$

(c) $U_\phi(a) := \{x \in [[\phi]] : \neg\exists y \in [[\phi]] . y \prec_a x\}$ (thus, $U_\phi(a)$ is the set of $\phi-$worlds closest to a)

(d) Finally, we define as usual: $\mathcal{W} \models \phi :\leftrightarrow [[\phi]] = W$

Strictly speaking, we should index $[[.]]$ and U by $\mathcal{W}$, but this will be done only when needed for clarity.

Definition 4.3.2

We say that $\prec_a$ is determined by a metric d_a on W, iff for all $x, y \in W$ $x \prec_a y \leftrightarrow d_a(a,x) < d_a(a,y)$. We say that all $\prec_a$ for $a \in W$ are determined by a common metric d iff for all $a, x, y \in W$, $x \prec_a y \leftrightarrow d(a,x) < d(a,y)$.

4.3.2 The results

We state right away the main result of Section 4.3.

Proposition 4.3.1

Let $\mathcal{W} =< W, \{\prec_a : a \in W\} >$ be a model for counterfactual conditionals such that each $\prec_a$ is determined by a metric d_a on W. Then there is a model $\mathcal{X} =< X, \{\prec_x : x \in X\} >$ for counterfactuals and a metric d on X such that:

(1) $\mathcal{W}$ and $\mathcal{X}$ validate exactly the same formulas of the language of counterfactual conditionals, indeed:

For all $a \in W$ there is $x_a \in X$ such that for all $\phi \in \mathcal{L}$ $a \models_{\mathcal{W}} \phi$ iff $x_a \models_{\mathcal{X}} \phi$.

and

For all $x \in X$ there is $a_x \in W$ such that for all $\phi \in \mathcal{L}$ $x \models_{\mathcal{X}} \phi$ iff $a_x \models_{\mathcal{W}} \phi$.

(2) Each $\prec_x$ is determined by the common metric d.

Remark: $\mathcal{X}$ will be ranked and respect 0, as all $\prec_x$ are determined by a metric.

The construction used to prove Proposition 4.3.1 is somewhat complex. For this reason, we first give an outline sketch, and then the formal details.

4.3.2.1 Outline of the construction for Proposition 4.3.1

Given a model with set W of worlds and relations $\prec_a$, each of which is determined by a metric d_a, we take the worlds in W and put them "very far from each other". For each world $a \in W$, we make copies of all the other worlds in W, and put those copies in a cluster relatively close to a, ordered among themselves by the relation $\prec_a$. Each such cluster is like a galaxy, with the separate galaxies far apart. This construction is iterated ω many times. Thus if b is in the cluster around a, we make fresh copies of all the other worlds and subcluster them tightly around b, internally ordered by the relation $\prec_b$ and all very close to b compared to their distance from anything outside the subcluster. And so on, ω many times.

To give this rough idea precise content, we shall take the elements of the metric space to be finite sequences of elements from W — for simplicity of construction, all beginning with some fixed element $* \notin W$ (Definition 4.3.4). The propositional properties of such a sequence will be inherited from its last element (see "Construction", (d)). The distance between two sequences is measured by "climbing" from the common intial segment to both ends and adding up the distances encountered on the way (Definition 4.3.4). Those latter distances depend on the position in the sequence — the later the position, the smaller the distance — but will preserve the relative sizes, "Construction", (b).

More precisely, as we are interested only in the comparison of distances, we define two metrics to be equivalent, $d \sim d'$, iff the resulting relations are the same (see Definition 4.3.3). Lemma 4.3.2, the proof of which is a straightforward construction from elementary calculus, says that we can choose the range of a metric almost ad libitum: For any metric d and any constant $c > 0$, there is an equivalent metric d' such that $d'(x, x') = 0$ or $\frac{3}{4}c \leq d'(x, x') \leq c$ for all x, x'. We use this result to make distances ever smaller along the sequences — but not too small — preserving the relative arrangement of worlds.

The main consequences of this construction are:

(1) the set $U(s)$ of sequences closest to a sequence s consists of s and its continuations by one further element (Lemma 4.3.4, (a)), and:

(2) $U(s)$ is arranged in the same way as the old universe was, as seen from the last element of s (Lemma 4.3.4, (b)).

It is then straightforward to show that s (in the new universe) and its last element (in the old universe) satisfy exactly the same formulas in the language of counterfactual conditionals (Lemma 4.3.4, (c)).

4.3.2.2 Detailed proof of Proposition 4.3.1

Definition 4.3.3

Two metrics $d, d' : X \times X \to \Re$ are called equivalent ($d \sim d'$) iff $d(a,b) < d(c,e) \leftrightarrow d'(a,b) < d'(c,e)$ for all $a,b,c,e \in X$.

Lemma 4.3.2

For each $c > 0$ and metric $d : X \times X \to \Re$, there is $d' \sim d$ such that $range(d') \subseteq \{0\} \cup [c * \frac{3}{4}, c]$. (For readability, we use $*$ for ordinary multiplication.)

Proof (elementary, but tedious):

We first show the following:

(a) Let $f : [0,\infty) \subseteq \Re \to \Re$ be such that for all a, b, c

(1) $f(a) \geq 0$

(2) f is monotonic, i.e. $a \leq b \to f(a) \leq f(b)$

(3) f is not concave, i.e. if $a < b < c$, then $\frac{f(b)-f(a)}{b-a} \geq \frac{f(c)-f(b)}{c-b}$.

Then $a \leq b + c \to f(a) \leq f(b) + f(c)$.

Proof of (a):

By $a \leq b+c$ and monotony, $f(a) \leq f(b+c)$, we show $f(b+c) \leq f(b)+f(c)$. If $b = 0$, then $f(b+c) = f(c) \leq f(b)+f(c)$ by $f(x) \geq 0$. So assume $b > 0$. If $b = c$, then $\frac{f(b+c)-f(b)}{b} \leq \frac{f(b)-f(0)}{b}$, so $f(b+c) \leq 2*f(b)-f(0) \leq 2*f(b)$. So assume without loss of generality $0 < b < c$. Then $\frac{f(c+b)-f(c)}{b} \leq \frac{f(c)-f(b)}{c-b}$ $\leq \frac{f(b)-f(0)}{b}$, so $f(c+b) - f(c) \leq f(b) - f(0) \leq f(b)$.

(b) If f satisfies the conditions of (a), and $f(x) = 0$ iff $x = 0$, and $d : X \times X \to \Re$ is a metric, then so is $f \circ d : X \times X \to \Re$.

Proof of (b):

(1) $f \circ d(x,y) = 0 \leftrightarrow d(x,y) = 0 \leftrightarrow x = y$.

(2) $d(x,z) \leq d(x,y) + d(y,z)$ implies by the above $f \circ d(x,z) \leq f \circ d(x,y) + f \circ d(y,z)$.

(c) If f satisfies the conditions of (b), and is in addition strictly monotone, i.e. $a < b \rightarrow f(a) < f(b)$, and $d : X \times X \rightarrow \Re$ is a metric, then $f \circ d$ is a metric equivalent to d. (Obvious)

The following functions satisfy the conditions of (b):

(1) for $c > 0$ let $f_c(x) := c * x$

(2) for $c \geq 0$, let

$$g_c(x) := \begin{cases} 0 & \text{iff} \quad x = 0 \\ x + c & \text{iff} \quad x > 0 \end{cases}$$

(3) Let

$$h(x) := \begin{cases} x & \text{iff} \quad 0 \leq x \leq 1 \\ 2 - \frac{1}{x} & \text{iff} \quad 1 < x. \end{cases}$$

So d' defined by $d' := f_c \circ f_{\frac{1}{8}} \circ g_6 \circ h \circ d$ will be a metric as desired. □

Definition 4.3.4

(a) Let, for any finite sequence $s =< s_0 \ldots s_n >$, $l(s)$ be its length, and s_∞ its last element. Let W be any set, and assume without loss of generality $\emptyset \notin W$ (if not, take, e.g. W instead of $\emptyset$).

Consider

$X := \{s$: $s =< s_0 \ldots s_n >$ is a finite sequence in $W \cup \{\emptyset\}$ such that

(1) $2 \leq l(s)$, (2) $s_0 = \emptyset$, (3) for $0 < i < l(s)$ $s_i \in W$, (4) s contains no direct repetitions, i.e. for $0 < i < l(s) - 1$ $s_i \neq s_{i+1}$ — but, e.g. $s_i = s_{i+2}$ is permitted $\}$.

We use the following notation: For $s, t \in X$, let the root of s and t be the maximal common initial segment of s and t, denoted $s \downarrow t$. By condition (2) above, $s \downarrow t$ will contain at least $\emptyset$. For $s \in X$, $a \in W$ let $< s; a >$ be the sequence resulting from appending a to s.

For $i \leq l(s)$, let $s\lceil i :=< s_0 \ldots s_{i-1} >$.

(b) For each $s \in X$ let a metric $d_s : W \times W \rightarrow \Re$, and for $s =< \emptyset >$ let a metric $d_s : (W \cup \{\emptyset\}) \times (W \cup \{\emptyset\}) \rightarrow \Re$ be defined. For $s \in X$ and $0 \leq i < l(s) - 1$ let $\delta(s, i) := d_{s\lceil i+1}(s_i, s_{i+1})$.

(Note: As direct repetitions are not allowed, $\delta(s, i) > 0$.)

(c) We define $d : X \times X \to \Re$ by

$$d(s,t) := \begin{cases} 0 & \text{iff} \quad s = t \\ \Sigma\{\delta(s,i) : l(s \downarrow t) - 1 \leq i < l(s) - 1\} + & \text{otherwise} \\ \Sigma\{\delta(t,i) : l(s \downarrow t) - 1 \leq i < l(t) - 1\} & \end{cases}$$

(If s is an initial segment of t, the first sum is 0, etc.)

Lemma 4.3.3

d as just defined in (c) is a metric on X.

Proof:

1. $d(s,t) \geq 0$

2. $d(s,s) = 0$

3. $d(s,t) = d(t,s)$

4. $d(s,t) > 0$ for $s \neq t$ are all trivial (note that we do not allow direct repetitions).

5. $d(s,u) \leq d(s,t) + d(t,u)$: There is not much to show: A look at the different cases $l(s \downarrow t) <$ or $=$ or $> l(s \downarrow u)$ will give the result. □

The construction of the metric space $\mathcal{X}'$:

Recall that the original structure $\mathcal{W}$ was given by W and a metric d_a for each $a \in W$.

(a) Define X as in Definition 4.3.4 (a) from W.

(b) Choose by Lemma 4.3.2 for each $s \in X$ a metric d_s on W such that

1. $\frac{3}{4} * \frac{1}{2^{l(s)}} \leq d_s(x,y) \leq \frac{1}{2^{l(s)}}$ for all $x, y \in W$, $x \neq y$

2. d_s is equivalent to d_{s_∞}.

Moreover, for $s =< \emptyset >$, define $d_{<\emptyset>} : (W \cup \{\emptyset\}) \times (W \cup \{\emptyset\}) \to \Re$ by

$$d_{<\emptyset>}(x,y) := \begin{cases} 0 & \text{iff} \quad x = y \\ 1 & \text{otherwise} \end{cases}$$

(c) Define a metric d on X as in Definition 4.3.4 (c) from the individual metrics d_s on W (or $W \cup \{\emptyset\}$), and let $u \prec_s t :\leftrightarrow d(s,u) < d(s,t)$. For $s \in X$, let $U(s) := \{< s; x >: x \in W\} \cup \{s\}$.

(d) Finally, construct a model for counterfactuals from X:

Set $\mathcal{X} :=< X, \{\prec_s : s \in X\} >$ and define classical validity at s as at s_∞: $s \models_\mathcal{X} \phi :\leftrightarrow s_\infty \models_\mathcal{W} \phi$ for classical ϕ.

Lemma 4.3.4

(a) $U(s)$ contains the elements closest to s, more precisely: for $t \in U(s)$ and $u \in X - U(s)$ $d(s,t) < d(s,u)$.

(b) For $< s;a >, < s;a' > \in U(s)$, we have $< s;a > \prec_s < s;a' > \leftrightarrow a \prec_{s_\infty} a'$

(c) $\mathcal{W}$ and $\mathcal{X}$ are logically equivalent in the language $\mathcal{L}$ of counterfactuals: for all $s \in X$, $\phi \in \mathcal{L}$ $s \models_\mathcal{X} \phi \leftrightarrow s_\infty \models_\mathcal{W} \phi$.

Proof:

(a) Let $n := l(s)$. Note that for all $t \in U(s)$ $d(s,t) \leq \frac{1}{2^n}$. Let $u \notin U(s)$.

Case 1: s is an initial segment of u, with $l(u) > l(s) + 1$: $d(u,s) \geq d_{u\lceil n}(u_{n-1}, u_n) + d_{u\lceil n+1}(u_n, u_{n+1}) \geq \frac{3}{4} * \frac{1}{2^n} + \frac{3}{4} * \frac{1}{2^{n+1}} = \frac{9}{4} * \frac{1}{2^{n+1}} > \frac{1}{2^n}$.

Case 2: u is an initial segment of s: $d(u,s) \geq d(s\lceil n-1, s) = d_{s\lceil n-1}(s_{n-2}, s_{n-1}) \geq \frac{3}{4} * \frac{1}{2^{n-1}} > \frac{1}{2^n}$.

Case 3: Neither s nor u is an initial segment of the other: Then $d(u,s) > d(s\lceil n-1, s)$.

(b) $< s;a > \prec_s < s;a' > :\leftrightarrow d(s, < s;a >) < d(s, < s;a' >) \leftrightarrow d_s(s_\infty, a) < d_s(s_\infty, a') \leftrightarrow d_{s_\infty}(s_\infty, a) < d_{s_\infty}(s_\infty, a') \leftrightarrow a \prec_{s_\infty} a'$.

(c) We show by a straightforward simultaneous induction on the complexity of ϕ:

(1) for all $s \in X$, $\phi \in \mathcal{L}$, we have $s \models_\mathcal{X} \phi \leftrightarrow s_\infty \models_\mathcal{W} \phi$

(2) If $s \not\models_\mathcal{X} \phi$, then $U_{\mathcal{X},\phi}(s) = \{< s;a >: a \in U_{\mathcal{W},\phi}(s_\infty)\}$.

(1) ϕ is a propositional variable: trivial by prerequisite. The cases $\phi = \neg\psi$ and $\phi = \psi \wedge \sigma$ are straightforward. Consider now $\phi = \psi \Rightarrow \sigma$, then $s \models_\mathcal{X} \psi \Rightarrow \sigma :\leftrightarrow U_{\mathcal{X},\psi}(s) \subseteq [[\sigma]]_\mathcal{X}$ and $s_\infty \models_\mathcal{W} \psi \Rightarrow \sigma :\leftrightarrow U_{\mathcal{W},\psi}(s_\infty) \subseteq [[\sigma]]_\mathcal{W}$.

Case 1: $s \models_\mathcal{X} \psi$. Then by induction hypothesis $s_\infty \models_\mathcal{W} \psi$, and $s \models_\mathcal{X} \psi \Rightarrow \sigma$ iff $s \models_\mathcal{X} \sigma$ iff $s_\infty \models_\mathcal{W} \sigma$ iff $s_\infty \models_\mathcal{W} \psi \Rightarrow \sigma$.

Case 2: $s \not\models_\mathcal{X} \psi$: "$\rightarrow$": Let $t' \in U_{\mathcal{W},\psi}(s_\infty) \rightarrow$ (by induction hypothesis) $< s;t' > \in U_{\mathcal{X},\psi}(s) \rightarrow$ (by prerequisite) $< s;t' > \models_\mathcal{X} \sigma \rightarrow$ (by induction hypothesis) $t' \models_\mathcal{W} \sigma$. "$\leftarrow$": Let $t \in U_{\mathcal{X},\psi}(s) \rightarrow$ (by induction hypothesis) $t =< s;t' >$ and $t' \in U_{\mathcal{W},\psi}(s_\infty) \rightarrow$ (by prerequisite) $t' \models_\mathcal{W} \sigma \rightarrow$ (by induction hypothesis) $t \models_\mathcal{X} \sigma$.

(2) Let $t \in U_{\mathcal{X},\phi}(s)$. We first show $t \in U(s)$: Note that $< s; t_\infty > \models_{\mathcal{X}} \phi$ by (1), and if $t \notin U(s)$, then $< s; t_\infty > \prec_s t$ by (a). So $t =< s; a >$ for some $a \in W$. As $t \models_{\mathcal{X}} \phi$, by (1) $a \models_{\mathcal{W}} \phi$, so $a \in [\![\phi]\!]_{\mathcal{W}}$. If $a \notin U_{\mathcal{W},\phi}(s_\infty)$, there must be some $a' \prec_{s_\infty} a$ such that $a' \models_{\mathcal{W}} \phi$, but then $< s; a' > \models_{\mathcal{X}} \phi$ by (1), and $< s; a' > \prec_s < s; a >$ by (b).

Conversely, let $a \in U_{\mathcal{W},\phi}(s_\infty)$, then $< s; a > \in U(s)$, $< s; a > \models_{\mathcal{X}} \phi$ by (1). Suppose now that there is $t \prec_s < s; a >$, $t \models_{\mathcal{X}} \phi$, then by (1) and (a), $t =< s; a' >$ for some $a' \in W$, and $a' \models_{\mathcal{W}} \phi$, but then $a' \prec_{s_\infty} a$ by (b), contradiction.

□ (Lemma 4.3.4)

Clearly by construction X is a set and the $\prec_x$ are ranked and respect 0 for all $x \in X$. By Lemma 4.3.3, d is a metric on X, and by Lemma 4.3.4, $\mathcal{W}$ and $\mathcal{X}$ validate exactly the same formulas of conditional logic. So Proposition 4.3.1 is proven.

□ (Proposition 4.3.1)

4.3.2.3 The limit variant

We conclude this section on counterfactuals with the following result: It might well be that $Con(\phi)$, but that, seen from m, there is no nearest $\phi-$model. In this case, the standard definition of counterfactuals trivializes to $m \models \phi > \bot$.

There is an obvious way out: We do not consider the closest $\phi-$worlds, but say that $m \models \phi > \psi$ iff "from a certain distance onward, the closer we get to m in the set of $\phi-$models, ψ will always hold". This is just the analogue of the limit variant for ranked preferential models, and we will call it the limit variant of the counterfactual semantics.

We now have the following nice result:

Proposition 4.3.5

The limit variant of the counterfactual semantics is equivalent to the minimal variant — provided we start with one copy each, and admit in the result many copies, and we do not have too many models to put into the reals.

Proof:

(Sketch) Start with the original structure and the limit approach. Fix m. Then the result on the limit variant for ranked structures applies (we consider formulas on the left), so, for this m, there is a minimal variant equivalent to the limit one. Do this for all m. Finally glue the resulting structures together to have one metric, with the technique described above. The final construction will have the same logical properties as the original one. □

(We do not know whether we can immediately create an equivalent minimal variant without using additional copies. This seems to be an open problem.)

Chapter 5

Definability preservation

5.1 Introduction

This chapter is addressed to the advanced reader. In particular, we assume familiarity with the basic concepts and definitions of preferential models and of theory revision (see Sections 2.3.1 and 2.3.2), but also with the results for the definability preserving case (Chapters 3 and 4). Explanations and proofs will be more succinct than in preceding (and following) chapters.

5.1.1 The problem

Many representation results for nonmonotonic and similar logics are valid only under certain restrictions. E.g. theories have to be equivalent to single formulas, or certain operators on the model sets have to preserve definability, i.e. applying the operator to the set of models of a theory has to result in the set of models of another theory. We address here this problem, show how to solve it, but will also show that, sometimes, such solutions will be necessarily different from normal characterizations — they have to speak about arbitrary sets of conditions.

The results presented in Sections 3.4 and 4.2.3 (and in [FL94], [Sch92], [Sch96-1], [Sch00-1], and [LMS01]) were all under the caveat that the operators μ and $\mid$ were definability preserving: If $M(T)$ is the set of (propositional) models of a theory T, then $\mu(M(T))$ has to be again exactly the set of models of some theory T'. Likewise, for model sets $M(S)$ and $M(T)$ $M(S) \mid M(T)$ has to be again exactly the set of models of some theory T'.

In the finite case, this will of course trivially hold, but not necessarily in the infinite case. The language of logic is too coarse to describe all possible operators. Counterexamples to the usual characterizations when definability is not preserved were given in [Sch92] and [LMS01], and are repeated here (see Example 5.1.2 below and Example 4.2.3).

More generally, a function $f : \mathcal{P}(M_{\mathcal{L}}) \to \mathcal{P}(M_{\mathcal{L}})$ is called definability preserving iff $f(M(T)) = M(T')$ for some theory T'. In the finite case, this condition holds trivially, as all sets of models are definable by a theory (even a formula). In the infinite case, this need not be the case, as the following, perhaps simplest, example shows:

Example 5.1.1

Let m be any $\mathcal{L}-$model of an infinite language $\mathcal{L}$. Then $M_{\mathcal{L}} - \{m\}$ is not definable, as shown in Fact 1.6.2. Thus, if we define f by $f(M_{\mathcal{L}}) := M_{\mathcal{L}} - \{m\}$, and $f(X) := X$ for any other set $X \subset M_{\mathcal{L}}$, f is not definability preserving. □

In a certain way, the situation is the same as with vector spaces and generating sets. One subspace has many generators, and we cannot identify the two.

When we translate f back into logic, by defining $F(T) := Th(f(M(T)))$, we do not "see" the missing m any more — we just see identity. Of course, there cannot be many such exceptions from identity — otherwise we would see them on the logics side, too. Such exceptions can blur the picture, in the sense that a structure might behave well "grosso modo", just as any definability preserving structure does, but still there might be "small" sets of exceptions. Yet, in cardinality, such "small" sets might be arbitrarily big — this just depends on the size of the language.

Finally, a remark on history and terminology: In [KLM90], definability preservation is called "fullness". The author first described the problem in the context of preferential structures in his [Sch92], and it reappeared in the joint paper [LMS01] on revision. Yet, in the author's opinion, the importance of definability preservation has been largely neglected so far, and the present chapter (and the related paper [Sch00-2]) are, to his knowledge, the first more systematic treatments of the problem.

We now describe shortly the positive results of Chapter 5, and then turn to the negative one.

The positive results of Chapter 5:

We characterize in this chapter not necessarily definability preserving operators, first for preferential structures, then for distance based revision. The basic idea is the same in both cases. We approximate a given choice function or set operator up to a (logically) small set of exceptions. Suppose that $T' = Th(\mu(M(T)))$, the set of formulas valid in the minimal models of T. If μ is definability preserving, then $M(T') = \mu(M(T))$, and there is no model m of T' s.t. there is some model m' of T with $m' \prec m$. If μ is not definability preserving, there might be a model m of T', not in $\mu(M(T))$ and thus a model m' of T s.t. $m' \prec m$. But there may not be many such models m, many in the logical sense, i.e. that there is ϕ s.t. $T' \not\vdash \phi$, $T' \not\vdash \neg\phi$, and $M(T \cup \{\phi\})$ consists of such models — otherwise $\mu(M(T)) \models \neg\phi$. In this sense, the set of such exceptional models is small. Small sets of exceptions can thus be tolerated, they correspond to the coarseness of the underlying language, which cannot describe all sets of models. The quantity of such models can, however, be arbitrarily big, when we measure it by cardinality.

We make this formal, and show two new representation results, one for smooth preferential structures, the other for distance based revision, valid in the absence of such definability preservation. The solution is, roughly, to admit small sets of exceptions to the usual representation results. We also give a new proof to the case of general preferential structures without definability preservation, already solved in [Sch00-2], following now the same strategy as in the other cases, i.e. re-using central parts of the proofs for the definability preserving case. The technique is quite similar in all the cases we describe, so it can certainly be re-applied in other situations, too.

It is in a way ironical that we find formal laws of representation with small sets of exceptions in a domain where we formalize reasoning with and about exceptions.

We now give the reader a concrete example of what can happen to the laws we have investigated so far, when definability preservation is not true any more.

The strategy to patch our previews results will be described in Section 5.1.2.

Example 5.1.2

This example was first given in [Sch92]. It shows that condition (PR) may fail in models which are not definability preserving.

Let $v(\mathcal{L}) := \{p_i : i \in \omega\}$, $n, n' \in M_{\mathcal{L}}$ be defined by $n \models \{p_i : i \in \omega\}$, $n' \models \{\neg p_0\} \cup \{p_i : 0 < i < \omega\}$. Let $\mathcal{M} :=< M_{\mathcal{L}}, \prec>$ where only $n \prec n'$, i.e. just two models are comparable. Let $\mu := \mu_{\mathcal{M}}$, and $\mid\sim$ be defined as usual by μ.

Set $T := \emptyset$, $T' := \{p_i : 0 < i < \omega\}$. We have $M_T = M_{\mathcal{L}}$, $\mu(M_T) = M_{\mathcal{L}} - \{n'\}$, $M_{T'} = \{n, n'\}$, $\mu(M_{T'}) = \{n\}$. So by the result of Example 5.1.1, $\mathcal{M}$ is not definability preserving, and, furthermore, $\overline{\overline{\overline{T}}} = \overline{T}$, $\overline{\overline{\overline{T'}}} = \overline{\{p_i : i < \omega\}}$, *so* $p_0 \in \overline{\overline{T \cup T'}}$, but $\overline{\overline{\overline{T}} \cup T'} = \overline{\overline{T} \cup T'} = \overline{T'}$, so $p_0 \notin \overline{\overline{\overline{T}} \cup T'}$, contradicting (PR).
□

When we work with formulas only (not real theories), some problems with definability preservation will probably not pose themselves: Let $f(M(\phi))$ be some set which is not definable, and $M(T')$ its closure. Then, we can find a theory T s.t. $M(T) \subseteq M(T') - f(M(\phi))$. But there is no such formula ψ, if there were such, then $M(T') \cap M(\neg\psi)$ were to contain $f(M(\phi))$, a contradiction. So, examples showing s.t. with the small set of exceptions cannot be captured — as a matter of fact, any ψ s.t. $M(\psi) \cap (M(T') - f(M(\phi))) \neq \emptyset$, will have $M(\psi) \cap f(M(\phi)) \neq \emptyset$, too. This will probably work in many cases, where we construct counterexamples with such sets $M(T') - f(M(\phi))$. It is thus tempting to conjecture: As long as we consider just model sets definable by a formula, there is no difference between definability preserving and not definability preserving structures, more precisely: If an axiomatization A characterizes the definability preserving instances of a certain class C of structures, then the same A will characterize all instances of C — be they definability preserving or not — as long as we restrict ourselves to sets definable by a single formula as argument. But this is wrong, as the following trivial, though very artificial, example shows:

Example 5.1.3

Fix some arbitrary $x \in U = M_{\mathcal{L}}$ for some infinite language $\mathcal{L}$. Define

$$f(X) := \begin{cases} X \\ \text{or} \\ U - \{x\} \end{cases}$$

Let C be the class of all structures whose choice function obeys this law. Obviously, the law $\overline{\overline{T}} = \overline{T}$ defines the definability preserving structures, but not the other ones. But the law $\overline{\overline{\phi}} = \overline{\phi}$ is not any better. □

The negative results of Chapter 5:

We will use "small" sets of exceptions to solve the representation problem. Yet, as mentioned above, such small sets can be arbitrarily big in cardinality. This has a negative consequence: We cannot describe general, preferential structures without definability preservation by normal logical means, even if we admit infinite expressions — but bounded in size in advance by some fixed cardinal. The necessary size depends on the language. Our characterizations use arbitrary unions, and are thus not bounded in cardinality. To prove the negative result, we construct in Section 5.2.3 a logic which is "almost" preferential, and show that we really use only a small fragment of the structure, and that this fragment can also be obtained in a true preferential structure. This negative result extends then trivially to the general limit variant, which, in nontrivial cases, is not definability preserving — see Fact 3.4.6 for precision and details. An analogous result holds for arbitrary ranked preferential structures and for arbitrary distance defined revision — see Section 5.2.3.

We now describe the remedy in more detail.

5.1.2 The remedy

We first define what a "small" subset is — in purely algebraic terms. There will be no particular properties (apart from the fact that small is downward closed), as long as we do not impose any conditions on $\mathcal{Y}$. (Intuitively, $\mathcal{Y}$ is the set of theory definable sets of models.)

Let $\mathcal{Y} \subseteq \mathcal{P}(Z)$. If $B \in \mathcal{Y}$, $A \subseteq B$ is called a small subset of B iff there is no $X \in \mathcal{Y}$, $B - A \subseteq X \subset B$. If $\mathcal{Y}$ is closed under arbitrary intersections, $Z \in \mathcal{Y}$, $A \subseteq Z$, $\widehat{A}$ will be the smallest $X \in \mathcal{Y}$ with $A \subseteq X$ — the closure, hull, or whatever you like. In the intended application, $\widehat{A}$ is $M(Th(A))$, see Definition 5.1.1.

We will show that our laws hold up to such small sets of exceptions. This is reflected, e.g. in condition (PR) for preferential structures without definability preservation:

($\mid\sim$ 4) Let T, T_i, $i \in I$ be theories s.t. $\forall i\ T_i \vdash T$, then there is no ϕ s.t. $\phi \notin \overline{\overline{T}}$ and $M(\overline{\overline{T}} \cup \{\neg\phi\}) \subseteq \bigcup\{M(T_i) - M(\overline{\overline{T_i}}) : i \in I\}$ (see Conditions 5.2.4), by the nonexistence of ϕ — which corresponds to the nonexistence of intermediate definable subsets. Note that I may be arbitrary big; this depends on the size of the language.

We could also have shown our basic results of Chapters 3 and 4 immediately for the general situation without definability preservation, covering the simpler case by the condition that $\mu = \mu'$ (see below), etc., but, we think, this would have blurred the picture too much and for too long, and introduced complications which are unnecessary in many cases.

The problem and the remedy for preferential structures and distance based revision are very similar. We begin with

5.1.2.1 Preferential structures

We present now the technique used to show the results in outline. The results are, literally and abstractly, very close to those used to obtain the results for the definability preserving case.

Let $\mathcal{Y} := \boldsymbol{D}_{\mathcal{L}}$. For an arbitrary, i.e. not necessarily definability preserving, preferential structure $\mathcal{Z}$ of $\mathcal{L}-$models, let for $X \in \mathcal{Y}$

$\mu'_{\mathcal{Z}}(X) := \mu_{\mathcal{Z}}(X) - \{x : \exists Y \in \mathcal{Y},\ Y \subseteq X,\ x \in Y - \mu(Y)\} = \{x \in X : \neg\exists Y \in \mathcal{Y}(Y \subseteq X \text{ and } x \in Y - \mu_{\mathcal{Z}}(Y)\}$ (see Definition 5.2.2). μ' (we omit the index $\mathcal{Z}$, when this does not create any ambiguity), and its modification adequate for the smooth case, are the central definitions, and will replace μ in the technical development.

Note that, $\mu(X) = \widehat{\mu'(X)}$, i.e. that $\mu(X) - \mu'(X)$ is small, and, if $\mathcal{Z}$ is definability preserving, then $\mu' = \mu$.

For representation, we consider now the Conditions 5.2.2:

$(\mu\emptyset)$ $U \neq \emptyset \rightarrow \mu(U) \neq \emptyset$,

$(\mu \subseteq)$ $\mu(U) \subseteq U$,

$(\mu 2)$ $\mu(U) - \mu'(U)$ is small, where $\mu'(U) := \{x \in U : \neg\exists Y \in \mathcal{Y}(Y \subseteq U$ and $x \in Y - \mu(Y))\}$

$(\mu 2s)$ $\mu(U) - \mu'(U)$ is small, where $\mu'(U) := \{x \in U : \neg\exists U' \in \mathcal{Y}(\mu(U \cup U') \subseteq U$ and $x \in U' - \mu(U'))\}$

(μCUM) $\mu(X) \subseteq Y \subseteq X \rightarrow \mu(X) = \mu(Y)$

for $X, Y, U \in \mathcal{Y}$.

and show that they — $(\mu \subseteq)$ and $(\mu 2)$ in the general case, $(\mu\emptyset)$, $(\mu \subseteq)$, $(\mu 2s)$, (μCUM) in the smooth case — imply a list of properties for μ' and $H(U) := \bigcup\{X \in \mathcal{Y} : \mu(X) \subseteq U\}$, described in Conditions 5.2.1 and 5.2.3. This is shown in Fact 5.2.2 and Fact 5.2.6.

We then show that such μ' can be represented by a (general or smooth) preferential structure, which can be chosen transitive. The strategy and execution is largely the same as for the definability preserving case. For the general case, this is formulated in Proposition 5.2.4, subsequent to an auxiliary lemma Fact 5.2.3. For the smooth case, this is formulated in Proposition 5.2.7. The proof of the transitive case is verbatim the same as for the definability preserving case. The first part, i.e. representation without transitivity, is done by an alternative construction, hinted at in Section 3.3, in Construction 5.2.1 and Fact 5.2.8.

It remains to replace, better approximate, μ' by μ to obtain representation, this is possible, as they differ only by small sets. E.g., in the general case, we obtain by putting our results together, Proposition 5.2.5:

Let Z be an arbitrary set, $\mathcal{Y} \subseteq \mathcal{P}(Z)$, $\mu : \mathcal{Y} \to \mathcal{Y}$, $\mathcal{Y}$ closed under arbitrary intersections and finite unions, and $\emptyset, Z \in \mathcal{Y}$, and let $\widehat{\cdot}$ be defined wrt. $\mathcal{Y}$.

(a) If μ satisfies $(\mu \subseteq)$ and $(\mu 2)$, then there is a transitive preferential structure $\mathcal{Z}$ over Z s.t. for all $U \in \mathcal{Y}$ $\mu(U) = \widehat{\mu_{\mathcal{Z}}(U)}$.

(b) If $\mathcal{Z}$ is a preferential structure over Z and $\mu : \mathcal{Y} \to \mathcal{Y}$ s.t. for all $U \in \mathcal{Y}$ $\mu(U) = \widehat{\mu_{\mathcal{Z}}(U)}$, then μ satisfies $(\mu \subseteq) - (\mu 2)$.

In the smooth case, we obtain a similar result (Proposition 5.2.9), the final proof is only slightly more complicated.

We turn to the logical counterpart.

The conditions are a little annoying, as we write down "small set of exceptions", or conditions like $(\mu 2)$ above in all logical detail. The proofs, however, are straightforward, even if they cover some pages.

5.1.2.2 Theory revision

Just as we have approximated μ by μ' for preferential structures, we approximate $\mid$ by $\mid'$ for revision. We consider $\mid'$ s.t. $A \mid B = \widehat{A \mid' B}$, formulate suitable conditions for $\mid'$, in particular a loop condition (see Conditions 5.3.1), show that $\mid'$ can be represented by a distance (Proposition 5.3.1, using the result on definability preserving distance based revision) and obtain an analogue of above Proposition 5.2.5 in Proposition 5.3.2, which describes the back and forth between $\mid$ and $\mid'$. The logical Conditions 5.3.2 describe the logical situation, Proposition 5.3.3 is the logical counterpart of Proposition 5.3.2.

The techniques are the same as those for the preferential case.

5.1.2.3 Summary

We show that definability preservation is an important property, the lack of which can make important laws fail. If we do not have definability preservation, we can still have (essentially) our old laws, we just have to admit small sets of exceptions — where small is relative to the expressiveness of classical propositional logic (in the infinite case, in the finite case, there is no problem).

We thus make precise what a small set is in the context of propositional logic. Of course, this is close to the concept of small set and exceptions in the rest of the book, but it is used in a special situation, if you like bombastic expressions, on the meta-level.

We then examine operators μ' and $|'$ sufficiently close to μ and $|$ respectively, to inherit still many of their properties, and show that these operators can be represented by a preference relation (a distance respectively). We then play the situation back to the original, very similar, μ and $|$. "Sufficiently close" has a somewhat nasty translation into logic, which makes the logical conditions cumbersome — and a usual characterization impossible, see Section 5.2.3.

Recommended reading:

The reader should first get familiar with the operation $\widehat{\ .\ }$ and consider the relation between μ and μ'.

We then suggest to follow the details of the general preferential case, the logical part included. This contains the main ideas.

He/she may then read the negative results in Section 5.2.3, which show that we are necessarily far from usual characterizations.

The rest of the chapter can then be read in any order.

We now give the basic definitions, and recall the corresponding representation results for definability preserving choice functions from Sections 3.4 and 4.2.3. We then discuss in detail the formal development to be presented in Sections 5.2 and 5.3.

5.1.3 Basic definitions and results

5.1.3.1 General part

Definition 5.1.1

Let $\mathcal{Y} \subseteq \mathcal{P}(Z)$ be given.

(1) For $B \in \mathcal{Y}$, we call $A \subseteq B$ a small subset of B (and $B - A$ dense in B) iff there is no $X \in \mathcal{Y}$ s.t. $B - A \subseteq X \subset B$.

(2) If $\mathcal{Y}$ is closed under arbitrary intersections, $Z \in \mathcal{Y}$, and $A \subseteq Z$, $\overbrace{A}$ will be the smallest $X \in \mathcal{Y}$ with $A \subseteq X$.

Intuitively, $\mathcal{Y}$ is $\boldsymbol{D}_{\mathcal{L}}$, and $\overbrace{A} = M(Th(A))$.

Fact 5.1.1

If $\mathcal{Y} \subseteq \mathcal{P}(Z)$ is closed under arbitrary intersections and finite unions, $Z \in \mathcal{Y}$, $X, Y \subseteq Z$, then condition

(~ 1) $\overbrace{X \cup Y} = \overbrace{X} \cup \overbrace{Y}$

holds.

Proof:

Let $\mathcal{Y}(U) := \{X \in \mathcal{Y} : U \subseteq X\}$. If $A \in \mathcal{Y}(X \cup Y)$, then $A \in \mathcal{Y}(X)$ and $A \in \mathcal{Y}(Y)$, so $\overbrace{X \cup Y} \supseteq \overbrace{X} \cup \overbrace{Y}$. If $A \in \mathcal{Y}(X)$ and $B \in \mathcal{Y}(Y)$, then $A \cup B \in \mathcal{Y}(X \cup Y)$, so $\overbrace{X \cup Y} \subseteq \overbrace{X} \cup \overbrace{Y}$. □

Recall the following central definitions:

Definition 5.1.2

A preferential structure $\mathcal{Z}$ is called definability preserving (for some fixed language $\mathcal{L}$) iff for all $\mathcal{L}-$theories T $\mu_{\mathcal{Z}}(M(T)) = M(T')$ for some $\mathcal{L}-$theory T'.

Definition 5.1.3

A binary function $|$ on the sets of models of some logic is called definability preserving iff $M(T) \mid M(T')$ is again the set of models of some theory S

for all theories T, T'. By abuse of language, a pseudo-distance d is called definability preserving iff $|_d$ is.

5.1.3.2 Discussion of the technical development

We first describe our approach in the case of general preferential structures. For simplicity, we neglect the use of multiple copies of models. If $\mathcal{M}$ is not a definability preserving preferential model, and if $\overline{\overline{T}}$ is the set of consequences of T in $\mathcal{M}$, not all models of $\overline{\overline{T}}$ need be minimal in the set of models of T. Thus, in such a structure, $M(\overline{\overline{T}})$ may contain nonminimal elements m, which we may identify through some T' s.t. $T' \vdash T$, $m \models T'$, $m \not\models \overline{\overline{T'}}$, i.e. $m \in M(T')$, but $m \notin \mu(M(T'))$. We eliminate these from μ, and consider $\mu'(M(T)) := \mu(M(T))$ - $\{m : \exists T'$ s.t. $T' \vdash T, m \in M(T') - \mu(M(T'))\}$. Note that we may take away here arbitrarily many elements (by cardinality), this is important in order to understand the negative results of Section 5.2.3. If we set $\mathcal{Y} := \boldsymbol{D}_{\mathcal{L}}$, we have $\mu : \mathcal{Y} \to \mathcal{Y}$, $\mu' : \mathcal{Y} \to \mathcal{P}(Z)$, and $\mu(U) = \widehat{\mu'(U)}$. To summarize, given a preferential structure, we have μ and μ' as defined above, and $\mu(U) \subseteq U$ and $\mu(U) = \widehat{\mu'(U)}$ will hold. Conversely, given such μ and μ', we can show that μ' satisfies the conditions

$(\mu' \subseteq)$ $\mu'(U) \subseteq U$

and

$(\mu'2)$ $x \in \mu'(U)$, $x \in Y - \mu'(Y) \to Y \not\subseteq U$.

But such μ' can be represented by a preferential structure (see Proposition 5.2.4), i.e. there is $\mathcal{Z}$ s.t. $\mu'(U) = \mu_{\mathcal{Z}}(U)$ for all $U \in \mathcal{Y}$, and thus $\mu(U) = \widehat{\mu'(U)} = \widehat{\mu_{\mathcal{Z}}(U)}$, the representation result we were looking for. If we examine the completeness proof for the definability preserving case, we see that we did use exactly the same properties, i.e. $\mu(U) \subseteq U$ and $x \in \mu(U)$, $x \in Y - \mu(Y) \to Y \not\subseteq U$ of μ to construct a representing preferential structure for μ. The only difference is that $\mu : \mathcal{Y} \to \mathcal{Y}$, and $\mu' : \mathcal{Y} \to \mathcal{P}(Z)$, but the proof does not use the fact that $\mu(U) \in \mathcal{Y}$ for $U \in \mathcal{Y}$. Consequently, we can use verbatim the same proof to show representation of μ', as we used to show representation of μ in the definability preserving case.

The cases of smooth preferential structures and distance based revision are analogous:

Given a smooth preferential structure, we define again μ' from μ (in a slightly different way to take care of the "semi-transitivity" of smooth structures), show that μ and μ' have certain properties, which, conversely, suffice to construct a representing structure. Again, we first show that μ' has the properties we needed for μ to construct a representing structure in the definability preserving case, so we can again use the main part of the old proof for the definability preserving case.

A distance d on a set U defines a revision operator $*$ by $T*T' := Th(M(T) \mid M(T'))$, where $A \mid B$ is the set of elements of B, closest to the set A. If $\mid$ is definability preserving, we have $\mid: \mathcal{Y} \times \mathcal{Y} \rightarrow \mathcal{Y}$. If not, we define again $\mid': \mathcal{Y} \times \mathcal{Y} \rightarrow \mathcal{P}(U)$, by $A \mid' B := \{b \in B : \forall B' \in \mathcal{Y}(B' \subseteq B, b \in B' \rightarrow b \in A \mid B')\}$ (see Proposition 5.3.2). We show that any operator $\mid$ defined by a distance, and $\mid'$ defined this way from $\mid$ have certain conditions, which, conversely, suffice to construct a representing distance. Again, the proof works via showing that $\mid'$ has properties which $\mid$ has in the definability preserving case, and which were used (and sufficient) to prove representation of $\mid$ in the latter case.

Those parts of the proofs for the definability preserving cases which we can re-use verbatim are omitted and the reader is referred to Sections 3.2, 3.3, 3.4, 4.2.2, 4.2.3 for details. We will also give a simple and complete proof for the nontransitive, smooth case, hinted at in Section 3.3, which is contained in Fact 5.2.6, Construction 5.2.1, and Fact 5.2.8.

The logical parts follow closely the development in Sections 3.4 and 4.2.3.

Comments on the negative result:

We discuss now the negative result, that there is no normal characterization of general preferential structures possible. Similar results hold for the ranked case, and for distance defined theory revision.

Let κ be any infinite cardinal. We show that there is no characterization Φ of general (i.e. not necessarily definability preserving) preferential structures which has size $\leq \kappa$. We suppose there were one such characterization Φ of size $\leq \kappa$, and construct a counterexample. The idea of the proof is very simple.

Take the language $\mathcal{L}$ defined by $p_i : i < \kappa$. We show that it suffices to consider for any given instantiation of $\Phi \leq \kappa$ many pairs $m \prec m^-$ in a case not representable by a preferential structure, and that $\leq \kappa$ many such pairs give the same result in a true preferential structure. Thus, every instantiation is true in an "illegal" and a "legal" example, so Φ cannot discern between legal and illegal examples. The main work is to show that $\leq \kappa$ many pairs suffice in the illegal example. This is, again, in principle, easy, we show that there is a "best" set of size $\leq \kappa$ which calculates $\overline{\overline{T}}$ for all T considered in the instantiation.

For any model m with $m \models p_0$, let m^- be exactly like m with the exception that $m^- \models \neg p_0$.

Define the logic $\mid\sim$ as follows in two steps:

(1) $\overline{\overline{Th(\{m, m^-\})}} := Th(\{m\})$. (Speaking preferentially, $m \prec m^-$, for all such m, m^-, this will be the entire relation. The relation is thus extremely simple, $\prec$ –paths have at most length 1, so $\prec$ is automatically transitive.)

We now look at (in terms of preferential models only some!) consequences:

(2) $\overline{\overline{T}} :=$ Th ($\bigcap\{M(Th(M(T) - A)) : card(A) \leq \kappa,\ A \subseteq M(T),\ \forall n(n \in A \rightarrow n = m^-$ and $m, m^- \in M(T))\}$).

This, without the size condition, would be exactly the preferential consequence of part (1) of the definition, but this logic as it stands is not preferential.

Suppose there were a characterization of size $\leq \kappa$. It has to say "no" for at least one instance of the universally quantified condition Φ. We will show that we find a true preferential structure where this instance of Φ has the same truth value, a contradiction. To demonstrate it, we consider the preferential structure where we do not make all $m \prec m^-$, but only the κ many of them we have used in the instance of Φ. We will see that the expression Φ still fails with our instances.

5.1.4 A remark on definability preservation and modal logic

We conclude this introductory Section 5.1 with a remark on definability preservation in classical modal logic, which shows that local evaluation can circumvent the problem.

Let R be an arbitrary binary relation.

Set $f(X) := \{y \in U : \forall z.yRz \rightarrow z \in X\}$,

i.e. (almost) the inverse image of R or $\Box$, as used by modal logic.

Set $T \mid\sim \phi$ iff $f(M(T)) \models \phi$.

Fact 5.1.2

(1) If $\forall x \exists y(xRy)$, then $f(\emptyset) = \emptyset$,

(2) $f(A \cap B) = f(A) \cap f(B)$,

(3) $Th(M(T) \cap M(T')) = \overline{T \cup T'}$,

(4) If f is definability preserving, then $\overline{\overline{T \cup T'}} = \overline{\overline{\overline{T}} \cup \overline{\overline{T'}}}$.

Proof:

(1) trivial.

(2) $x \in f(A \cap B) \leftrightarrow \forall y(xRy \to y \in A \cap B) \leftrightarrow \forall y(xRy \to y \in A) \wedge \forall y(xRy \to y \in B) \leftrightarrow x \in f(A) \wedge x \in f(B)$.

(3) $M(T) \cap M(T') = M(T \cup T')$, so $Th(M(T) \cap M(T')) = Th(M(T \cup T')) = \overline{T \cup T'}$.

(4) $f(M(T \cup T')) = f(M(T) \cap M(T')) = f(M(T)) \cap f(M(T')) =$ (by definability preservation) $M(T_0) \cap M(T_0')$ for some T_0, T_0', so $\overline{\overline{T \cup T'}} = Th(f(M(T \cup T'))) = Th(M(T_0) \cap M(T_0')) = \overline{T_0 \cup T_0'} = \overline{\overline{\overline{T}} \cup \overline{\overline{T'}}}$. □

Consider now the following

Example 5.1.4

Take the language: p, q, p_i, $i < \omega$. For any model m and propositional variable r, let m_r be the same model as m, only r has the opposite value.

The relation: for any $q-$model m mRm_q, i.e. a $q-$model is in $R-$relation with the analog $\neg q-$model. Let $n \models p, \neg q, p_i$, all i, and $M' := M(\neg q) - \{n\}$. Let n be in $R-$relation with all $p \wedge q-$models, and all $m \in M'$ in $R-$relation with all $\neg p \wedge q-$models.

Then $f(\emptyset) = \emptyset$ holds, $f(M(p \wedge q)) = \{n\}$, $f(M(\neg p \wedge q)) = M'$, $M(\{p \wedge q\} \cup \{\neg p \wedge q\}) = \emptyset$, so $\overline{\overline{\{p \wedge q\} \cup \{\neg p \wedge q\}}} = Th(\emptyset) = \mathcal{L}$. On the other hand, $\overline{\overline{\{p \wedge q\}}} = Th(\{n\})$ and $\overline{\overline{\{\neg p \wedge q\}}} = \overline{\neg q}$, *so* $\overline{\overline{\overline{\{p \wedge q\}}} \cup \overline{\overline{\{\neg p \wedge q\}}}} = Th(\{n\})$. Consequently, (4) above does not hold, and the definability preservation property can also be felt in the context of classical modal logic. □

Comments:

(1) The definability problem occurs here already with formulas, we do not need full theories to see the problem.

(2) In modal logic, the local evaluation prevents definability preservation problems, as we do not look at $M(Th(X))$, but work directly with single models, thus, we do not see the "blurring" effect between X and $M(Th(X))$.

5.2 Preferential structures

5.2.1 The algebraic results

We assume that the reader is familiar with the basic concepts, definitions, and results on preferential structures (see Section 2.3.1 and Chapter 3).

5.2.1.1 The conditions

We define in this introductory part first the hull H, and then two versions of the approximation μ' to μ (for the general and the smooth case), and formulate the Conditions 5.2.2 for μ, which we will use for characterization. We then give another set of conditions, Conditions 5.2.3 for μ', which, as we will show in Fact 5.2.2 and 5.2.6, are implied by the first set in Conditions 5.2.2. We will then work with the conditions for μ' in the proofs, and represent μ' just as we represented μ in Chapter 3, using suitable conditions. This is done in Proposition 5.2.4 and 5.2.7. Proposition 5.2.5 and 5.2.9 bridge the gap between μ and μ', and state the results we worked for, i.e. representation of μ.

Let in the following $\mathcal{Y} \subseteq \mathcal{P}(Z)$ be closed under arbitrary intersections and finite unions, $\emptyset, Z \in \mathcal{Y}$, and let $\widehat{\cdot}$ be defined wrt. $\mathcal{Y}$. So Fact 5.1.1 holds. Let $\mu : \mathcal{Y} \to \mathcal{Y}$. Smoothness will also be wrt. $\mathcal{Y}$.

Definition 5.2.1

Define $H : \mathcal{Y} \to \mathcal{P}(Z)$ by $H(U) := \bigcup\{X \in \mathcal{Y} : \mu(X) \subseteq U\}$ for $U \in \mathcal{Y}$.

Condition 5.2.1

(H1) $U \subseteq H(U)$,

(H2) $U \subseteq U' \to H(U) \subseteq H(U')$

for $U, U' \in \mathcal{Y}$.

Fact 5.2.1

Conditions (H1) and (H2) hold for H as defined in Definition 5.2.1, if $\mu(U) \subseteq U$. □

We will define for $U \in \mathcal{Y}$ in the general case:

$\mu'(U) := \{x \in U : \neg\exists Y \in \mathcal{Y}(Y \subseteq U \text{ and } x \in Y - \mu(Y))\}$

and in the smooth case:

$\mu'(U) := \{x \in U : \neg\exists U' \in \mathcal{Y}(\mu(U \cup U') \subseteq U \text{ and } x \in U' - \mu(U'))\}$.

Note that even if $U \in \mathcal{Y}$, $\mu'(U)$ is not necessarily in $\mathcal{Y}$.

$\mu(U) - \mu'(U)$ will be required to be small. Consider first the general case: Any $x \in Y \subseteq U$ s.t. $x \in Y - \mu(Y)$ cannot be minimal in U, as any element smaller than x, which is in Y, will also be present in U. (This is the fundamental condition for preferential structures, see Proposition 3.2.2.) If μ is definability preserving, no exceptions are possible. Consider now the smooth case: If $x \in U' - \mu(U')$, then x cannot be minimal in U', and for the same reason, not in $U \cup U'$. But, if $\mu(U \cup U') \subseteq U$, then, by smoothness, there must be y minimal in $U \cup U'$, i.e. in U, smaller than x, so x is not minimal in U. This captures the "semi-transitivity" of smooth structures, which can perhaps best be described by the following simple situation: If $x \in X$ is not minimal in X, then there is $y < x$, $y \in X$. Suppose that y is not minimal either, but there is $z < y$, z minimal in X. Then $z < x$ need not hold, but there will be $z' < x$, z' minimal in X. If there is just one minimal element in X, then we have full transitivity, we have only "almost" transitivity.

We make now the conditions for μ, μ', and H precise.

Condition 5.2.2

$(\mu\emptyset)$ $U \neq \emptyset \rightarrow \mu(U) \neq \emptyset$,

$(\mu \subseteq)$ $\mu(U) \subseteq U$,

$(\mu 2)$ $\mu(U) - \mu'(U)$ is small, where

$\mu'(U) := \{x \in U : \neg\exists Y \in \mathcal{Y}(Y \subseteq U \text{ and } x \in Y - \mu(Y))\}$,

$(\mu 2s)$ $\mu(U) - \mu'(U)$ is small, where

$\mu'(U) := \{x \in U : \neg\exists U' \in \mathcal{Y}(\mu(U \cup U') \subseteq U \text{ and } x \in U' - \mu(U'))\}$,

(μCUM) $\mu(X) \subseteq Y \subseteq X \rightarrow \mu(X) = \mu(Y)$

for $X, Y, U \in \mathcal{Y}$.

Note that $(\mu 2)$ contains essentially the fundamental condition $X \subseteq Y \rightarrow \mu(Y) \cap X \subseteq \mu(X)$ of preferential structures. To see this, it suffices to take $\emptyset$ as the only small set, or $\mu(U) = \mu'(U)$. We re-emphasize that "small" does not mean "small by cardinality" — see Section 5.2.3.

Condition 5.2.3

$(\mu' \subseteq)$ $\mu'(U) \subseteq U$,

$(\mu'2)$ $x \in \mu'(U)$, $x \in Y - \mu'(Y) \rightarrow Y \not\subseteq U$,

$(\mu'\emptyset)$ $U \neq \emptyset \rightarrow \mu'(U) \neq \emptyset$,

$(\mu'4)$ $\mu'(U \cup Y) - H(U) \subseteq \mu'(Y)$,

$(\mu'5)$ $x \in \mu'(U)$, $x \in Y - \mu'(Y) \rightarrow Y \not\subseteq H(U)$,

$(\mu'6)$ $Y \not\subseteq H(U) \rightarrow \mu'(U \cup Y) \not\subseteq H(U)$

for $Y, U \in \mathcal{Y}$.

Note that $(\mu'5)$ implies $(\mu'2)$ if (H1) holds.

Outline of the proofs:

In both cases, i.e. the general and the smooth case, we follow the same strategy: First, we show from the conditions on μ – $(\mu \subseteq)$, $(\mu 2)$ in the general case, $(\mu\emptyset)$, $(\mu \subseteq)$, $(\mu 2s)$, (μCUM) in the smooth case — that certain conditions hold for μ' (and for H in the smooth case) — $(\mu' \subseteq)$, $(\mu'2)$ in the general case, $(\mu' \subseteq)$, $(\mu'\emptyset)$, $(\mu'4)$–$(\mu'6)$ in the smooth case. We then show that any $\mu' : \mathcal{Y} \rightarrow \mathcal{P}(Z)$ satisfying these conditions can be represented by a (smooth) preferential structure $\mathcal{Z}$, and that the structure can be chosen transitive. As the proof for the not necessarily transitive case is easier, we do this proof first, and then the transitive case. The basic ideas are mostly the same as those used in Sections 3.2 and 3.3. Finally, we show that if $\mathcal{Z}$ is a [smooth] preferential structure, $(\mu \subseteq)$, $(\mu 2)$ [$(\mu\emptyset)$, $(\mu \subseteq)$, $(\mu 2s)$, (μCUM) in the smooth case] will hold for $\widehat{\mu_{\mathcal{Z}}}$. Moreover, if μ' was defined from μ as indicated, and if in addition $(\mu 2)$ [or $(\mu 2s)$] holds, then $\mu = \widehat{\mu_{\mathcal{Z}}}$. Putting all these things together results in representation results for the general and the smooth case, Proposition 5.2.5 and 5.2.9.

Our basic idea to construct representing preferential structures is to consider (modifications of) the following construction: $\mathcal{X} := \{< x, f >: x \in Z \wedge f \in \Pi_x\}$, $< x', f' > \prec < x, f > :\leftrightarrow x' \in ran(f)$, and $\mathcal{Z} := < \mathcal{X}, \prec >$. where $\Pi_x := \Pi\{Y \in \mathcal{Y} : x \in Y - \mu'(Y)\}$ - see Sections 3.2 and 3.3.

5.2.1.2 The general case

We show here the representation result for the general (and transitive) case (Proposition 5.2.5). It is an easy consequence of the main auxiliary result, Proposition 5.2.4, which has verbatim the same proof as the corresponding result for definability preserving preferential structures (Propositions 3.2.2 and 3.2.4).

We first construct μ' from μ.

Definition 5.2.2

$\mu'(U) := \{x \in U : \neg\exists Y \in \mathcal{Y}(Y \subseteq U \text{ and } x \in Y - \mu(Y))\}$

We now show the properties of μ' which will be used for representation in Proposition 5.2.4.

Fact 5.2.2

Let μ' be defined as in Definition 5.2.2, then:

(a) $(\mu' \subseteq)$ and $(\mu'2)$ hold,

(b) $\mu'(U) \subseteq \mu(U)$.

Proof:

(a) $(\mu' \subseteq)$ is trivial. For $(\mu'2)$: If $x \in Y - \mu'(Y)$, then, by definition, there is $Y' \in \mathcal{Y}$, $Y' \subseteq Y$ and $x \in Y' - \mu(Y')$. If, in addition, $Y \subseteq U$, then $Y' \subseteq U$, so $x \notin \mu'(U)$.

(b) Take $Y := U$. □

In preparation of the proof of Proposition 5.2.4, we show the following Fact 5.2.3. The rest of the proof is verbatim the same as the proof of Propositions 3.2.2 and 3.2.4, and the reader is referred there.

Set $\mathcal{Y}_x := \{Y \in \mathcal{Y} : x \in Y - \mu'(Y)\}$ and $\Pi_x := \Pi\mathcal{Y}_x$.

Fact 5.2.3

If μ' satisfies $(\mu' \subseteq)$ and $(\mu'2)$, then $x \in \mu'(U) \leftrightarrow x \in U \wedge \exists f \in \Pi\{Y \in \mathcal{Y} : x \in Y - \mu'(Y)\}.ran(f) \cap U = \emptyset$.

Proof:

"$\rightarrow$": $x \in \mu'(U) \rightarrow x \in U$ by $(\mu' \subseteq)$. Let $x \in \mu'(U)$, $Y \in \mathcal{Y}$, $x \in Y - \mu'(Y)$, then $Y \not\subseteq U$ by $(\mu'2)$, so there is $f \in \Pi_x.ran(f) \cap U = \emptyset$.

"$\leftarrow$": $x \notin \mu'(U)$, $x \in U$ implies $\exists Y \in \mathcal{Y}(Y \subseteq U, x \in Y - \mu(Y))$, so by Fact 5.2.2, (b) $\exists Y \subseteq U.Y \in \mathcal{Y}_x$, so for all $f \in \Pi_x$ $ran(f) \cap U \neq \emptyset$. □

The following Proposition 5.2.4 shows representation for μ'.

Proposition 5.2.4

(a) If μ' satisfies $(\mu' \subseteq)$ and $(\mu' 2)$, then there is a preferential structure $\mathcal{Z}$ over Z s.t. $\mu' = \mu_{\mathcal{Z}}$.

(b) $\mathcal{Z}$ can be chosen transitive.

The proof is the same as the proof of Propositions 3.2.2 and 3.2.4. □

Proposition 5.2.5 uses the approximation of μ by μ' and the representation of μ' demonstrated in Proposition 5.2.4, to obtain a representation for μ.

Proposition 5.2.5

Let Z be an arbitrary set, $\mathcal{Y} \subseteq \mathcal{P}(Z)$, $\mu : \mathcal{Y} \to \mathcal{Y}$, $\mathcal{Y}$ closed under arbitrary intersections and finite unions, and $\emptyset, Z \in \mathcal{Y}$, and let $\widehat{\cdot}$ be defined wrt. $\mathcal{Y}$.

(a) If μ satisfies $(\mu \subseteq)$, $(\mu 2)$, then there is a transitive preferential structure $\mathcal{Z}$ over Z s.t. for all $U \in \mathcal{Y}$ $\mu(U) = \widehat{\mu_{\mathcal{Z}}(U)}$.

(b) If $\mathcal{Z}$ is a preferential structure over Z and $\mu : \mathcal{Y} \to \mathcal{Y}$ s.t. for all $U \in \mathcal{Y}$ $\mu(U) = \widehat{\mu_{\mathcal{Z}}(U)}$, then μ satisfies $(\mu \subseteq)$, $(\mu 2)$.

Proof:

(a) Let μ satisfy $(\mu \subseteq)$, $(\mu 2)$. μ' as defined in Definition 5.2.2 and in $(\mu 2)$ satisfies properties $(\mu' \subseteq)$, $(\mu' 2)$ by Fact 5.2.2. Thus, by Proposition 5.2.4, there is a transitive structure $\mathcal{Z}$ over Z s.t. $\mu' = \mu_{\mathcal{Z}}$, but by $(\mu 2)$ $\mu(U) = \widehat{\mu'(U)} = \widehat{\mu_{\mathcal{Z}}(U)}$ for $U \in \mathcal{Y}$.

(b) $(\mu \subseteq) : \mu_{\mathcal{Z}}(U) \subseteq U$, so by $U \in \mathcal{Y}$ $\mu(U) = \widehat{\mu_{\mathcal{Z}}(U)} \subseteq U$.

$(\mu 2)$: If $(\mu 2)$ is false, there is $U \in \mathcal{Y}$ s.t. for $U' := \bigcup\{Y' - \mu(Y') : Y' \in \mathcal{Y}, Y' \subseteq U\}$ $\widehat{\mu(U) - U'} \subset \mu(U)$. *By* $\mu_{\mathcal{Z}}(Y') \subseteq \mu(Y')$, $Y' - \mu(Y') \subseteq Y' - \mu_{\mathcal{Z}}(Y')$. No copy of any $x \in Y' - \mu_{\mathcal{Z}}(Y')$ with $Y' \subseteq U$, $Y' \in \mathcal{Y}$ can be minimal in $\mathcal{Z} \lceil U$. Thus, by $\mu_{\mathcal{Z}}(U) \subseteq \mu(U)$, $\mu_{\mathcal{Z}}(U) \subseteq \mu(U) - U'$, so $\widehat{\mu_{\mathcal{Z}}(U)} \subseteq \widehat{\mu(U) - U'} \subset \mu(U)$, contradiction. □

5.2.1.3 The smooth case

The main result here is Proposition 5.2.9. It is a direct and easy consequence of Proposition 5.2.7. The construction in the proof of part (a) of Proposition 5.2.7 is an adaptation of a simplification hinted at in Section 3.3. It uses the properties of μ' demonstrated in Fact 5.2.6, in particular validity of the Conditions 5.2.3. Part (b) has again verbatim the same proof as the corresponding result for definability preserving smooth and transitive structures.

Generally, in smooth preferential structures $\mu(U)$ may be empty, even if U is not — it suffices to consider the trivial empty structure as an example. This introduces unnecessary and rather uninteresting complications. To avoid this, we shall assume that $U \neq \emptyset \rightarrow \mu(U) \neq \emptyset$, and the structures $< \mathcal{X}, \prec >$ considered will be over Z, i.e. for all $z \in Z$ there is some $< z, i > \in \mathcal{X}$.

Again, we first construct μ' from μ.

Definition 5.2.3

$\mu'(U) := \{x \in U : \neg \exists U' \in \mathcal{Y}(x \in U' - \mu(U') \text{ and } \mu(U \cup U') \subseteq U)\}.$

We show the important properties of μ' in the following Fact 5.2.6. They will be used for the representation of μ' in Proposition 5.2.7.

Fact 5.2.6

Let μ satisfy $(\mu\emptyset)$, $(\mu \subseteq)$, $(\mu 2s)$, (μCUM) and H and μ' be defined from μ as in Definitions 5.2.1 and 5.2.3 (and $(\mu 2s)$). Let A, B, U, U', X, Y, A_i for $i \in I$ be elements of $\mathcal{Y}$. Then:

(1) $\mu(U \cup U') \subseteq U \leftrightarrow \mu(U \cup U') = \mu(U)$,

(2) $\mu'(U) \subseteq \mu(U)$, and $\mu'(U) \subseteq U$,

(3) $\mu(A \cup B) \subseteq \mu(A) \cup \mu(B)$,

(4) $\mu'(U) \subseteq U' \leftrightarrow \mu(U) \subseteq U'$,

(5) if $\bigcup\{A_i : i \in I\} \in \mathcal{Y}$, and for all i $\mu(U \cup A_i) \subseteq U$, then $\mu(U \cup \bigcup A_i) \subseteq U$,

(6) we can define equivalently $H(U)$ via μ',

(7) $X \subseteq Y$, $\mu(X \cup U) \subseteq X \rightarrow \mu(Y \cup U) \subseteq Y$,

(8) $X \subseteq Y$, $x \in X$, $x \notin \mu'(X) \rightarrow x \notin \mu'(Y)$,

(9) $\mu'(X) \subseteq Y \subseteq X \to \mu'(X) = \mu'(Y)$,

(10) $X \neq \emptyset \to \mu'(X) \neq \emptyset$,

(11) $\mu'(A_1 \cup A_2) \subseteq \mu'(A_1) \cup \mu'(A_2)$,

(12) $\mu'(U \cup Y) - H(U) \subseteq \mu'(Y)$,

(13) $U \subseteq A$, $\mu'(A) \subseteq H(U) \to \mu'(A) \subseteq U$,

(14) $\mu'(Y) \subseteq H(U) \to Y \subseteq H(U)$ and $\mu'(U \cup Y) = \mu'(U)$,

(15) if $Y \subseteq H(U)$, then $\mu(U \cup Y) \subseteq U$,

(16) $x \in \mu'(U)$, $x \in Y - \mu'(Y) \to Y \not\subseteq H(U)$,

(17) $Y \not\subseteq H(U) \to \mu'(U \cup Y) \not\subseteq H(U)$.

Proof:

(1) $\mu(U \cup U') \subseteq U \subseteq U \cup U' \to_{(\mu CUM)} \mu(U \cup U') = \mu(U)$.

(2) $\mu'(U) \subseteq U$ by definition. If $x \in U - \mu(U)$, take $U' := U$ in the definition of $\mu'(U)$, this shows $x \notin \mu'(U)$.

(3) By definition of μ', we have $\mu'(A \cup B) \subseteq A \cup B$, $\mu'(A \cup B) \cap (A - \mu(A)) = \emptyset$, $\mu'(A \cup B) \cap (B - \mu(B)) = \emptyset$, so $\mu'(A \cup B) \cap A \subseteq \mu(A)$, $\mu'(A \cup B) \cap B \subseteq \mu(B)$, and $\mu'(A \cup B) \subseteq \mu(A) \cup \mu(B)$. By the prerequisites about μ and $\mathcal{Y}$, $\mu(A) \cup \mu(B) \in \mathcal{Y}$. Moreover, by (2) $\mu'(A \cup B) \subseteq \mu(A \cup B)$, so $\mu'(A \cup B) \subseteq (\mu(A) \cup \mu(B)) \cap \mu(A \cup B)$, but if $\mu(A \cup B) \not\subseteq \mu(A) \cup \mu(B)$, then $(\mu(A) \cup \mu(B)) \cap \mu(A \cup B) \subset \mu(A \cup B)$, contradicting $(\mu 2s)$.

(4) "$\leftarrow$" by (2). "$\to$": By $(\mu 2s)$, $\mu(U) - \mu'(U)$ is small, so there is no $X \in \mathcal{Y}$ s.t. $\mu'(U) \subseteq X \subset \mu(U)$. If there were $U' \in \mathcal{Y}$ s.t. $\mu'(U) \subseteq U'$, but $\mu(U) \not\subseteq U'$, then for $X := U' \cap \mu(U) \in \mathcal{Y}$, $\mu'(U) \subseteq X \subset \mu(U)$, contradiction.

(5) Set $U_i := U \cup A_i$. Then $\mu(U \cup \bigcup A_i \cup U_i) = \mu(U \cup \bigcup A_i) \subseteq U \cup \bigcup A_i$, so for all i $(U_i - \mu(U_i)) \cap \mu'(U \cup \bigcup A_i)) = \emptyset$ by definition of μ'. But $(U \cup A_i) - \mu(U \cup A_i) \supseteq A_i$–U, so $(A_i - U) \cap \mu'(U \cup \bigcup A_i) = \emptyset$ for all i, so $\mu'(U \cup \bigcup A_i) \subseteq U$, so by (4) $\mu(U \cup \bigcup A_i) \subseteq U$.

(6) By (4) (note that $U \in \mathcal{Y}$).

(7) $\mu(Y \cup U) = \mu(Y \cup X \cup U) \subseteq_{(3)} \mu(Y) \cup \mu(X \cup U) \subseteq Y \cup X = Y$.

(8) By (7) and the definition of μ'.

(9) "$\subseteq$": Let $x \in \mu'(X)$, so $x \in Y$, and $x \in \mu'(Y)$ by (8). "$\supseteq$": Let $x \in \mu'(Y)$, so $x \in X$. Suppose $x \notin \mu'(X)$, so there is $U' \in \mathcal{Y}$ s.t. $x \in U' - \mu(U')$ and $\mu(X \cup U') \subseteq X$. Note that by $\mu(X \cup U') \subseteq X$ and (1), $\mu(X \cup U') = \mu(X)$. Now, $\mu'(X) \subseteq Y$, so by (4) $\mu(X) \subseteq Y$, thus $\mu(X \cup U') = \mu(X) \subseteq Y \subseteq Y \cup U' \subseteq X \cup U'$, so $\mu(Y \cup U') = \mu(X \cup U') = \mu(X) \subseteq Y$, so

$x \notin \mu'(Y)$, contradiction.

(10) By $(\mu\emptyset)$, $\emptyset \in \mathcal{Y}$, and $(\mu 2s)$.

(11) Let, e.g. $x \in A_2$, $x \notin \mu'(A_2)$, so by (8) $x \notin \mu'(A_1 \cup A_2)$.

(12) Suppose $x \in \mu'(U \cup Y) - H(U)$, $x \notin \mu'(Y)$. As $x \notin H(U)$, $x \notin U$, so $x \in Y$, so by (8) $x \notin \mu'(U \cup Y)$, contradiction.

(13) Suppose there is $x \in H(U)$–U, $x \in \mu'(A)$. Then there is B with $x \in B - \mu(B)$ and $\mu(B) \subseteq U$. But $\mu(A \cup B) \subseteq_{(3)} \mu(A) \cup \mu(B) \subseteq A \cup U \subseteq A$, so $x \notin \mu'(A)$, contradiction.

(14) Let $\mu'(Y) \subseteq H(U)$, then by $\mu'(U) \subseteq H(U)$ and (11) $\mu'(U \cup Y) \subseteq \mu'(U) \cup \mu'(Y) \subseteq H(U)$, so by (13) $\mu'(U \cup Y) \subseteq U$, so by (4) $\mu(U \cup Y) \subseteq U$ and $U \cup Y \subseteq H(U)$. Moreover, $\mu'(U \cup Y) \subseteq U \subseteq U \cup Y$ so by (9) $\mu'(U \cup Y) = \mu'(U)$.

(15) Let $H(U) = \bigcup\{A_i : i \in I\}$ with $A_i \in \mathcal{Y}$, $\mu(A_i) \subseteq U$, so $Y = \bigcup\{Y \cap A_i : i \in I\}$. By $\mu(A_i) \subseteq U$ and (3), $\mu(A_i \cup U) \subseteq \mu(A_i) \cup \mu(U) \subseteq U$. So $\mu(U \cup A_i) \subseteq U \subseteq U \cup (Y \cap A_i) \subseteq U \cup A_i$, so by (μCUM) $\mu(U \cup (Y \cap A_i)) = \mu(U \cup A_i) \subseteq U$ for all i. So by (5) $\mu(U \cup Y) \subseteq U$.

(16) Suppose $Y \subseteq H(U)$, so by (15) $\mu(U \cup Y) \subseteq U$. By definition of μ', there is Y' s.t. $x \in Y' - \mu(Y')$, $\mu(Y \cup Y') \subseteq Y$. Now, $\mu(U \cup Y \cup Y') \subseteq_{(3)} \mu(U \cup Y) \cup \mu(Y') \subseteq U \cup Y' \subseteq U \cup Y \cup Y'$, so $\mu(U \cup Y') = \mu(U \cup Y \cup Y')$. Moreover, $\mu(U \cup Y \cup Y') \subseteq \mu(U) \cup \mu(Y \cup Y') \subseteq U \cup Y \subseteq U \cup Y \cup Y'$, so $\mu(U \cup Y') = \mu(U \cup Y \cup Y') = \mu(U \cup Y) \subseteq U$, so $x \notin \mu'(U)$.

(17) $\mu'(U \cup Y) \subseteq H(U) \rightarrow_{(14)} U \cup Y \subseteq H(U)$.

□

In the following Proposition 5.2.7, showing the representation of μ', we first construct a not necessarily transitive structure, which contains many elements of the somewhat more difficult construction for the transitive case, and then address the transitive construction.

Proposition 5.2.7

Let $\mu' : \mathcal{Y} \rightarrow \mathcal{P}(Z)$ and $H : \mathcal{Y} \rightarrow \mathcal{P}(Z)$ be two operations satisfying (H1) and (H2) of Conditions 5.2.1, and $(\mu' \subseteq)$, $(\mu'\emptyset)$, $(\mu'4) - (\mu'6)$. Then

(a) there is a smooth preferential structure $\mathcal{Z}$ over Z s.t. $\mu' = \mu_{\mathcal{Z}}$,

(b) $\mathcal{Z}$ can be chosen transitive.

Proof:

The proof of part (b) is verbatim the same as the proof of Proposition 3.3.8, and the reader is referred there.

(a) The following Construction 5.2.1 and Fact 5.2.8 are adaptations of the simplified construction hinted at, but not carried out in Section 3.3 for the smooth, but not necessarily transitive case.

Construction 5.2.1

Let $U \in \mathcal{Y}$.

1. Let $x \in \mu'(U)$, $x \in Z$. Construct for this x and U :

$\{< x, f, U >: f \in \Pi\{\mu'(U \cup Y) - H(U) : x \in Y \not\subseteq H(U), Y \in \mathcal{Y}\}\ \}$.

2. Let $x \in Z$ be arbitrary. Construct for this x (independent from U):

$\{< x, f, 0 >: f \in \Pi\{\mu'(Y) : x \in Y, Y \in \mathcal{Y}\}\ \}$.

3. Let $< x, f, U > \succ < y, g, V > :\leftrightarrow y \in ran(f)$,

let $< x, f, 0 > \succ < y, g, V > :\leftrightarrow y \in ran(f)$.

(No $< x, f, 0 >$ will be smaller than any other element.)

We note:

Fact 5.2.8

(1) If $Y \not\subseteq H(U)$, then $\mu'(U \cup Y) - H(U) \neq \emptyset$, so in Construction 5.2.1, 1., there is always some f (which may be $\emptyset$),

(2) $< x, f, U > \succ < y, g, V > \rightarrow y \notin H(U)$,

(3) $< x, f, U >$ is $\mathcal{Z}-$minimal in $\mathcal{Z}\lceil U$,

(4) no $< x, f, 0 >$ is $\mathcal{Z}-$minimal in any U,

(5) smoothness is respected for the elements of the form $< x, f, U >$,

(6) smoothness is respected for the elements of the form $< x, f, 0 >$,

(7) $\mu' = \mu_{\mathcal{Z}}$.

Proof of Fact 5.2.8:

(1): Trivial by $(\mu'6)$.

(2), (3): trivial, the latter by (H1).

(4): Let $x \in Y$, then $\mu'(Y) \neq \emptyset$ by $(\mu'\emptyset)$, take $y \in ran(f) \cap \mu'(Y)$, then there is $< y, g, Y >$, and $< x, f, 0 > \succ < y, g, Y >$.

(5) Fix $< x, f, U >$, let $x \in A \in \mathcal{Y}$.

Case 1: $A \subseteq H(U)$. If $< y, g, V > \prec < x, f, U >$, then $y \notin A$, so $< x, f, U >$ is minimal in $\mathcal{Z} \lceil A$.

Case 2: $A \not\subseteq H(U)$. Then A is one of the Y considered, take $y \in ran(f) \cap (\mu'(U \cup A) - H(U))$, then $y \in \mu'(A) \subseteq A$ by $(\mu'4)$ and $(\mu' \subseteq)$. Consider the construction for y, $U \cup A$, take $g \in \Pi\{\mu'(U \cup A \cup Y) - H(U \cup A) : y \in Y \not\subseteq H(U \cup A)\}$, then $< y, g, U \cup A >$ is minimal in $\mathcal{Z} \lceil U \cup A$ by (3), so by $y \in A$ minimal in $\mathcal{Z} \lceil A$, and $< y, g, U \cup A > \prec < x, f, U >$.

(6) Fix $< x, f, 0 >$, let $x \in A$. Take $y \in ran(f) \cap \mu'(A)$, then any $< y, g, A >$ is minimal in $\mathcal{Z} \lceil A$ by (3) and $< y, g, A > \prec < x, f, 0 >$.

(7) "$\subseteq$": $x \in \mu'(U) \rightarrow$ any $< x, f, U >$ is minimal in $\mathcal{Z} \lceil U$.
"$\supseteq$": Let $x \in U - \mu'(U)$. No $< x, f, 0 >$ is minimal in $\mathcal{Z} \lceil U$. Consider now some $< x, f, V >$, $x \in \mu'(V)$, $V \in \mathcal{Y}$, so $f \in \Pi\{\mu'(V \cup Y) - H(V) : x \in Y \not\subseteq H(V), Y \in \mathcal{Y}\}$. As $x \in U - \mu'(U)$, $U \not\subseteq H(V)$ by $(\mu'5)$, so U was considered as one of the Y, so $ran(f) \cap \mu'(U) \neq \emptyset$, as $\mu'(V \cup U) - H(V) \subseteq \mu'(U)$ by $(\mu'4)$, so any $< y, g, U >$ with $y \in ran(f) \cap \mu'(U)$ will be smaller than $< x, f, V >$ and in $\mathcal{Z} \lceil U$.

□ (Fact 5.2.8 and Proposition 5.2.7)

Proposition 5.2.9

Let Z be an arbitrary set, $\mathcal{Y} \subseteq \mathcal{P}(Z)$, $\mu : \mathcal{Y} \rightarrow \mathcal{Y}$, $\mathcal{Y}$ closed under arbitrary intersections and finite unions, and $\emptyset, Z \in \mathcal{Y}$, and let $\widehat{\cdot}$ be defined wrt. $\mathcal{Y}$.

(a) If μ satisfies $(\mu\emptyset)$, $(\mu \subseteq)$, $(\mu 2s)$, (μCUM), then there is a transitive smooth preferential structure $\mathcal{Z}$ over Z s.t. for all $U \in \mathcal{Y}$ $\mu(U) = \widehat{\mu_{\mathcal{Z}}(U)}$.

(b) If $\mathcal{Z}$ is a smooth preferential structure over Z and $\mu : \mathcal{Y} \rightarrow \mathcal{Y}$ s.t. for all $U \in \mathcal{Y}$ $\mu(U) = \widehat{\mu_{\mathcal{Z}}(U)}$, then μ satisfies $(\mu\emptyset)$, $(\mu \subseteq)$, $(\mu 2s)$, (μCUM).

Proof:

(a) If μ satisfies $(\mu\emptyset)$, $(\mu \subseteq)$, $(\mu 2s)$, (μCUM), then μ' and H as defined from μ in Definition 5.2.1 and 5.2.3 (and $(\mu 2s)$) satisfy properties (H1), (H2), $(\mu' \subseteq)$, $(\mu'\emptyset)$, $(\mu'4) - (\mu'6)$ by Fact 5.2.1 and Fact 5.2.6, (2), (10), (12), (16), (17). Thus, by Proposition 5.2.7, there is a smooth transitive preferential structure $\mathcal{Z}$ over Z s.t. $\mu' = \mu_{\mathcal{Z}}$, but by $(\mu 2s)$ $\mu(U) = \widehat{\mu'(U)} = \widehat{\mu_{\mathcal{Z}}(U)}$.

(b) $(\mu\emptyset)$: If $U \neq \emptyset$, $U \in \mathcal{Y}$, then there is $< x, i > \in \mathcal{X}$, $x \in U$. If $< x, i >$ is minimal in $\mathcal{Z}\lceil U$, $\mu_{\mathcal{Z}}(U) \neq \emptyset$. If not, there is by smoothness $< x', i' > \prec < x, i >$, $x' \in U$, $< x', i' >$ minimal in $\mathcal{Z}\lceil U$, and again $\mu_{\mathcal{Z}}(U) \neq \emptyset$. But $\mu_{\mathcal{Z}}(U) \subseteq \mu(U)$.

$(\mu \subseteq)$: $\mu_{\mathcal{Z}}(U) \subseteq U \rightarrow \mu(U) = \overbrace{\mu_{\mathcal{Z}}(U)} \subseteq U$ by $U \in \mathcal{Y}$.

$(\mu 2s)$: If $(\mu 2s)$ fails, then there is $U \in \mathcal{Y}$ s.t. for $U' := \bigcup\{Y' - \mu(Y') : Y' \in \mathcal{Y},\ \mu(U \cup Y') \subseteq U\}$ $\overbrace{\mu(U) - U'} \subset \mu(U)$. By $\mu_{\mathcal{Z}}(Y') \subseteq \mu(Y')$, $Y' - \mu(Y') \subseteq Y' - \mu_{\mathcal{Z}}(Y')$. But no copy of any $x \in Y' - \mu_{\mathcal{Z}}(Y')$ with $\mu_{\mathcal{Z}}(U \cup Y') \subseteq \mu(U \cup Y') \subseteq U$ can be minimal in $\mathcal{Z}\lceil U$: As $x \in Y' - \mu_{\mathcal{Z}}(Y')$, if $< x, i >$ is any copy of x, then there is $< y, j > \prec < x, i >$, $y \in Y'$. Consider now $U \cup Y'$. As $< x, i >$ is not minimal in $\mathcal{Z}\lceil U \cup Y'$, by smoothness of $\mathcal{Z}$ there must be $< z, k > \prec < x, i >$, $< z, k >$ minimal in $\mathcal{Z}\lceil U \cup Y'$. But all minimal elements of $\mathcal{Z}\lceil U \cup Y'$ must be in $\mathcal{Z}\lceil U$, so there must be $< z, k > \prec < x, i >$, $z \in U$, thus $< x, i >$ is not minimal in $\mathcal{Z}\lceil U$. Thus by $\mu_{\mathcal{Z}}(U) \subseteq \mu(U)$, $\mu_{\mathcal{Z}}(U) \subseteq \mu(U) - U'$, so $\overbrace{\mu_{\mathcal{Z}}(U)} \subseteq \overbrace{\mu(U) - U'} \subset \mu(U)$, contradiction.

(μCUM) : Let $\mu(X) \subseteq Y \subseteq X$. Now $\mu_{\mathcal{Z}}(X) \subseteq \overbrace{\mu_{\mathcal{Z}}(X)} = \mu(X)$, so by smoothness of $\mathcal{Z}$ $\mu_{\mathcal{Z}}(Y) = \mu_{\mathcal{Z}}(X)$, thus $\mu(X) = \overbrace{\mu_{\mathcal{Z}}(X)} = \overbrace{\mu_{\mathcal{Z}}(Y)} = \mu(Y)$.
□

5.2.2 The logical results

We turn to (propositional) logic.

The main result here is Proposition 5.2.11. Fact 5.2.10 shows some trivial but useful properties we will need later on in this Section. The conditions are formulated or recalled in Conditions 5.2.4, the auxiliary Lemma 5.2.12 is the main step in the proof of Proposition 5.2.11.

Fact 5.2.10

Let $\mu : \boldsymbol{D}_{\mathcal{L}} \rightarrow \boldsymbol{D}_{\mathcal{L}}$, and $\overline{\overline{T}} = Th(\mu(M(T)))$ for all theories T. Then:

(0) $\mu(M(T)) = M(\overline{\overline{T}})$,

(1) $M(T \vee T') = M(T) \cup M(T')$,

(2) $Th(X \cup Y) = \overline{Th(X) \vee Th(Y)}$,

(3) If $\widehat{U} := M(Th(U))$, then $\widehat{X \cup Y} = \widehat{X} \cup \widehat{Y}$ for $X, Y \in \boldsymbol{D}_{\mathcal{L}}$,

(4) $\mu(M(T) \cup M(T')) \subseteq M(T) \Leftrightarrow T \subseteq \overline{\overline{T \vee T'}}$,

(5) $M(\overline{\overline{T \vee T'}}) \subseteq M(\overline{\overline{T}} \vee \overline{\overline{T'}}) \Leftrightarrow \overline{\overline{T}} \vee \overline{\overline{T'}} \subseteq \overline{\overline{T \vee T'}}$.

Proof:

(0) By prerequisite, there is T' s.t. $\mu(M(T)) = M(T')$, so $Th(M(T')) = \overline{\overline{T}}$, so $\mu(M(T)) = M(T') = M(\overline{\overline{T}})$.

(1) If, e.g. $m \in M(T)$, then $m \models T$, so $m \models T \vee T'$. If $m \notin M(T)$, $m \notin M(T')$, then there are $\phi \in T$, $\phi' \in T'$, $m \not\models \phi$, $m \not\models \phi'$, so $m \not\models \phi \vee \phi'$ and $m \notin M(T \vee T')$.

(2) "$\supseteq$": Let $\sigma \in Th(X) \vee Th(Y)$, so $\sigma = \phi \vee \psi$, $X \models \phi$, $Y \models \psi$, so $X \cup Y \models \sigma$, and $\sigma \in Th(X \cup Y)$, but $Th(X \cup Y)$ is deductively closed. "$\subseteq$": $X \cup Y \models \phi$ $\rightarrow X \models \phi$, $Y \models \phi \rightarrow \phi \vee \phi \in Th(X) \vee Th(Y) \rightarrow \phi \in \overline{Th(X) \vee Th(Y)}$.

(3) Trivial by Fact 5.1.1.

(4) $M(\overline{\overline{T \vee T'}}) =_{(0)} \mu(M(T \vee T')) =_{(1)} \mu(M(T) \cup M(T'))$. So $\mu(M(T) \cup M(T')) \subseteq M(T)$ iff $M(\overline{\overline{T \vee T'}}) \subseteq M(T)$ iff $T \subseteq \overline{\overline{T \vee T'}}$ (as $\overline{\overline{T \vee T'}}$ is deductively closed).

(5) $M(\overline{\overline{T \vee T'}}) \subseteq M(\overline{\overline{T}} \vee \overline{\overline{T'}}) \Leftrightarrow \overline{\overline{T}} \vee \overline{\overline{T'}} \subseteq \overline{\overline{T \vee T'}}$, again as $\overline{\overline{T \vee T'}}$ is deductively closed.

□

We work now in $\mathcal{Y} := \boldsymbol{D}_{\mathcal{L}}$, so $\widehat{U} = M(Th(U))$ for $U \subseteq M_{\mathcal{L}}$ and (~ 1) in Fact 5.1.1 will hold.

Condition 5.2.4

(CP) $Con(T) \rightarrow Con(\overline{\overline{T}})$,

(LLE) $\overline{T} = \overline{T'} \rightarrow \overline{\overline{T}} = \overline{\overline{T'}}$,

(CCL) $\overline{\overline{T}}$ is classically closed,

(SC) $T \subseteq \overline{\overline{T}}$,

($\mid\!\sim 4$) Let T, T_i, $i \in I$ be theories s.t. $\forall i\ T_i \vdash T$, then there is no ϕ s.t. $\phi \notin \overline{\overline{T}}$ and $M(\overline{\overline{T}} \cup \{\neg\phi\}) \subseteq \bigcup\{M(T_i) - M(\overline{\overline{T_i}}) : i \in I\}$,

($\mid\!\sim 4s$) Let T, T_i, $i \in I$ be theories s.t. $\forall i\ T \subseteq \overline{\overline{T_i \vee T}}$, then there is no ϕ s.t.

$\phi \notin \overline{\overline{T}}$ and $M(\overline{\overline{T}} \cup \{\neg\phi\}) \subseteq \bigcup\{M(T_i) - M(\overline{\overline{T_i}}) : i \in I\}$,

$(\mid\!\sim 5)$ $\overline{\overline{T}} \vee \overline{\overline{T'}} \subseteq \overline{\overline{T \vee T'}}$,

(CUM) $T \subseteq \overline{T'} \subseteq \overline{\overline{T}} \rightarrow \overline{\overline{T}} = \overline{\overline{T'}}$

for all T, T', T_i.

Note:

Condition (CP) is auxiliary and corresponds to the nonemptiness condition $(\mu\emptyset)$ of μ in the smooth case: $U \neq \emptyset \rightarrow \mu(U) \neq \emptyset$ — or to the fact that all models occur in the structure.

We formulate now the logical representation theorem for not necessarily definability preserving preferential structures.

Proposition 5.2.11

Let $\mid\!\sim$ be a logic for $\mathcal{L}$. Then:

(a.1) If $\mathcal{M}$ is a classical preferential model over $M_{\mathcal{L}}$ and $\overline{\overline{T}} = Th(\mu_{\mathcal{M}}(M(T)))$, then (LLE), (CCL), (SC), $(\mid\!\sim 4)$ hold for the logic so defined.

(a.2) If (LLE), (CCL), (SC), $(\mid\!\sim 4)$ hold for a logic, then there is a transitive classical preferential model over $M_{\mathcal{L}}$ $\mathcal{M}$ s.t. $\overline{\overline{T}} = Th(\mu_{\mathcal{M}}(M(T)))$.

(b.1) If $\mathcal{M}$ is a smooth classical preferential model over $M_{\mathcal{L}}$ and $\overline{\overline{T}} = Th(\mu_{\mathcal{M}}(M(T)))$, then (CP), (LLE), (CCL), (SC), $(\mid\!\sim 4s)$, $(\mid\!\sim 5)$, (CUM) hold for the logic so defined.

(b.2) If (CP), (LLE), (CCL), (SC), $(\mid\!\sim 4s)$, $(\mid\!\sim 5)$, (CUM) hold for a logic, then there is a smooth transitive classical preferential model $\mathcal{M}$ over $M_{\mathcal{L}}$ s.t. $\overline{\overline{T}} = Th(\mu_{\mathcal{M}}(M(T)))$.

The proof is an easy consequence of Propositions 5.2.5, 5.2.9, and Lemma 5.2.12, and will be shown after the proof of the latter.

Lemma 5.2.12

(a) If $\mu : \boldsymbol{D}_{\mathcal{L}} \rightarrow \boldsymbol{D}_{\mathcal{L}}$ satisfies $(\mu \subseteq)$, $(\mu 2)$ (for $\mathcal{Y} = \boldsymbol{D}_{\mathcal{L}}$), then $\mid\!\sim$ defined by $\overline{\overline{T}} := Th(\mu(M(T)))$ satisfies (LLE), (CCL), (SC), $(\mid\!\sim 4)$.

(b) If $\mu : \boldsymbol{D}_{\mathcal{L}} \rightarrow \boldsymbol{D}_{\mathcal{L}}$ satisfies $(\mu\emptyset)$, $(\mu \subseteq)$, $(\mu 2s)$, (μCUM) (for $\mathcal{Y} = \boldsymbol{D}_{\mathcal{L}}$), then $\mid\!\sim$ defined by $\overline{\overline{T}} := Th(\mu(M(T)))$ satisfies (CP), (LLE), (CCL), (SC), $(\mid\!\sim 4s)$, $(\mid\!\sim 5)$, (CUM).

(c) If $\mid\!\sim$ satisfies (LLE), (CCL), (SC), ($\mid\!\sim$ 4), then there is $\mu : \boldsymbol{D}_{\mathcal{L}} \to \boldsymbol{D}_{\mathcal{L}}$ such that $\overline{\overline{T}} = Th(\mu(M(T)))$ for all T and μ satisfies ($\mu \subseteq$), ($\mu 2$) (for $\mathcal{Y} = \boldsymbol{D}_{\mathcal{L}}$).

(d) If $\mid\!\sim$ satisfies (CP), (LLE), (CCL), (SC), ($\mid\!\sim 4s$), ($\mid\!\sim$ 5), (CUM), then there is $\mu : \boldsymbol{D}_{\mathcal{L}} \to \boldsymbol{D}_{\mathcal{L}}$ such that $\overline{\overline{T}} = Th(\mu(M(T)))$ for all T and μ satisfies ($\mu\emptyset$), ($\mu \subseteq$), ($\mu 2s$), (μCUM) (for $\mathcal{Y} = \boldsymbol{D}_{\mathcal{L}}$).

Proof:

We show (a) and (b) together.

Let $\overline{\overline{T}} = Th(\mu(M(T)))$, thus by Fact 5.2.10, (0) $\mu(M(T)) = M(\overline{\overline{T}})$.

(CP): $Con(T) \to M(T) \neq \emptyset \to$ (by ($\mu\emptyset$)) $\mu(M(T)) \neq \emptyset \to Con(\overline{\overline{T}})$.

(LLE): If $\overline{T} = \overline{T'}$, then $M(T) = M(T')$, so $\mu(M(T)) = \mu(M(T'))$, and $\overline{\overline{T}} = \overline{\overline{T'}}$.

(CCL) is trivial by definition, and (SC) is trivial by ($\mu \subseteq$).

($\mid\!\sim$ 4) : Suppose there are T, T_i s.t. $\forall i \in I$ $T_i \vdash T$ and ϕ s.t. $\phi \notin \overline{\overline{T}}$, and $M(\overline{\overline{T}} \cup \{\neg\phi\}) \subseteq X := \bigcup\{M(T_i) - M(\overline{\overline{T_i}}) : i \in I\}$. Then $M(\overline{\overline{T}}) - X \subseteq M(\overline{\overline{T}} \cup \{\phi\}) \subset M(\overline{\overline{T}})$, by $\phi \notin \overline{\overline{T}}$. Now $M(\overline{\overline{T}}) = \mu(M(T))$, and $X = \bigcup\{M(T_i) - \mu(M(T_i)) : i \in I\} \subseteq \bigcup\{M(T') - \mu(M(T')) : T' \vdash T\} = \bigcup\{M(T') - \mu(M(T')) : M(T') \subseteq M(T)\}$. Thus $\mu'(M(T)) = \mu(M(T)) - \bigcup\{M(T') - \mu(M(T')) : M(T') \subseteq M(T)\} \subseteq \mu(M(T)) - X = M(\overline{\overline{T}}) - X \subseteq M(\overline{\overline{T}} \cup \{\phi\}) \subset M(\overline{\overline{T}}) = \mu(M(T))$, contradicting ($\mu 2$).

($\mid\!\sim 4s$) : The proof is almost the same as for ($\mid\!\sim$ 4). Suppose there are T, T_i s.t. $\forall i \in I$ $T \subseteq \overline{\overline{T \vee T_i}}$ and ϕ s.t. $\phi \notin \overline{\overline{T}}$, and $M(\overline{\overline{T}} \cup \{\neg\phi\}) \subseteq X := \bigcup\{M(T_i) - M(\overline{\overline{T_i}}) : i \in I\}$. Then $M(\overline{\overline{T}}) - X \subseteq M(\overline{\overline{T}} \cup \{\phi\}) \subset M(\overline{\overline{T}})$, by $\phi \notin \overline{\overline{T}}$. Now $M(\overline{\overline{T}}) = \mu(M(T))$, and $X = \bigcup\{M(T_i) - \mu(M(T_i)) : i \in I\}$ $\subseteq \bigcup\{M(T') - \mu(M(T')) : T'$ s.t. $T \subseteq \overline{\overline{T \vee T'}}\}$ = (by Fact 5.2.10, (4)) $\bigcup\{M(T') - \mu(M(T')) : T'$ s.t. $\mu(M(T) \cup M(T')) \subseteq M(T)\}$. Thus $\mu'(M(T))$ = $\mu(M(T))$ - $\bigcup\{M(T') - \mu(M(T')) : T'$ s.t. $\mu(M(T) \cup M(T')) \subseteq M(T)\} \subseteq \mu(M(T)) - X = M(\overline{\overline{T}}) - X \subseteq M(\overline{\overline{T}} \cup \{\phi\}) \subset M(\overline{\overline{T}}) = \mu(M(T))$, contradicting ($\mu 2s$).

($\mid\!\sim$ 5) : $M(\overline{\overline{T \vee T'}}) = \mu(M(T \vee T')) = \mu(M(T) \cup M(T')) \subseteq$ (by Fact 5.2.6, (3)) $\mu(M(T)) \cup \mu(M(T')) = M(\overline{\overline{T}}) \cup M(\overline{\overline{T'}}) = M(\overline{\overline{T}} \vee \overline{\overline{T'}})$, so $\overline{\overline{T}} \vee \overline{\overline{T'}} \subseteq \overline{\overline{T \vee T'}}$ by Fact 5.2.10, (5).

(CUM): $T \subseteq \overline{T'} \subseteq \overline{\overline{T}} \to \mu(M(T)) = M(\overline{\overline{T}}) \subseteq M(\overline{T'}) \subseteq M(T) \to_{(\mu CUM)}$

$\mu(M(T')) = \mu(M(T)) \rightarrow \overline{\overline{T}} = \overline{\overline{T'}}$.

We now show (c) and (d) together:

Let $\mid\sim$ satisfy the mentioned properties. We define $\mu : \boldsymbol{D}_{\mathcal{L}} \rightarrow \boldsymbol{D}_{\mathcal{L}}$, show $\overline{\overline{T}} = Th(\mu(M(T)))$, and that the corresponding properties for μ hold.

Set $\mu(M(T)) := M(\overline{\overline{T}})$ for all T. If $M(T) = M(T')$, then $\overline{T} = \overline{T'}$, thus $\overline{\overline{T}} = \overline{\overline{T'}}$ by (LLE), so $M(\overline{\overline{T}}) = M(\overline{\overline{T'}})$, and μ is well-defined. As $\overline{\overline{T}}$ is classically closed, $Th(M(\overline{\overline{T}})) = \overline{\overline{T}}$, so $\overline{\overline{T}} = Th(\mu(M(T)))$, and Fact 5.2.10 holds.

$(\mu\emptyset)$ is trivial by (CP), so is $(\mu \subseteq)$ by (SC).

$(\mu 2)$: Suppose there were $U \in \boldsymbol{D}_{\mathcal{L}}$ s.t. $\bigcup\{U' - \mu(U') : U' \in \boldsymbol{D}_{\mathcal{L}}, U' \subseteq U\} \cap \mu(U)$ is not a small (in $\boldsymbol{D}_{\mathcal{L}}$) subset of $\mu(U)$. Let $U = M(T)$, $U' = M(T')$. So $\mu(U) = M(\overline{\overline{T}})$. Then there must be some $X \in \boldsymbol{D}_{\mathcal{L}}$ with $\mu(U) - \bigcup\{U' - \mu(U') : U' \in \boldsymbol{D}_{\mathcal{L}}, U' \subseteq U\} \subseteq X \subset \mu(U)$. So there must be some $\phi \notin \overline{\overline{T}}$ with $X \subseteq M(\overline{\overline{T}} \cup \{\phi\})$, so $M(\overline{\overline{T}} \cup \{\neg\phi\}) \subseteq \bigcup\{M(T') - M(\overline{\overline{T'}}) : T' \vdash T\}$, contradicting $(\mid\sim 4)$.

$(\mu 2s)$: (This is alsmost the same proof as for $(\mu 2)$.) Suppose there were $U \in \boldsymbol{D}_{\mathcal{L}}$ s.t. $\bigcup\{U' - \mu(U') : U' \in \boldsymbol{D}_{\mathcal{L}}, \mu(U \cup U') \subseteq U\} \cap \mu(U)$ is not a small (in $\boldsymbol{D}_{\mathcal{L}}$) subset of $\mu(U)$. Let $U = M(T)$, $U' = M(T')$. So $\mu(U) = M(\overline{\overline{T}})$. Then there must be some $X \in \boldsymbol{D}_{\mathcal{L}}$ with $\mu(U)$ - $\bigcup\{U' - \mu(U') : U' \in \boldsymbol{D}_{\mathcal{L}}, \mu(U \cup U') \subseteq U\} \subseteq X \subset \mu(U)$. So there must be some $\phi \notin \overline{\overline{T}}$ with $X \subseteq M(\overline{\overline{T}} \cup \{\phi\})$, so $M(\overline{\overline{T}} \cup \{\neg\phi\}) \subseteq \bigcup\{M(T') - M(\overline{\overline{T'}}) : T' \text{ s.t. } T \subseteq \overline{\overline{T \vee T'}}\}$, contradicting $(\mid\sim 4s)$.

$(\mu CUM) : M(\overline{\overline{T}}) = \mu(M(T)) \subseteq M(T') \subseteq M(T) \rightarrow T \subseteq \overline{T'} \subseteq \overline{\overline{T}} \rightarrow_{(CUM)}$ $\overline{\overline{T}} = \overline{\overline{T'}} \rightarrow \mu(M(T)) = M(\overline{\overline{T}}) = M(\overline{\overline{T'}}) = \mu(M(T'))$.

□

Proof of Proposition 5.2.11:

We show (a.1) and (b.1) together.

Let for some [smooth] classical preferential model $\mathcal{M}$ over $M_{\mathcal{L}}$ $\overline{\overline{T}} = Th(\mu_{\mathcal{M}}(M(T)))$ for all T. For the function $\mu : \boldsymbol{D}_{\mathcal{L}} \rightarrow \boldsymbol{D}_{\mathcal{L}}$ defined by $\mu(M(T)) := M(\overline{\overline{T}})$ $\widehat{\mu_{\mathcal{M}}(M(T))} = \mu(M(T))$ (*in* $\boldsymbol{D}_{\mathcal{L}}$) holds, so by Proposition 5.2.5 (b) [5.2.9 (b)] μ satisfies $(\mu \subseteq)$, $(\mu 2)$ [$(\mu\emptyset)$, $(\mu \subseteq)$, $(\mu 2s)$, (μCUM)].

Thus, by Lemma 5.2.12, $\mid\!\sim'$ defined by $\overline{\overline{T}}' := Th(\mu(M(T)))$ satisfies (LLE), (CCL), (SC), ($\mid\!\sim$ 4) [(CP), (LLE), (CCL), (SC), ($\mid\!\sim$ 4s), ($\mid\!\sim$ 5), (CUM)], but $\overline{\overline{T}}' = Th(\mu(M(T))) = Th(M(\overline{\overline{T}})) = \overline{\overline{T}}$, as $\overline{\overline{T}}$ is deductively closed.

We now show (a.2) and (b.2) together.

Let $\mid\!\sim$ satisfy (LLE), (CCL), (SC), ($\mid\!\sim$ 4) [(CP), (LLE), (CCL), (SC), ($\mid\!\sim$ 4s), ($\mid\!\sim$ 5), (CUM)]. By Lemma 5.2.12, there is $\mu : \boldsymbol{D}_{\mathcal{L}} \to \boldsymbol{D}_{\mathcal{L}}$ s.t. $\overline{\overline{T}} = Th(\mu(M(T)))$ and μ satisfies ($\mu \subseteq$), ($\mu 2$) [($\mu\emptyset$), ($\mu \subseteq$), ($\mu 2s$), (μCUM)]. So by Proposition 5.2.5 (a) [5.2.9 (a)], there is a [smooth] transitive classical preferential model $\mathcal{M}$ over $M_{\mathcal{L}}$ s.t. $\mu(M(T)) = \widehat{\mu_{\mathcal{M}}(M(T))}$ for all T, so $\overline{\overline{T}} = Th(\mu(M(T))) = Th(\mu_{\mathcal{M}}(M(T)))$. □

5.2.3 The general case and the limit version cannot be characterized

Introduction:

We show in this section more than what the headline announces:

- general, not necessarily definability preserving preferential structures,
- the general limit version of preferential structures,
- not necessarily definability preserving ranked preferential structures,
- the limit version of ranked preferential structures,
- general, not necessarily definability preserving distance based revision,
- the general limit version of distance based revision

all have no "normal" characterization by logical means of any size.

The limit version will in all cases be a trivial consequence of the minimal version: On the one hand, the constructed structures will give the same results in the minimal and the limit reading (this is due to the simplicity of the relation, where paths will have length at most 1). On the other hand, the logics we define will not be preferentially or distance representable in both readings — this is again trivial.

We have seen that distance based revision has no finite characterization, but a countable set of finite conditions suffices, as transitivity speaks about arbitrarily long finite chains. The case of not necessarily definability preserving preferential structures (and, as a consequence, of the limit version of preferentials structures) is much worse, as we will see now in Proposition 5.2.15. This proposition shows that there is no normal characterization of any size of general preferential structures, and consequently of the limit variant.

As said in Section 2.3.1, this negative result, together with the reductory results of Sections 3.4.1 and 3.10.3, casts a heavy doubt on the utility of the limit version as a reasoning tool. It seems either hopelessly, or unnecessarily, complicated. But it seems useful as a tool for theoretical investigations, as it separates finitary from infinitary versions, see Section 3.4.1.

A similar result is shown in Proposition 5.2.16 for not necessarily definability preserving ranked structures, and in Proposition 5.2.17 for not necessarily definability preserving distance based revision. It was at first sight a little surprising to the author that the generally simpler ranked case revealed itself as more difficult. But a second look shows why: In a ranked structure, we cannot only take pairs of the relation away, as we do in the proof of Proposition 5.2.15 for the general case, we then also have to put new pairs in, in order to keep the structure ranked. Still, the same technique works with some modifications. The result on revision is then a simple corollary of the ranked case. For this reason, we treat the revision case here, too, instead of, e.g. in Section 5.3.

We first introduce some notation and simple facts, which will be used throughout this Section 5.2.3. We then state the technical Lemma 5.2.14, which shows that for each theory T in our language of size κ whose set of consequences $\overline{\overline{T}}$ is defined in a certain way, there is an "optimal" model set A_T of size $\leq \kappa$ which generates $\overline{\overline{T}}$. This lemma might also prove useful in other contexts, so it has some interest on its own. The reader should perhaps look first at the proof of Proposition 5.2.15, and only then at that of Proposition 5.2.16. The former proof is simpler, and we also use small facts from the former in the latter. Proposition 5.2.17 is an easy corollary to Proposition 5.2.16, as revision is essentially rankedness with mobile left hand side. We will take care that all nontrivial cases involve just one left hand point.

The reader may perhaps best skip the Lemma 5.2.14 and its proof, retain the intuitive idea just described, read the propositions and their proofs, and come back to Lemma 5.2.14 afterwards.

Notation 5.2.1

(1) We will always work in a propositional language $\mathcal{L}$ with κ many (κ an

infinite cardinal) propositional variables $p_i : i < \kappa$. As p_0 will have a special role, we will set $p := p_0$. In the revision case, we will use another special variable, which we will call q. (This will just avoid excessive indexing.)

(2) In all cases, we will show that there is no normal characterization of size $\leq \kappa$. As κ was arbitrary, we will have shown the results. We will always assume that there is such a characterization Φ of size κ, and derive a contradiction. For this purpose, we construct suitable logics which are not representable, and show that for any instantiation of Φ (i.e. with at most κ theories T or formulas ϕ) in these logics, we find a "legal" structure where these instances have the same value as in the original logic, a contradiction to the assumed discerning power of Φ. (By hypothesis, at least one instance has to have negative value in the not representable logics, but then it has the same negative value in a legal structure, a contradiction.) To simplify notation, we assume wlog. that the characterization works with theories only, we can always replace a formula ϕ by the theory $\{\phi\}$, etc. The structures to be constructed depend of course on the particular instantiation of Φ, a set of theories of size $\leq \kappa$, we will denote this set $\mathcal{T}$, and construct the structures from $\mathcal{T}$ and the "illegal" original logic.

(3) Given any model set $X \subseteq M_{\mathcal{L}}$, we define again $\overbrace{X} := M(Th(X))$ — the closure of X in the standard topology.

We then have:

Fact 5.2.13

(1) $X \subseteq \overbrace{X}$.

(2) Let T be any $\mathcal{L}-$theory, and $A \subseteq M_{\mathcal{L}}$, then $\overbrace{M(T) - A} = M(T \cup T_A)$ for some T_A. Of course, T_A may be empty or a subset of $\overline{T}$, if $\overbrace{M(T) - A} = M(T)$. Thus, for $\mathcal{X} \subseteq \mathcal{P}(M_{\mathcal{L}})$ $\bigcap\{\overbrace{M(T) - A} : A \in \mathcal{X}\} = \bigcap\{M(T \cup T_A) : A \in \mathcal{X}\} = M(\bigcup\{T \cup T_A : A \in \mathcal{X}\})$ for suitable T_A.

(3) If $\overbrace{M(T) - A} \neq M(T)$, then $Th(M(T) - A) \supset \overline{T}$, so $\overbrace{M(T) - A} = M(T \cup T_A)$ for some T_A s.t. $T \not\vdash T_A$.

(Trivial).

□

We now state and prove our main technical lemma.

Lemma 5.2.14

Let $\mathcal{L}$ be a language of κ many (κ an infinite cardinal) propositional variables.

Let a theory T be given, $\mathcal{E}_T \subseteq \{X \subseteq M_{\mathcal{L}} : card(X) \leq \kappa\}$ be closed under unions of size $\leq \kappa$ and subsets, and $\overline{\overline{T}}$ be defined by $\overline{\overline{T}} := Th(\bigcap\{\widehat{M(T) - A} : A \in \mathcal{E}_T\})$.

Then there is an (usually not unique) "optimal" $A_T \in \mathcal{E}_T$ s.t.

(1) $\overline{\overline{T}} = Th(M(T) - A_T)$,

(2) for all $A \in \mathcal{E}_T$ $\widehat{M(T) - A_T} \subseteq \widehat{M(T) - A}$.

Proof:

Before we give the details, we describe the (simple) idea. The proof shows essentially how to do the right counting.

We cannot work directly with the $A \in \mathcal{E}_T$, and take the union, there might be too many of them, and the resulting set might be too big. But the $A \in \mathcal{E}_T$ give mostly the same results $\widehat{M(T) - A}$, and there are not very many interesting ones of them, or of their corresponding theories: To each A corresponds a theory T_A with $\widehat{M(T) - A} = M(T \cup T_A)$. As we are only interested in those A or T_A which change $M(T)$, we will successively add formulas to some initial T_A, until we have found a maximal $T_{A'}$ s.t. $\overline{\overline{T}} = Th(M(T \cup T_{A'})$, and A' will be the A_T. Thus, we work neither directly with all A, nor with all T_A, but count formulas, and there are only $\leq \kappa$ many of them. We will then take the union A_T of the corresponding A (i.e. which add new formulas), this will have size $\leq \kappa$ again.

Now the details.

By Fact 5.2.13, $\bigcap\{\widehat{M(T) - A} : A \in \mathcal{E}_T\} = M(\bigcup\{T \cup T_A : A \in \mathcal{E}_T\})$, so $\overline{\overline{T}}$ $= Th(M(\bigcup\{T \cup T_A : A \in \mathcal{E}_T\})) = \overline{\bigcup\{T \cup T_A : A \in \mathcal{E}_T\}}$ for suitable T_A.

We have to show that we can obtain $\overline{\overline{T}}$ with one single $A_T \in \mathcal{E}_T$, i.e. $\overline{\overline{T}} = Th(M(T) - A_T)$.

Let $\mathcal{E} := \mathcal{E}_T$, and let Ψ_i be an (arbitrary) enumeration of $\{T_A : A \in \mathcal{E}\}$.

We define an increasing chain $\Gamma_i : i \leq \mu$ ($\mu \leq \kappa$) of sets of formulas by induction, and show that for each Γ_i there is $A_i \in \mathcal{E}$ s.t. $\widehat{M(T) - A_i} = M(T \cup \Gamma_i)$, and $\overline{\overline{T}} = \overline{T \cup \bigcup\{\Gamma_i : i \leq \mu\}}$.

$\Gamma_0 := \Psi_0$.

$\Gamma_{i+1} := \Gamma_i \cup \Psi_j$, where Ψ_j is the first $\Psi_l \not\subseteq \Gamma_i$ — if this does not exist, as Γ_i contains already all Ψ_l, we stop the construction.

$\Gamma_\lambda := \bigcup\{\Gamma_i : i < \lambda\}$ for limits λ.

Note that the chain of $\Gamma's$ has length $\leq \kappa$, as we always add at least one of the κ many formulas of $\mathcal{L}$ in the successor step (the construction will stop at a successor step).

We now show that there is $A_i \in \mathcal{E}$ s.t. $\overbrace{M(T) - A_i} = M(T \cup \Gamma_i)$ by induction.

By construction,

$\overbrace{M(T) - A_0} = M(T \cup \Gamma_0)$ – where $A_0 \in \mathcal{E}$ is one of the $A \in \mathcal{E}$ which correspond to Ψ_0 (usually, there are many of them).

Suppose $\overbrace{M(T) - A_i} = M(T \cup \Gamma_i)$ by induction, and $\overbrace{M(T) - A_j} = M(T \cup \Psi_j)$. Then $M(T) - (A_i \cup A_j) \models T \cup \Gamma_i \cup \Psi_j$, so there is a subset A_{i+1} of $A_i \cup A_j$, thus of size $\leq \kappa$, and $A_{i+1} \in \mathcal{E}$, s.t. $\overbrace{M(T) - A_{i+1}} = M(T \cup \Gamma_i \cup \Psi_j) = M(T \cup \Gamma_{i+1})$, as $\Gamma_{i+1} = \Gamma_i \cup \Psi_j$.

Suppose $\overbrace{M(T) - A_i} = M(T \cup \Gamma_i)$ for $i < \lambda \leq \kappa$ by induction. Then $M(T) - \bigcup\{A_i : i < \lambda\} \models T \cup \bigcup\{\Gamma_i : i < \lambda\}$, so there is a subset A_λ of $\bigcup\{A_i : i < \lambda\}$, i.e. of size $\leq \kappa$, and $A_\lambda \in \mathcal{E}$, s.t. $\overbrace{M(T) - A_\lambda} = M(T \cup \bigcup\{\Gamma_i : i < \lambda\}) = M(T \cup \Gamma_\lambda)$.

This is also true for the last element Γ_μ, as the entire chain has length $\leq \kappa$.

Consequently, there is $A_T := A_\mu \in \mathcal{E}$ s.t. $\overbrace{M(T) - A_T} = M(T \cup \Gamma_\mu) = M(\overline{\overline{T}})$, as $\overline{\overline{T}} = \overline{\bigcup\{T \cup T_A : A \in \mathcal{E}\}}$ and $\Gamma_\mu = \bigcup\{T_A : A \in \mathcal{E}\}$, and for each $A \in \mathcal{E}$ $\overbrace{M(T) - A} \supseteq \overbrace{M(T) - A_T}$, and $\overline{\overline{T}} = Th(\overbrace{M(T) - A_T}) = Th(M(T) - A_T)$, as Γ_μ contains all Ψ corresponding to some $A \in \mathcal{E}$. Thus, (1) and (2) hold.

(Loosely speaking, $A_T := A_\mu$ is a maximal element of $\mathcal{E}_T$, more precisely, its Ψ is maximal. The important fact is that such A_T exists, and still has size $\leq \kappa$.)

□

We are now ready to state and prove the negative result for general, not necessarily definability preserving preferential structures and the general limit variant.

Proposition 5.2.15

(1) There is no "normal" characterization of any fixed size of not necessarily definability preserving preferential structures.

(2) There is no "normal" characterization of any fixed size of the general limit variant of preferential structures.

Proof:

Before we begin the proof, we recall that the "small sets of exceptions" we speak about in this Chapter 5 can be arbitrarily big unions of exceptions, this depends essentially on the size of the language. So there is no contradiction in our results. If you like, the "small" of the "small sets of exceptions" is relative, the κ discussed here is absolute.

(2) It is easy to see that (2) is a consequence of (1): Any minimal variant of suitable preferential structures can also be read as a degenerate case of the limit variant: There is a smallest closed minimizing set, so both variants coincide. This is in particular true for the structurally extremely simple cases we consider here — the relation will be trivial, as the paths in the relation have length at most 1, we work with quantity. On the other hand, it is easily seen that the logic we define first is not preferential, neither in the minimal, nor in the limit reading.

Proof of (1):

Let then κ be any infinite cardinal. We show that there is no characterization of general (i.e. not necessarily definability preserving) preferential structures which has size $\leq \kappa$. We suppose there were one such characterization Φ of size $\leq \kappa$, and construct a counterexample.

The idea of the proof is very simple. We show that it suffices to consider for any given instantiation of $\Phi \leq \kappa$ many pairs $m \prec m^-$ in a case not representable by a preferential structure, and that $\leq \kappa$ many such pairs give the same result in a true preferential structure for this instantiation. Thus, every instantiation is true in an "illegal" and a "legal" example, so Φ cannot discern between legal and illegal examples. The main work is to show that $\leq \kappa$ many pairs suffice in the illegal example, this was done in Lemma 5.2.14.

We first note some auxiliary facts and definitions, and then define the logic, which, as we show, is not representable by a preferential structure. We then use the union of all the "optimal" sets A_T guaranteed by Lemma 5.2.14 to define the preferential structure, and show that in this structure $\overline{\overline{T}}$ for $T \in \mathcal{T}$ is the same as in the old logic, so the truth value of the instantiated expression is the same in the old logic and the new structure.

Writing down all details properly is a little complicated.

As any formula ϕ in the language has finite size, ϕ uses only a finite number of variables, so ϕ has 0 or 2^κ different models.

For any model m with $m \models p$, let m^- be exactly like m with the exception that $m^- \models \neg p$. (If $m \not\models p$, m^- is not defined.)

Let $\mathcal{A} := \{X \subseteq M(\neg p) : card(X) \leq \kappa\}$. For given T, let $\mathcal{A}_T := \{X \in \mathcal{A} : X \subseteq M(T) \wedge \forall m^- \in X. m \in M(T)\}$. Note that $\mathcal{A}_T$ is closed under subsets and under unions of size $\leq \kappa$. For T, let $\mathcal{B}_T := \{X \in \mathcal{A}_T : \widehat{M(T) - X} \neq M(T)\}$, the (in the logical sense) "big" elements of $\mathcal{A}_T$. For $X \subseteq M_\mathcal{L}$, let $X \lceil\lceil M(T) := \{m^- \in X : m^- \in M(T) \wedge m \in M(T)\}$. Thus, $\mathcal{A}_T = \{X \lceil\lceil M(T) : X \in \mathcal{A}\}$.

Define now the logic $\mid\!\sim$ as follows in two steps:

(1) $\overline{\overline{Th(\{m, m^-\})}} := Th(\{m\})$

(Speaking preferentially, $m \prec m^-$, for all pairs m, m^-, this will be the entire relation. The relation is thus extremely simple, $\prec$ −paths have length at most 1, so $\prec$ is automatically transitive.)

We now look at (in terms of preferential models only some!) consequences:

(2) $\overline{\overline{T}} := Th(\bigcap\{\widehat{M(T) - A} : A \in \mathcal{B}_T\}) = Th(\bigcap\{\widehat{M(T) - A} : A \in \mathcal{A}_T\})$.

We note:

(a) This — with exception of the size condition — would be exactly the preferential consequence of part (1) of the definition.

(b) (1) is a special case of (2), we have seperated them for didactic reasons.

(c) The prerequisites of Lemma 5.2.14 are satisfied for $\overline{\overline{T}}$ and $\mathcal{A}_T$.

(d) It is crucial that we close before intersecting.

(Remark: We discussed a similar idea — better "protection" of single models by bigger model sets — in Section 3.5, where we gave a counterexample to the KLM characterization.)

This logic is not preferential. We give the argument for the minimal case, the argument for the limit case is the same.

Take $T := \emptyset$. Take any $A \in \mathcal{A}_T$. Then $Th(M_\mathcal{L}) = Th(M_\mathcal{L} - A)$, as any ϕ, which holds in A, will have 2^κ models, so there must be a model of ϕ in $M_\mathcal{L} - A$, so we cannot separate A or any of its subsets. Thus, $\widehat{M(\emptyset) - A} =$

$M(\emptyset)$ for all A of size $\leq \kappa$, so $\overline{\overline{\emptyset}} = \overline{\emptyset}$, which cannot be if $\mid\!\sim$ is preferential, for then $\overline{\overline{\emptyset}} = \overline{p}$.

Suppose there were a characterization Φ of size $\leq \kappa$. It has to say "no" for at least one instance $\mathcal{T}$ (i.e. a set of size $\leq \kappa$ of theories) of the universally quantified condition Φ. We will show that we find a true preferential structure where this instance $\mathcal{T}$ of Φ has the same truth value, more precisely, where all $T \in \mathcal{T}$ have the same $\overline{\overline{T}}$ in the old logic and in the preferential structure, a contradiction, as this instance evaluates now to "false" in the preferential structure, too.

Suppose $T \in \mathcal{T}$.

If $\overline{\overline{T}} = \overline{T}$, we do nothing (or set $A_T := \emptyset$). When $\overline{\overline{T}}$ is different from $\overline{T}$, this is because $\mathcal{B}_T \neq \emptyset$.

By Lemma 5.2.14, for each of the $\leq \kappa$ $T \in \mathcal{T}$, it suffices to consider a set A_T of size $\leq \kappa$ of suitable models of $\neg p$ to calculate $\overline{\overline{T}}$, i.e. $\overline{\overline{T}} = Th(M(T) - A_T)$, so, all in all, we work just with at most κ many such models. More precisely, set

$B := \bigcup\{A_T : \overline{\overline{T}} = Th(M(T) - A_T) \neq \overline{T}, T \in \mathcal{T}\}$.

Note that for each T with $\overline{\overline{T}} \neq \overline{T}$, $B\lceil\lceil M(T) \in \mathcal{B}_T$, as B has size $\leq \kappa$, and B contains A_T, so $\overbrace{M(T) - B\lceil\lceil M(T)} \neq M(T)$. But we also have $\overline{\overline{T}} = Th(M(T) - A_T) = Th(M(T) - B\lceil\lceil M(T))$, as A_T was optimal in $\mathcal{B}_T$.

Consider now the preferential structure where we do not make all $m \prec m^-$, but only the κ many of them featuring in B, i.e. those we have used in the instance $\mathcal{T}$ of Φ. We have to show that the instance $\mathcal{T}$ of Φ still fails in the new structure. But this is now trivial. Things like $\overline{T}$, etc. do not change, the only problem might be $\overline{\overline{T}}$. As we work in a true preferential structure, we now have to consider not subsets of size at most κ, but all of $B\lceil\lceil M(T)$ at once — which also has size $\leq \kappa$. But, by definition of the new structure, $\overline{\overline{T}} = Th(M(T) - B\lceil\lceil M(T)) = Th(M(T) - A_T)$. On the other hand, if $\overline{\overline{T}} = \overline{T}$ in the old structure, the same will hold in the new structure, as $B\lceil\lceil M(T)$ is one of the sets considered, and they did not change $\overline{\overline{T}}$.

Thus, the $\overline{\overline{T}}$ in the new and in the old structure are the same. So the instance $\mathcal{T}$ of Φ fails also in a suitable preferential structure, contradicting its supposed discriminatory power.

The limit reading of this simple structure gives the same result.

□

We turn to the ranked case. Lemma 5.2.14 will again play a central role.

Before we state and prove the result, we describe a small additional difficulty we have to solve in the ranked case. When we try to adapt above proof to the ranked case, we meet a problem which we will describe shortly before modifying the construction. We begin the construction as above, but let every $p-$model minimize every $\neg p-$model, not only its counterpart. We then continue as above the construction of κ size sets. This works fine. The problem is at the end. We have omitted part of the relation, but, to ensure rankedness, we have to put the $\neg p-$models we did not use somewhere without changing the $\overline{\overline{T}}$. By rankedness, we cannot just take away part of the relation, we have to add something, too. If we put them all on the upper level, we risk to run into a contradiction, as then, e.g. $\overline{\overline{\emptyset}} = \overline{p}$, which did not hold in the original structure, and $\overline{\overline{\emptyset}}$ might be one of the parameters. If we put one $\neg p-$model m on the lower level, then it might happen that $Th(m, m')$ — for m' another $\neg p-$model — is one of the parameters, and by $\overline{\overline{Th(m, m')}} = \overline{Th(m, m')}$ in the original structure, we have to put m' down too, etc. Thus, we have to invest a little more work.

Note that the representation conditions for limit ranked structures in Section 3.10.3 contain implicitly arbitrarily big sets — see ($\Lambda 6$) there. Moreover, the characterization of arbitrary ranked structures we sketch there concerns the algebraic side, here we work on the logical part.

Proposition 5.2.16

(1) There is no "normal" characterization of any fixed size of not necessarily definability preserving ranked preferential structures.

(2) There is no "normal" characterization of any fixed size of the general limit version of ranked preferential structures.

Proof:

The proof follows closely the proof of Proposition 5.2.15. We omit some simple facts shown already there.

Let as above κ be any infinite cardinal. We show that there is no characterization of general (i.e. not necessarily definability preserving) ranked preferential structures which has size $\leq \kappa$. We suppose there were one such characterization Φ of size $\leq \kappa$, and construct a counterexample. Again, we assume wlog. that the characterization works with theories only.

We introduce again some notation:

$\mathcal{A} := \{X \subseteq M(\neg p) : card(X) \leq \kappa\}.$

If $Con(T, p)$ or if ($T \vdash \neg p$ and $card(M(T)) > \kappa$) define $\mathcal{A}_T := \{X \in \mathcal{A} : X \subseteq M(T)\}$ and $\mathcal{B}_T := \{X \in \mathcal{A}_T : \overbrace{M(T) - X} \neq M(T)\}$.

(If $T \vdash \neg p$ and $card(M(T)) \leq \kappa$, $\mathcal{A}_T$ and $\mathcal{B}_T$ are undefined.)

Note that $\mathcal{A}_T$ is closed under subsets and under unions of size $\leq \kappa$.

Define now the logic $\mid\!\sim$ as follows in three steps:

(1) $\overline{\overline{Th(\{m, m'\})}} := Th(\{m\})$ if $m \models p$, $m' \models \neg p$.

This corresponds to a ranked structure with two levels, on the top level all $\neg p-$models, on the bottom level all $p-$models.

We now look again at (in terms of preferential models only some!) consequences of (1) in the first case of the following definition, and add more information in the second case:

(2) For T s.t. $Con(T, p)$ or ($T \vdash \neg p$ and $card(M(T)) > \kappa$), we define

$\overline{\overline{T}} := Th(\bigcap\{\overbrace{M(T) - A} : A \in \mathcal{A}_T\})$.

Thus, (1) is again a special case of (2), and in case (2) the prerequisites of Lemma 5.2.14 are satisfied for $\overline{\overline{T}}$ and $\mathcal{A}_T$.

(3) For T s.t. $T \vdash \neg p$ and $card(M(T)) \leq \kappa$, define $\overline{\overline{T}} := \overline{T}$.

As in the proof of Proposition 5.2.15, we see that this logic is not preferential, neither in the minimal, nor in the limit reading.

Suppose there were a characterization Φ of size $\leq \kappa$. It has to say "no" for at least one instance $\mathcal{T}$ of the universally quantified condition Φ. We will again show that we find a true preferential structure where all $T \in \mathcal{T}$ evaluate to the same value as just described, a contradiction.

By Lemma 5.2.14, for each of the $\leq \kappa$ $T \in \mathcal{T}$ of case (2), it suffices to consider a set A_T of size $\leq \kappa$ of suitable models of $\neg p$, so, all in all, we work just with at most κ many such models. So we set again

$B := \bigcup\{A_T : \overline{\overline{T}} = Th(M(T) - A_T) \neq \overline{T}, T \text{ as in case (2)}, T \in \mathcal{T}\}.$

Define now B' by induction:

$B_0 := B,$

$B_\lambda := \bigcup\{B_i : i < \lambda\}$ for limit λ,

$B_{i+1} := B_i \cup \bigcup\{M(T) : T \vdash \neg p, card(M(T)) \leq \kappa, M(T) \cap B_i \neq \emptyset, T \in \mathcal{T}\}.$

$B' := \bigcup\{B_i : i < \lambda\}$, $\lambda \leq \kappa$ sufficiently big.

Thus, B' still has card $\leq \kappa$, as there were only $\leq \kappa$ many $T \in \mathcal{T}$ involved, and each $M(T)$ has size $\leq \kappa$.

Consider now the ranked preferential structure with two levels where we put in the top level all $b \in B'$ (i.e. not all $\neg p-$models, but only $\leq \kappa$ many of them), and all other $\mathcal{L}-$models on the bottom level.

We have to show that the expression Φ with our instances still fails in the new structure, or, that it has still the same truth value. For this purpose, we show that $\overline{\overline{T}}$ is the same in the first definition and in the new structure for all $T \in \mathcal{T}$.

Case 1: T is s.t. $Con(T, p)$ or ($T \vdash \neg p$ and $card(M(T)) > \kappa$) : If $\overline{\overline{T}} \neq \overline{T}$ in the old definition, then this is an immediate consequence of the fact that $card(B') \leq \kappa$, and that the $A_T \subseteq B \subseteq B'$ were optimal. If $\overline{\overline{T}} = \overline{T}$ in the old definition, then there is no $X \in \mathcal{A}_T$ s.t. $\overbrace{M(T) - X} \neq M(T)$, so a fortiori there is no $Y \subseteq B'$ s.t. $\overbrace{M(T) - Y} \neq M(T)$, as $card(B') \leq \kappa$.

Case 2: $T \vdash \neg p$ and $card(M(T)) \leq \kappa$: Thus, $\overline{\overline{T}} = \overline{T}$ in the old definition, but in the new structure all $x \in M(T)$ for every such $T \in \mathcal{T}$ are either all on the top level, or they are all on the bottom level, so $\overline{\overline{T}} = \overline{T}$ in the new structure, too.

Thus, the $\overline{\overline{T}}$ for $T \in \mathcal{T}$ in the new and in the old structure are the same. So the instance $\mathcal{T}$ of Φ fails also in a suitable preferential structure, contradicting its supposed discriminatory power.

□

We turn to revision.

Familiarity with concepts, definitions, and results of theory revision are assumed.

We have done the main work in the ranked case, and just have to apply our result to revision. The idea is as follows: Take a language as above, and add for simplicity one new variable q. We make almost everything trivial, with one exception: We single out one $q-$model m, and look from m to the $\neg q-$models, which we treat as in the ranked case. The $\neg q \wedge p-$models will play the role of the $p-$models in the ranked case, and the $\neg q \wedge \neg p-$models will play the role of the $\neg p-$models in the ranked case. Thus, all $\neg q \wedge p-$models will be closer to m than some $\neg q \wedge \neg p-$models, e.g. the first distance is 0.5, the second one 0.6. All other distances will be made bigger, e.g. 1.0. (See Figure 5.2.1.)

To avoid trivial complications, we make the distance not symmetric, but it is easy to construct a similar case with a symmetric distance. (If $m \models T'$ and $T \vdash \neg q$, then $T * T' := Th(m)$, else $T * T' := T'$, etc.)

Note again that the representation conditions for limit revision also contain arbitrarily big sets — see Section 4.2.5, and that we worked there on the algebraic side.

some $\neg q \wedge \neg p-$models $d(m, x) = 0.6$

all $\neg q \wedge p-$models and most $\neg q \wedge \neg p-$models $d(m, x) = 0.5$

$m \models q$

all other $q-$models $d(m, x) = 1$

All other distances are 1.0 or 0

Figure 5.2.1

We thus have:

Proposition 5.2.17

(1) There is no normal characterization of not necessarily definability preserving distance defined revision. The distance can be chosen symmetric or not.

(2) There is no normal characterization of the limit version of distance defined revision. The distance can be chosen symmetric or not.

Proof:

Choose p, q, m as in the discussion preceding this proposition.

The following definition is motivated by the distance defined structures as described above, see also Figure 5.2.1.

Define a revision operator as follows:

Case 1, $Con(T,T') : T * T' := \overline{T \cup T'}$.

Case 2, $\neg Con(T,T')$:

Case 2.1, $m \notin M(T)$ or $T' \vdash q$: $T * T' := T'$.

Case 2.2, $m \in M(T)$ and $Con(T', \neg q)$:

Only the $\neg q-$models are interesting, as all $q-$models have distance 1. For this reason, we now define T'', and $T * T'$ will be $\overline{\overline{T''}}$ as constructed in the proof of Proposition 5.3.16, using the similarity between ranked preferential structures and distance defined structures — ranked structures are distance defined structures with fixed left hand point.

Let $T'' := T' \cup \{\neg q\}$. As in the proof of Proposition 5.2.16, define $\mathcal{A}_{T''} := \{X \subseteq M(\neg q \wedge \neg p) : card(X) \leq \kappa$ and $X \subseteq M(T'')\}$.

Case 2.2.1, T'' is s.t. $Con(T'', p)$ or $(T'' \vdash \neg p$ and $card(M(T'')) > \kappa)$:

$T * T' := Th(\bigcap\{\widehat{M(T'') - A} : A \in \mathcal{A}_{T''}\})$.

Case 2.2.2, T'' is s.t. $T'' \vdash \neg p$ and $card(M(T'')) \leq \kappa$: $T * T' := T''$.

Obviously, this is not distance definable — argue as in the proof of Propositions 5.3.15 or 5.3.16.

We turn to the construction of the distance, to obtain a positive case for small (size $\leq \kappa$) subsets of the information for the chosen set $\mathcal{T}$ of parameters.

We argue now just as in the proof of Proposition 5.2.16, and define a set of $\neg q \wedge \neg p-$models B' (depending, of course, on $\mathcal{T}$), which we move farther away from m than the other $\neg q-$models. Thus, we construct the distance relation as follows:

$d(x,x) := 0,$

$d(m,x) := 0.5$ iff $x \in M(\neg q) - B'$,

$d(m,x) := 0.6$ iff $x \in B'$,

$d(x,y) := 1.0$ iff $x \neq m$ or $y \models q$.

This distance gives the same revision results for the $\leq \kappa$ parameters in $\mathcal{T}$ (as in the proof of Proposition 5.2.16), and we are finished.

$\square$

5.3 Revision

The reader should be familiar with the basic concepts and definitions of theory revision (see Section 2.3.2), and the representation results of Section 4.2.

Recall that we consider here only the symmetric case of Section 4.2, as the not necessarily symmetric case was treated only in the finitary version, where definability preservation is trivially true.

5.3.1 The algebraic result

The proof of the main result of this part, Proposition 5.3.2 follows the same lines as the corresponding proof for preferential structures. The auxiliary result Proposition 5.3.1 has again verbatim the same proof as the version for definability preserving distance based revision, Proposition 4.2.2.

Let, in this section, $U \neq \emptyset$, $\mathcal{Y} \subseteq \mathcal{P}(U)$ be closed under finite $\cap$ and finite $\cup$, $\emptyset \in \mathcal{Y}$. Recall that for all $A, B \in \mathcal{Y}$, and all distances d considered $A, B \neq \emptyset$ $\rightarrow A \mid_d B \neq \emptyset$ is assumed to hold.

We consider the following conditions:

Condition 5.3.1

Let $\mid: \mathcal{Y} \times \mathcal{Y} \rightarrow \mathcal{Y}$, $\mid': \mathcal{Y} \times \mathcal{Y} \rightarrow \mathcal{P}(U)$, and $A, B, X_i \in \mathcal{Y}$.

$(\mid 0)$ $A \mid B = \widehat{A \mid' B}$,

$(\mid 1)$ $A \mid B \subseteq B$,

$(\mid 2)$ $A \cap B \neq \emptyset \rightarrow A \mid B = A \cap B$,

$(\mid 3)$ $A, B \neq \emptyset \rightarrow A \mid B \neq \emptyset$,

$(\mid' 1)$ $A \mid' B \subseteq B$,

$(\mid' 2)$ $A \cap B \neq \emptyset \rightarrow A \mid' B = A \cap B$,

$(\mid' 3)$ $A, B \neq \emptyset \rightarrow A \mid' B \neq \emptyset$,

$(\mid' L)$ (Loop):

$(X_1 \mid' (X_0 \cup X_2)) \cap X_0 \neq \emptyset$,

$(X_2 \mid' (X_1 \cup X_3)) \cap X_1 \neq \emptyset$,

$(X_3 \mid' (X_2 \cup X_4)) \cap X_2 \neq \emptyset, \ldots.$

$(X_k \mid' (X_{k-1} \cup X_0)) \cap X_{k-1} \neq \emptyset$

imply

$(X_0 \mid' (X_k \cup X_1)) \cap X_1 \neq \emptyset$.

We follow the same strategy as for preferential structures in Section 5.2, and represent first $\mid'$.

Proposition 5.3.1

Let $\mid' : \mathcal{Y} \times \mathcal{Y} \rightarrow \mathcal{P}(U)$. Then $\mid'$ is representable by an [identity respecting], consistency preserving symmetric pseudo-distance $d : U \times U \rightarrow Z$ if $\mid'$ satisfies $(\mid' 3)$, $(\mid' 1)$, $(\mid' L)$ [and $(\mid' 2)$].

Note that $(\mid' L)$ corresponds to: $d(X_1, X_0) \leq d(X_1, X_2)$, $d(X_2, X_1) \leq d(X_2, X_3)$, $d(X_3, X_2) \leq d(X_3, X_4) \leq \ldots \leq d(X_k, X_{k-1}) \leq d(X_k, X_0) \rightarrow d(X_0, X_1) \leq d(X_0, X_k)$, and, by symmetry, $d(X_0, X_1) \leq d(X_1, X_2) \leq \ldots \leq d(X_0, X_k) \rightarrow d(X_0, X_1) \leq d(X_0, X_k)$, i.e. transitivity, or to absence of loops involving $<$.

The proof is verbatim the same as for Proposition 4.2.2 in Section 4.2.2, and the reader is referred there for details. □

We obtain essentially as corollary the corresponding result for $\mid$, with a suitable definition of $\mid'$ from $\mid$.

Definition 5.3.1

For $\mid : \mathcal{Y} \times \mathcal{Y} \rightarrow \mathcal{Y}$, define $A \mid' B := A \mid B$ - $\{b \in B : \exists B' \in \mathcal{Y}(b \in B' \subseteq B$ and $b \notin A \mid B')\}$.

Proposition 5.3.2

Let $U \neq \emptyset$, $\mathcal{Y} \subseteq \mathcal{P}(U)$ be closed under finite $\cap$ and finite $\cup$, $\emptyset \in \mathcal{Y}$.

(a) Let $|: \mathcal{Y} \times \mathcal{Y} \to \mathcal{Y}$. If $|$ and $|'$ as defined in Definition 5.3.1 satisfy $(| 3)$, $(| 0)$, $(| 1)$, and $(|' L)$ [and $(| 2)$], then there is an [identity respecting] consistency preserving symmetric pseudo-distance $d : U \times U \to Z$ s.t.

$A \mid B = \overbrace{A \mid_d B}$ holds.

(b) If d is an [identity respecting] consistency preserving symmetric pseudo-distance $d : U \times U \to Z$ and $A \mid B := \overbrace{A \mid_d B}$, $A \mid' B := A \mid_d B$, then $|: \mathcal{Y} \times \mathcal{Y} \to \mathcal{Y}$ and $|': \mathcal{Y} \times \mathcal{Y} \to \mathcal{P}(U)$ satisfy $(| 3)$, $(| 0)$, $(| 1)$, and $(|' L)$ [and $(| 2)$].

Proof:

(a) We show that the prerequisites of Proposition 5.3.1 hold.

$(|' 1)$: $|'$ satisfies $(|' 1)$ by definition and $(| 1)$.

$(|' 3)$: If $A, B \neq \emptyset$, then $A \mid B \neq \emptyset$ by $(| 3)$, and as $\emptyset \in \mathcal{Y}$, $A \mid' B \neq \emptyset$ by $(| 0)$.

$(|' 2)$: If $(| 2)$ holds, then $(|' 2)$ holds: Let $A \cap B \neq \emptyset$. By definition, $A \mid' B \subseteq A \mid B = A \cap B$. Suppose $b \in (A \cap B) - (A \mid' B)$. Then there is $B' \in \mathcal{Y}$, $B' \subseteq B$ s.t. $b \in B'$, $b \notin A \mid B'$, but by $b \in B'$ $A \cap B' \neq \emptyset$, so $A \mid B' = A \cap B'$, contradiction.

So Proposition 5.3.1 applies, and there is d representing $|'$. By $(| 0)$,

$A \mid B = \overbrace{A \mid' B} = \overbrace{A \mid_d B}$.

(b) $(| 3)$, $(| 0)$, and $(| 1)$ are trivial.

$(| 2)$: If d is identity respecting, then, if $A \cap B \neq \emptyset$, $A \mid_d B = A \cap B$, so by $A \cap B \in \mathcal{Y}$ $A \mid B = A \cap B$, and $(| 2)$ holds.

$(|' L)$: Define for two sets $A, B \neq \emptyset$ $d(A, B) := d(a_b, b)$, where $b \in A \mid_d B$, and $a_b \in b \mid_d A$. Then $d(A, B) = d(B, A)$ by $d(a, b) = d(b, a)$ for all a,b. Loop amounts thus to $d(X_1, X_0) \leq \ldots \leq d(X_k, X_0) \to d(X_0, X_1) \leq d(X_0, X_k)$, which is now obvious. □

5.3.2 The logical result

We turn to (propositional) logic. The conditions for representation are formulated in Conditions 5.3.2. The result, Proposition 5.3.3 will be shown directly.

First, just as we had defined $|'$ for $|$ in Definition 5.3.1, we define now $*'$ for $*$ — a suitable approximation to $*$.

Definition 5.3.2

If $*$ is a revision function, we define $S *' T := M(S * T)$ - $\{m \in M(T) : \exists T'(m \models T', T' \vdash T, m \not\models S * T')\}$

We consider the following conditions for a revision function $*$ defined for arbitrary consistent theories on both sides.

Condition 5.3.2

$(*0)$ If $\models T \leftrightarrow S$, $\models T' \leftrightarrow S'$, then $T * T' = S * S'$,

$(*1)$ $T * T'$ is a consistent, deductively closed theory,

$(*2)$ $T' \subseteq T * T'$,

$(*3)$ If $T \cup T'$ is consistent, then $T * T' = \overline{T \cup T'}$,

$(*5)$ $Th(T *' T') = T * T'$,

$(*'L)$ $M(T_0) \cap (T_1 *' (T_0 \vee T_2)) \neq \emptyset$, $M(T_1) \cap (T_2 *' (T_1 \vee T_3)) \neq \emptyset$, $M(T_2) \cap (T_3 *' (T_2 \vee T_4)) \neq \emptyset$, $\ldots M(T_{k-1}) \cap (T_k *' (T_{k-1} \vee T_0)) \neq \emptyset$ imply $M(T_1) \cap (T_0 *' (T_k \vee T_1))$.

Proposition 5.3.3

Let $\mathcal{L}$ be a propositional language. A revision function $*$ is representable by a symmetric consistency preserving [identity respecting] pseudo-distance iff $*$ satisfies $(*0)$–$(*2)$, $(*5)$, $(*'L)$ [and $(*3)$].

Proof:

T, T', etc. will be consistent theories, and let

$\mathcal{Y} := \{M(T) : T \text{ a } \mathcal{L}\text{−theory }\}$.

"$\Leftarrow$":

Define $M(T) \mid M(T') := M(T * T')$. By $(*0)$, this is well-defined. We show the prerequisites of Proposition 5.3.2, (a).

$(\mid 3)$: By $(*1)$, $M(T) \mid M(T') \neq \emptyset$ if $M(T), M(T') \neq \emptyset$.

Note that by the definition of $\mid'$ from $\mid$ in Proposition 5.3.2 $M(S) \mid' M(T) := M(S) \mid M(T) - \{m \in M(T)\colon \exists T'(m \in M(T') \subseteq M(T), m \notin M(S) \mid M(T'))\}$. By Definition 5.3.2, $S*'T := M(S*T) - \{m \in M(T)\colon \exists T'(m \models T', T' \vdash T, m \not\models S*T')\} = M(S) \mid M(T) - \{m \in M(T)\colon \exists T'(m \models T', T' \vdash T, m \notin M(S) \mid M(T'))\}$. Thus, $M(S) \mid' M(T) = S *' T$.

$(\mid 0)$: By $(*5)$, $Th(T *' T') = T * T'$, so $M(T * T') = \overbrace{T *' T'}$. So by the above, $M(T) \mid M(T') = M(T * T') = \overbrace{T *' T'} = \overbrace{M(T) \mid' M(T')}$.

$(\mid 1)$ holds by $(*2)$, if $(*3)$ holds, so will $(\mid 2)$.

$(\mid' L)$ holds by $(*'L)$: $(T_i *' (T_j \vee T_k)) \cap M(T_j) \neq \emptyset$ iff $(M(T_i) \mid' (M(T_j) \cup M(T_k))) \cap M(T_j) \neq \emptyset$ by the above remark, so the loop conditions $(\mid' L)$ and $(*'L)$ are equivalent.

By Proposition 5.3.2, (a), there is a — if $(\mid 2)$ holds, identity respecting — symmetric, consistency preserving pseudo-distance d on $M_{\mathcal{L}}$ s.t. $M(T) \mid M(T') = \overbrace{M(T) \mid_d M(T')}$, so $M(T * T') = \overbrace{M(T) \mid_d M(T')}$, and $Th(M(T * T')) = Th(\overbrace{M(T) \mid_d M(T')}) = Th(M(T) \mid_d M(T'))$. As $T*T'$ is deductively closed, $T * T' = Th(M(T * T'))$, so $T * T' = Th(M(T) \mid_d M(T'))$.

"$\rightarrow$":

Let d be such a pseudo-distance on $M_{\mathcal{L}}$.

Define $T * T' := Th(M(T) \mid_d M(T'))$. We use Proposition 5.3.2, (b).

Note: if $y \notin X \mid_d Y$, then there is $Y' \subseteq Y$ finite (and thus in $\mathcal{Y}$) s.t. $y \in Y'$, $y \notin X \mid_d Y'$. Moreover, for finite Y $X \mid_d Y = \overbrace{X \mid_d Y}$. Let $\mathcal{F}$ be the set of $\mathcal{L}-$theories with finitely many models. Thus, by $M(S * T) = \overbrace{M(S) \mid_d M(T)}$ we have $S *' T := M(S * T)$ - $\{m \in M(T) : \exists T'(m \models T', T' \vdash T, m \not\models S*T')\} = \overbrace{M(S) \mid_d M(T)} - \{m \in M(T) : \exists T'(m \models T', T' \vdash T, m \notin \overbrace{M(S) \mid_d M(T')})\} = \overbrace{M(S) \mid_d M(T)} - \{m \in M(T) : \exists T' \in \mathcal{F}(m \models T', T' \vdash T, m \notin M(S) \mid_d M(T'))\} = M(S) \mid_d M(T)$, as $y \in X \mid_d Y$ iff $y \in X \mid_d Y'$ for all Y' s.t. $y \in Y' \subseteq Y$.

$(*5)$: $Th(T*'T') = Th(M(T) \mid_d M(T')) = T*T'$. $(*0)$ and $(*1)$ will trivially hold. By $(\mid 1)$, $(*2)$ holds, if $(\mid 2)$ holds, so will $(*3)$.

We see that $(*'L)$ holds by $(|'\ L)$: By the above, the two loop conditions are again equivalent. □

Chapter 6

Sums

6.1 Introduction

We consider in this Chapter 6 minimal sums (mostly of distances).

In Chapter 4 on revision and counterfactuals, we looked at just one distance at a time, and we were interested in the question whether it was minimal. Here, we will consider (finite) sums, and will be interested in those objects, which correspond to minimal sums (of some entity), like shortest trajectories. Perhaps the easiest way to see their importance is the example of iterated update. Given some principle of maximal inertia, we are interested in those sequences or developments, which have smallest change, i.e. the smallest sum of individual changes from time t to time t'.

More precisely, we look at several representation problems:

- first, the problem of update characterization as just described,
- second, the BFH Markov characterization problem — where "BFH" stands for the three authors Boutilier/Friedman/Halpern, or their paper [BFH95], or their approach,
- third, the problem to characterize "between" and "behind".

In all cases, the solution will be given by an old algorithm which goes back to Farkas at the beginning of the 20th century. And, in all cases, we show that it is impossible to find a finite characterization in the general case.

6.1.1 The general situation and the Farkas algorithm

In a certain way, this chapter carries the consideration about distances one step further. Whereas we were interested in Chapter 4 in minimal distances, we look here at something like the sum or mean of several values — not just one value decides, but many values together, and, we can add them. In short, we are interested in minimal sums. We will, for instance, consider developments over several steps, and will be interested in those which use the least resources, change the least, etc.

Thus, we have to compare sums, to determine whether sum S is smaller than sum S'. Conversely, we will try to determine whether certain choices — supposed to be achieved by minimization of certain sums — are really determined by sums. This amounts to determine whether systems of inequalities have a solution. Suppose, e.g. that sum S consists of a and b, S' of a' and b', the development corresponding to S is preferred to S'. We know then that $a + b < a' + b'$ has to hold, if this preference is really determined by comparing sums. If we know all preferences, we have a complete (for this situation) system of inequalities, and we can try to see whether the system has a solution. If so, the preferences are determined by (or at least equivalent to) minimizing sums, if not, this cannot be.

Thus, completeness can be determined by solving systems of inequalities: if we impose sufficiently many conditions on preference, we can show that the resulting systems of inequalities have a solution. Our system of conditions is then complete for being determined by minimizing sums. If this is not the case, then the set of conditions is not complete.

Basic addition seems rather difficult to characterize, especially under lack of suitable closure properties of the domain. More abstract forms — where, e.g. the biggest element carries it all — seem better suited. This is no surprise, as in the basic form, one big element can be compensated by many small ones, so it is plausible that representation becomes more difficult.

Solving systems of inequalities is an ubiquous problem, so it is not surprising that there is an old solution to the problem, an algorithm due to Farkas at the beginning of the 20th century. (S. Koppelberg, Berlin, pointed out this algorithm to the author.) We present it now (in slight modification).

The algorithm eliminates variables by induction, it instantiates them, and it roughly works like this:

Let wlog. all inequalitis be of the form $a + a' + \ldots \leq b + b' + \ldots$

We eliminate variables successively:

If some variable b occurs only on the right, we can omit it. If the remaining, smaller, system has a solution, the original one will have one too, we just choose b big enough.

The same holds if a occurs only on the left — we make it sufficiently small.

Suppose c occurs on the left and twice on the right in a simple example:

For instance, for $c \preceq x_1$, $c \preceq x_2$, $x_3 \prec c$, $x_3 + x_4 \preceq c + x_5$, we consider then $c+x_3 \prec x_1+c$, $c+x_3+x_4 \preceq x_1+c+x_5$, $c+x_3 \prec x_2+c$, $c+x_3+x_4 \preceq x_2+c+x_5$. We cancel c on both sides, solve the system of the remaining variables, and then "squeeze" a suitable value for c in.

In this way, we eliminate all variables down to the last one, and it will be evident whether the final system has a solution or not.

Now, solvability by an algorithm is a funny condition: the system is representable by inequalities of sums iff the Farkas algorithm terminates with a solution. It is natural to look for other, more usual and compact conditions. Moreover, we see here again a problem of domain closure: if we can observe directly the transformations of the Farkas algorithm in the domain, it suffices to give the transformation steps as conditions. E.g., if, for the last step, we can reshuffle the variables in the manner indicated, i.e. if we can observe whether $x_3 \prec x_1$, etc., we can work with the new system. If such sequences are not directly observable, it will not be sufficient to give the transformation steps, as we cannot observe their results.

On the other hand, we can sometimes show that finite characterizations are not possible under weak closure conditions. This is trivial in the BFH situation, as we have there no strong closure conditions, easy for the case of update by minimal sums, and, in my opinion, most interesting in the case of "between" and "behind". We discuss there a class of examples which remind the author strangely of switching diagrams for independent switching of one lightbulb by more than two switches — a problem the reader might have pondered over as a kid (like the author). The strategy will always be the same one as for revision: we show that the difference between "good" and "bad" cases, i.e. those representable by inequalities of sums and those not so representable, can be made arbitrarily "small", and we need arbitrarily much information to discern the cases.

We present now the three problems.

6.1.2 Update by minimal sums

The basic idea is a formalization of the principle of maximal inertia. We reason about the states of a dynamic system at time t, t', t'', ... We have only partial information, usually partial in two ways: first, we are not given information about all moments s, second, even when we have information about moment s, this information usually does not allow a full determination of the state the system is in. For instance, the system's states may be described by the propositional language p, q, r. Then, e.g., for time t we know that p holds, at t', we know that $q \wedge \neg r$ holds, and we know nothing about t''. The task is to give the best guess for t''. Once we have a distance between models, based on the principle of maximal inertia, or minimal change, we can calculate the sums of all possible trajectories, i.e. starting with $p-$models, then going through $q \wedge \neg r-$models, and undetermined for the last step, choose the trajectories with minimal sums, and the (disjunction of the) possible endpoints give us the best guess. Such distance based reasoning will have certain properties, in particular, the Farkas algorithm will terminate positively for it.

The representation problem is the converse: We are given a language, a number of incomplete descriptions together with the best guesses, and we have to determine whether this can be a rational guess based on maximal inertia or minimal sums of distances between models. If so, we have to present one such distance measure. For this purpose, we apply the Farkas algorithm to determine whether it is distance based, and to calculate a suitable distance in the positive case.

We have in all cases examined a positive result — a criterion for representability by minimal sums, the Farkas algorithm — and a negative result: we cannot do better than give arbitrarily long conditions.

The positive result:

We first show an abstract representation result (Proposition 6.3.1), which shows how to extend a relation $\prec$ to a ranked relation $<$, essentially representing a choice function μ — at least up to some equivalence relation (which has nothing to do with $\prec$ or $<$), and the same for an abstract distance (see Section 6.3.2 for the exact formulation). The proof is more tedious than difficult. This proposition will be used later on in the proof of the main result.

We then show that knowledge of sufficiently many intermediate results allows us to completely reconstruct the best sequences (Fact 6.3.4). Again, the argument is more tedious than really complicated.

In the conditions we use, we impose quite strong closure of the domain. We can thus avoid to just cite the Farkas algorithm. We make the domain sufficiently rich to be able to make the transformations in the domain itself, and are not obliged to leave it with "just the values in our pockets", and perform the transformations "in the air". This is a matter of taste — or of the actual problem at hand.

An inconvenient is that the conditions themselves become quite complicated, as we have to put all necessary transformations into them.

In Definition 6.3.1, we show how to define the relation $\prec$: (Here $\prec$ means that the smallest sums of distances in sequences of the left hand side are smaller than the smallest sums of distances in sequences of the right hand side.)

We will be given some (perhaps incomplete) information at various time points, this corresponds to a cartesian product of model sets.

Assume that all sets of sequences written are "observable" sets. E.g., in (R4) below, $\Sigma' \cup \bigcup \Sigma_i$ will be some product Σ of sets (such products are observable).

For the right hand side:

(R1) $\Sigma \times B \preceq \Sigma \times B'$ if $\mu(\Sigma \times (B \cup B')) \cap B \neq \emptyset$,

(R2) $\Sigma \times B \prec \Sigma \times B'$ if $\mu(\Sigma \times (B \cup B')) \cap B' = \emptyset$.

For the left hand side:

(R3) $\Sigma \times B \preceq \Sigma' \times B$ if $\Sigma' \subseteq \Sigma$,

(R4) (for all i $\Sigma' \times B \preceq \Sigma_i \times B$) if $\mu((\Sigma' \cup \bigcup \Sigma_i) \times B) \not\subseteq \bigcup \mu(\Sigma_i \times B)$,

(R5) (for all i $\Sigma' \times B \prec \Sigma_i \times B$) if $\bigcap \mu(\Sigma_i \times B) \not\subseteq \mu((\Sigma' \cup \bigcup \Sigma_i) \times B)$

(i ranges over some I in (R4) and (R5)).

The role of 0:

(R6) If $\bigcap \Sigma(i) \neq \emptyset$, then $\mu(\Sigma) = \bigcap \Sigma(i)$.

Addition:

(Here, a, b, etc. are sums of distances, corresponding to one sequence. When they are compared, they are of equal length. $\approx$ stands for $\preceq$ and $\succeq$ simultaneously).

(+1) $a + b \approx b + a$,

(+2) $(a + b) + c \approx a + (b + c)$,

(+3) $a \approx a' \rightarrow (b \approx b' \leftrightarrow a + b \approx a' + b')$,

(+4) $a \approx a' \rightarrow (b \prec b' \leftrightarrow a + b \prec a' + b')$,

(+5) $a \prec a' \wedge b \prec b' \rightarrow a + b \prec a' + b'$

and in Condition 6.3.1 we give the necessary and sufficient conditions for update by minimal sums:

(C1) $\mu(\Sigma) \subseteq \Sigma(n)$, if $\Sigma(n)$ is the last component of Σ,

(C2) $\Sigma \neq \emptyset \rightarrow \mu(\Sigma) \neq \emptyset$,

(C3) $\mu((\bigcup \Sigma_i) \times B) \subseteq \bigcup \mu(\Sigma_i \times B)$,

(C4) Loop: The relation defined by $(R1)$–$(R6)$, $(+1)$–$(+5)$ contains no loops involving $\prec$.

They translate then into the Farkas algorithm, using two lemmas (Fact 6.3.6 and Fact 6.3.7), which demonstrate the existence of suitable "witnesses", sequences $\sigma \in \Sigma$, for the results of μ described there.

The negative result:

Now to the negative result: There is no finite characterization in the general case.

As said already above, the domain has some closure conditions, resulting from the fact that we are always given snapshots at different time points, so we always consider whole cartesian products of model sets.

We will consider trajectories of length 2, e.g. from a to b to c, and the distances (e.g. from a to b, from b to c) will be small, medium, or big. Small will be much smaller than medium, which will be much smaller than big. In this way, if one component of the path is small, this path will always be smaller than one with two middle size components, and a path with a big component will always be bigger than one with two middle size components. Small and big will be constant sizes, and we fiddle with the medium ones. A suitable choice of the distances shows that only particular products of two-element sets A, B, C, $A \times B \times C$, give us any information about the middle sizes — in all other cases, there will be some small distance which interferes, or, the case is trivial for some other reason. Thus, the only information we get is from very particular products. It is now easy to give update results which do not fit with any choice of the middle values, but it suffices to change just one of the results to have a distance representable update — and all the other information stays the same. Thus, we have the same situation as for revision: We can choose arbitrarily big "bad" examples which differ from "good" ones by just one bit of information, and the same reasoning applies: it is impossible to find a finite characterization of the "good" cases.

6.1.3 Comments on "Belief revision with unreliable observations"

The definitions are unfortunately somewhat involved. Consequently, we will describe in this overview only the general outline in an informal way, and refer the reader immediately to Section 6.4 for all formal details.

BFH give a careful motivation for examining exactly these systems, and the present chapter is mostly on representation results. We therefore invite the reader interested in motivation to consult the original article [BFH95].

(We have not discussed this article in earlier chapters, as it is a little off from our main line of more analytical interests. Yet, it fits fully into the rest of this formal chapter, having a representation via Farkas, and no finite representation. The reader is asked to be lenient with this conceptually, but not formally, somewhat isolated iceberg.)

Seen abstractly, BFH discuss systems of "runs", i.e. systems of sequences, with a ranking κ of the sequences, and, by the natural extension, of sets of runs. We first give the basic definitions, show some basic results taken from the original BFH article, and then turn to "Markov" systems (definition of BFH, see Definition 6.4.5). Markov systems have, essentially, no memory, or, their ranking can be calculated component-wise by sums of values, see Fact 6.4.3. This makes the connection to the rest of the chapter obvious. We correct a small mistake in the original article (Example 6.4.1), show that the conditions (O1) and (O2) given in BFH hold in Markov systems, but give an example (Example 6.4.2) of a non-Markov system, where they also hold — the latter result goes beyond the original article.

We then show our main results for this section:

First, we can have a complete characterization using the Farkas algorithm, see Proposition 6.4.5, and the preparing Fact 6.4.4. To obtain the result, we use a (harmless) closure condition — see the remarks after Fact 6.4.4. Example 6.4.3 illustrates problems when we try to extend the result to infinite runs.

Second, we show in Example 6.4.4 that there is no finite characterization. This is easy, as there are no nontrivial closure conditions for the domain to respect.

6.1.4 "Between" and "behind"

Recall that we define "b is between a and c", or, equivalently "c is behind b, seen from a" iff $d(a,c) = d(a,b) + d(b,c)$. Thus, given a distance with addition, we can easily determine the elements which are in the "between" relation. Conversely, given a relation of "between" or "behind", i.e. a set of triples, we can ask the question again whether there is a distance d, which determines the same relation. Using Farkas, this is easy. And, again, the question arises whether it is possible to find an essentially simpler characterization, i.e. a finite one. The answer is negative, which might be somewhat surprising for such a simple problem.

We follow again the same strategy, i.e. we produce arbitrarily big "bad", i.e. not distance representable examples which differ from "good" ones by just one "bit" of information — changing just one triple from "yes" to "no" transforms the negative example to a positive one. The idea is indicated in Example 6.5.1, which shows by an easy calculation that the following situation cannot be represented by a distance (where $< a, b, c >$ stands for "b is between a and c"):

Take x and y as endpoints, and a_1, a_2, a_3, b_1, b_2, b_3 as intermediate, and $< x, a_1, a_2 >$, $< x, b_1, b_2 >$, $< a_1, b_2, b_3 >$, $< b_1, a_2, a_3 >$, $< a_2, b_3, y >$, $< b_2, a_3, y >$. All other triples are not in the "between" relation.

This cannot correspond to a distance (see the discussion of Example 6.5.2), but we can make any proper subset work: If we omit any of the triples, we can find a distance which generates the relation.

The general example is the same, only with arbitrarily many points. We show that eliminating just one of the triples transforms the example into a "good" one, examining by cases a triple in the middle or a triple at the end. The details are again a little fiddly, we have to determine the right numerical values.

6.1.5 Summary

We treat in this chapter representation for systems, where the choice is determined by minimality of certain sums. There is a general approach to such situations, using an old algorithm, due to Farkas, which determines whether a system of inequalities has a solution or not. If we make sufficiently strong assumptions about domain closure (see, e.g. in the section on update by minimal sums), we can put the Farkas transformations into the domain, so they become observable, if not, we have to treat the algorithm as a black box which churns out "yes" or "no". In all three cases we show that we cannot do really better: there is no finite characterization possible (unless, again, we have sufficiently strong domain closure properties). The most interesting of these last results is perhaps the one on "between", as this is a seemingly very simple problem — and the example is somewhat funny.

Recommended reading

It might be best to begin with the Farkas algorithm. You might then turn immediately to the negative result about "between" and "behind", as this requires no further definitions and preliminaries. You might then read the positive result about update defined by minimal sums, and then the negative one, or vice versa, while the "between" result is still fresh in your mind. At the end, you might turn to the BFH version of update.

6.2 The Farkas algorithm

The following (old) algorithm will be used in this chapter several times in various forms. It is a modification of an algorithm communicated by S. Koppelberg, Berlin. The original version seems to be due to Farkas [Far02].

We have a system of inequalities and equalities of the type

$x_{1,1}+\ldots+x_{1,m} \prec x_{2,1}+\ldots+x_{2,m}$ or

$x_{1,1}+\ldots+x_{1,m} \preceq x_{2,1}+\ldots+x_{2,m}$ or

$x_{1,1}+\ldots+x_{1,m} \approx x_{2,1}+\ldots+x_{2,m}$,

where m can differ.

The last one can be transformed to

$x_{1,1}+\ldots+x_{1,m} \preceq x_{2,1}+\ldots+x_{2,m}$ and

$x_{2,1}+\ldots+x_{2,m} \preceq x_{1,1}+\ldots+x_{1,m}$.

Let the system contain $x_1 \ldots x_n$. We eliminate by induction all but one of the $x_{i,k}$. In the positive case, the procedure will be successful (by the loop condition), and tells us how to assign positive rationals to the $x_{i,k}$. The procedure eliminates the x_n by induction, and the simplified system of inequalities Π' has a solution iff the original one Π has.

In the cases which interest us, all x_n will be ≥ 0. Fix now x_n.

Assume without loss of generality that the left hand side is always less or equal the right hand side.

Without loss of generality, x_n does not occur on both sides of the same inequality, otherwise, subtract one each on both sides repeatedly. See Remark (1) below.

Case 1: x_n does not occur in Π — we are done.

Case 2: x_n occurs only on the right hand side. Let $\Pi' \subseteq \Pi$ be the set of those inequalities, where x_n does not occur. If Π' has a solution, choose x_n big enough to make Π true.

Case 3: x_n occurs only on the left hand side. Replace, e.g. $x_m + x_n \preceq x_k + x_l$ by $x_m + 0 \preceq x_k + x_l$, $x_m + x_n \prec x_k + x_l$ by $x_m + 0 \prec x_k + x_l$, let the other inequalities unchanged. Let Π' have a solution, then the difference in the modified inequalities is some minimum or 0, where we can put x_n in.

Case 4: x_n occurs on both the left and the right hand side. Let Π_l be the set of inequalities, where x_n occurs on the left hand side, let Π_r be the set of inequalities, where x_n occurs on the right hand side.

Informally, we isolate x_n and transform all $\delta_i \in \Pi_l$ into $x_n \preceq R$ or $x_n \prec R$, and all $\delta_j \in \Pi_r$ into $L \preceq x_n$ or $L \prec x_n$, e.g. $x_3 + x_4 \preceq x_n + x_5$ will become $x_3 + x_4 - x_5 \preceq x_n$. We then consider all inequalities of the form $L \preceq R$ or $L \prec R$ resulting from $L \preceq x_n \preceq R$, etc., and "squeeze" x_n into a solution of the system of $L \preceq R$ and $L \prec R$. In general, this procedure will use subtraction, which is not observable in our cases, and will not figure among the conditions. Instead we consider the sums $\delta_i + \delta_j$ where $\delta_i \in \Pi_l$, $\delta_j \in \Pi_r$, and eliminate x_n from both sides. We then solve this system, have numbers, and "squeeze" x_n into the inequalities.

Let Π' be the set of inequalities where x_n does not occur, and all sums $\delta_i + \delta_j$, $\delta_i \in \Pi_l$, $\delta_j \in \Pi_r$. See again Remark (1) below.

For instance, for $x_n \preceq x_1$, $x_n \preceq x_2$, $x_3 \prec x_n$, $x_3 + x_4 \preceq x_n + x_5$, we consider $x_n + x_3 \prec x_1 + x_n$, $x_n + x_3 + x_4 \preceq x_1 + x_n + x_5$, $x_n + x_3 \prec x_2 + x_n$, $x_n + x_3 + x_4 \preceq x_2 + x_n + x_5$.

Now we can eliminate x_n on both sides, where it still occurs.

In our example, $x_3 \prec x_1$, $x_3 + x_4 \preceq x_1 + x_5$, $x_3 \prec x_2$, $x_3 + x_4 \preceq x_2 + x_5$.

Let this Π' have a solution. Then $x_3 \prec x_1$, $x_3 + x_4 - x_5 \preceq x_1$, $x_3 \prec x_2$, $x_3 + x_4 - x_5 \preceq x_2$. We then fit x_n into $[max(x_3 + x_4 - x_5, x_3), min(x_1, x_2)]$.

In the end, we have

$0 + \ldots + 0 + x_1 + \ldots + x_1 \prec 0 + \ldots + 0$ (sums of equal length on both sides), or

$0 + \ldots + 0 + x_1 + \ldots + x_1 \preceq 0 + \ldots + 0$, or

$0 + \ldots + 0 \prec 0 + \ldots + 0 + x_1 + \ldots + x_1$, or

$0 + \ldots + 0 \preceq 0 + \ldots + 0 + x_1 + \ldots + x_1$.

By $0 \preceq x_1$, the first possibility is excluded. The other three show that the final system has a solution, which can be transformed into one for the original system as indicated.

Remark 6.2.1

(1) The less we can observe, the more we have to impose closure conditions on the values. For instance, usually we will not have differences, therefore we have avoided them in above algorithm. Whenever we can determine whether some $x_i = 0$, we will add this to the system. Whenever we can observe the results of the transformations (e.g. by concatenation), we can put this into the conditions, otherwise, we will have to impose it directly on the values: that the operations on the values, and not on the universe, preserve freedom from loop.

For example, if we have commutativity and adding the same value built in, we can code the elimination of the same x_n on both sides into the system: $S \subseteq S'$ iff $T \subseteq T'$, where T and T' result from S and S' by adding x_n — expressed in conditions about the systems. E.g. S corresponds to a sequence preferred to the sequence corresponding to S' iff the same applies to the $T-$ and $T'-$sequences. Similarly for Case 4, addition of inequalities.

(2) The algorithm will usually result in a finite system of rationals ≥ 0, which we can transform by suitable multiplication into a system of positive integers, by finiteness.

(3) We do not see how to extend the algorithm for infinite sets of inequalities.

6.3 Representation for update by minimal sums

6.3.1 Introduction

Reasoning about developments or changing situations is an important problem in Artificial Intelligence (AI), as has been recognized very early. Much of human reasoning about these problems is based on the assumption that the world is relatively static. We will, for instance, hesitate to accept an explanation as plausible which involves many and unmotivated changes. This assumption of inertia is part of the AI folklore.

More formally, we can describe a possible development of the world as a sequence of models, a situation as a theory (or set of models), and say that the development $\sigma =< \sigma(0) \ldots \sigma(n) >$ explains the change from situation S to situation S' iff $\sigma(0) \in S$ and $\sigma(n) \in S'$. Usually, several different developments can explain the change from S to S', and the criterion of inertia may permit to choose a subset of these developments as containing the most plausible ones. A natural way to use this criterion is the following: We assume a measure of difference or distance d between the models to be given, and consider those developments as most plausible whose sum of differences $\delta(\sigma) := d(\sigma(0), \sigma(1)) + \ldots + d(\sigma(n-1), \sigma(n))$ is minimal among those which explain the change from S to S'. Conversely, given a preference relation between possible explanations, it is an interesting question whether there exists a measure of difference d between the models s.t. $\delta(\sigma) < \delta(\sigma')$ iff σ is preferred over σ'. The aim of this Section 6.3 is to solve this question, i.e. to characterize those preference relations which can be generated by such a distance, in other words, to give a representation result. We use (an adaptation of) the Farkas algorithm, given in Section 6.2.

We now describe precisely the situation we work in, and the assumptions we make.

We assume discrete time, and that we have (incomplete) information about the state of affairs at time $t_0 \ldots t_n$. This information is given by a sequence of sets of models, $\Sigma = < \Sigma(0), \ldots, \Sigma(n) >$, or, equivalently, by the product $\Sigma(0) \times \ldots \times \Sigma(n)$. We further assume that our (fixed) propositional language $\mathcal{L}$ is finite, and that we consider only sequences of finite length. $\Sigma(M_{\mathcal{L}})$ will denote the set of finite products of sets of models of $\mathcal{L}$.

If there is a distance d defined on $M_{\mathcal{L}}$, we can determine the set of those sequences σ with $\sigma(i) \in \Sigma(i)$ for which $\delta(\sigma)$ (as defined above) is minimal. We call such sequences ($d-$) preferred sequences and denote by $\nu_d(\Sigma)$ the set of $d-$preferred sequences in σ. We can then determine the set $\mu_d(\Sigma) \subseteq \Sigma(n)$, the set of endpoints of $d-$preferred sequences in σ. Formally, we have defined a function $F_d : \Sigma(M_{\mathcal{L}}) \to \mathcal{P}(M_{\mathcal{L}})$ by $F_d(\Sigma) := \mu_d(\Sigma)$. This function F_d has certain properties, like $F_d(\Sigma) \subseteq \Sigma(n)$, if $\Sigma(n)$ is the last element of Σ. We look for a complete list of properties, which characterize such F_d , i.e. if some $F : \Sigma(M_{\mathcal{L}}) \to \mathcal{P}(M_{\mathcal{L}})$ satisfies these properties, then there is some distance d on $M_{\mathcal{L}}$ s.t. $F = F_d$.

Note that, in general, the scarcity of information we dispose of — we know only $\mu(\Sigma(n))$, and nothing about the intermediate $\nu_i(\Sigma) \subseteq \Sigma(i)$ through which the preferred sequences pass — will not allow us to reconstruct $\nu(\Sigma)$, the set of the ($d-$) preferred sequences. Knowledge of these $\nu_i(\Sigma)$, on the other hand, allows to reconstruct completely $\nu(\Sigma)$, as Fact 6.3.4 will show.

We assume that we have the information $\mu(\Sigma)$ only about products Σ of sets of models, but not about arbitrary sets of sequences. Thus, $\mu(\{a, a'\} \times \{b, b'\})$ will be given, but, if $a \neq a'$ and $b \neq b'$, not $\mu(\{< a, b >, < a', b' >\})$ — the sequences $< a, b' >$, $< a', b >$ are missing. On the other hand, we assume that we can reason about unions of sets of sequences, in particular say that a union of products of sets is itself a product of sets, like $\{a, a'\} \times \{b, b'\} = (\{a\} \times \{b\}) \cup (\{a\} \times \{b'\}) \cup (\{a'\} \times \{b\}) \cup (\{a'\} \times \{b'\})$. We will call products of sets "legal" sets of sequences. Thus, we can reason about arbitrary sets of sequences, but "the world" does not give us information about arbitrary, only about legal sets of sequences. It seems a natural hypothesis that the language of reasoning may be stronger that the language of observation.

Obviously, the $\Sigma(n)$ are in a stronger position than the other $\Sigma(i)$, by definition of $\mu(\Sigma)$. This corresponds to the fact that, considering a development into the future, we are probably most interested in the final outcome. Conversely, given a development from the past to the present, we might have most information about the present.

There are, however, other directions of interest possible, and the reader will see how to adapt our conditions and proofs to the case which interests him. We have examined the two extremes — all $\nu_i(\Sigma)$ are known, and, only one ν_i is known — it should not be too difficult to modify our results and techniques accordingly.

We start our formal exposition with a "higher abstract nonsense" result in Section 6.3.2, which has proved useful in many situations. The result itself, Proposition 6.3.1, is neither conceptually nor technically deep, but it serves very well as a guideline to prove representation theorems (in the finite case) for operations based on distances: it shows that, essentially, it suffices to show the properties ($\mu 1$) and ($\mu 2$). These two properties are sufficiently close to the operation considered to give an idea how to build the proof (or to see which properties one still has to add for completeness).

6.3.2 An abstract result

As announced, we start the development with a very abstract result, which will be useful later. The conditions and statements might seem rather obscure, but they will be exactly what we need later.

The general prerequisite is that we have witnesses in some abstract equivalence classes, demonstrating the cooperation between an operation μ and a relation $\prec$. Part (a) says (very roughly) that, if $\prec$ is free from cycles, we can extend $\prec$ to a ranked relation $<$, s.t. the resulting minimality function $\mu_<$ is the same as μ. Part (b) says, again very roughly, that the same holds for a distance function.

The reader should note that in part (a), $\prec$ is an arbitrary relation between elements of Z, whereas in part (b), $\prec$ is also a relation on Z, but this time

Z is a set of abstract distances between elements of a set W, so we have statements like $(a, b) \prec (c, d)$, where $a, b, c, d \in W$, and $(a, b), (c, d) \in Z$ are abstract distances between a and b, etc.

The proof of this abstract result is a little tedious, but not difficult. In a first reading, the reader might skip the result and its proof, and come back to it later when it is needed.

We work in a universe Z, with a function $\mu : \mathcal{P}(Z) \to \mathcal{P}(Z)$, and two relations $\prec$ and $\preceq$ on Z, with $\prec \subseteq \preceq$, for which the conditions $(\mu 0)$ – $(\mu 2)$ below hold. $\preceq^*$ will denote the transitive closure of $\preceq$, we will also write $\prec^*$, if at least once $\prec$ is involved.

Moreover, we have an equivalence relation $\equiv$ on Z, (which has nothing to do with $\preceq$ or the relation to be constructed). Let $[\![a]\!]$ be the equivalence class under $\equiv$ of a.

In the first part, we construct a ranked relation $<$ on Z by extending the relation $\prec$ (and $\preceq$), and show that (essentially) $\mu = \mu_<$, where $\mu_<$ is the minimality operation defined as usual by the relation $<$, i.e. $\mu_<(X) := \{x \in X : \exists x' \in X.x' < x\}$. In the second part, we show an analogous result for a suitably defined distance function.

Proposition 6.3.1

Let $[\![a]\!]$ be finite for all a.

Let the following conditions $(\mu 0)$ – $(\mu 2)$ hold for μ and $\prec / \preceq$:

$(\mu 0)$ $\mu(A) \subseteq A$, and $A \neq \emptyset \to \mu(A) \neq \emptyset$,

$(\mu 1)$ if $a \in A$, $[\![a]\!] \cap \mu(A) = \emptyset$, $[\![b]\!] \cap \mu(A) \neq \emptyset$, then there is $b' \in [\![b]\!] \cap A$, $b' \prec^* a$,

$(\mu 2)$ if $a \in A$, $[\![a]\!] \cap \mu(A) \neq \emptyset$, $[\![b]\!] \cap \mu(A) \neq \emptyset$, then there is $b' \in [\![b]\!] \cap A$, $b' \preceq^* a$.

(1) If the relation $\preceq$ is free from cycles containing $\prec$, then $\prec / \preceq$ can be extended to a ranked order $<$ s.t. for all $A \subseteq Z$, $a \in A$ $[\![a]\!] \cap \mu(A) = \emptyset$ iff $[\![a]\!] \cap \mu_<(A) = \emptyset$.

(2) If Z is a set of abstract distances over some space W (written (x, y) for $x, y \in W$) s.t., in addition

(d1) $\forall x, y \in W$ $x \neq y \to (x, x) \prec (x, y)$,

(d2) $\forall x, y \in W$ $(x, x) \preceq (y, y)$

hold, and the relation $\preceq$ is free from cycles containing $\prec$, then there is a distance function $d : W \times W \to \mathcal{Z}$ and a total order $<$ on $\mathcal{Z}$ s.t.

(α) $0 = d(x,x)$ for any $x \in W$,

(β) $(u,v) \prec (x,y) \rightarrow d(u,v) < d(x,y)$, $(u,v) \preceq (x,y) \rightarrow d(u,v) \leq d(x,y)$,

(γ) for all $A \subseteq Z$, $a \in A$ $[[a]] \cap \mu(A) = \emptyset$ iff $[[a]] \cap \mu_<(A) = \emptyset$.

Proof:

The proofs of (1) and (2) are very close, and have a common beginning.

Let $\prec^+$ and $\preceq^+$ be the closures of $\preceq$ and $\prec$ under reflexivity and transitivity of $\prec$ or $\preceq$, more precisely:

$a \preceq^+ a$,

$a \preceq b$ implies $a \preceq^+ b$, and $a \prec b$ implies $a \prec^+ b$,

$a \preceq^+ b \preceq^+ c$ imply $a \preceq^+ c$,

($a \preceq^+ b \prec^+ c$ or $a \prec^+ b \preceq^+ c$ or $a \prec^+ b \prec^+ c$) implies ($a \prec^+ c$ and $a \preceq^+ c$).

Define $a \approx b$ iff $a \preceq^+ b$ and $b \preceq^+ a$. This is an equivalence relation. Let $\ll a \gg$ denote the $\approx$ $-$equivalence class of a, and $\mathcal{Z} := \{\ll a \gg : a \in Z\}$.

Define $\prec$ on $\mathcal{Z}$ by $\ll a \gg \prec \ll b \gg$ iff $a \preceq^+ b$, but $\ll a \gg \neq \ll b \gg$ (thus $b \not\preceq^+ a$). This is well-defined, and $\prec$ on $\mathcal{Z}$ is transitive and free of cycles too. (For the latter, e.g. $\ll a \gg \prec \ll b \gg \prec \ll a \gg$ implies $\ll a \gg = \ll b \gg$.)

Extend $\prec$ on $\mathcal{Z}$ to a strict total order $<$ on $\mathcal{Z}$.

We turn to the proof of the first part.

(1) Define $a < b$ iff $\ll a \gg < \ll b \gg$. $<$ is a ranked order on Z, e.g. $a \bot b < c$ implies $\ll a \gg = \ll b \gg < \ll c \gg$, so $a < c$.

Fact 6.3.2

Let $a, b \in Z$.

(a) $a \prec^* b \rightarrow a < b$,

(b) $a \preceq^* b \rightarrow a < b$ or $a \bot b$ or $a = b$.

Proof:

(a) $a \prec^* b \rightarrow a \preceq^+ b$, but not $b \preceq^+ a$ (otherwise there is a cycle involving $\prec$), so $\ll a \gg \prec \ll b \gg$, so $\ll a \gg < \ll b \gg$, so $a < b$.

(b) $a \preceq^* b \rightarrow a \preceq^+ b$. If $b \preceq^+ a$ too, then $a \approx b$, and $a \bot b$ or $a = b$. Otherwise $\ll a \gg < \ll b \gg$, so $a < b$. □ (Fact 6.3.2)

We now show $[\![a]\!] \cap \mu(A) = \emptyset$ iff $[\![a]\!] \cap \mu_<(A) = \emptyset$.

Let $a \in A$.

(a) Let $[\![a]\!] \cap \mu(A) = \emptyset$. By $(\mu 0)$, $\exists b \in A.[\![b]\!] \cap \mu(A) \neq \emptyset \rightarrow_{(\mu 1)} \exists b' \in [\![b]\!] \cap A$, $b' \prec^* a \rightarrow$ (by Fact 6.3.2) $b' < a$ and $a \notin \mu_<(A)$, so $[\![a]\!] \cap \mu_<(A) = \emptyset$.

(b) Suppose $[\![a]\!] \cap \mu_<(A) = \emptyset$, but $[\![a]\!] \cap \mu(A) \neq \emptyset$. Choose $a' \in [\![a]\!] \cap A$ $<-$minimal in $[\![a]\!] \cap A$ (i.e. there is no $a'' \in [\![a]\!] \cap A$ $a'' < a'$). This is possible, as $[\![a]\!] \cap A$ is finite and $<$ is free from cycles. As $[\![a]\!] \cap \mu_<(A) = \emptyset$, there is $b \in A$, $b < a'$. If $[\![b]\!] \cap \mu(A) = \emptyset$, then by $(\mu 1)$ there is $a'' \in [\![a]\!] \cap A$, $a'' \prec^* b$, so by Fact 6.3.2 $a'' < b < a'$, contradiction. If $[\![b]\!] \cap \mu(A) \neq \emptyset$, then by $(\mu 2)$ there is $a'' \in [\![a]\!] \cap A$, $a'' \preceq^* b$. By Fact 6.3.2, $a'' < b$ or $a'' \perp b$ or $a'' = b$. If $a'' \perp b$, then by rankedness $a'' < a'$. If $a'' < b$, then $a'' < a'$ by transitivity, contradiction.

We turn to the proof of the second part.

(2) Define $d(x, y) := \ll (x, y) \gg$.

Fact 6.3.3

(a) $a \prec^* b \rightarrow d(a) < d(b)$,

(b) $a \preceq^* b \rightarrow d(a) \leq d(b)$,

(c) $0 := d(x, x)$ for any $x \in W$ is well defined, and $0 \leq d(x, y)$ for any $x, y \in W$, and $0 < d(x, y)$ iff $x \neq y \in W$.

Proof:

(a) $a \prec^* b \rightarrow a \preceq^+ b$, but not $b \preceq^+ a$ (otherwise there is a cycle involving $\prec$), so $\ll a \gg \prec \ll b \gg$, so $\ll a \gg < \ll b \gg$, so $d(a) < d(b)$.

(b) $a \preceq^* b \rightarrow a \preceq^+ b$. If $b \preceq^+ a$ too, then $a \approx b$, and $d(a) = d(b)$. Otherwise $\ll a \gg < \ll b \gg$, so $d(a) < d(b)$.

(c) 0 is well defined by (d2) and (b). $0 \leq d(x, y)$ holds by (d1), (d2), (a), (b). $0 < d(x, y)$ iff $x \neq y$ holds for the same reasons. □ (Fact 6.3.3)

The rest of the proof for (2) is almost verbatim the same as the one for (1):

It remains to show $[\![a]\!] \cap \mu(A) = \emptyset$ iff $[\![a]\!] \cap \mu_<(A) = \emptyset$.

Let $a \in A$.

(a) Let $[\![a]\!] \cap \mu(A) = \emptyset$. By $(\mu 0)$, $\exists b \in A.[\![b]\!] \cap \mu(A) \neq \emptyset \rightarrow_{(\mu 1)} \exists b' \in [\![b]\!] \cap A$, $b' \prec^* a \rightarrow d(b') < d(a)$ and $a \notin \mu_<(A)$, so $[\![a]\!] \cap \mu_<(A) = \emptyset$.

(b) Suppose $[[a]] \cap \mu_<(A) = \emptyset$, but $[[a]] \cap \mu(A) \neq \emptyset$. Choose $a' \in [[a]] \cap A$ $<-$minimal in $[[a]] \cap A$ (i.e. there is no $a'' \in [[a]] \cap A$ $d(a'') < d(a')$). This is possible, as $[[a]] \cap A$ is finite and $<$ is free from cycles. As $[[a]] \cap \mu_<(A) = \emptyset$, there is $b \in A$, $d(b) < d(a')$. If $[[b]] \cap \mu(A) = \emptyset$, then by $(\mu 1)$ there is $a'' \in [[a]] \cap A$, $a'' \prec^* b$, so $d(a'') < d(b) < d(a')$, contradiction. If $[[b]] \cap \mu(A) \neq \emptyset$, then by $(\mu 2)$ there is $a'' \in [[a]] \cap A$, $a'' \preceq^* b$. So by Fact 6.3.3, (b) $d(a'') \leq d(b)$, contradiction. □ (Proposition 6.3.1)

Note that we use in both cases for the representation result essentially Fact 6.3.2 or Fact 6.3.3 respectively, independent of the details of the construction of the orders $\leq$ or $<$ from $\preceq$ and $\prec$.

6.3.3 Representation

6.3.3.1 Introduction

We will work here with (finite) sequences σ of points. In this introduction, we present some notation, and show (in Fact 6.3.4) how to reconstruct the set of $\nu-$minimal sequences from its coordinates.

Notation 6.3.1

We will work in the finite case, and with models.

σ, etc. will denote sequences of models, or, more generally, of arbitrary points. Σ, etc. will denote products of sets of points, thus special sets of sequences. We call such sets of sequences legal sets of sequences. a, etc. will denote points, A, etc. sets of points. If σ is a sequence, a a point, σa will be the concatenation of σ with a. $\sigma \times A$ will denote the set of all sequences σa, $a \in A$. $\Sigma \times A$ will denote the set of all sequences σa, $\sigma \in \Sigma$, $a \in A$, likewise $\Sigma \times a$ by abuse of notation. $\mu(\Sigma)$ will denote the set of endpoints through which preferred sequences in Σ pass.

$\sigma(i)$ will be the $i-$th element of the sequence σ, $\Sigma(i)$ the $i-$th component of the product.

If σ, σ' have same length, then $[\sigma, \sigma'] := \{\sigma'' : \sigma''(i) \in \{\sigma(i), \sigma'(i)\}$ for all $i\}$. Note that $[\sigma, \sigma']$ is thus a (legal) product of sets (of size ≤ 2). Likewise, if Σ is a legal set of sequences, and σ a sequence, both of same length, then $[\Sigma, \sigma] := \{\ \sigma' : \sigma'(i) \in \Sigma(i) \cup \{\sigma(i)\}\ \}$.

If Σ is a set of sequences, σ a sequence, both of the same length, then the Hamming distance $h(\Sigma, \sigma)$ will be the minimum of the Hamming distances

$h(\sigma', \sigma)$, $\sigma' \in \Sigma$.

Define $\sigma \equiv \sigma'$ iff σ and σ' have the same endpoint ($\equiv$ is the relation of Section 6.3.2).

Recall the definition $\delta(\sigma) := \Sigma\{d(\sigma(i), \sigma(i+1)) : 0 \leq i < n\}$ (Σ is here the sum), d a distance between models of a language $\mathcal{L}$, σ a sequence of models $\sigma := < \sigma(0), \sigma(1), \ldots, \sigma(n) >$.

Given a product of sets $\Sigma(0) \times \ldots \times \Sigma(n)$, $\nu(\Sigma)$ shall be the set of sequences $\sigma \in \Sigma$ s.t. $\delta(\sigma)$ is minimal in Σ (the "lazy" sequences with minimal changes). $\mu(\Sigma)$ shall be the set of endpoints of sequences in $\nu(\Sigma)$, $\nu_i(\Sigma)$ shall be the $i-$th projection of $\nu(\Sigma)$. Thus, if $\Sigma(n)$ is the last component of Σ, then $\mu(\Sigma) = \nu_n(\Sigma)$.

If σ, σ' are two sequences with $\sigma(n) = \sigma'(0)$, then $\sigma\sigma'$ will be their concatenation $< \sigma(0), \ldots, \sigma(n), \sigma'(1), \ldots, \sigma'(n') >$. If $\Sigma(n) = \Sigma'(0)$, let $\Sigma\Sigma' := \Sigma(0) \times \ldots \times \Sigma(n) \times \Sigma'(1) \times \ldots \times \Sigma'(n')$.

We look for conditions on μ which guarantee that we can find a distance with suitable order and addition on the values, which singles out exactly $\mu(\Sigma)$ for all legal Σ.

Remark: Note that, in general, we cannot "observe" sums: if $x = d(a, b)$, $y = d(c, e)$, we cannot be sure to see a sequence σ s.t. $\delta(\sigma) = x + y$. It is, e.g. not certain that there is f s.t. $d(b, f) = y$. This is the reason that we pack the conditions $(+i)$ into the relation and Loop, and do not use conditions like (for $\Sigma(n) = \Sigma'(0)$, $\sigma(n) = \sigma'(0)$) :

- If $\sigma \in \mu$, $\sigma' \in \mu(\Sigma')$, then $\sigma\sigma' \in \mu(\Sigma\Sigma')$, and, if $\sigma \in \mu(\Sigma)$, $\sigma' \notin \mu(\Sigma')$, then $\sigma\sigma' \notin \mu(\Sigma\Sigma')$,

- if $\sigma \notin \mu$, $\sigma' \in \mu(\Sigma')$, then $\sigma\sigma' \notin \mu(\Sigma\Sigma')$, and, if $\sigma \notin \mu(\Sigma)$, $\sigma' \notin \mu(\Sigma')$, then $\sigma\sigma' \notin \mu(\Sigma\Sigma')$.

Such conditions are much weaker, because they apply only to those sums which are really observable. We could, of course, stipulate a general condition of homogenousness of the space: that we can perform sufficient translations to guarantee concatenability. This, however, would impose a restriction on the models we consider.

We have only very limited information, the endpoints of preferred sequences. The following Fact 6.3.4 is a side remark. It illustrates that, if we also know the intermediate points of preferred sequences, we can determine the preferred sequences much better:

Fact 6.3.4

Assume $\nu(\Sigma)$ given by a distance d. Then $\nu(\Sigma)$ is reconstructible from the $\nu_i(\Sigma')$ for suitable Σ' with $\Sigma'(i) \subseteq \Sigma(i)$.

Proof:

Fix i.

Case 1: $\nu_i(\Sigma) = \{a_i\}$. Then for all $x \in \nu_{i-1}(\Sigma)$ there is a preferred sequence containing $< x, a_i >$ as a subsequence. Likewise for $y \in \nu_{i+1}(\Sigma)$.

Case 2: $Card(\nu_i(\Sigma)) > 1$. If, e.g. $card(\nu_{i-1}(\Sigma)) = 1$, we apply Case 1 to $i-1$. So suppose $card(\nu_{i-1}(\Sigma)) > 1$, $card(\nu_{i+1}(\Sigma)) > 1$. Fix $a_i \in \nu_i(\Sigma)$, and consider Σ_{a_i}, where $\Sigma(i)$ has been replaced by $\{a_i\}$, i.e. $\Sigma_{a_i} := \Sigma(0) \times \ldots \times \{a_i\} \times \ldots \times \Sigma(n)$.

If $a_{i-1} \notin \nu_{i-1}(\Sigma_{a_i})$, then there is no preferred sequence through $< a_{i-1}, a_i >$ in Σ : Any such sequence σ' through $< a_{i-1}, a_i >$ is already in $\Sigma_{a_i} \subseteq \Sigma$, and there is a better one in $\Sigma_{a_i} \subseteq \Sigma$.

Suppose $a_{i-1} \in \nu_{i-1}(\Sigma_{a_i})$. As $a_i \in \nu_i(\Sigma)$, there is a preferred sequence in Σ through a_i. It is already in Σ_{a_i}. But in Σ_{a_i}, there is one through all $a_{i-1} \in \nu_{i-1}(\Sigma_{a_i})$. By rankedness, all are preferred in Σ. So there is a preferred sequence in Σ through $< a_{i-1}, a_i >$ for all $a_{i-1} \in \nu_{i-1}(\Sigma_{a_i})$. The same argument applies to $i+1$.

Suppose now $\sigma, \sigma' \in \nu(\Sigma)$, and $\sigma(i) = \sigma'(i)$. Let $\sigma = \sigma_0\sigma_1$, where $\sigma_0 = \sigma(0)\ldots\sigma(i)$, $\sigma_1 = \sigma(i+1)\ldots\sigma(n)$, likewise $\sigma' = \sigma'_0\sigma'_1$. Then also $\sigma_0\sigma'_1$ and $\sigma'_0\sigma_1 \in \nu(\Sigma)$. For if not, then, e.g. $\delta(\sigma_0) > \delta(\sigma'_0)$, as $\sigma' \in \nu(\Sigma)$, but then $\delta(\sigma_0) + \delta(\sigma_1) > \delta(\sigma'_0) + \delta(\sigma_1)$, contradicting $\sigma \in \nu(\Sigma)$.

Thus, any sequence constructed as follows:

$a_i \in \nu_i(\Sigma)$, $a_{i-1} \in \nu_{i-1}(\Sigma_{a_i})$, $a_{i+1} \in \nu_{i+1}(\Sigma_{a_i})$ belongs to $\nu(\Sigma)$, and no others. □

6.3.3.2 The result

We recall that we put some of the Farkas machinery into the conditions and the domain properties. The conditions are consequently quite complicated. On the other hand, this avoids a black box use of the Farkas algorithm. Depending on the case, the reader might prefer or need the opposite: a small set of conditions, especially for the domain, and a more nasty completeness check (which, by the way, can be easy for an automatic demonstration).

Definition 6.3.1 will give the construction of the relations $\prec$ and $\preceq$. Condition 6.3.1 will give the conditions for representability, using a loop condition for the relation just defined. Facts 6.3.5 – 6.3.7 are auxiliary lemmas, the latter two show the essential prerequisites of the abstract representation result of Section 6.3.2. Proposition 6.3.8 states the representation result, the only thing still to show is the treatment of sums, this is done in its proof.

In the following, we assume that all sets of sequences written are legal sets, i.e. cartesian products. E.g., in (R4) of Definition 6.3.1, $\Sigma' \cup \bigcup \Sigma_i$ will be some product Σ of sets.

The conditions (R1) and (R2) contain the main ideas, so the reader should look at them in the first place.

Definition 6.3.1

(Here $\prec$ means that the smallest sums of distances in sequences of the left hand side are smaller than the smallest sums of distances in sequences of the right hand side.)

For the right hand side:

(R1) $\Sigma \times B \preceq \Sigma \times B'$ if $\mu(\Sigma \times (B \cup B')) \cap B \neq \emptyset$,

(R2) $\Sigma \times B \prec \Sigma \times B'$ if $\mu(\Sigma \times (B \cup B')) \cap B' = \emptyset$.

For the left hand side:

(R3) $\Sigma \times B \preceq \Sigma' \times B$ if $\Sigma' \subseteq \Sigma$,

(R4) (for all i $\Sigma' \times B \preceq \Sigma_i \times B$) if $\mu((\Sigma' \cup \bigcup \Sigma_i) \times B) \not\subseteq \bigcup \mu(\Sigma_i \times B)$,

(R5) (for all i $\Sigma' \times B \prec \Sigma_i \times B$) if $\bigcap \mu(\Sigma_i \times B) \not\subseteq \mu((\Sigma' \cup \bigcup \Sigma_i) \times B)$

(i ranges over some I in (R4) and (R5).)

The role of 0:

(R6) If $\bigcap \Sigma(i) \neq \emptyset$, then $\mu(\Sigma) = \bigcap \Sigma(i)$.

Addition:

(Here, a, b, etc. are sums of distances, corresponding to one sequence. When they are compared, they are of equal length. $\approx$ stands for $\preceq$ and $\succeq$ simultaneously.)

(+1) $a + b \approx b + a$,

(+2) $(a + b) + c \approx a + (b + c)$,

(+3) $a \approx a' \rightarrow (b \approx b' \leftrightarrow a + b \approx a' + b')$,

(+4) $a \approx a' \rightarrow (b \prec b' \leftrightarrow a + b \prec a' + b')$,

(+5) $a \prec a' \wedge b \prec b' \rightarrow a + b \prec a' + b'$.

Condition 6.3.1

(The conditions)

(C1) $\mu(\Sigma) \subseteq \Sigma(n)$, if $\Sigma(n)$ is the last component of Σ,

(C2) $\Sigma \neq \emptyset \rightarrow \mu(\Sigma) \neq \emptyset$,

(C3) $\mu((\bigcup \Sigma_i) \times B) \subseteq \bigcup \mu(\Sigma_i \times B)$,

(C4) Loop: The relation defined by $(R1)$–$(R6)$, (+1)–(+5) contains no loops involving $\prec$.

The loop condition is the main condition here.

Fact 6.3.5

(1) $B' \subseteq B \rightarrow \Sigma \times B \preceq \Sigma \times B'$,

(2) $b \in \mu(\Sigma \times B) \rightarrow \Sigma \times b \preceq \Sigma \times B$,

(3) $b \in B$, $b \notin \mu(\Sigma \times B) \rightarrow (\Sigma \times B) \prec (\Sigma \times b)$,

(4) If $b \in B$, $b \notin \mu(\Sigma \times B)$, then there is $\sigma' \in \Sigma$ s.t. $\forall \Sigma' \subseteq \Sigma$ (Σ' legal, $\sigma' \in \Sigma' \rightarrow b \notin \mu(\Sigma' \times B)$),

(5) If $b \in \mu(\Sigma \times B)$, then there is $\sigma' \in \Sigma$ s.t. $\forall \Sigma' \subseteq \Sigma$ (Σ' legal, $\sigma' \in \Sigma' \rightarrow b \in \mu(\Sigma' \times B)$).

Proof:

(1) trivial by (R1), (C1), (C2).

(2) trivial by (R1).

(3) trivial by (R2).

(4) If not, then $\forall \sigma' \in \Sigma \exists \Sigma' \subseteq \Sigma$ (Σ' legal, $\sigma' \in \Sigma'$, $b \in \mu(\Sigma' \times B)$). Let then $\sigma \in \Sigma$. For $\sigma' \neq \sigma$, $\sigma' \in \Sigma$, there is $\Sigma'_{\sigma'}$ with $\sigma' \in \Sigma'_{\sigma'} \subseteq \Sigma$, $b \in \mu(\Sigma'_{\sigma'} \times B)$. Then $\Sigma = \{\sigma\} \cup \bigcup \{\Sigma'_{\sigma'} : \sigma' \neq \sigma\}$, but $b \notin \mu(\Sigma \times B)$. Thus $\sigma \times B \prec_{(R5)} \Sigma'_{\sigma'} \times B \preceq_{(R3)} \sigma' \times B$ for all $\sigma' \neq \sigma$. Using the argument twice shows that $\prec$ contains a cycle.

(5) If not, then $\forall \sigma' \in \Sigma \exists \Sigma'_{\sigma'} \subseteq \Sigma$ ($\Sigma'_{\sigma'}$ legal, $\sigma' \in \Sigma'_{\sigma'}$, $b \notin \mu(\Sigma'_{\sigma'} \times B)$), but $\Sigma = \bigcup \Sigma'_{\sigma'}$, so this contradicts (C3).

□

Fact 6.3.6

$b, b' \in \mu(\Sigma \times B)$, $\sigma' \in \Sigma$ imply $\exists \sigma \in \Sigma . \sigma b \preceq^* \sigma' b'$.

Proof:

If $b \in \mu(\sigma' \times B)$, then $\sigma' b \preceq_{\text{Fact 6.3.5,(2)}} \sigma' \times B \preceq_{\text{Fact 6.3.5,(1)}} \sigma' b'$.

Suppose $b \notin \mu(\sigma' \times B)$. By Fact 6.3.5, (5), there is σ s.t. $\sigma \in \Sigma' \subseteq \Sigma \rightarrow b \in \mu(\Sigma' \times B)$. Thus the set of σ s.t. $b \in \mu([\sigma', \sigma] \times B)$ is not empty. Choose such σ with minimal Hamming distance from σ'. Then $[\sigma', \sigma] = \bigcup\{[\sigma', \sigma''] : \sigma'' \in [\sigma', \sigma], \sigma'' \neq \sigma\} \cup \{\sigma\}$. Moreover, for each $\sigma'' \in [\sigma', \sigma]$, $\sigma'' \neq \sigma$ $b \notin \mu([\sigma', \sigma''] \times B)$. Thus, $\sigma \times B \preceq_{(R4)} [\sigma', \sigma'] \times B = \sigma' \times B \preceq_{\text{Fact 6.3.5,(1)}} \sigma' b'$. As $b \in \mu([\sigma', \sigma] \times B)$, there must be by Fact 6.3.5, (5) $\sigma'' \in [\sigma', \sigma]$ s.t. $\forall \Sigma'' \subseteq [\sigma', \sigma]$ $(\sigma'' \in \Sigma'' \rightarrow b \in \mu(\Sigma'' \times B))$. Choice of σ shows that this σ'' can only be σ. Thus, in particular, $b \in \mu(\sigma \times B)$. Thus $(\sigma, b) \preceq_{\text{Fact 6.3.5,(2)}} (\sigma \times B)$. □

Fact 6.3.7

$b \in \mu(\Sigma \times B)$, $b' \notin \mu(\Sigma \times B)$, $\sigma' \in \Sigma$ imply $\exists \sigma \in \Sigma . \sigma b \prec^* \sigma' b'$.

Proof:

(a) If $b \in \mu(\sigma' \times B)$, $b' \notin \mu(\sigma' \times B)$, then $\sigma' b \preceq_{\text{Fact 6.3.5,(2)}} \sigma' \times B \prec_{\text{Fact 6.3.5,(3)}} \sigma' b'$.

(b) If $b' \in \mu(\sigma' \times B)$, then there is by Fact 6.3.5, (4) $\sigma \in \Sigma$ s.t. $\sigma \in \Sigma' \subseteq \Sigma \rightarrow b' \notin \mu(\Sigma' \times B)$. Thus, the set of $\sigma \in \Sigma$ s.t. $b' \notin \mu([\sigma', \sigma] \times B)$ is not empty, let σ be such with minimal Hamming distance from σ'. Then, as in the proof of Fact 6.3.6, $\sigma \times B \prec [\sigma', \sigma''] \times B$ by (R5) for any $\sigma'' \in [\sigma', \sigma]$, $\sigma'' \neq \sigma$, so $\sigma \times B \prec [\sigma', \sigma'] \times B = \sigma' \times B \preceq_{\text{Fact 6.3.5,(1)}} \sigma' b'$. If $b \in \mu(\sigma \times B)$, then $\sigma b \preceq_{\text{Fact 6.3.5,(2)}} \sigma \times B$, and we are done. If $b \notin \mu(\sigma \times B)$, then by an argument as above, using Fact 6.3.5, (5), we find σ^+ with minimal Hamming distance from σ s.t. $b \in \mu([\sigma, \sigma^+] \times B)$. As above, we see that $\sigma^+ \times B \preceq [\sigma, \sigma] \times B$, and as in the proof of Fact 6.3.6, we see that $b \in \mu(\sigma^+ \times B)$, so $\sigma^+ b \preceq_{\text{Fact 6.3.5,(2)}} \sigma^+ \times B$. Thus, we have $\sigma^+ b \preceq \sigma^+ \times B \preceq \sigma \times B \prec \sigma' \times B \preceq \sigma' b'$.

(c) If $b, b' \not\in \mu(\sigma' \times B)$, then $\sigma' \times B \prec_{\text{Fact } 6.3.5,(3)} \sigma' b'$. Choose as above σ with least Hamming distance from σ' s.t. $b \in \mu([\sigma', \sigma] \times B)$. As above, we see $b \in \mu(\sigma \times B)$, and $\sigma b \preceq_{\text{Fact } 6.3.5,(2)} \sigma \times B \preceq \sigma' \times B \prec \sigma' b'$. □

Proposition 6.3.8

The properties of Condition 6.3.1 hold iff there is a distance with values in an ordered abelian group (which can be assumed to be $\boldsymbol{Q}$), and best sequences are those with minimal sums of distances between their elements.

Proof:

Outline: Consider the relations $\preceq / \prec$ restricted to σ's, i.e. neglect the Σ's. The algorithm below shows, by loop, that the resulting system of inequalities has a solution, and constructs this solution, with all $d(m, m')$ and thus all $\delta(\sigma)$ in $\boldsymbol{Q}$. In particular, the original $\preceq / \prec$ between $\sigma's$ are respected by the assignment of values. It remains to show that $\mu(\Sigma) = \mu_<(\Sigma)$, where $<$ is defined from the natural order on $\boldsymbol{Q}$. We use Fact 6.3.6 and Fact 6.3.7 and the strategy of the proof of Proposition 6.3.1.

Note that by (R6) the system of inequalities contains $0 \prec d(a, b)$ for all $a \neq b$.

We use now the adapted Farkas algorithm, and note the following comments:

(1) As we can determine when $x_{i,k}$ is 0 ($x_{i,k} = d(a, a)$ for some a in some sequence σ), we can note 0 as 0.

Without loss of generality, x_n does not occur on both sides of the same inequality.

(2) Subtraction of x_n on both sides is justified by (+3) and (+4).

(3) We consider the sums $\delta_i + \delta_j$ where $\delta_i \in \Pi_l$, $\delta_j \in \Pi_r$, and eliminate x_n from both sides. These are "legal" operations covered by the $(+i)$.

(4) As the transformations we did from Π to Π' were legal, covered by the conditions $(+i)$ in Definition 6.3.1, and thus preserved freedom from loop, and by $0 \prec x_1$, the first two possibilities lead to a cycle — which was excluded. The latter two show that the final system has a solution, which can be transformed into one for the original system as indicated.

So the algorithm defines a distance compatible with +. 0 does what it should. It remains to show that the distance represents μ, i.e. that $\mu(\Sigma) = \mu_<(\Sigma)$. For better readability, we separate the last component from Σ.

Let $b \in \mu(\Sigma \times B) \rightarrow$ (by Fact 6.3.6) $\forall \sigma' b' \exists \sigma . \sigma b \preceq \sigma' b' \rightarrow \forall \sigma' b' \exists \sigma . \delta(\sigma b) \leq \delta(\sigma' b') \rightarrow$ (by finiteness) $b \in \mu_{<}(\Sigma \times B)$.

Let $b \notin \mu(\Sigma \times B)$, $\sigma \in \Sigma$. So there are (by $\mu(\Sigma \times B) \neq \emptyset$ and Fact 6.3.7) $b' \in \mu(\Sigma \times B)$, $\sigma' \in \Sigma$ s.t. $\sigma' b' \prec \sigma b$, so $\delta(\sigma' b') < \delta(\sigma b)$, so $b \notin \mu_{<}(\Sigma \times B)$. □

6.3.4 There is no finite representation for our type of update possible

6.3.4.1 Outline

We have shown in Section 4.2.4 on theory revision that there cannot be a finite characterization of distance generated revision. It was essential in the example that the "good" and the "bad" variant differed only marginally, and in a well controlled way. For this purpose, we took care that most revision results are trivial, and do not depend on the critical values — so it was impossible to conclude on good or bad by indirect means. We follow here the same strategy. We will choose distances in a way that results will change only in a carefully controlled way when we change the relevant values.

We start by giving certain conditions on possible distances between points, which determine to a large degree the outcome of the updates — only a small set of particular updates, involving nontrivial sums, will not be fixed. We then give update structures which are not distance representable, but which agree on all update results but the nontrivial sums with the distance defined update structures as just defined. Moreover, changing just one update result involving the nontrivial sums will make them again distance definable, respecting above conditions on the sums. We will thus generate update systems, which are distance compatible on all results but nontrivial sums, and we need to know all nontrivial sums to see that they are not fully distance compatible.

Again, as in the case of revision, we will construct the examples such that most results are the same in the representable and the not representable case, and the cases where the results differ involve only small sets (again of size 2 each). Much effort goes into avoiding too many changes at a time, in other words, we will need the full picture to determine whether it is a good or a bad example, i.e. distance representable or not.

Finally, there are several ways to interpret our results. We can look at update as finding shortest paths, or as looking at the projections (e.g. on the end point) of shortest paths. We will choose the latter, and show that there is no finite characterization in a suitable language.

6.3.4.2 The details

Definition 6.3.2

Call a path $a - a' - a''$ through the points a, a', a'', a 2-step path.

Example 6.3.1

We will define for given n a set of n pairs of paths with two steps, i.e. of the form $a - -- > a' - -- > a''$, which essentially compare among themselves but not to anything else in a nontrivial way. Recall that we consider here products of sets, so, usually, we have to consider many paths at the same time to find the shortest ones, so we will carefully isolate the ones we are interested in from the rest. Part of the solution is to make, within those pairs, cross-overs between the sequences of one pair very long, so they do not interfere. The other part is to make, outside of those pairs, paths very short, so we do not see the pairs any more — cross-overs between sequences of different pairs will be all that is visible. The distances which interest us will have medium size. To give an example:

Consider the two-step paths $a_0 - a_1 - a_2$, $b_0 - b_1 - b_2$, $a_1 - a_2 - a_3$, $b_1 - b_2 - b_3$. The first and the second form a pair, so do the third and the fourth. We will then have the distances: $d(a_0, b_1) = d(b_0, a_1) = b$ (for big), $d(a_1, b_1) = d(b_0, a_2) = s$ (for small), $d(a_0, a_1)$ and $d(b_0, b_1)$, etc. variable medium size m, where b is constant big, s constant small, and m a variable medium size. The details will be made clear in a moment.

To have a structure which is not distance representable, we will make a kind of circular comparison, taking care that all comparisons of single distances present no problem, and addition really plays an important role.

We will choose distances s,b,m short, big, medium (many mediums, one short, one big) s.t. $m > \frac{s+b}{2}$, in this way, if one of the two distances in a path of length two is short, it will win over any path composed of two medium distances. Moreover, if m, m', m'' are medium size, then $m+m' < m''+b$, so cross-overs do not interfere in neighboring paths. s and b are now suitable, fixed values.

We can, e.g. choose: $s := 1$, $b := 2$, $1.51 \leq m \leq 1.6$, then $(s + b)/2 = 1.5$, and $m + m' \leq 3.2 < 1.5 + 2 = 3.5$.

Distances will not necessarily be symmetric.

Choose wlog. n even, sufficiently big. We describe the distances, first the crucial medium ones, then the big and small ones. The medium ones will be fixed here only up to some inequalities, e.g. $1.51 \leq m \leq 1.6$ as above, which will already determine most of the update results, the precise values will only be given below when we construct a legal, i.e. distance representable, update formalism from an illegal one, which would contain cycles.

The medium distances:

We consider a "channel" which we will chop up again into 2-step pieces. The set U consists of the points a_i and b_i, $1 \leq i \leq n$. The interesting distances are shown in Figure 6.3.1.

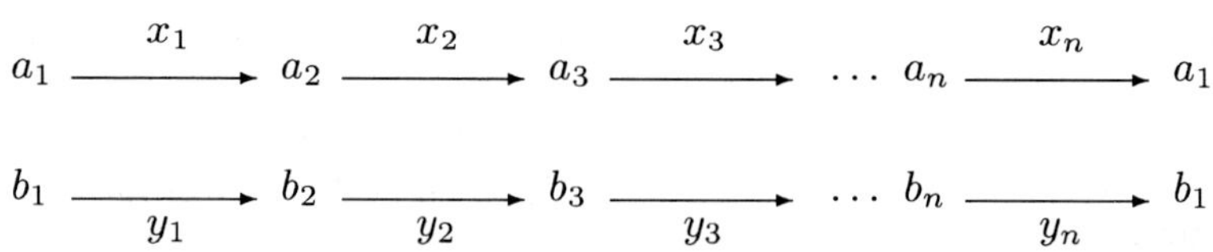

Figure 6.3.1

Formally, $x_i := d(a_i, a_{i+1})$, $x_n := d(a_n, a_1)$, $y_i := d(b_i, b_{i+1})$, $y_n := d(b_n, b_1)$, with $1.51 < y_1 < x_2 < y_3 < \ldots < y_{2i+1} < x_{2i+2} < y_{2i+3} < \ldots < x_n < 1.55$, $1.56 < x_1 < y_2 < x_3 < \ldots < x_{2i+1} < y_{2i+2} < x_{2i+3} < \ldots < y_n < 1.60$.

Thus, for i uneven $1.51 < y_i < x_{i+1} < y_{i+2} < 1.55$ and for i even $1.56 < x_i < y_{i+1} < x_{i+2} < 1.6$.

The exact sizes of the x_i and y_i will be left open for the moment, we will see that, in many cases, they do not matter.

The big distances:

Choose furthermore $d(a_i, b_{i+1}) = d(b_i, a_{i+1}) = d(a_n, b_1) = d(b_n, a_1) = b$, we call them cross-overs.

The small distances:

All other distances are chosen as s (also the other directions), or 0 for $d(z, z)$.

We consider now three layers of U and look at all update results for 2-step paths through the set. We will show that almost all update results (those where we do not consider a comparison of two sums of the type $m_1 + m_2/m_3 + m_4$) are determined for all choices of the x_i and y_i which obey above conditions, and the complicated comparisons only arise in very specific circumstances.

It is important to note the following: We will compare sums of the type $e+f$ and $g+h$. We have chosen the values such that $0 < s < m < b < s+s$, $s+b < m+m$. Thus, anytime a sum contains 0 or s, it will be smaller than any sum composed of two medium values. By above inequalities, for any element z, there is exactly one element z' s.t. $d(z', z)$ is 0 ($z' = z$), is m (e.g. $z' = a_{i-1}$ and $z = a_i$), is b (e.g. $z' = b_{i-1}$, $z = a_i$), and, likewise, there is exactly one element z' s.t. $d(z, z')$ is 0, m, or b. All other distances are s. Thus, in many cases, one of the sums to consider will be of the form $0 + e$ or $s + e$, where e is any value, which is smaller than any sum of the form $m + m'$. Thus, in many cases, we might have to compare, e.g. $s + m$ with $s + m'$, i.e. m with m', but not sums of the form $m_1 + m_2$ with $m_3 + m_4$. But all comparisons of the first type are already decided by the inequalities we gave above.

Suppose then that we consider the set product $A \times B \times C$, where $A, B, C \subseteq U$, and we examine the shortest paths of the product.

Properties:

We distinguish several cases, in all but the last one the exact choice of the values for the m will not matter, as long as all values respect above inequalities. Again, the interesting cases will involve small sets (of cardinality 2), the situation is thus quite parallel to the one in revision, see Section 4.2.4.

Case 1, one of the three sets A, B, C has just one element: Suppose, e.g. that $A = \{a\}$. The other cases are similar. As there is only one element with $d(a, b) = m$, we will never have to compare $m_1 + m_2$ with $m_3 + m_4$ for $m_1 \neq m_3$. Consequently, all comparisons are already decided by the inequalities we already know.

In subsequent Cases 2 and part of 3, one of the distances will be s or even 0, and sums of type $m + m'$ are no match: paths containing 0 or s will always win.

Case 2, one of the three sets A, B, C has at least three elements: Suppose again, e.g. that this set is A. Then for any $b \in B$ there is $a \in A$ s.t.

$d(a,b) = 0$ or $d(a,b) = s$. Consequently, one of the sums is of type $0+e$ or $s+e$, which is smaller than any $m+m'$, so the latter are unimportant for the outcome, and again the known inequalities decide already.

Case 3, all three sets A, B, C have two elements: Assume we have to compare two sums $m_1 + m_2$ and $m_3 + m_4$, with $m_1 \neq m_3$, $m_2 \neq m_4$. Consider $A = \{r,s\}$, $B = \{t,u\}$. Say $d(r,t)$ and $d(s,u)$ are medium size (as discussed above, it cannot be that $d(r,t)$ and $d(s,t)$ are medium size, etc.). Then, e.g. $r = a_i$, and $t = a_{i+1}$. But, if $u \neq b_{i+1}$, then $d(r,u) = 0$ or $d(r,u) = s$, and then the sums m_1+m_2 and m_3+m_4 cannot be minimal, and the inequalities known so far decide already again.

Consequently, the only cases perhaps not yet decided are of the following form: $\{a_i, b_i\} \times \{a_{i+1}, b_{i+1}\} \times \{a_{i+2}, b_{i+2}\}$.

On the other hand, as $d(a_i, b_{i+1}) = b$, etc., the sums $m_1 + m_2$ and $m_3 + m_4$ with four medium size values really decide the outcome.

We construct now a not distance representable update structure as follows:

For all trivial updates of 2-step length, choose the results as they are determined by the individual relations between the m-values and s and b, as already decided by above inequalities.

Consider now the following n update results:

For uneven i, choose the path $< a_i, a_{i+1}, a_{i+2} >$ over the path $< b_i, b_{i+1}, b_{i+2} >$, by the result $\mu(\{a_i, b_i\} \times \{a_{i+1}, b_{i+1}\} \times \{a_{i+2}, b_{i+2}\}) = \{a_{i+2}\}$ — recall that we look only at the last coordinate. By above analysis, we know that then $x_i + x_{i+1} < y_i + y_{i+1}$.

For i even, choose the path $< b_i, b_{i+1}, b_{i+2} >$ over the path $< a_i, a_{i+1}, a_{i+2} >$, i.e. $y_i + y_{i+1} < x_i + x_{i+1}$.

(We calculate modulo n.)

If this situation were distance representable, we would have for uneven i $x_i + x_{i+1} < y_i + y_{i+1}$, for i even $y_i + y_{i+1} < x_i + x_{i+1}$, a contradiction, as all distances occur exactly once on each side of $<$.

But we can transform the structure into one which is distance representable by changing just one of the nontrivial update results, e.g. by preferring the path $< a_n, a_1, a_2 >$ over the path $< b_n, b_1, b_2 >$, i.e. $x_n + x_1 < y_n + y_1$. We show next how to achieve this result by choosing appropriate values for the x_i and y_i.

Choose c small enough, e.g. $2n * c < 0.04$.

We increase the "right" values so the balance in favor of x or y will change from one side to the other. As $x_1 + x_2 < y_1 + y_2$, we choose $y_1 := 1.51 + c$,

$x_1 := 1.56 + c$, $y_2 := 1.56 + 3 * c$, $x_2 := 1.51 + 2 * c$ As $x_2 + x_3 > y_2 + y_3$, we choose $y_3 := 1.51 + 3 * c$, $x_3 := 1.56 + 5 * c$, etc., formally:

For i uneven, $y_i := 1.51 + i * c$, $x_i := 1.56 + (2i - 1) * c$, for i even, $x_i := 1.51 + i * c$, $y_i := 1.56 + (2i - 1) * c$.

It is easily seen that the inequalities for i uneven $1.51 < y_i < x_{i+1} < y_{i+2} < 1.55$ and for i even $1.56 < x_i < y_{i+1} < x_{i+2} < 1.6$ hold.

Then, for i uneven: $x_i + x_{i+1} = 1.56 + (2i - 1) * c + 1.51 + (i + 1) * c = 3.07 + 3i * c$, $y_i + y_{i+1} = 1.51 + i * c + 1.56 + (2(i+1) - 1) * c = 3.07 + (3i + 1) * c$,

for i even: $x_i + x_{i+1} = 1.51 + i * c + 1.56 + (2(i+1) - 1) * c = 3.07 + (3i + 1) * c$, $y_i + y_{i+1} = 1.56 + (2i - 1) * c + 1.51 + (i + 1) * c = 3.07 + 3i * c$,

finally: $x_n + x_1 = 1.51 + n * c + 1.56 + c = 3.07 + (n + 1) * c$, $y_n + y_1 = 1.56 + (2n - 1) * c + 1.51 + c = 3.07 + 2n * c$.

We preserve thus the original inequalities, while modifying the increases, thus all trivial sums are unchanged, and so are the trivial update results, but not the nontrivial ones.

Now, any characterization of distance generated update has to fail for the illegal example somewhere. As all but the nontrivial sums are distance compatible — they were generated by a distance, respecting the original inequalities — any true subset of above nontrivial update results is also distance compatible, so this characterization has to speak about all the above n update results.

We give an example.

Example 6.3.2

(Construction for $n = 4$)

Recall that $d(a_i, b_{i+1}) = d(b_i, a_{i+1}) = b$ and all other distances are 0 or s.

This does not work, as then $x_1 + x_2 + y_2 + y_3 + x_3 + x_4 + y_4 + y_1 < y_1 + y_2 + x_2 + x_3 + y_3 + y_4 + x_4 + x_1$.

The choice of the x_i and y_i :

$y_1 = 1.511$, $x_1 = 1.561$,

$y_2 = 1.563$, $x_2 = 1.512$, y_2 has to win over the rest for update 1,

$y_3 = 1.513$, $x_3 = 1.565$, x_3 has to win over the rest for update 2,

$y_4 = 1.567$, $x_4 = 1.514$, y_4 has to win over the rest for update 3.

Changing x_4 to 1.516 will turn the decision for update 3 the other way, while preserving the relations between singletons.

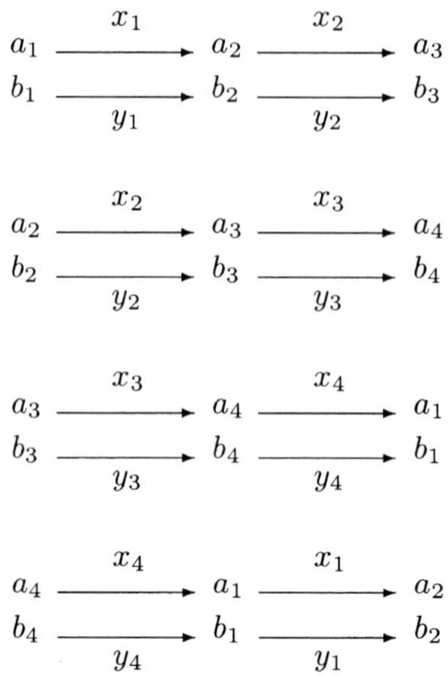

$x_1 + x_2 < y_1 + y_2$ update 1

$x_2 + x_3 > y_2 + y_3$ update 2

$x_3 + x_4 < y_3 + y_4$ update 3

$x_4 + x_1 > y_4 + y_1$ update 4

Figure 6.3.2

Summary:

For all n, we can create a 2-step update structure A which:

(1) is not distance representable,

(2) there are distance-representable update structures which agree on all trivial sums (i.e. of type $a+b < a+c$ with fixed a) with A, and which agree for all fragments of size $n-1$ with A — we have to look at arbitrarily many nontrivial sums to see that the structure is not distance representable.

(3) The example shows how we can achieve arbitrary consistent update results, by manipulating the increases, while respecting the conditions for the x_i and y_i.

We turn to interpretation.

We consider as base operator a ternary one on sets: $< A, B, C >$ is the set of elements of C through which paths of minimal length go. Suppose there is a characterization of distance representable structures by a formula $\phi = \forall x_1, \ldots, x_k \phi'(x_1, \ldots, x_k)$, where ϕ' is quantifier free, and contains only set operators and relations, and above update operator $< .,.,. >$. Let ϕ' contain n update operators, and look at structures with more than $6 * n$ elements. ϕ has to fail in a counterexample, say $C \models \neg\phi'[a_1, \ldots, a_k]$. As in the case of revision, we determine the set of relevant elements, there are at most $2 * n$ many, as all sets of size 1 or ≥ 3 evaluate to the same in the representable and the not representable cases. Choose now in $L(C)$, the set of of "legal" structures almost equivalent to C, one T which agrees with C on all relevant elements (and on the irrelevant ones, too, as all $S \in L(C)$ agree on the irrelevant ones), then $T \models \neg\phi'[a_1, \ldots, a_k]$, so $T \models \neg\phi$, a contradiction.

(If we had worked directly with shortest sequences, and not with their projections, we would have introduced suitable operators on set triples into the language.)

6.4 Comments on "Belief revision with unreliable observations"

6.4.1 Introduction

We discuss the article "Belief Revision with Unreliable Observations" by C. Boutilier, N. Friedman, and J. Halpern, [BFH95], and give a characterization of (a finite variant) of Markov systems, using again the Farkas algorithm. Moreover, we also show that the problem has no finite representation.

We present in Section 6.4.1 the basic definitions and results, so the reader can understand the subsequent main results. Unfortunately, the definitions are a little complicated, and the reader will need some patience.

Most of the results in this Introduction 6.4.1 are trivial and/or contained already in [BFH95]. Definitions and intuition are due to [BFH95], for motivation, the reader is referred to the original article.

6.4.1.1 The situation

We shall work in propositional logic, in some fixed language $\mathcal{L}$, which we identify with its formulas.

We have an evolving scenario of an environment e_i and an agent who has information ϕ_i (about the environment), where ϕ_i is an $\mathcal{L}-$formula. We assume discrete time, starting at 1. A development, also called a run, has now the form $\{< e_i, \phi_i >: i \in \omega\}$. We assume that the agent has perfect recall, which we may express by $\phi_{i+1} = \phi_i \bullet \phi'$, where $\bullet$ denotes appending an element to a sequence. ϕ' is the new information arriving at moment $i+1$. We simplify a little: As we are interested here in theory revision in contrast to theory update, we assume the environment to be constant over a run. To avoid excessive notation, we write down only the new information arriving at the individual moments. In our above example, ϕ_{i+1} will then just be ϕ'. Runs have now the form $< e, \sigma >$, where $\sigma =< \phi_1, \phi_2, \ldots >$, without any a priori connections between the different ϕ_i's. The environment e will now be a classical $\mathcal{L}-$model. Such runs will also be called runs over $\mathcal{L}$, as e and the ϕ_i are $\mathcal{L}-$models ($-$formulas).

6.4.1.2 Basic definitions and results

Definition 6.4.1

If $r =< e, \sigma >$ is a run, then

$r_e := e$, $r_\sigma := \sigma$, $r_n := \sigma_n$, $r\lceil m := \sigma\lceil m :=< \sigma_1, \ldots, \sigma_{m-1} >$.

All this will be used unambigously.

$\mathcal{R}$ will be a set of runs over $\mathcal{L}$ — no closure properties are required for the moment.

Given a set X, a function $\kappa : X \to \omega + 1$ is called a ranking on X. $\kappa(x) < \kappa(x')$ means intuitively that x is more plausible than x'. We set $\kappa(\emptyset) := \omega$, and for $\emptyset \neq U \subseteq X$ $\kappa(U) := min\{\kappa(x) : x \in U\}$. If $U, V \subseteq X$, then $\kappa(V \mid U) := \kappa(V \cap U) - \kappa(U)$, where $\omega - \omega$ will be undefined, and $\omega - n := \omega$ for $n \in \omega$.

A pair $\mathcal{I} =< \mathcal{R}, \kappa >$ with $\mathcal{R}$ a set of runs over $\mathcal{L}$, and κ a ranking on $\mathcal{R}$ will be called a ranked or interpreted system.

$\mathcal{I} :=< \mathcal{R}, \kappa >$ will now be a fixed interpreted system over $\mathcal{L}$.

The rest of this section will be elementary definitions, necessary for the comprehension, and some simple facts.

Definition 6.4.2

(1) For $w \in M_\mathcal{L}$, set $[[w]] := \{r \in \mathcal{R} : r_e = w\}$.

(2) For $\phi \in \mathcal{L}$, set $[[\phi]] := \{r \in \mathcal{R} : r_e \models \phi\}$.

(3) For $\phi_i \in \mathcal{L}$, set
$[[Obs = \phi_1, \ldots, \phi_n]] := \{r \in \mathcal{R} : r\lceil n+1 =< \phi_1, \ldots, \phi_n >\}$.
(4) For $\phi \in \mathcal{L}$, set $[[Obs_n = \phi]] := \{r \in \mathcal{R} : r_n = \phi\}$.
(5) For $w \in M_{\mathcal{L}}$, $\phi_i \in \mathcal{L}$, set
$[[w, Obs = \phi_1, \ldots, \phi_n]] := [[w]] \cap [[Obs = \phi_1, \ldots, \phi_n]]$.
(6) For $U \subseteq \mathcal{R}$, set $\mu(U) := \{r \in U : \forall r' \in U . \kappa(r) \leq \kappa(r')\}$.

Condition 6.4.1

If $r \in \mathcal{R}$ and $n < \omega$, then $\kappa([[Obs = r_1, \ldots, r_n]]) < \omega$, i.e. for any $r \in \mathcal{R}$ and $n \in \omega$ there is $r' \in \mathcal{R}$ s.t. $r\lceil n = r'\lceil n$ and $\kappa(r') < \omega$.

Definition 6.4.3

(1) For $r \in \mathcal{R}$, $m \in \omega$, $U \subseteq \mathcal{R}$ we define inductively
$\kappa^{r,0} := \kappa$,
$\kappa^{r,m+1}(U) := \kappa^{r,m}(U \mid [[Obs = r_1, \ldots, r_{m+1}]])$.
(2) $\kappa^{\phi_1,\ldots,\phi_n} := \kappa^{r,n}$ for any $r \in \mathcal{R}$ s.t. $r\lceil n+1 =< \phi_1, \ldots, \phi_n >$ (and undefined if there is no such r).

Fact 6.4.1

(1) $\kappa^{r,m}(U) = \kappa(U \cap [[Obs = r_1, \ldots, r_m]])$ - $\kappa([[Obs = r_1, \ldots, r_m]])$ for all $U \subseteq \mathcal{R}$ and for $m > 0$.

(2) $\kappa^{r,m}(U \cap [[Obs_{m+1} = r_{m+1}]]) = \kappa^{r,m}(U \cap [[Obs = r_1, \ldots, r_{m+1}]])$ for all $U \subseteq \mathcal{R}$.

(3) $\kappa^{r,m}(U \mid [[Obs_{m+1} = r_{m+1}]]) = \kappa^{r,m}(U \mid [[Obs = r_1, \ldots, r_{m+1}]])$ for all $U \subseteq \mathcal{R}$.

(4) $\kappa^{r,m}(r') = 0$ iff $r' \in \mu([[Obs = r_1, \ldots, r_m]])$ and $\kappa(r') < \infty$, for $m > 0$.

Proof:

(1) (By induction)

$m = 1 : \kappa^{r,1}(U) := \kappa^{r,0}(U \mid [[Obs = r_1]]) := \kappa(U \cap [[Obs = r_1]]) - \kappa([[Obs = r_1]])$.

$m \to m+1 : \kappa^{r,m+1}(U) := \kappa^{r,m}(U \mid [[Obs = r_1, \ldots, r_{m+1}]]) = \kappa^{r,m}(U \cap [[Obs = r_1, \ldots, r_{m+1}]]) - \kappa^{r,m}([[Obs = r_1, \ldots, r_{m+1}]])$ = (by induction hypothesis) $[\kappa(U \cap [[Obs = r_1, \ldots, r_{m+1}]] \cap [[Obs = r_1, \ldots, r_m]]) - \kappa([[Obs = r_1, \ldots, r_m]])] - [\kappa([[Obs = r_1, \ldots, r_{m+1}]] \cap [[Obs = r_1, \ldots, r_m]]) -$

$\kappa([\![Obs = r_1,\ldots,r_m]\!])] = \kappa(U \cap [\![Obs = r_1,\ldots,r_{m+1}]\!]) - \kappa([\![Obs = r_1,\ldots,r_{m+1}]\!])$.

(2) $m = 0$: trivial. $m > 0$: straightforward, e.g. by (1).

(3) Trivial by (2).

(4) $\kappa^{r,m}(r') = 0$ iff (by (1)) $\kappa(\{r'\} \cap [\![Obs = r_1,\ldots,r_m]\!]) = \kappa([\![Obs = r_1,\ldots,r_m]\!])$, and $\kappa([\![Obs = r_1,\ldots,r_m]\!]) < \infty$ iff $r' \lceil m+1 = r \lceil m+1$ and $\kappa(r')$ is minimal among the $\kappa(r'')$ with $r'' \lceil m+1 = r \lceil m+1$ and $\kappa([\![Obs = r_1,\ldots,r_m]\!]) < \infty$. □

Definition 6.4.4

(1) $Bel(\mathcal{I}, r, n) := \{w : \exists r' \in \mu([\![Obs = r_1,\ldots,r_n]\!]).r'_e = w\}$ for $r \in \mathcal{R}$.

(2)

$$B_{\mathcal{I}}(\phi_1,\ldots,\phi_n) := \begin{cases} \{w : \exists r' \in \mu([\![Obs = \phi_1,\ldots,\phi_n]\!]).r'_e = w\} \text{ iff} \\ \kappa([\![Obs = \phi_1,\ldots,\phi_n]\!]) \neq \omega \\ \text{and} \\ \emptyset \text{ iff } \kappa([\![Obs = \phi_1,\ldots,\phi_n]\!]) = \omega. \end{cases}$$

Condition 6.4.2

(O1): For $\phi_1,\ldots,\phi_n$ there is $P(\phi_1,\ldots,\phi_n) \subseteq \mathcal{L}$ s.t. $B_{\mathcal{I}}(\phi_1,\ldots,\phi_n) = \bigcup\{B_{\mathcal{I}}(\phi_1,\ldots,\phi_n,\psi) : \psi \in P(\phi_1,\ldots,\phi_n)\}$.

(O2): If ρ is a permutation of $\{1,\ldots,n\}$, then $B_{\mathcal{I}}(\phi_1,\ldots,\phi_n) = B_{\mathcal{I}}(\phi_{\rho(1)},\ldots,\phi_{\rho(n)})$.

Fact 6.4.2

(1) $w \in B_{\mathcal{I}}(\phi_1,\ldots,\phi_n)$ iff $\kappa([\![w, Obs = \phi_1,\ldots,\phi_n]\!]) \leq \kappa([\![w', Obs = \phi_1,\ldots,\phi_n]\!])$ for all $w' \in M_{\mathcal{L}}$ (and $\kappa([\![Obs = \phi_1,\ldots,\phi_n]\!]) \neq \omega$).

(2) (O1) holds in all ranked systems.

Proof:

(1) is trivial.

(2) Define $P(\phi_1,\ldots,\phi_n) := \{\psi : \exists r \in \mu([\![Obs = \phi_1,\ldots,\phi_n]\!]).r_{n+1} = \psi\}$.

Case 1: $B_{\mathcal{I}}(\phi_1,\ldots,\phi_n) = \emptyset$: Then $\kappa([\![Obs = \phi_1,\ldots,\phi_n]\!]) = \omega$, so for all ψ $\kappa([\![Obs = \phi_1,\ldots,\phi_n,\psi]\!]) = \omega$.

Case 2: $B_{\mathcal{I}}(\phi_1,\ldots,\phi_n) \neq \emptyset$: Let $w \in B_{\mathcal{I}}(\phi_1,\ldots,\phi_n)$, so there is $r \in \mu([\![Obs = \phi_1,\ldots,\phi_n]\!]).r_e = w$. Thus $\psi := r_{n+1} \in P(\phi_1,\ldots,\phi_n)$ and $w \in B_{\mathcal{I}}(\phi_1,\ldots,\phi_n,\psi)$. Conversely, let $w \in B_{\mathcal{I}}(\phi_1,\ldots,\phi_n,\psi)$ with $\psi \in P(\phi_1,\ldots,\phi_n)$. Thus there is $r \in \mu([\![Obs = \phi_1,\ldots,\phi_n,\psi]\!]).r_e = w$. By definition of $P(\phi_1,\ldots,\phi_n)$, $\exists r' \in \mu([\![Obs = \phi_1,\ldots,\phi_n]\!]).r'_{n+1} = \psi$. By choice of r, $\kappa(r') \geq \kappa(r)$, so $r \in \mu([\![Obs = \phi_1,\ldots,\phi_n]\!])$ and $w \in B_{\mathcal{I}}(\phi_1,\ldots,\phi_n)$.
□

Definition 6.4.5

$\mathcal{I}$ is called a Markov system iff the following hold:

(a) $\kappa(\{r \in \mathcal{R} : r_{m+1} = \phi\} \mid \{r \in \mathcal{R} : r\lceil m+1 = \phi_1,\ldots,\phi_m \text{ and } r_e = w\}) = \kappa(\{r \in \mathcal{R} : r_{m+1} = \phi\} \mid \{r \in \mathcal{R} : r_e = w\})$ for all m,

and

(b) $\kappa(\{r \in \mathcal{R} : r_m = \phi\} \mid \{r \in \mathcal{R} : r_e = w\}) = \kappa(\{r \in \mathcal{R} : r_{m'} = \phi\} \mid \{r \in \mathcal{R} : r_e = w\})$ for any m, m'.

Thus, in Markov systems $\kappa([\![Obs_i = \phi]\!] \mid [\![w]\!]) = \kappa([\![Obs_j = \phi]\!] \mid [\![w]\!])$ for all $i, j \in \omega$, and we define $\lambda(\phi, w) := \kappa([\![Obs_* = \phi]\!] \mid [\![w]\!])$, where $*$ is any i.

Moreover, we set $\lambda(w) := \kappa([\![w]\!])$.

Fact 6.4.3

(1) Let $\mathcal{I}$ be Markov. Then $\kappa([\![w, Obs = \phi_1,\ldots,\phi_n]\!]) = \lambda(w) + \Sigma\{\lambda(\phi_i, w) : 1 \leq i \leq n\}$.

(2) If $r = < w, < \phi_1, \phi_2, \ldots >>$, then $\lambda(w) + \Sigma\{\lambda(\phi_i, w) : 1 \leq i < \omega\} \leq \kappa(r)$.

(3) (O2) holds in Markov systems.

Proof:

(1) Case 1: $\lambda(w) < \omega$ and all $\lambda(\phi_i, w) < \omega$.

$\lambda(\phi_i, w) =_{\text{Markov}} \kappa([\![Obs_i = \phi_i]\!] \mid [\![w]\!]) =_{\text{Markov}} \kappa([\![Obs_i = \phi_i]\!] \mid [\![w, Obs = \phi_1,\ldots,\phi_{i-1}]\!]) := \kappa([\![w, Obs = \phi_1,\ldots,\phi_i]\!])$ - $\kappa([\![w, Obs = \phi_1,\ldots,\phi_{i-1}]\!])$. (If $i = 1$, then $\kappa([\![Obs_i = \phi_i]\!] \mid [\![w]\!]) = \kappa([\![w, Obs = \phi_1]\!])$ - $\lambda(w)$.) Thus, $\lambda(w) + \Sigma\{\lambda(\phi_i, w) : 1 \leq i \leq n\} = \lambda(w) + \Sigma\{[\kappa([\![w, Obs = \phi_1,\ldots,\phi_i]\!]) - \kappa([\![w, Obs = \phi_1,\ldots,\phi_{i-1}]\!])] : 1 \leq i \leq n\} = \kappa([\![w, Obs = \phi_1,\ldots,\phi_n]\!])$.

Case 2: $\lambda(w) = \omega$.
Then $\kappa([\![w, Obs = \phi_1, \ldots, \phi_n]\!]) = \omega$, as $[\![w, Obs = \phi_1, \ldots, \phi_n]\!] \subseteq [\![w]\!]$.

Case 3: $\lambda(w) < \omega$, but one of the $\lambda(\phi_i, w) = \omega$:
Then $\kappa([\![w, Obs_i = \phi_i]\!]) - \lambda(w) = \omega$, so $\kappa([\![w, Obs_i = \phi_i]\!]) = \omega$, but $[\![w, Obs = \phi_1, \ldots, \phi_n]\!] \subseteq [\![w, Obs_i = \phi_i]\!]$.

(2) For all $n < \omega$ $\kappa(r) \geq \kappa([\![w, Obs = \phi_1, \ldots, \phi_n]\!]) =_{(1)} \lambda(w) + \Sigma\{\lambda(\phi_i, w) : 1 \leq i \leq n\}$.

(3) Trivial, e.g. by Facts 6.4.2, (1) and 6.4.3, (1). □

Example 6.4.1

$\lambda(w) + \Sigma\{\lambda(\phi_i, w) : 1 \leq i < \omega\} = \kappa(r)$ is in general wrong in Markov systems (and thus, Lemma 4.2 in [BFH95] is wrong).

Proof:

Take any language, and let $\mathcal{R}$ be the set of all possible runs, fix any $r \in \mathcal{R}$. Define $\kappa(r) := 1$, and $\kappa(r') := 0$ for any $r' \neq r$.

Obviously, Condition 6.4.1 is satisfied, and κ is Markov, as for any w, ϕ_i $\lambda(w) = \kappa([\![Obs = \phi_1, \ldots, \phi_n]\!]) = \kappa([\![Obs_n = \phi]\!]) = \kappa([\![w, Obs = \phi_1, \ldots, \phi_n]\!]) = 0$, so (a) and (b) of the Markov Definition will trivially hold. If the statement were correct, then $\kappa(r) = 0$, contradiction. □

Example 6.4.2

There are systems where (O1) and (O2) hold, but which are not Markov.

Proof:

Let $\mathcal{L}$ be the language with one propositional variable, p.

Let $\mathcal{R}$ be the set of all runs over $\mathcal{L}$, define κ by: $\kappa(< w, \sigma >) = 0$ if p does not occur in σ, $\kappa(< w, \sigma >) = 2$ if p does occur in σ, $\kappa(< w', \sigma >) = 1$ if p occurs at most once in σ, $\kappa(< w', \sigma >) = 3$ if p occurs at least twice in σ.

Then $B_{\mathcal{I}}(\phi_1, \ldots, \phi_n) = w'$ iff p occurs exactly once in $\phi_1, \ldots, \phi_n$, and $B_{\mathcal{I}}(\phi_1, \ldots, \phi_n) = w$ otherwise.

Thus, (O2) holds.

But $\mathcal{I}$ is not Markov: $\lambda(w) = 0$, $\lambda(p, w) = 2$, $\kappa([\![w, Obs = p, p]\!]) = 2$, contradicting Fact 6.4.3, (1). □

6.4.2 A characterization of Markov systems in the finite case

6.4.2.1 Outline and introduction

We first present some weak restrictions. We then formally connect the Markov property to systems of inequalities in Fact 6.4.4, show in Example 6.4.3 that an extension to the infinite case has some difficulties, and summarize representation in Proposition 6.4.5.

We will work in the finite case (this will be made precise below).

Given a Markov system $\mathcal{I} =< \mathcal{R}, \kappa >$, we have defined $B_{\mathcal{I}}$ and λ. We first show that there is a connection between $B_{\mathcal{I}}$ and λ in Markov systems. More precisely, an element w belongs to $B_{\mathcal{I}}(\phi_1, \ldots, \phi_n)$ iff certain inequalities hold between sums of $\lambda(v)$'s and $\lambda(\phi, v)$'s. (This is trivial.) Conversely, given ($\mathcal{R}$ and) B, we can write a system of inequalities between sums of $\lambda(v)$'s and $\lambda(\phi, v)$'s. If this system has a solution, we can define κ s.t. $B = B_{\mathcal{I}}$ for $\mathcal{I} =< \mathcal{R}, \kappa >$, and the λ for this κ corresponds to the solution of the system of inequalities. This is again trivial. We have thus transformed the question whether B is the B of a Markov system to the question whether a certain system of inequalities has a solution. This can be decided by the (adapted) Farkas algorithm. The complexity of the question whether such a system of inequalities has a solution made it somewhat doubtful whether there is a much simpler characterization of Markov systems, e.g. in the spirit of the conditions (O1) and (O2), so Example 6.4.2 came as no surprise. And we see that, indeed, there is no finite characterization possible — see Section 6.4.3 below.

The restrictions (finiteness and others):

We will assume the following conditions for finiteness:

1. the number of propositional variables of the language is finite,
2. the length of the sequences r_σ is some finite, fixed N,
3. $\lambda(\phi, w) = \lambda(\phi', w)$ for all w, ϕ, ϕ' s.t. $\models \phi \leftrightarrow \phi'$.

Moreover, we assume that for all w there is ϕ s.t. $\lambda(\phi, w) = 0$. This is motivated by the following: If κ satisfies Condition 6.4.1, and $\mathcal{I}$ is Markov, and there is $r \in \mathcal{R}$, then there is ϕ s.t. $\lambda(\phi, w) = 0$ for some w. (Proof: Let $r' \in \mathcal{R}$ be s.t. $r\lceil 1 = r'\lceil 1$ and $\kappa(r') < \omega$. Let $r' =< w, < \phi_1, \phi_2, \ldots >>$. Thus $\lambda(w) + \Sigma\{\lambda(\phi_i, w) : 1 \leq i < \omega\} \leq \kappa(r) < \omega$, but as $\lambda(\phi_i, w) \in \omega$, only finitely many $\lambda(\phi_i, w)$ can be > 0, so some $\lambda(\phi_i, w) = 0$.)

Note that Condition 6.4.1 only guarantees that for some w $\lambda(\phi, w) = 0$, we assume that for all w there is such ϕ. Now, "true" seems the least surprising fact about the world, so we will assume that for all w $\lambda(\boldsymbol{T}, w) = 0$.

Moreover, we will assume that any initial segment of a sequence in $\mathcal{R}$ can be continued by "true", i.e. if $r =< w, < \phi_1, \ldots, \phi_N >> \in \mathcal{R}$, and $1 \leq i \leq N$, then $r' =< w, < \phi_1, \ldots, \phi_{i-1}, true, \ldots, true >> \in \mathcal{R}$.

A consequence of our finiteness assumption is, that Fact 6.4.3, (1) suffices to construct $\kappa(r)$ from λ, as $[\![w, Obs = \phi_1, \ldots, \phi_N]\!]$ is a singleton or empty (N is the length of the sequences).

Before we explain why the existence of ϕ s.t. $\lambda(\phi, w) = 0$ is desirable, we show the connection between the belief system B and systems of inequalities for Markov systems.

Fact 6.4.4

Let, for simplicity, $\mathcal{R}$ be the set of all runs over $\mathcal{L}$ of length N.

(1) Let $\mathcal{I} =< \mathcal{R}, \kappa >$ be Markov, then

(1.1) $w, w' \in B_{\mathcal{I}}(\phi_1, \ldots, \phi_N)$ iff $\lambda(w) + \lambda(\phi_1, w) + \ldots + \lambda(\phi_N, w) = \lambda(w') + \lambda(\phi_1, w') + \ldots + \lambda(\phi_N, w') < \infty$,

(1.2) $w \in B_{\mathcal{I}}(\phi_1, \ldots, \phi_N)$, $w' \notin B_{\mathcal{I}}(\phi_1, \ldots, \phi_N)$ iff $\lambda(w) + \lambda(\phi_1, w) + \ldots + \lambda(\phi_N, w) < \lambda(w') + \lambda(\phi_1, w') + \ldots + \lambda(\phi_N, w')$.

(2) Let $B(\phi_1, \ldots, \phi_N) \subseteq M_{\mathcal{L}}$ for all ϕ_i.

Let $\lambda : M_{\mathcal{L}} \cup (M_{\mathcal{L}} \times \mathcal{L}) \to \omega + 1$ s.t. for all $\phi_1, \ldots, \phi_N$, w, w'.

(2.1) $w, w' \in B_{\mathcal{I}}(\phi_1, \ldots, \phi_N)$ iff $\lambda(w) + \lambda(\phi_1, w) + \ldots + \lambda(\phi_N, w) = \lambda(w') + \lambda(\phi_1, w') + \ldots + \lambda(\phi_N, w') < \infty$,

(2.2) $w \in B_{\mathcal{I}}(\phi_1, \ldots, \phi_N)$, $w' \notin B_{\mathcal{I}}(\phi_1, \ldots, \phi_N)$ iff $\lambda(w) + \lambda(\phi_1, w) + \ldots + \lambda(\phi_N, w) < \lambda(w') + \lambda(\phi_1, w') + \ldots + \lambda(\phi_N, w')$.

Then there is $\kappa : \mathcal{R} \to \omega + 1$ s.t. $\kappa(r) = \lambda(w) + \Sigma\{\lambda(\phi_i, w) : 1 \leq i \leq N\}$ for $r =< w, < \phi_1, \ldots, \phi_N >>$ and for $\mathcal{I} :=< \mathcal{R}, \kappa >$ $B(\phi_1, \ldots, \phi_N) = B_{\mathcal{I}}(\phi_1, \ldots, \phi_N)$, and $\mathcal{I}$ is Markov.

Proof:

(1) Trivial by Facts 6.4.2 (1) and 6.4.3 (1).

(2) Define $\kappa(r) := \lambda(w) + \Sigma\{\lambda(\phi_i, w) : 1 \leq i \leq N\}$ for $r =< w, < \phi_1, \ldots \phi_N >>$. □

We now argue why it is desirable to have for all w some ϕ s.t. $\lambda(\phi, w) = 0$. Consider the information we get. Let N again be the length of the runs. For $n < N$, e.g. for $n := 1$ with $N > 1$, $B_{\mathcal{I}}(\phi_1)$ gives us only information of the type $\lambda(w) + \lambda(\phi_1, w) + a < \lambda(w') + \lambda(\phi_1, w') + b$, etc., as we do not know how the best (comparing between different w's) sequences behave. For this reason, it is very desirable to know that the best sequences continue by $0 = \lambda(true, w)$, and a, b will be 0. For the same reason, we will assume that any sequence can be continued by "true".

We now use the Farkas algorithm.

Comments:

We see that the algorithm works — as it is — only for finite sets of inequalities. In particular, it does not seem trivial how to extend the result to infinitely long sequences, as the following example shows. The described belief sets can be belief sets for a Markov system of sequences of any fixed finite length, but cannot be belief sets for a Markov system with infinite sequences. (We will make the assumption that $\mathcal{R}$ is the set of all runs over $\mathcal{L}$, and that $\lambda(\boldsymbol{T}, w) = 0$ for all $w \in M_{\mathcal{L}}$.)

Example 6.4.3

Take as $\mathcal{L}$ the language with one propositional variable, p. Let $\mathcal{R}$ be the set of all runs over $\mathcal{L}$, and let $\lambda(\boldsymbol{T}, w) = 0$ for all $w \in M_{\mathcal{L}}$.

Let $\phi := p$, $\psi := \neg p$. Set

(1) $B(<>) = \{w, w'\}$,

(2) $B(< \phi, \psi >) = \{w'\}$,

(3) $B(< \phi^m, \psi^{m+1} >) = \{w\}$ for all $m \geq 1$ — where $< \phi^m, \psi^n >$ is the sequence of m times ϕ, followed by n times ψ.

Assume now B to be defined by a Markov system. We conclude:

By (1), $\lambda(w) = \lambda(w')$. Set now $x := \lambda(\phi, w)$, $x' := \lambda(\phi, w')$, $y := \lambda(\psi, w)$, $y' := \lambda(\psi, w')$, and $u := x - x'$, $v := y' - y$. By (2) $x' + y' < x + y$, so $y' - y < x - x'$, and $v < u$. Setting $m := 1$ in (3), we see $x + y + y < x' + y' + y'$, so $x - x' < 2 * (y' - y)$, i.e. $u < 2 * v$. As $v < u$, we see $0 < v$. In general, we have $m * x + (m+1) * y < m * x' + (m+1) * y'$, thus $m * (x - x') < (m+1) * (y' - y)$, i.e. $u * m < v * (m+1)$.

We note:

(a) If u, v are integers, $u > v > 0$, $u * v < v * (m + 1)$, then $v > m$. Proof: $u > v \rightarrow u \geq v+1$. $v*(m+1) = v*m+v > u*m \geq (v+1)*m = v*m+m$ $\rightarrow v > m$.

(b) $v \geq m + 1$, $u = v + 1$ are solutions for the conditions: $m \geq 1$, u, v integers, $u > v > 0$, $u * m < v * (m + 1)$. Proof: $u * m = v * m + m < v * m + v = v * (m + 1)$.

Consequences:

1. (1) + (2) + (3) for all $m \geq 1$ has no solution. Proof: By (a), $v = y' - y = \lambda(\psi, w') - \lambda(\psi, w)$ has to be arbitrarily big, so $\lambda(\psi, w')$ has to be arbitrarily big.

2. (1) + (2) + (3) for all $m_0 \geq m \geq 1$ has a solution. Proof: Set $v := m_0+1$, $u := m_0 + 2$. By (b), this is a solution for all $m \leq m_0$. □

6.4.2.2 The representation result for the finite case

We summarize:

Proposition 6.4.5

Let the number of propositional variables of the language be finite, the length of the sequences r_σ be some finite, fixed N, and $\lambda(\phi, w) = \lambda(\phi', w)$ for all w, ϕ, ϕ' s.t. $\models \phi \leftrightarrow \phi'$.

(1) Let $\mathcal{I} =< \mathcal{R}, \kappa >$ be Markov, then

(1.1) $w, w' \in B_{\mathcal{I}}(\phi_1, \ldots, \phi_N)$ iff $\lambda(w) + \lambda(\phi_1, w) + \ldots + \lambda(\phi_N, w) = \lambda(w') + \lambda(\phi_1, w') + \ldots + \lambda(\phi_N, w') < \infty$,

(1.2) $w \in B_{\mathcal{I}}(\phi_1, \ldots, \phi_N)$, $w' \notin B_{\mathcal{I}}(\phi_1, \ldots, \phi_N)$ iff $\lambda(w) + \lambda(\phi_1, w) + \ldots + \lambda(\phi_N, w) < \lambda(w') + \lambda(\phi_1, w') + \ldots + \lambda(\phi_N, w')$.

(2) Let $\mathcal{R}$ be a set of runs, such that any initial segment of a sequence in $\mathcal{R}$ can be continued by "true", i.e. if $r =< w, < \phi_1, \ldots, \phi_N >> \in \mathcal{R}$, and $1 \leq i \leq N$, then $r' =< w, < \phi_1, \ldots, \phi_{i-1}, true, \ldots, true >> \in \mathcal{R}$. Let $B(\phi_1, \ldots, \phi_n) \subseteq M_{\mathcal{L}}$ for all ϕ_i ($n \leq N$). Let $\lambda : M_{\mathcal{L}} \cup (M_{\mathcal{L}} \times \mathcal{L}) \rightarrow \omega + 1$ s.t. for all $\phi_1, \ldots, \phi_N$, w, w' $\lambda(\boldsymbol{T}, w) = 0$, and the system of inequalities generated as follows:

(2.1) $w, w' \in B_{\mathcal{I}}(\phi_1, \ldots, \phi_N)$ iff $\lambda(w) + \lambda(\phi_1, w) + \ldots + \lambda(\phi_N, w) = \lambda(w') + \lambda(\phi_1, w') + \ldots + \lambda(\phi_N, w') < \infty$,

(2.2) $w \in B_{\mathcal{I}}(\phi_1, \ldots, \phi_N)$, $w' \notin B_{\mathcal{I}}(\phi_1, \ldots, \phi_N)$ iff $\lambda(w) + \lambda(\phi_1, w) + \ldots + \lambda(\phi_N, w) < \lambda(w') + \lambda(\phi_1, w') + \ldots + \lambda(\phi_N, w')$

have a solution (by the Farkas algorithm).

Then there is $\kappa : \mathcal{R} \to \omega + 1$ s.t. $\kappa(r) = \lambda(w) + \Sigma\{\lambda(\phi_i, w) : 1 \leq i \leq N\}$ for $r =< w, < \phi_1, \ldots, \phi_N >>$ and for $\mathcal{I} :=< \mathcal{R}, \kappa > B(\phi_1, \ldots, \phi_N) = B_{\mathcal{I}}(\phi_1, \ldots, \phi_N)$, and $\mathcal{I}$ is Markov. □

6.4.3 There is no finite representation possible

We show here that there is no finite characterization of Markov developments. The technique is the same as in the other cases, and facilitated by the fact that we have no strong closure conditions.

We present again an example of an arbitrarily complicated situation, where "good" and "bad" cases differ by just one "bit" of information. It will be evident how to generalize this example.

Example 6.4.4

We work with four points, w_1, w_2, w_3, w_4, and two formulas, ϕ_1, ϕ_2. We give the impossible conditions in two groups.

Group 1 (see Fact 6.4.2 and 6.4.3):

Information	Consequence
$B(\phi_1, \phi_2) = w_1$	$\lambda(w_1) + \lambda(\phi_1, w_1) + \lambda(\phi_2, w_1) < \lambda(w_2) + \lambda(\phi_1, w_2) + \lambda(\phi_2, w_2)$
$B(\phi_2, \phi_3) = w_2$	$\lambda(w_2) + \lambda(\phi_2, w_2) + \lambda(\phi_3, w_2) < \lambda(w_3) + \lambda(\phi_2, w_3) + \lambda(\phi_3, w_3)$
$B(\phi_3, \phi_4) = w_3$	$\lambda(w_3) + \lambda(\phi_3, w_3) + \lambda(\phi_4, w_3) < \lambda(w_4) + \lambda(\phi_3, w_4) + \lambda(\phi_4, w_4)$
$B(\phi_4, \phi_1) = w_4$	$\lambda(w_4) + \lambda(\phi_4, w_4) + \lambda(\phi_1, w_4) < \lambda(w_1) + \lambda(\phi_4, w_1) + \lambda(\phi_1, w_1)$

and

Group 2 (see Proposition 6.4.5):

Information	Consequence
$w_2 \in B(\phi_1)$, $w_4 \notin B(\phi_1)$	$\lambda(w_2) + \lambda(\phi_1, w_2) < \lambda(w_4) + \lambda(\phi_1, w_4)$
$w_3 \in B(\phi_2)$, $w_1 \notin B(\phi_2)$	$\lambda(w_3) + \lambda(\phi_2, w_3) < \lambda(w_1) + \lambda(\phi_2, w_1)$
$w_4 \in B(\phi_3)$, $w_2 \notin B(\phi_3)$	$\lambda(w_4) + \lambda(\phi_3, w_4) < \lambda(w_2) + \lambda(\phi_3, w_2)$
$w_1 \in B(\phi_4)$, $w_3 \notin B(\phi_4)$	$\lambda(w_1) + \lambda(\phi_4, w_1) < \lambda(w_3) + \lambda(\phi_4, w_3)$

By adding all inequalities, we see that this is contradictory.

We can generalize the example as follows: Lines 1, 2, 3, 4 in group 2 correspond to the following in group 1. "$*$" will stand for any expression, "1l" for the left hand side of line 1 in group 2, etc.

$*$ 2r $<$ 1l $*$

$*$ 3r $<$ 2l $*$

$*$ 4r $<$ 3l $*$

$*$ 1r $<$ 4l $*$

To simplify further, we may assume all $\lambda(w_i) = 0$.

It remains to create a full example:

1. We do not assume any domain closure, this simplifies the construction considerably.

2. To construct runs of length ω, we append $\boldsymbol{T} = true$.

3. We take $\kappa(< w_i, \boldsymbol{T}, \boldsymbol{T}, \ldots >) = 0$, thus $\lambda(w_i) = 0$ and $\lambda(w_i, \boldsymbol{T}) = 0$, so we have to put $< w_i, \boldsymbol{T}, \boldsymbol{T}, \ldots >$ into $\mathcal{R}$.

4. The set of runs, $\mathcal{R}$, is now:

(a) all $< w_i, \boldsymbol{T}, \boldsymbol{T}, \ldots >$

(b) all completions with $\boldsymbol{T}$ of above initial segments, i.e.
$< w_1, \phi_1, \phi_2, \boldsymbol{T}, \boldsymbol{T}, \ldots >$, etc. of group 1, and $< w_2, \phi_1, \boldsymbol{T}, \boldsymbol{T}, \ldots >$, etc. of group 2.

We thus have a set of runs $\mathcal{R}$, and $B(\phi_i)$, $B(\phi_i, \phi_j)$ as information. But we need the complete information to see that it is not Markov, as any true subset has a Markov solution. This case is simpler than the revision case discussed above in Section 4.2.4, as we have no completions of the domain — again an illustration of the importance of closure conditions of the domain.

We conclude with a word on the language. Condition (O1) above seems complicated, as a matter of fact, it is not, as we can speak about arbitrary points instead of formulas, taking the usual characterization for ranked structures. (Note that the only way we can cut up a domain is by considering longer segments.)

We can simplify further by setting almost everywhere 0, this was not possible in the other cases.

It is left to the reader to fill in the details.

6.5 "Between" and "Behind"

We introduce the (trivial) notation, and go immediately to the (class of) example(s) which show that there is no finite characterization possible. This case is particularly interesting by the simplicity of the situation.

In the case of "between" and "behind" we have information of the type $< a, b, c >$ or $\neg < a, b, c >$, which mean that b is (is not) between a and c, or that c is (is not) behind b, as seen from a. As said in the introduction, b is between a and c iff $d(a, c) = d(a, b) + d(b, c)$ for some fixed distance d. A (finite) amount of such information can be solved using the Farkas algorithm, this solution is evident. We shall also see below that there is no finite representation of the problem.

There are, however, also easy properties for between, like: $< a, b, c >$, $< a, x, b >$, $< x, b, c >$ imply $< a, x, c >$: $d(a, x) + d(x, c) = d(a, x) + d(x, b) + d(b, c) = d(a, b) + d(b, c) = d(a, c)$.

Considering problems like B is between A and C, where A, B, C are sets of points is probably very complicated, we have not looked into the question.

6.5.1 There is no finite representation for "between" and "behind"

Definition 6.5.1

Given a distance d, we define "x is between a and b", in symbols $< a, x, b >$, iff $d(a, b) = d(a, x) + d(x, b)$ (and consequently $\neg < a, x, b >$ iff $d(a, b) < d(a, x) + d(x, b)$, by the triangle inequality). For simplicity, we assume the distance to be symmetrical.

It is then an interesting question to characterize the relation "between" thus defined, i.e. to give sound and sufficient conditions for "between" so it can be generated by a distance.

The following class of examples shows that for any such characterization, we need arbitrarily much information.

Before we consider the general picture, we give one example.

Example 6.5.1

Take x and y as endpoints, and a_1, a_2, a_3, b_1, b_2, b_3 as intermediate, and $< x, a_1, a_2 >$, $< x, b_1, b_2 >$, $< a_1, b_2, b_3 >$, $< b_1, a_2, a_3 >$, $< a_2, b_3, y >$, $< b_2, a_3, y >$. All other triples are not in the "between" relation.

This cannot correspond to a distance (see the discussion of Example 6.5.2 below), but we can make any proper subset work: If we omit any one of the $< \ldots . >$ −triples, we can find a distance which generates the relation. If, e.g. we omit $< a_1, b_2, b_3 >$, we have an upper part $< x, a_1, a_2 >$, $< x, b_1, b_2 >$, and a lower part $< a_2, b_3, y >$, $< b_2, a_3, y >$, and in the middle just $< b_1, a_2, a_3 >$, with a_2 a "hinge" connecting the upper and the lower part.

(The situation reminds me strangely of "cross switches" for independent switching of electric current.)

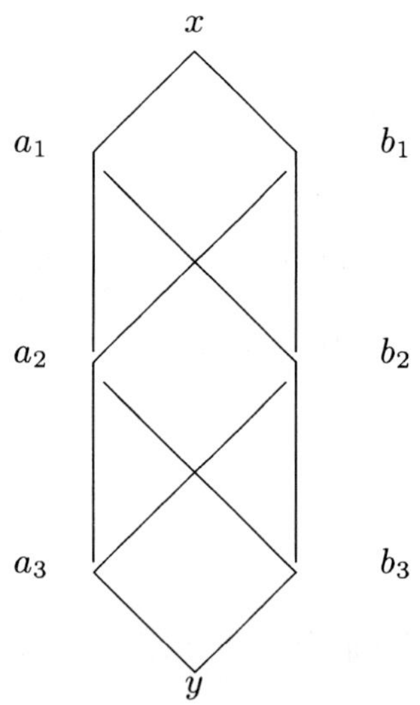

Figure 6.5.1

□

Example 6.5.2

A systematic construction:

The points: x, $a_1, \ldots, a_n$, $b_1, \ldots, b_n$, y.

The full diagram consists of the "between" relations:

$< x, a_1, a_2 >$, $< x, b_1, b_2 >$, $< a_i, b_{i+1}, b_{i+2} >$, $< b_i, a_{i+1}, a_{i+2} >$, for $1 \leq i \leq n-2$, $< a_{n-1}, b_n, y >$, $< b_{n-1}, a_n, y >$, and cannot be represented by a distance.

This is easy to see: We have

$d(x, a_1) + d(a_1, a_2) < d(x, b_1) + d(b_1, a_2)$,

$d(x, b_1) + d(b_1, b_2) < d(x, a_1) + d(a_1, b_2)$,

$d(a_i, b_{i+1}) + d(b_{i+1}, b_{i+2}) < d(a_i, a_{i+1}) + d(a_{i+1}, b_{i+2})$, for $1 \leq i \leq n-2$,

$d(b_i, a_{i+1}) + d(a_{i+1}, a_{i+2}) < d(b_i, b_{i+1}) + d(b_{i+1}, a_{i+2})$, for $1 \leq i \leq n-2$,

$d(a_{n-1}, b_n) + d(b_n, y) < d(a_{n-1}, a_n) + d(a_n, y)$,

$d(b_{n-1}, a_n) + d(a_n, y) < d(b_{n-1}, b_n) + d(b_n, y)$,

but any $d(s, t)$ occurs exactly once on the left and on the right of $<$, so adding all inequalities results in $A < A$.

We now give for any one missing triple a symmetrical distance which generates exactly the remaining triples. Thus, as in the case of distance representable revision, a minimal change transforms again an "illegal" into a "legal" situation.

Case 1:

The missing triple is in the middle.

Say wlog. that $< a_{m-1}, b_m, b_{m+1} >$ is missing (but $< b_{m-1}, a_m, a_{m+1} >$ is present).

The upper part consists of all points and triples involving only x and a_i, b_i with $1 \leq i \leq m$, the lower part consists of all points and triples involving only y and a_i, b_i with $m+1 \leq i \leq n$, the triple $< b_{m-1}, a_m, a_{m+1} >$ is the "connecting rod", and a_m the "hinge".

The distances between a_m and a_{m+1}, a_m and b_{m+1}, b_m and b_{m+1}, b_m and a_{m+1} will have a special treatment, we determine the other distances now.

We construct from the endpoints x and y, downwards in the upper part, upwards in the lower part. Attention: in the upper part, the interior (in the diagram) distances increase when going downwards, in the lower part, the exterior distances increase when going upward.

Let $c := 1/(8 * n)$. In all triples of both parts, let the second distance be 1. Let $d(x, a_1) = d(x, b_1) = d(y, a_n) = d(y, b_n) := 1$. In all triples of both parts, we increase the first distances each time by c, e.g. $d(a_1, b_2) = d(b_1, a_2) = d(a_n, a_{n-1}) = d(b_n, b_{n-1}) = 1 + c$, $d(a_2, b_3) = d(b_2, a_3) = d(a_{n-1}, a_{n-2}) = d(b_{n-1}, b_{n-2}) = 1 + 2 * c$, etc. In detail: for $1 \leq i \leq m - 1$ $d(a_i, a_{i+1}) = d(b_i, b_{i+1}) = 1$, $d(a_i, b_{i+1}) = d(b_i, a_{i+1}) = 1 + i * c$, for $m + 1 \leq i \leq n - 1$ $d(a_i, b_{i+1}) = d(b_i, a_{i+1}) = 1$, $d(a_i, a_{i+1}) = d(b_i, b_{i+1}) = 1 + (n - i) * c$.

We still have to connect the two parts, and define $d(u_m, v_{m+1})$, where u, v are a or b. $d(b_{m+1}, a_m) := 1$ will be as usual for a second part of a triple, $d(a_{m+1}, a_m)$ will be as usual for a first part from below, i.e. $1 + (n - m) * c$, this makes the length of the "connecting rod" $1 + (m - 1) * c + 1 + (n - m) * c = 2 + (n - 1) * c$. As $b_{m-1} - b_m - a_{m+1}$ has to be a detour, and $d(b_{m-1}, b_m) = 1$, we choose $d(b_m, a_{m+1}) := 1 + nc$. As $< a_{m+2}, a_{m+1}, b_m >$ and $a_{m+2} - b_{m+1} - b_m$ is a detour, and $d(a_{m+2}, a_{m+1}) = 1 + (n - m - 1) * c$, we choose $d(b_{m+1}, b_m) := 1 + (2n - m) * c$.

All the distances defined so far are between 1 and $1 + 2n * c = 1.25$.

Let now the distances between the endpoints of the triples in the upper and lower part be as "between" dictates: the sum of the first and the second distance. Note that increasing by c assures that "detours are detours", e.g. $d(x, a_2) = 2 < d(x, b_1) + d(b_1, a_2) = 1 + 1 + c$. But also around the center, detours are respected:

$d(a_{m+2}, b_m) = d(a_{m+2}, a_{m+1}) + d(a_{m+1}, b_m) = 2 + (2n - m - 1) * c < d(a_{m+2}, b_{m+1}) + d(b_{m+1}, b_m) = 2 + (2n - m) * c$,

$d(b_{m+2}, a_m) = d(b_{m+2}, b_{m+1}) + d(b_{m+1}, a_m) = 2 + (n - m - 1) * c < d(b_{m+2}, a_{m+1}) + d(a_{m+1}, a_m) = 2 + (n - m) * c$,

$d(a_{m+1}, b_{m-1}) = d(a_{m+1}, a_m) + d(a_m, b_{m-1}) = 2 + (n-1) * c < d(a_{m+1}, b_m) + d(b_m, b_{m-1}) = 2 + n * c$.

Choose all other distances as 1.9. This cannot introduce new triples, as we will show now. Suppose $< u, v, w >$ holds, i.e. $d(u, w) = d(u, v) + d(v, w)$. $d(u, w) = 1.9$ is impossible, as all distances are ≥ 1. The same holds for all individual distances defined above like $d(a_i, b_{i+1})$, so u and w can only be endpoints of triples, as those distances were between 2 and 2.5. v cannot have distance 1.9 from u or w, so v must be on the direct or indirect path from u to w, but we took care that only the desired triples satisfy this condition.

It remains to check the triangle inequality. The only problem can again be with distances $d(u, w)$ between 2 and 2.5. But again for any v with

$d(u, v) = 1.9$ or $d(v, w) = 1.9$, there is no problem. The remaining cases are precisely the sequences we looked at in detail, and they present no problem either.

Case 2:

The missing triple is at the end.

Say wlog. that $< a_{n-1}, b_n, y >$ is missing (but $< b_{n-1}, a_n, y >$ is present).

Construct everything as above (from above) down to a_n, b_n. Then $d(a_{n-1}, a_n) = d(b_{n-1}, b_n) = 1$, $d(a_{n-1}, b_n) = d(b_{n-1}, a_n) = 1 + (n-1) * c$. Let $d(a_n, y) := 1$, and choose $d(b_n, y) := 1 + nc$. Finish as above.

We show the absence of a finite characterization.

Let now $\phi = \forall x_1, \ldots, x_k \phi'(x_1, \ldots, x_n)$ be a universally quantified formula containing just the ternary relations $< ., ., . >$ described above. Suppose it characterizes the distance representable structures. Let n be the number of triples in ϕ.

Take a sufficiently big counterexample C, so for some $a_1, \ldots, a_k$ ϕ has to fail: $C \models \neg\phi'[a_1, \ldots, a_n]$. But, there is a legal structure S which gives exactly the same information on the n triples involved, so $S \models \neg\phi'[a_1, \ldots, a_n]$, and $S \models \neg\phi$, contradiction.

□

Discussion:

When we look at the "illegal" cases, which cannot be represented by a distance, we note the following:

Each distance $d(x, y)$ occurs on one side as part of a direct path, and on the other side, as part of an indirect path. But this cannot be: Each inequality $A < B$ is a strict one, so if we add them all up, this results in one big strict inequality. So, if each $d(x, y)$ occurs as part of a direct and of an indirect path (and, more precisely, the same number of times), we have a contradiction.

Thus, we can describe the situation a little sloppily as follows: It must not be possible to recombine all $d(x, y)$ which form part of direct paths in a way that they all form part of indirect paths. If we could, we would have exactly the same terms on the left and on the right of a strict inequality.

Chapter 7

Size

7.1 Introduction

In this chapter, we will

- give an interpretation of "almost all" or "big subsets" in first order logic (FOL) by a generalized quantifier, for which we give a sound and complete axiomatization,
- compare coherent systems of filters with certain order relations,
- show how to construct epistemic entrenchment relations, and thus revision functions from model size.

In the first part, we assume no coherence conditions, we just give a bare bones system based semantically on weak filters. It is straightforward to add full filter or coherence conditions as we like — it suffices to write them down, the corresponding semantical conditions are obvious. Section 7.2 is, if you wish, just a formal version of basic intuitions about abstract size in a first oder setting. It shows that what you think to be right is really so: the straightforward axiomatization works.

In the second part, on the contrary, the emphasis is on coherence. We compare the coherent filter systems of S. Ben-David and R. Ben-Eliyahu with that of the author and the order relation of N. Friedman and J. Halpern. The latter correspondence is not exact, but the systems are very very close. In a certain way, such orders as the one by Friedman/Halpern are already implicit in completeness proofs in [KLM90] and [LM92]. What appears there as an auxiliary notion is put into the foreground here. This approach has the advantage of being very general, abstract, and thus free from the technicalities of the completeness proofs in [KLM90] and similar approaches. Note that the completeness proofs by the author (see Chapter 3) are quite different, they concentrate more on the algebraic side and the minimal constructions needed. Consequently, they are more technical, more specific to preferential structures, but also more general within this framework.

In the third part, we show a construction of epistemic entrenchment relations from model size, published by the author quite long ago. In hindsight, this is exactly the construction of stable sets from point size described in Chapter 2. So, this construction seems quite natural, as well from an abstract as from a more concrete viewpoint. The idea is to assign weight to models, and to prefer those formulas which — better whose models — have more weight. This very simple idea does already half of the work. But, as epistemic entrenchment relations are stable under finite intersection ($\alpha \wedge \beta \leq \alpha$ or $\alpha \wedge \beta \leq \beta$), we have to take care to make this construction sufficiently robust. This is done via the construction of "stable" sets, where, roughly, each single element in the set is stronger than all elements outside the set together. Stable sets make really big leaps in the size of elements. We go quite into detail for the construction of such sets and the relation. The size of formulas resulting from the size of models can be read as well on the formula side as on the side of model sets, there is no difference. Conceptually, this part is perhaps the most interesting of the present Chapter 7, as it shows how to go from size of points in a quite robust way essentially to a ranked order of sets.

7.1.1 The details

Defaults as generalized quantifiers (Section 7.2):

In Section 7.2 on defaults as generalized quantifiers in a FOL setting, we use weak filters on the semantical side, and add the following axioms on the syntactical side to a FOL axiomatization:

1. $\nabla x\phi(x) \wedge \forall x(\phi(x) \rightarrow \psi(x)) \rightarrow \nabla x\psi(x)$,
2. $\nabla x\phi(x) \rightarrow \neg\nabla x\neg\phi(x)$,
3. $\forall x\phi(x) \rightarrow \nabla x\phi(x)$ and $\nabla x\phi(x) \rightarrow \exists x\phi(x)$.

A model is now a pair, consisting of a classical FOL model M, and a weak filter over its universe. Both sides are connected by the following definition, where $\mathcal{N}(M)$ is the weak filter on the universe of the classical model M :

$< M, \mathcal{N}(M) > \models \nabla x \phi(x)$ iff there is $A \in \mathcal{N}(M)$ s.t.
$\forall a \in A$ $(< M, \mathcal{N}(M) > \models \phi[a])$.

To show soundness and completeness, we prove the central Lemma 7.2.4:

Let T be a $\nabla - \mathcal{L}-$theory. Then T is consistent under our axioms iff T has a model as defined above.

The main formal work is to prove this lemma.

The extension to defaults with prerequisites by restricted quantifiers is straightforward.

Three abstract coherent systems (Section 7.3):

We present and compare in this section the abstract systems of Ben-David/Ben-Eliyahu, the author, and of Friedman/Halpern. The fact that they are very close, though in different disguise, can be seen as an argument for their naturalness.

They work either with filter systems, or with abstract ordering relations.

(1) The system of S. Ben-David and R. Ben-Eliyahu (see Proposition 7.3.2):

Let $\mathcal{N}' := \{\mathcal{N}'(A) : A \subseteq U\}$ be a system of filters for $\mathcal{P}(U)$, i.e. each $\mathcal{N}'(A)$ is a filter over A. The conditions are (in slight modification):

UC′: $B \in \mathcal{N}'(A) \to \mathcal{N}'(B) \subseteq \mathcal{N}'(A)$,

DC′: $B \in \mathcal{N}'(A) \to \mathcal{N}'(A) \cap \mathcal{P}(B) \subseteq \mathcal{N}'(B)$,

RBC′: $X \in \mathcal{N}'(A)$, $Y \in \mathcal{N}'(B) \to X \cup Y \in \mathcal{N}'(A \cup B)$,

SRM′: $X \in \mathcal{N}'(A)$, $Y \subseteq A \to A - Y \in \mathcal{N}'(A) \vee X \cap Y \in \mathcal{N}'(Y)$,

GTS′: $C \in \mathcal{N}'(A)$, $B \subseteq A \to C \cap B \in \mathcal{N}'(B)$.

(2) The system of the author (see Definition 7.3.4, and Definition 2.3.6) :

We consider a system of ideals over $\mathcal{P}(U)$, i.e. let for $A \subseteq U$ an ideal $\mathcal{I}(A) \subseteq \mathcal{P}(A)$ be given.

($\emptyset$) If $A \neq \emptyset$, then $\emptyset \in \mathcal{I}(A)$,

(Coh0) if $B \subseteq C \subseteq D$, $B \in \mathcal{I}(C)$, then $B \in \mathcal{I}(D)$,

(CohCUM) if $A, C \in \mathcal{I}(B)$, then $A - C \in \mathcal{I}(B - C)$,

(CohRM) if $A \in \mathcal{I}(B)$, $C \subseteq B$, $B - C \notin \mathcal{I}(B)$, then $A - C \in \mathcal{I}(B - C)$.

(3) The system of N. Friedman and J. Halpern (see Definition 7.3.5):

Let U be a set, $<$ a strict partial order on $\mathcal{P}(U)$, (i.e. $<$ is transitive, and contains no cycles). Consider the following conditions for $<$:

(B1) $A' \subseteq A < B \subseteq B' \rightarrow A' < B'$,

(B2) if A, B, C are pairwise disjoint, then $C < A \cup B$, $B < A \cup C \rightarrow B \cup C < A$,

(B3) $\emptyset < X$ for all $X \neq \emptyset$,

(B4) $A < B \rightarrow A < B - \mathrm{A}$,

(B5) Let $X, Y \subseteq A$. If $A - X < X$, then $Y < A - Y$ or $Y - X < X \cap Y$.

We then show the following equivalences:

(1) Equivalence of the first two systems (see Proposition 7.3.9):

$\mathcal{N}'$ satisfies UC′, DC′, RBC′, SRM′, iff the corresponding system of ideals $\mathcal{I}$ defined by $\mathcal{I}(A) := \{X : A - X \in \mathcal{N}'(A)\}$ satisfies $(\emptyset)$–(CohRM).

(2) Equivalence of the first and third system (see Proposition 7.3.11):

Let, on the one hand, $<$ on $\mathcal{P}(U)$ satisfy (B1)–(B4), and, on the other hand, $\mathcal{N}'$ be a coherent system of proper filters on U (i.e. for $A \subseteq U$ $\mathcal{N}'(A) \neq \mathcal{P}(A)$), satisfying UC′, DC′, RBC′.

Define for $X \neq \emptyset$ $\mathcal{N}'_<(X) := \{B \subseteq X : X - B < B\}$, and
$A <_{\mathcal{N}'} B :\leftrightarrow \exists X, Y (A \subseteq X - Y, Y \subseteq B, Y \in \mathcal{N}'(X), Y \neq \emptyset)$.

Then:

(1) Setting $\mathcal{N}'(X) := \mathcal{N}'_<(X)$, $\mathcal{N}'(X)$ is a proper filter, and UC′, DC′, RBC′, hold for $\mathcal{N}'$.

(2) Setting $< := <_{\mathcal{N}'}$, $<$ will be transitive, cycle-free, and satisfy (B1)–(B4).

(3) The operations are inverse: $\mathcal{N}'(X) = \mathcal{N}'_{<_{\mathcal{N}'}}(X)$ and $< = <_{\mathcal{N}'_<}$.

(4) If (B5) holds for $<$, then SRM′ holds for $\mathcal{N}'_<$. Conversely, if SRM′ holds for $\mathcal{N}'$, then (B5) holds for $<_{\mathcal{N}'}$.

The details and proofs are more complicated than difficult.

Theory revision based on model size (Section 7.4):

We base here theory revision a la AGM on model size. The main reason to repeat this old construction of the author in this context is, that it illustrates well the use of individual (model) size to obtain a ranking of the models, which is quite robust, and thus supports the strong property of epistemic entrenchment relations that $A \cap B$ has the same size as A or as B.

As discussed already in Section 2.3.3.1, the main idea is to group elements by size, so that new groups contain really much bigger elements: Any element in the stronger group has to be bigger than all elements of the weaker groups together. This procedure is, of course, highly context dependent. The more we have elements, the more weaker elements can form "coalitions" to beat stronger elements, and pull them into their group of weaker elements. This is asymmetrical, as stronger elements cannot from coalitions, they have to stand on their own. If you wish, this is the opposite of preferential structures, where we may need many stronger elements to beat one weaker element, but then we can beat as many weak elements of the same kind as we like, our forces do not get used up.

This procedure reflects well the strong property of epistemic entrenchment — and thus of revision — that A or B has the same size as $A \cap B$. Speaking in terms of distance, if C is somewhat excentric around X, the farthest elements do not interest us, only the closest ones, so, in a way, we can form intersections until we have the (Grove-) sphere around X. Below this sphere begins something new, and we slide down to the next sphere. This is not exactly the same, but to some degree, and may help the intuition.

We first introduce pre-EE relations (Definition 7.4.1) on the powerset of some set U, which are (essentially) just total orders compatible with the subset relation. We then give, and this is the central idea of this Section 7.4, a method to construct epistemic entrenchment relations relative to some fixed, arbitrary set $X \subseteq U$ (Definition 7.4.2 and Proposition 7.4.2). This construction allows to recover all epistemic entrenchment relations — see Proposition 7.4.3. Note that we did not speak about X before, so pre-EE relations are — in theory revision terms — universal for all K, K intervenes only when we make the full epistemic entrenchment relation conrete. We then look at the special case where the pre-EE relation is constructed in a natural way from size.

To summarize:

We have three essays on size in this Chapter 7. The first one translates abstract size (weak filters) to a generalized quantifier in FOL, the second compares three largely equivalent systems of abstract size and coherence, and the third shows how to group elements by size in a robust and useful way, leading to a size based semantics for theory revision.

Recommended reading:

The sections are independent. Section 7.3 is the longest, most abstract, but perhaps the most straigthforward one. Section 7.2 is one of the few moments where we go into FOL in this book. Section 7.4 has close ties with other sections on theory revision, so readers interested in this subject should perhaps read this section first.

Technically, none of the sections of this chapter is very involved.

7.2 Generalized quantifiers

7.2.1 Introduction

We have discussed weak filters in Chapter 2, and also said that a reasonable abstract notion of size without the properties of weak filters seems difficult to imagine. The full set seems the best candidate for a "big" subset, "big" should cooperate with inclusion, and, finally, no set should be big and small at the same time.

If defaults have something to do with "big" subsets — however we measure them — then weak filters should give some kind of minimal semantics to defaults. This is what we do in this section for FOL. We introduce a new, generalized, quantifier to which we give exactly the desired properties: it should express that a property holds almost everywhere. In particular, the property should hold somewhere if it does so almost everwhere, and, if it holds everywhere, then it holds almost everywhere (so read it, if you like, better: at least almost everywhere), and it cannot be that ϕ and $\neg\phi$ hold almost everywhere at the same time. The latter gives a notion of consistency, we cannot write down just anything any more and pretend that it is still a reasonable default theory. Note that the new quantifier is fully in the object language, so we can negate it, nest it, mix it with classical quantifiers, everything we can do in usual FOL. We will recover the same advantage again in Chapter 8, where we put (almost) everything together in a generalized modal logic.

The essential axioms are now

1. $\nabla x\phi(x) \wedge \forall x(\phi(x) \to \psi(x)) \to \nabla x\psi(x)$,
2. $\nabla x\phi(x) \to \neg\nabla x\neg\phi(x)$,
3. $\forall x\phi(x) \to \nabla x\phi(x)$ and $\nabla x\phi(x) \to \exists x\phi(x)$.

We show here formally that these axioms correspond exactly to weak filters, i.e. we prove a soundness and completeness theorem for weak filter models, where the essential supplementary definition of validity for ∇ and the weak filter $\mathcal{N}(M)$ is (with M a classical FOL model):

$< M, \mathcal{N}(M) > \models \nabla x\phi(x)$ iff there is $A \in \mathcal{N}(M)$ s.t.
$\forall a \in A\ (< M, \mathcal{N}(M) > \models \phi[a])$.

The main definitions are Definitions 7.2.1, 7.2.2, 7.2.3.

The proof shows that in our axiomatization consistent theories have a model, and uses a normal form to construct the structure. For this, we need a dual of ∇, noted $\heartsuit$, to be able to pass negation through. This is done in Definition 7.2.4 and Lemma 7.2.3. The construction of the structure, i.e. the main result, is in Lemma 7.2.4, and the result is summarized in Theorem 7.2.5.

We then extend the approach to restricted quantifiers — corresponding to defaults with prerequisites — this is straightforward, and done at the end of this section.

Recall that this is a bare bones system — weak filters instead of filters, and no coherence whatsoever. Such very weak systems have the advantage to allow easily multiple extensions, we just have to add on both sides (semantics and proof theory) what we think desirable, it is not necessary to start more or less difficult completeness proofs anew, the correspondence will be obvious.

7.2.2 Results

We give first the main definitions of this section, the extension of the language in Definition 7.2.1, the definition of the (extended) FOL model in Definition 7.2.2, and the corresponding axiomatization in Definition 7.2.3. These three definitions contain the essential concepts of our approach.

Definition 7.2.1

We augment the language of first order logic by the new quantifiers: If ϕ and ψ are formulas, then so are $\nabla x\phi(x)$, $\heartsuit x\phi(x)$,
$\nabla x\phi(x) : \psi(x)$, $\heartsuit x\phi(x) : \psi(x)$ for any variable x. Intuitively, $\heartsuit$ means: "for at least a medium size set", and the $:$ –versions are the restricted variants. We call any formula of $\mathcal{L}$, possibly containing ∇ or $\heartsuit$ a $\nabla - \mathcal{L}$–formula.

Definition 7.2.2

($\mathcal{N}$–Model)

Let $\mathcal{L}$ be a first order language, and M be a $\mathcal{L}$–structure. Let $\mathcal{N}(M)$ be a weak filter, or $\mathcal{N}$–system — $\mathcal{N}$ for normal — over M. Define $< M, \mathcal{N}(M) > \models \phi$ for any $\nabla - \mathcal{L}$–formula inductively as usual, with two additional induction steps: $< M, \mathcal{N}(M) > \models \nabla x\phi(x)$ iff there is $A \in \mathcal{N}(M)$ s.t. $\forall a \in A$ ($< M, \mathcal{N}(M) > \models \phi[a]$), and $< M, \mathcal{N}(M) > \models \heartsuit x\phi(x)$ iff
$\{a \in M$: $< M, \mathcal{N}(M) > \models \neg\phi[a]\} \notin \mathcal{N}(M)$.

Lemma 7.2.1

$< M, \mathcal{N}(M) > \models \heartsuit x\phi(x)$ iff $\forall A \in \mathcal{N}(M) \exists a \in A(< M, \mathcal{N}(M) > \models \phi[a])$. □

Proof theory:

Definition 7.2.3

Let any axiomatization of predicate calculus be given. Augment this with the axiom schemata

(1) $\nabla x\phi(x) \wedge \forall x(\phi(x) \rightarrow \psi(x)) \rightarrow \nabla x\psi(x)$,

(2) $\nabla x\phi(x) \rightarrow \neg\nabla x\neg\phi(x)$,

(3) $\forall x\phi(x) \rightarrow \nabla x\phi(x)$ and $\nabla x\phi(x) \rightarrow \exists x\phi(x)$,

(4) $\heartsuit x\phi(x) :\leftrightarrow \neg\nabla x\neg\phi(x)$,

(5) $\nabla x\phi(x) \leftrightarrow \nabla y\phi(y)$ if x does not occur free in $\phi(y)$ and y does not occur free in $\phi(x)$

(for all ϕ, ψ).

We also denote the corresponding notion of derivability by $\vdash_\nabla$.

The following lemmas and definition prepare the central Lemma 7.2.4.

Lemma 7.2.2

The following formulas are derivable:

(1) $\nabla x\phi(x) \wedge \nabla x\psi(x) \rightarrow \exists x(\phi \wedge \psi)(x)$,

(2) $\nabla x\phi(x) \wedge \neg\nabla x\psi(x) \rightarrow \exists x(\phi \wedge \neg\psi)(x)$,

(3) $\neg\nabla x\neg\phi(x) \rightarrow \exists x\phi(x)$,

(4) $\heartsuit x\phi(x) \rightarrow \exists x\phi(x)$,

(5) $\nabla x\phi(x) \wedge \heartsuit x\psi(x) \rightarrow \exists x(\phi \wedge \psi)(x)$,

(6) $\forall x(\phi(x) \leftrightarrow \psi(x)) \rightarrow (\nabla x\phi(x) \leftrightarrow \nabla x\psi(x)) \wedge (\heartsuit x\phi(x) \leftrightarrow \heartsuit x\psi(x))$,

(7) $\forall x\phi(x) \rightarrow \heartsuit x\phi(x)$.

It is usually <u>not</u> derivable: $\heartsuit x\phi(x) \wedge \heartsuit x\psi(x) \rightarrow \exists x(\phi \wedge \psi)(x)$. (To see this, use Theorem 7.2.5 below and argue semantically.) □

Soundness and completeness:

To prepare the proof of completeness, we introduce $\nabla-$normal forms ($\nabla-$NF).

Definition 7.2.4

ϕ is in $\nabla-$normal form ($\nabla-$NF) iff

1. ϕ contains only $\neg$, $\wedge$, $\vee$ as propositional operators
2. only atomic FOL formulas are in the scope of $\neg$.

Lemma 7.2.3

For every ϕ there is ϕ' in $\nabla-$NF s.t. $\vdash_\nabla \phi \leftrightarrow \phi'$.

Proof:

By induction on the depth of $\nabla + \heartsuit$ - nesting.

Case 1: depth= 0: This is a classical result, take, e.g. the disjunctive prenex normal form (PNF).

Case 2: Let the depth of ϕ be $n+1$, and the result be proven up to depth n. We take an $\vdash_\nabla$ $-$equivalent ϕ'' in, e.g. disjunctive PNF, treating the subformulas within the outmost ∇ and $\heartsuit$ quantifiers like classical atomic formulas, so ϕ'' is of the form $Q_1 \ldots Q_n[\phi_1 \vee \ldots \vee \phi_m]$, the Q_i classical quantifiers, the ϕ_i of the form $\phi_{i1} \wedge \ldots \wedge \phi_{ik_i}$, where the ϕ_{ij} are either classical (negated) atomic formulas, or of the form $\nabla x \psi(x, \overline{y})$, $\neg \nabla x \psi(x, \overline{y})$, $\heartsuit x \psi(x, \overline{y})$, or $\neg \heartsuit x \psi(x, \overline{y})$. The negation can be passed through by $\vdash_\nabla$ $\neg \nabla x \psi(x, \overline{y}) \leftrightarrow \heartsuit x \neg \psi(x, \overline{y})$ and $\vdash_\nabla \neg \heartsuit x \psi(x, \overline{y}) \leftrightarrow \nabla x \neg \psi(x, \overline{y})$. By induction hypothesis, the $(\neg)\psi(x, \overline{y})$ can be transformed into an $\vdash_\nabla$ $-$equivalent $\psi'(x, \overline{y})$ in $\nabla-$NF. Axioms 1 and 5 in Definition 7.2.3 give the result. □

We come to the central result of this section: every consistent theory has a model, which we construct now.

Lemma 7.2.4

Let T be a $\nabla - \mathcal{L}-$theory. Then T is consistent under the axioms of Definition 7.2.3 iff T has a model as defined in Definition 7.2.2.

Proof:

The consistency of T when it has a model is trivial.

Let T be a $\vdash_\nabla$ −consistent $\nabla - \mathcal{L}$−theory. We have to show that it has a model. Throughout the proof, let "$\vdash_\nabla$ −consistent" be abbreviated by "consistent". We give a constructive proof, to make the reader comfortable with the new logic. By the above, assume wlog. that all $\phi \in T$ are in ∇−NF.

We first construct a consistent $T' \supseteq T$.

We add $c_\alpha : \alpha < \kappa$ new constants to $\mathcal{L}$, where κ is the size of $\mathcal{L}$, and inductively construct $T' = \bigcup\{T_\gamma : \gamma < \beta\}$ (T_γ ascending, β large enough) with $T_0 := T$, by adding new formulas to T, preserving consistency. (For simplicity, we omit the exact enumeration process — it does not matter anyway.) Let $\phi \in T_\gamma$, depending on the topmost operator, we add 0, 1, or several new formulas. It should be noted that all added formulas are in ∇−NF too.

Case 1: $\phi = \neg\psi$: We do nothing, by ∇−NF, ψ is a classical atomic formula.

Case 2: $\phi = \psi \wedge \psi'$: We add ψ, ψ', obviously preserving consistency.

Case 3: $\phi = \psi \vee \psi'$: Both $T_\gamma + \psi$ and $T_\gamma + \psi'$ cannot be inconsistent, as $\phi \in T_\gamma$, so add one (or both) which preserves consistency.

Case 4: $\phi = \forall x \psi(x)$: Add all $\psi(c_\alpha)$, $\alpha < \kappa$.

Case 5: $\phi = \exists x \psi(x)$: Add some $\psi(c_\alpha)$ which preserves consistency.

Case 6: $\phi = \nabla x \psi(x)$: Add $\exists x \psi(x)$, and for each $\nabla y \psi'(y) \in T_\gamma$ $\exists x(\psi \wedge \psi')(x)$ and for each $\heartsuit y \psi'(y) \in T_\gamma$ $\exists x(\psi \wedge \psi')(x)$ (after suitable renaming, preserving consistency by Lemma 7.2.2).

Case 7: $\phi = \heartsuit x \psi(x)$: Add $\exists x \psi(x)$, and for each $\nabla y \psi'(y) \in T_\gamma$ $\exists x(\psi \wedge \psi')(x)$ (after suitable renaming, preserving consistency by Lemma 7.2.2).

In case 6 and 7, we mark all new $\exists x \psi(x)$ / $\exists x(\psi \wedge \psi')(x)$ as children of $\phi = \nabla x \psi(x)$ / $\phi = \nabla x \psi(x)$ and $\phi = \nabla x \psi'(x)$, etc.

Let $T' := \bigcup\{T_\gamma : \gamma < \beta\}$, β large enough, and $T'' \subseteq T'$ be the set of FOL-formulas of T'. By FOL-completeness, T' has a model M with universe U, where each $u \in U$ is denoted by some c_α.

Next, we define the weak filter $\mathcal{N}(U)$ over U.

Case 1: T' contains no $\nabla x \psi(x)$: Set $\mathcal{N}(U) := \{U\}$.

Case 2: Otherwise. Let $\nabla x \psi(x)$ be in T', and its children be $\exists x \psi(x)$, $\exists x(\psi \wedge \psi_i)(x)$, $i \in I$ (with $\nabla y \psi_i(y)$ / $\heartsuit y \psi_i(y) \in T'$), so there are $\psi(c_\alpha)$, $(\psi \wedge \psi_i)(c_{\alpha_i}) \in T'$. Let $X_{\nabla x \psi(x)} := \{c_\alpha\} \cup \{c_{\alpha_i} : i \in I\}$ (we identify the c_α with their interpretation), and set $\mathcal{N}(U) := \{V \subseteq U\colon X_{\nabla x \psi(x)} \subseteq V$ for some $\nabla x \psi(x) \in T'\}$ Obviously, for $\nabla x \psi(x)$, $\nabla x \psi'(x) \in T'$, $X_{\nabla x \psi(x)} \cap X_{\nabla x \psi'(x)} \neq$

$\emptyset$, as they have the common child $\exists x(\psi \wedge \psi')(x)$, so $\mathcal{N}(U)$ is a $\mathcal{N}-$system.

It remains to show that T holds in $\mathcal{M} :=< M, \mathcal{N}(U) >$. We show by induction on the complexity of ϕ that all $\phi \in T'$ hold in $\mathcal{M}$.

The atomic case is trivial, so are the cases $\neg$, $\wedge$, $\vee$, $\forall$, $\exists$. Consider now $\nabla x\psi(x)$. Note that for each $c_\alpha \in X_{\nabla x\psi(x)}$, $\mathcal{M} \models \psi(c_\alpha)$ by induction hypothesis. But $X_{\nabla x\psi(x)} \in \mathcal{N}(U)$, so $\mathcal{M} \models \nabla x\psi(x)$. Finally, consider $\heartsuit x\psi(x)$.

Case 1: $\mathcal{N}(U) = \{U\}$. $\heartsuit x\psi(x)$ has the child $\exists x\psi(x)$, so $\mathcal{M} \models \psi(c_\alpha)$ for some c_α, so $\mathcal{M} \models \heartsuit x\psi(x)$ by Lemma 7.2.1.

Case 2: $\mathcal{N}(U) \neq \{U\}$. Let $V \in \mathcal{N}(U)$, so there is some $X_{\nabla x\psi'(x)} \subseteq V$, $\nabla x\psi'(x) \in T'$. $\heartsuit x\psi(x)$ and $\nabla x\psi'(x)$ have the common child $\exists x(\psi \wedge \psi')(x)$, so there is some $c_\alpha \in X_{\nabla x\psi'(x)}$ with $\mathcal{M} \models (\psi \wedge \psi')(c_\alpha)$ by induction hypothesis. As this holds for all such V, $\mathcal{M} \models \heartsuit x\psi(x)$ by Lemma 7.2.1 again.
□

The following Theorem 7.2.5 summarizes soundness and completeness and is an easy consequence of Lemma 7.2.4.

Theorem 7.2.5

The axioms given in Definition 7.2.3 are sound and complete for the semantics of Definition 7.2.2.

Proof:

Let $T \not\models \phi$. Then there is a model M, s.t. $M \models T \wedge \neg\phi$. Thus, $Con(T \wedge \neg\phi)$, so $T \not\vdash \phi$. The other direction is analogous. □

Extension to normal defaults with prerequisites

We follow exactly the development in the case without prerequisites, i.e. of unrestricted quantifiers. The reader may safely skip this extension in a first reading, and come back to it later. Neither definitions nor results are surprising. We define again an axiomatization (Definition 7.2.6), the semantics (Definition 7.2.5), and the notion of validity (Definition 7.2.7). Theorem 7.2.7 formulates soundness and completeness.

Definition 7.2.5

Call $\mathcal{N}^+(M) =< \mathcal{N}(N) : N \subseteq M >$ a $\mathcal{N}^+-$system or system of weak filters over M iff for each $N \subseteq M$ $\mathcal{N}(N)$ is a weak filter or $\mathcal{N}-$system over N. (It suffices to consider the definable subsets of M.)

Definition 7.2.6

Extend the logic of first order predicate calculus by adding the axiom schemata

(1) $a.$ $\nabla x\phi(x) \leftrightarrow \nabla x(x = x) : \phi(x)$, $b.$ $\forall x(\sigma(x) \leftrightarrow \tau(x)) \wedge \nabla x\sigma(x) : \phi(x) \rightarrow \nabla x\tau(x) : \phi(x)$,

(2) $\nabla x\phi(x) : \psi(x) \wedge \forall x(\phi(x) \wedge \psi(x) \rightarrow \vartheta(x)) \rightarrow \nabla x\phi(x) : \vartheta(x)$,

(3) $\exists x\phi(x) \wedge \nabla x\phi(x) : \psi(x) \rightarrow \neg\nabla x\phi(x) : \neg\psi(x)$,

(4) $\forall x(\phi(x) \rightarrow \psi(x)) \rightarrow \nabla x\phi(x) : \psi(x)$ and $\nabla x\phi(x) : \psi(x) \rightarrow [\exists x\phi(x) \rightarrow \exists x(\phi(x) \wedge \psi(x))]$,

(5) $\heartsuit x\phi(x) : \psi(x) \leftrightarrow \neg\nabla x\phi(x) : \neg\psi(x)$,

(6) $\nabla x\phi(x) : \psi(x) \leftrightarrow \nabla y\phi(y) : \psi(y)$ (under the usual caveat for substitution)

(for all ϕ, ψ, ϑ, σ, τ).

Lemma 7.2.6

The following are derivable:

a) the axioms of Definition 7.2.3, and the formulas of Lemma 7.2.2 (via Definition 7.2.6 (1) and the corresponding relativized versions).

b) the relativized versions of Lemma 7.2.2, where the existential statements have to be weakened by an existential assumption as in Definition 7.2.6 (4).

□

Definition 7.2.7

Let $\mathcal{L}$ be a first order language, and M a $\mathcal{L}-$structure. Let $\mathcal{N}^+(M)$ be a $\mathcal{N}^+-$system over M.

Define $< M, \mathcal{N}^+(M) > \models \phi$ for any formula inductively as usual, with the additional induction steps:

1. $< M, \mathcal{N}^+(M) > \models \nabla x\phi(x)$ iff there is $A \in \mathcal{N}(M)$ s.t. $\forall a \in A$ $(< M, \mathcal{N}^+(M) > \models \phi[a])$,

2. $< M, \mathcal{N}^+(M) > \models \heartsuit x\phi(x)$ iff $\{a \in M\colon < M, \mathcal{N}^+(M) > \models \neg\phi[a]\} \notin \mathcal{N}(M)$,

3. $< M, \mathcal{N}^+(M) > \models \nabla x\phi(x) : \psi(x)$ iff there is $A \in \mathcal{N}(\{x :< M, \mathcal{N}^+(M) > \models \phi(x)\})$ s.t. $\forall a \in A$ $(< M, \mathcal{N}^+(M) > \models \psi[a])$,

4. $< M, \mathcal{N}^+(M) > \models \heartsuit x\phi(x) : \psi(x)$ iff $\{a \in M$: $< M, \mathcal{N}^+(M) > \models \phi[a] \wedge \neg\psi[a]\} \notin \mathcal{N}(\{x :< M, \mathcal{N}^+(M) > \models \phi(x)\})$.

Theorem 7.2.7

The axioms of Definition 7.2.6 are sound and complete for the $\mathcal{N}^+-$semantics of ∇ as defined in Definition 7.2.7.

Proof:

Fix any $\phi(x)$. The proof of Lemma 7.2.4 shows how to construct a model for all $\nabla x\phi(x) : \psi_i(x)$. Axiom 7.2.6 (1) shows that equivalent ϕ will give the same construction of normal subsets of $\{x$: $M \models \phi(x)\}$. The additional assumption $\exists x\phi(x)$ in Definition 7.2.6 (4) was not needed in Definition 7.2.3 (3), because the domain of a classical model is always nonempty. □

7.3 Comparison of three abstract coherent systems based on size

7.3.1 Introduction

We compare here three abstract coherent systems based on size:

- The system of S. Ben-David and R. Ben-Eliyahu (see [BB94]),
- the system of the author, see also Section 2.3.3 for a discussion,
- the system of N. Friedman and J. Halpern (see [FH98]).

Our main interest is on the semantic side, so we will only compare this part of the articles [BB94] and [FH98] in detail. It would, however, do injustice to these articles to cite them without their essential parts of the proof theoretical side, so we will at least mention the main definitions and results of the logical side, too.

Now, we present first in some modification (for details see below) the three coherent systems:

(1) The system of S. Ben-David and R. Ben-Eliyahu (reformulated in Proposition 7.3.2 below, and called henceforth BB for brevity):

Let $\mathcal{N}' := \{\mathcal{N}'(A) : A \subseteq U\}$ be a system of filters for $\mathcal{P}(U)$, i.e. each $\mathcal{N}'(A)$ is a filter over A. The conditions are (in slight modification):

UC′: $B \in \mathcal{N}'(A) \rightarrow \mathcal{N}'(B) \subseteq \mathcal{N}'(A)$,

DC′: $B \in \mathcal{N}'(A) \rightarrow \mathcal{N}'(A) \cap \mathcal{P}(B) \subseteq \mathcal{N}'(B)$,

RBC′: $X \in \mathcal{N}'(A),\ Y \in \mathcal{N}'(B) \rightarrow X \cup Y \in \mathcal{N}'(A \cup B)$,

SRM′: $X \in \mathcal{N}'(A),\ Y \subseteq A \rightarrow A - Y \in \mathcal{N}'(A) \vee X \cap Y \in \mathcal{N}'(Y)$,

GTS′: $C \in \mathcal{N}'(A),\ B \subseteq A \rightarrow C \cap B \in \mathcal{N}'(B)$.

(2) The system of the author (see Definition 7.3.4 below, and Definition 2.3.6, called KS from now on for brevity):

We consider a system of ideals over $\mathcal{P}(U)$, i.e. let for $A \subseteq U$ an ideal $\mathcal{I}(A) \subseteq \mathcal{P}(A)$ be given.

($\emptyset$) If $A \neq \emptyset$, then $\emptyset \in \mathcal{I}(A)$,

(Coh0) if $B \subseteq C \subseteq D$, $B \in \mathcal{I}(C)$, then $B \in \mathcal{I}(D)$,

(CohCUM) if $A, C \in \mathcal{I}(B)$, then $A - C \in \mathcal{I}(B - C)$,

(CohRM) if $A \in \mathcal{I}(B)$, $C \subseteq B$, $B - C \notin \mathcal{I}(B)$, then $A - C \in \mathcal{I}(B - C)$.

(3) The system of N. Friedman and J. Halpern (see Definition 7.3.5 below, called FH in future):

Let U be a set, $<$ a strict partial order on $\mathcal{P}(U)$, (i.e. $<$ is transitive, and contains no cycles). Consider the following conditions for $<$:

(B1) $A' \subseteq A < B \subseteq B' \rightarrow A' < B'$,

(B2) if A, B, C are pairwise disjoint, then $C < A \cup B$, $B < A \cup C \rightarrow B \cup C < A$,

(B3) $\emptyset < X$ for all $X \neq \emptyset$,

(B4) $A < B \rightarrow A < B$–A,

(B5) Let $X, Y \subseteq A$. If $A - X < X$, then $Y < A - Y$ or $Y - X < X \cap Y$.

Our main results in this section are:

(1) Equivalence of the systems BB and KS (see Proposition 7.3.9 below):

$\mathcal{N}'$ satisfies UC′, DC′, RBC′, SRM′, iff the corresponding system of ideals $\mathcal{I}$ defined by $\mathcal{I}(A) := \{X : A - X \in \mathcal{N}'(A)\}$ satisfies ($\emptyset$)–(CohRM).

(2) Equivalence of the systems BB and FH (see Proposition 7.3.11 below):

Let, on the one hand, $<$ on $\mathcal{P}(U)$ satisfy (B1)–(B4), and, on the other hand, $\mathcal{N}'$ be a coherent system of proper filters on U (i.e. for $A \subseteq U$ $\mathcal{N}'(A) \neq \mathcal{P}(A)$), satisfying UC′, DC′, RBC′.

Define for $X \neq \emptyset$ $\mathcal{N}'_<(X) := \{B \subseteq X : X - B < B\}$, and $A <_{\mathcal{N}'} B :\leftrightarrow \exists X, Y(A \subseteq X - Y, Y \subseteq B, Y \in \mathcal{N}'(X), Y \neq \emptyset)$.

Then:

(1) Setting $\mathcal{N}'(X) := \mathcal{N}'_<(X)$, $\mathcal{N}'(X)$ is a proper filter, and UC′, DC′, RBC′, hold for $\mathcal{N}'$.

(2) Setting $< := <_{\mathcal{N}'}$, $<$ will be transitive, cycle-free, and satisfy (B1)–(B4).

(3) The operations are inverse: $\mathcal{N}'(X) = \mathcal{N}'_{<_{\mathcal{N}'}}(X)$ and $< \; = \; <_{\mathcal{N}'_<}$.

(4) If (B5) holds for $<$, then SRM′ holds for $\mathcal{N}'_<$. Conversely, if SRM′ holds for $\mathcal{N}'$, then (B5) holds for $<_{\mathcal{N}'}$.

There is another variant of this equivalence, stated in Proposition 7.3.12 below, which, however, is in the same spirit as the last one, so the reader is referred there.

As usual in such cases, the proofs are elementary, though a little long, so there seems no need for further comments.

The fact that we find essentially the same systems in several, at first sight quite different, dialects can probably be seen as an argument for the validity of the underlying intuition. Recall also that such systems are already implicit to a certain degree in the completeness proofs of [KLM90] and [LM92]. This intuition, is, of course, that filters are good — perhaps too idealistic — abstractions of size, but that this does not suffice, that we need a more abstract notion of size, which allows us to change the reference set, too.

If the reader is not interested in the accompanying logical systems, and just wants to see the correspondences on the semantical side, she or he might just skip them. In some cases, we have also modified the original systems, the reader who wants to see the common traits, might thus go directly to the modified systems (see Definition 7.3.9 for BB, and Definition 7.3.10 for FH) to see the main lines, and, perhaps, only later return to the original versions.

7.3.2 Presentation of the three systems

We now present the systems of Ben-David/Ben-Eliyahu, of the author and of Friedman/Halpern, and some of their main results. We have concentrated on those parts essential to understand our comparisons in Sections 7.3.3 and

7.3.4. For details, the reader is referred to the original papers.

7.3.2.1 The system of Ben-David/Ben-Eliyahu

Ben-David/Ben-Eliyahu consider a conditional language with a binary operator $\Rightarrow$, and their structures are as usual in the conditionals framework, i.e. relativized to all points in the structure. We will drop this relativization in the later development, as we are mainly interested in the nonmonotonic framework. Their language is the usual one for propositional conditionals, and admits full nestedness, etc. of $\Rightarrow$. ($\rightarrow$ will continue to denote classical implication.)

Definition 7.3.1

(Ben-David/Ben-Eliyahu)

$M =< U, l, \mathcal{N} >$ is a filter based model (FBM) for a set V of propositional variables

iff

(1) U (the universe) is a set (of worlds),

(2) $l : U \rightarrow \mathcal{P}(V)$ is a labelling function, which assigns as usual to each $w \in U$ the set of variables which hold in w,

(3) $\mathcal{N} : U \times \mathcal{P}(U) \rightarrow \mathcal{P}(\mathcal{P}(U))$ is a function s.t. $\mathcal{N}_w(A) := \mathcal{N}(w, A)$ is a filter over U with $A \in \mathcal{N}_w(A)$.

(We shall later modify equivalently so that $\mathcal{N}_w(A)$ will be a filter over A.)

Definition 7.3.2

(Ben-David/Ben-Eliyahu)

Given $M =< U, l, \mathcal{N} >$, define validity of an arbitrary $\Rightarrow$ −formula in M at a world w and the set $\| \phi \|$ of worlds in M where a formula ϕ holds by simultaneous induction:

(1) for $\phi \in V$ $M \models_w \phi$ iff $\phi \in l(w)$,

(2) classical propositional connectives are treated as usual, i.e. $M \models_w \neg\phi$ iff $M \not\models_w \phi$, etc.,

(3) $M \models_w \phi \Rightarrow \psi$ iff $\| \psi \| \in \mathcal{N}_w(\| \phi \|)$.

The authors then turn to proof theory, first consider a basic system of axioms and rules, called F (for filter), and various extensions.

We have seen some of the conditions already before, with $\mid\!\sim$ instead of $\Rightarrow$. For instance, reflexivity reads now $\alpha \Rightarrow \alpha$.

Definition 7.3.3

The system F consists of of the following two axioms and four rules:

all instances of classical tautologies and reflexivity,

Modus ponens: $\frac{\alpha \rightarrow \beta, \alpha}{\beta}$, and the rules (LLE), (RW), (AND).

Proposition 7.3.1

(Ben-David/Ben-Eliyahu)

The system F is sound and complete for the family of filter based models.

Ben-David/Ben-Eliyahu obtain among other results the following representation theorem:

Proposition 7.3.2

(Ben-David/Ben-Eliyahu)

Consider the following coherence properties for $\mathcal{N}$:

UC: $B \in \mathcal{N}_w(A) \rightarrow \mathcal{N}_w(A \cap B) \subseteq \mathcal{N}_w(A)$,

DC: $B \in \mathcal{N}_w(A) \rightarrow \mathcal{N}_w(A) \subseteq \mathcal{N}_w(A \cap B)$,

RBC: $\mathcal{N}_w(A) \cap \mathcal{N}_w(B) \subseteq \mathcal{N}_w(A \cup B)$,

SRM: (re-written) $X \in \mathcal{N}_w(A) \rightarrow X \in \mathcal{N}_w(A \cap Y) \vee \boldsymbol{C}(Y) \in \mathcal{N}_w(A)$,

GTS: $\mathcal{N}_w(A \cup B) \subseteq \mathcal{N}_w(A) \cap \mathcal{N}_w(B)$.

(Remark on our version of SRM: By $F_*(\| \alpha \|, \| \beta \|) = \{w : w \models \alpha \Rightarrow \beta\}$ and $\mathcal{N}_w(\alpha) := \{\| \beta \| : w \models \alpha \Rightarrow \beta\}$, $\mathcal{N}_w(\alpha) = \{\| \beta \| : w \in F_*(\| \alpha \|, \| \beta \|)\}$. So, essentially, $\mathcal{N}_w(A) = \{B : w \in F_*(A, B)\}$. Then $F_*(A, B) \subseteq F_*(A \cap Y, B) \cup F_*(A, U - Y)$ is equivalent with $X \in \mathcal{N}_w(A) \rightarrow X \in \mathcal{N}_w(A \cap Y) \vee \boldsymbol{C}(Y) \in \mathcal{N}_w(A)$.)

Then:

F+(CUT) is sound and complete for FBM's satisfying UC,

F+(CM) is sound and complete for FBM's satisfying DC,

P is sound and complete for FBM's satisfying UC, DC, RBC,

F+(RM) is sound and complete for FBM's satisfying SRM,

F+Monotony is sound and complete for FBM's satisfying GTS.

7.3.2.2 The system of the author

We repeat Definition 2.3.6 for easier reference. Recall that an extensive discussion of this system is in Section 2.3.3.

Definition 7.3.4

We consider a system of ideals over $\mathcal{P}(U)$, i.e. let for $A \subseteq U$ an ideal $\mathcal{I}(A) \subseteq \mathcal{P}(A)$ be given.

($\emptyset$) If $A \neq \emptyset$, then $\emptyset \in \mathcal{I}(A)$,

(Coh0) if $B \subseteq C \subseteq D$, $B \in \mathcal{I}(C)$, then $B \in \mathcal{I}(D)$,

(CohCUM) if $A, C \in \mathcal{I}(B)$, then $A - C \in \mathcal{I}(B - C)$,

(CohRM) if $A \in \mathcal{I}(B)$, $C \subseteq B$, $B - C \notin \mathcal{I}(B)$, then $A - C \in \mathcal{I}(B - C)$.

7.3.2.3 The system of Friedman/Halpern

The system of Friedman/Halpern does not work with filter systems, but with partial orders, and is a priori farther away from the first systems than they differ among each other. A detailed study of, e.g. [KLM90] will show that such order approaches are not new in the field, so there is a common intuition behind various masks.

Definition 7.3.5

(Friedman/Halpern)

Let U be a set, $\leq$ a partial order on some set D (i.e. $\leq$ is reflexive, transitive, anti-symmetric). Let $\perp, \boldsymbol{T} \in D$ with $\perp \leq d \leq \boldsymbol{T}$ for all $d \in D$. (Thus, $\perp$ and $\boldsymbol{T}$ have here a nonstandard meaning compared to the rest of this book, as we follow their notation.) Let $Pl : \mathcal{P}(U) \to D$ s.t. $Pl(U) = \boldsymbol{T}$, $Pl(\emptyset) = \perp$, and the following conditions hold:

(A1) $A \subseteq B \to Pl(A) \leq Pl(B)$,

(A2) If A, B, C are pairwise disjoint, then $Pl(C) < Pl(A \cup B)$, $Pl(B) < Pl(A \cup C) \to Pl(B \cup C) < Pl(A)$,

(A2′) $Pl(A - B) < Pl(A \cap B)$, $Pl(A - B') < Pl(A \cap B') \to Pl((A - B) \cup (A - B')) < Pl(A \cap B \cap B')$,

(A3) $Pl(A) = Pl(B) = \perp \to Pl(A \cup B) = \perp$.

Then Pl is called a qualitative plausibility measure, and (U, Pl) a qualitative plausibility space.

Fact 7.3.3

(Friedman/Halpern)

In the presence of (A1), (A2) and (A2′) are equivalent.

Definition 7.3.6

(Friedman/Halpern) Given a qualitative plausibility space (U, Pl), a propositional language $\mathcal{L}$ with set of variables $v(\mathcal{L})$, and a truth assignment function $\pi : U \to \mathcal{P}(v(\mathcal{L}))$, (U, Pl, π) is called a qualitative plausibility structure. For a classical formula ϕ, $[\![\phi]\!]$ is the set of $w \in U$, where ϕ holds — the latter defined as usual.

Given a flat conditional $\phi \Rightarrow \psi$, we define $(U, Pl, \pi) \models_{Pl} \phi \Rightarrow \psi$ iff $Pl([\![\phi]\!]) = \perp$ or $Pl([\![\phi \wedge \psi]\!]) > Pl([\![\phi \wedge \neg\psi]\!])$

Given a set $\boldsymbol{P}$ of qualitative plausibility structures, a set Δ of flat conditionals, and a flat conditional $\phi \Rightarrow \psi$, we define $\Delta \models_{\boldsymbol{P}} \phi \Rightarrow \psi$ iff for all $(U, Pl, \pi) \in \boldsymbol{P}$ ($\forall \delta \in \Delta((U, Pl, \pi) \models_{Pl} \delta)$ implies $(U, Pl, \pi) \models_{Pl} \phi \Rightarrow \psi$).

Definition 7.3.7

(Friedman/Halpern)

A set $\boldsymbol{P}$ of qualitative plausibility structures is called rich iff for all sets $\{\phi_1 \ldots \phi_n\}$ of mutually exclusive classical formulas, there is a plausibility structure $(U, Pl, \pi) \in \boldsymbol{P}$ s.t. $\perp = Pl([\![\phi_1]\!]) < Pl([\![\phi_2]\!]) < \ldots < Pl([\![\phi_n]\!])$.

Proposition 7.3.4

(Friedman/Halpern)

Let Δ be a set of flat conditionals, $\phi \Rightarrow \psi$ a flat conditional. Let further $\Delta \vdash_P \phi \Rightarrow \psi$ denote that $\phi \Rightarrow \psi$ follows from Δ in the system P (see Definition 1.6.5).

If $\boldsymbol{P}$ is a set of qualitative plausibility structures, and $\Delta \vdash_P \phi \Rightarrow \psi$, then $\Delta \models_{\boldsymbol{P}} \phi \Rightarrow \psi$.

Conversely, a set of qualitative plausibility structures $\boldsymbol{P}$ is rich, iff for all sets Δ of flat conditionals, and all flat conditionals $\phi \Rightarrow \psi$, $\Delta \models_{\boldsymbol{P}} \phi \Rightarrow \psi$ implies $\Delta \vdash_P \phi \Rightarrow \psi$.

Friedman/Halpern then show that a number of well-known systems, e.g. (essentially) that of preferential reasoning, give rise to equivalent sets of qualitative plausibility structures, which satisfy the richness condition. Consequently, they are all characterized by the system P, despite their other differences.

We have modified Definition 7.3.5 slightly — see Definition 7.3.10 and Fact 7.3.10 — and obtain with this modified version a very close connection between the systems of Ben-David/Ben-Eliyahu and Friedman/Halpern (see Proposition 7.3.11). The connection between the former and the original version of the latter is a bit looser (see Proposition 7.3.12).

7.3.3 Comparison of the systems of Ben-David/Ben-Eliyahu and the author

We remind the reader that we shall henceforth drop the indices w of the filter systems. We collect the modified conditions of Proposition 7.3.2 in the following

Definition 7.3.8

(Ben-David/Ben-Eliyahu)

Let $\mathcal{N} := \{\mathcal{N}(A) : A \subseteq U\}$, where each $\mathcal{N}(A)$ is a filter over U. We define the conditions:

UC: $B \in \mathcal{N}(A) \rightarrow \mathcal{N}(A \cap B) \subseteq \mathcal{N}(A)$,

DC: $B \in \mathcal{N}(A) \rightarrow \mathcal{N}(A) \subseteq \mathcal{N}(A \cap B)$,

RBC: $\mathcal{N}(A) \cap \mathcal{N}(B) \subseteq \mathcal{N}(A \cup B)$,

SRM: (re-written) $X \in \mathcal{N}(A) \rightarrow X \in \mathcal{N}(A \cap Y) \vee A - Y \in \mathcal{N}(A)$,

GTS: $\mathcal{N}(A \cup B) \subseteq \mathcal{N}(A) \cap \mathcal{N}(B)$.

We modify the system of Ben-David/Ben-Eliyahu slightly and obtain conditions, which are less elegant, but perhaps more intuitive. Their equivalence with the original version is shown in Proposition 7.3.6.

Fact 7.3.5

Let $A \subseteq U$, and $\mathcal{N}(A)$ be a filter over U, with $A \in \mathcal{N}(A)$. Then

(1) $\mathcal{N}'(A) := \{A \cap B : B \in \mathcal{N}(A)\}$ is a filter over A,

(2) $\mathcal{N}'(A) = \mathcal{N}(A) \cap \mathcal{P}(A)$,

(3) $\mathcal{N}(A) = \{C \subseteq U : \exists B \in \mathcal{N}'(A).B \subseteq C\}$.

Proof:

(1) $A \in \mathcal{N}'(A)$ by prerequisite. If $A \cap B \subseteq C \subseteq A$, $B \in \mathcal{N}(A)$, then by $A \in \mathcal{N}(A)$ $A \cap B \in \mathcal{N}(A)$, so $C \in \mathcal{N}(A)$, and $C \in \mathcal{N}'(A)$. $A \cap B$, $A \cap B' \in \mathcal{N}'(A) \rightarrow A \cap B \cap B' \in \mathcal{N}'(A)$, as $B \cap B' \in \mathcal{N}(A)$.

(2) $A \cap B \in \mathcal{N}'(A) \to A \cap B \in \mathcal{N}(A)$. $B \in \mathcal{N}(A)$, $B \subseteq A \to A \cap B = B \in \mathcal{N}'(A)$.

(3) "$\subseteq$": Let $C \in \mathcal{N}(A)$, then $C \cap A \in \mathcal{N}'(A)$ by definition. "$\supseteq$": Let $C \subseteq U$, $\exists B \in \mathcal{N}'(A).B \subseteq C$. As $\mathcal{N}'(A) \subseteq \mathcal{N}(A)$ by (2), $B \in \mathcal{N}(A)$, so $C \in \mathcal{N}(A)$. $\square$

Definition 7.3.9

We define the coherence conditions for a modified system $\mathcal{N}' := \{\mathcal{N}'(A) : A \subseteq U\}$, each $\mathcal{N}'(A)$ a filter over A :

UC′: $B \in \mathcal{N}'(A) \to \mathcal{N}'(B) \subseteq \mathcal{N}'(A)$,

DC′: $B \in \mathcal{N}'(A) \to \mathcal{N}'(A) \cap \mathcal{P}(B) \subseteq \mathcal{N}'(B)$,

RBC′: $X \in \mathcal{N}'(A)$, $Y \in \mathcal{N}'(B) \to X \cup Y \in \mathcal{N}'(A \cup B)$,

SRM′: $X \in \mathcal{N}'(A)$, $Y \subseteq A \to A - Y \in \mathcal{N}'(A) \vee X \cap Y \in \mathcal{N}'(Y)$,

GTS′: $C \in \mathcal{N}'(A)$, $B \subseteq A \to C \cap B \in \mathcal{N}'(B)$.

The following Proposition 7.3.6 shows that we did not really change anything.

Proposition 7.3.6

If $\mathcal{N}$ and $\mathcal{N}'$ are interdefinable as in Fact 7.3.5, i.e. for given $\mathcal{N}$, $\mathcal{N}'$ is as defined by (1) or (2), for given $\mathcal{N}'$, $\mathcal{N}$ is defined by (3), then:

(1) UC for $\mathcal{N} \leftrightarrow$ UC′ for $\mathcal{N}'$,

(2) DC for $\mathcal{N} \leftrightarrow$ DC′ for $\mathcal{N}'$,

(3) RBC for $\mathcal{N} \leftrightarrow$ RBC′ for $\mathcal{N}'$,

(4) SRM for $\mathcal{N} \leftrightarrow$ SRM′ for $\mathcal{N}'$,

(5) GTS for $\mathcal{N} \leftrightarrow$ GTS′ for $\mathcal{N}'$.

Proof:

We use Fact 7.3.5.

(1)

"$\to$": Let $B \in \mathcal{N}'(A) \to B \in \mathcal{N}(A) \to \mathcal{N}(A \cap B) \subseteq \mathcal{N}(A) \to \mathcal{N}'(B) = \mathcal{N}'(A \cap B) = \mathcal{N}(A \cap B) \cap \mathcal{P}(A \cap B) \subseteq \mathcal{N}(A \cap B) \cap \mathcal{P}(A) \subseteq \mathcal{N}(A) \cap \mathcal{P}(A) = \mathcal{N}'(A)$. "$\leftarrow$": Let $B \in \mathcal{N}(A) \to B \cap A \in \mathcal{N}'(A) \to \mathcal{N}'(A \cap B) \subseteq \mathcal{N}'(A)$.

Let now $C \in \mathcal{N}(A \cap B)$. Then $C \cap A \cap B \in \mathcal{N}'(A \cap B) \subseteq \mathcal{N}'(A) \subseteq \mathcal{N}(A)$, so $C \in \mathcal{N}(A)$.

(2)

"$\rightarrow$": $B \in \mathcal{N}'(A) \subseteq \mathcal{N}(A) \rightarrow \mathcal{N}(A) \subseteq \mathcal{N}(A \cap B) \rightarrow \mathcal{N}'(A) \cap \mathcal{P}(B) =_{B \subseteq A} \mathcal{N}(A) \cap \mathcal{P}(B) \subseteq \mathcal{N}(A \cap B) \cap \mathcal{P}(B) =_{B \subseteq A} \mathcal{N}(A \cap B) \cap \mathcal{P}(A \cap B) = \mathcal{N}'(A \cap B) = \mathcal{N}'(B)$. "$\leftarrow$": Let $B \in \mathcal{N}(A) \rightarrow B \cap A \in \mathcal{N}'(A) \rightarrow \mathcal{N}'(A) \cap \mathcal{P}(A \cap B) \subseteq \mathcal{N}'(A \cap B)$. Let $C \in \mathcal{N}(A)$, then $C \cap B \cap A \in \mathcal{N}'(A) \cap \mathcal{P}(A \cap B) \subseteq \mathcal{N}'(A \cap B)$ $\rightarrow C \in \mathcal{N}(A \cap B)$.

(3)

"$\rightarrow$": $X \in \mathcal{N}'(A)$, $Y \in \mathcal{N}'(B) \rightarrow X \in \mathcal{N}(A)$, $Y \in \mathcal{N}(B) \rightarrow X \cup Y \in \mathcal{N}(A) \cap \mathcal{N}(B) \rightarrow X \cup Y \in \mathcal{N}(A \cup B) \rightarrow_{X \subseteq A, Y \subseteq B} X \cup Y \in \mathcal{N}'(A \cup B)$. "$\leftarrow$": $C \in \mathcal{N}(A) \cap \mathcal{N}(B) \rightarrow C \cap A \in \mathcal{N}'(A)$, $C \cap B \in \mathcal{N}'(B) \rightarrow C \cap (A \cup B) \in \mathcal{N}'(A \cup B) \subseteq \mathcal{N}(A \cup B) \rightarrow C \in \mathcal{N}(A \cup B)$.

(4)

"$\rightarrow$": Let $X \in \mathcal{N}'(A)$, $Y \subseteq A$. Then $X \in \mathcal{N}(A)$, so $X \in \mathcal{N}(Y)$ or $\boldsymbol{C}(Y) \in \mathcal{N}(A)$, so $X \cap Y \in \mathcal{N}'(Y)$ or $A - Y \in \mathcal{N}'(A)$. "$\leftarrow$": Let $X \in \mathcal{N}(A)$, Y arbitrary. Then $X \cap A \in \mathcal{N}'(A)$, and by $Y \cap A \subseteq A$ $A - (Y \cap A) = A - Y \in \mathcal{N}'(A)$, so $\boldsymbol{C}(Y) \in \mathcal{N}(A)$, or $X \cap Y \cap A \in \mathcal{N}'(Y \cap A)$, so $X \in \mathcal{N}(A \cap Y)$.

(5)

"$\rightarrow$": Let $C \in \mathcal{N}'(A)$, $B \subseteq A$. Then $C \in \mathcal{N}(B \cup (A - B)) = \mathcal{N}(A) \rightarrow C \in \mathcal{N}(B) \rightarrow C \cap B \in \mathcal{N}'(B)$. "$\leftarrow$": $C \in \mathcal{N}(A \cup B) \rightarrow (C \cap A) \cup (C \cap B) \in \mathcal{N}'(A \cup B) \rightarrow C \cap A \in \mathcal{N}'(A)$, $C \cap B \in \mathcal{N}'(B) \rightarrow C \in \mathcal{N}(A) \cap \mathcal{N}(B)$. □

GTS/GTS′ express monotony (see Ben-David/Ben-Eliyahu), and will not be considered any further.

From now on, we work with $\mathcal{N}'$. Furthermore, we restrict our attention to those $\mathcal{N}'(A)$, where $A \neq \emptyset$.

Fact 7.3.7

(a) DC′ and RBC′ entail: $Z \in \mathcal{N}'(Y)$, $Z \subseteq B$, $X - B \subseteq Y - Z \rightarrow B \in \mathcal{N}'(X \cup B)$,

(b) RBC′ $\rightarrow$ UC′.

Proof:

(a) $Z \in \mathcal{N}'(Y) \rightarrow Z \subseteq B \cap Y \in \mathcal{N}'(Y)$. $B \cap Y \in \mathcal{N}'(Y)$, $B \in \mathcal{N}'(B) \rightarrow_{RBC'}$

$B = (B \cap Y) \cup B \in \mathcal{N}'(B \cup Y)$. Thus $B \subseteq B \cup (X - B) \in \mathcal{N}'(B \cup Y)$, and $\mathcal{N}'(B \cup Y) \cap \mathcal{P}(B \cup (X - B)) \subseteq \mathcal{N}'(B \cup (X - B)) = \mathcal{N}'(X \cup B)$ by DC′.

(b) Let $B \in \mathcal{N}'(A)$, $X \in \mathcal{N}'(B)$. $X \in \mathcal{N}'(B) \rightarrow_{RBC'} X \cup (A - B) \in \mathcal{N}'(B \cup (A - B)) = \mathcal{N}'(A)$. So by $B \in \mathcal{N}'(A)$, and $X \subseteq B \subseteq A$, $X = (X \cup (A - B)) \cap B \in \mathcal{N}'(A)$. □

Fact 7.3.8

UC′, DC′, RBC′ entail: $A, B \in \mathcal{N}'(A \cup B)$, $X \subseteq (A \cup C) \cap (B \cup C) \rightarrow$ $(X \in \mathcal{N}'(A \cup C) \leftrightarrow X \in \mathcal{N}'(B \cup C))$

Proof:

$A, B \in \mathcal{N}'(A \cup B) \rightarrow A \cap B \in \mathcal{N}'(A \cup B) \rightarrow_{A \in \mathcal{N}'(A \cup B), DC'} A \cap B \in \mathcal{N}'(A)$ $\rightarrow_{RBC'} C \cup (A \cap B) \in \mathcal{N}'(A \cup C) \rightarrow_{UC', DC'} (X \in \mathcal{N}'(A \cup C) \leftrightarrow X \in \mathcal{N}'(C \cup (A \cap B)))$. Likewise, by $A \cap B \in \mathcal{N}'(B)$, $(A \cap B) \cup C \in \mathcal{N}'(B \cup C)$ and $X \in \mathcal{N}'(B \cup C) \leftrightarrow X \in \mathcal{N}'(C \cup (A \cap B))$ □

7.3.3.1 Equivalence of both systems

Proposition 7.3.9

$\mathcal{N}'$ satisfies UC′, DC′, RBC′, SRM′, iff the corresponding system of ideals $\mathcal{I}$ defined by $\mathcal{I}(A) := \{X : A - X \in \mathcal{N}'(A)\}$ satisfies (∅)−(CohRM).

Proof:

"→":

(∅) by $A \in \mathcal{N}'(A)$

(Coh0) $A \subseteq B \subseteq C$, B small in $C \rightarrow$ A small in C by the filter properties. $B \subseteq C \subseteq D$, B small in $C \rightarrow C - B \in \mathcal{N}'(C)$, $D - C \in \mathcal{N}'(D - C) \rightarrow_{RBC'}$ D-B $= (C - B) \cup (D - C) \in \mathcal{N}'(D)$.

(CohCUM) Let $A, C \subseteq B$, $A \cap C = \emptyset$. $B - A, B - C \in \mathcal{N}'(B) \rightarrow (B - C) - A = (B - A) \cap (B - C) \in \mathcal{N}'(B)$, $(B - C) - A \in \mathcal{N}'(B - C)$ by DC′.

(CohRM) $B - A \in \mathcal{N}'(B)$, $C \notin \mathcal{N}'(B)$, $(B - C) - A = (B - A) \cap (B - C) \in \mathcal{N}'(B - C)$ by SRM′.

"$\leftarrow$":

$\mathcal{N}'(A)$ is a filter: $\emptyset \subseteq A$ is small by $(\emptyset)$, so $A \in \mathcal{N}'(A)$. If $B \subseteq C \subseteq A$, $B \in \mathcal{N}'(A)$, then by (Coh0), $A - C$ is small in A, so $C \in \mathcal{N}'(A)$. If $B, C \in \mathcal{N}'(A)$, then $B \cap C \in \mathcal{N}'(A)$ by the filter or ideal properties.

DC$'$: Let $B, C \in \mathcal{N}'(A)$, $C \subseteq B$, then $A - B$, $A - C$ are small in A, then $B - C$ is small in A, then $B - C$ is small in B by (CohCUM), so $C \in \mathcal{N}'(B)$.

RBC$'$: $X \in \mathcal{N}'(A)$, $Y \in \mathcal{N}'(B)$, so $A - X$ is small in A, thus in $A \cup B$, likewise, $B - Y$ is small in $A \cup B$, so by the filter or ideal properties $(A \cup B) - (X \cup Y) \subseteq (A - X) \cup (B - Y)$ is small in $A \cup B$, so $X \cup Y \in \mathcal{N}'(A \cup B)$.

SRM$'$: Let $X \in \mathcal{N}'(A)$, $Y \subseteq A$. Then $A - X$ is small in A, so $(A - X) \cap Y$ is small in A. If $A - Y \notin \mathcal{N}'(A)$, then $(A - X) \cap Y$ is small in Y by (CohRM), so $X \cap Y = Y - ((A - X) \cap Y) \in \mathcal{N}'(Y)$. □

7.3.4 Comparison of the systems of Ben-David/Ben-Eliyahu and of Friedman/Halpern

Definition 7.3.10

(Friedman/Halpern, modified)

Let U be a set, $<$ a strict partial order on $\mathcal{P}(U)$, (i.e. $<$ is transitive, and contains no cycles). Consider the following conditions for $<$:

(B1) $A' \subseteq A < B \subseteq B' \rightarrow A' < B'$,

(B2') $A - B < A \cap B$, $A - B' < A \cap B' \rightarrow (A - B) \cup (A - B') < A \cap B \cap B'$,

(B2) if A, B, C are pairwise disjoint, then $C < A \cup B$, $B < A \cup C \rightarrow B \cup C < A$,

(B3) $\emptyset < X$ for all $X \neq \emptyset$,

(B4) $A < B \rightarrow A < B$–A,

(B5) Let $X, Y \subseteq A$. If $A - X < X$, then $Y < A - Y$ or $Y - X < X \cap Y$.

Fact 7.3.10

(essentially Friedman/Halpern)

In the presence of (B1), (B2) and (B2$'$) are equivalent.

Proof:

(B2) $\rightarrow$ (B2'): Assume without loss of generality $B, B' \subseteq A$. Set $A'' := B \cap B'$, $B'' := (A - B) \cap B'$, $C'' := A - B'$. A'', B'', C'' are pairwise disjoint. We have $B'' = (A-B) \cap B' \subseteq A-B < A \cap B = B = (B \cap B') \cup (B \cap (A-B'))$ $\subseteq (B \cap B') \cup (A - B') = A'' \cup C''$. $C'' = A - B' < A \cap B' = B' = A'' \cup B''$. So by (B2), $(A - B) \cup (A - B') = B'' \cup C'' < A'' = B \cap B'$.

(B2') $\rightarrow$ (B2): Let A, B, C be pairwise disjoint, $C < A \cup B$, $B < A \cup C$. Set $A'' := A \cup B \cup C$, $B'' := A \cup B$, $C'' := A \cup C$. Then $A'' - B'' = C < A \cup B = B''$, $A'' - C'' = B < A \cup C = C''$. Thus $B \cup C = (A'' - B'') \cup (A'' - C'')$ $< B'' \cap C'' = A$. □

Proposition 7.3.11 is perhaps the main result of this Section 7.3. It shows that the systems BB and FH are essentially equivalent, so the statement about equivalent intuitions in different disguises finds its formal proof.

Proposition 7.3.11

Let $<$ on $\mathcal{P}(U)$ satisfy (B1)–(B4), and $\mathcal{N}'$ be a coherent system of proper filters on U (i.e. for $A \subseteq U$ $\mathcal{N}'(A) \neq \mathcal{P}(A)$), satisfying UC', DC', RBC'.

Define for $X \neq \emptyset$ $\mathcal{N}'_<(X) := \{B \subseteq X : X - B < B\}$, and $A <_{\mathcal{N}'} B :\leftrightarrow \exists X, Y (A \subseteq X - Y, Y \subseteq B, Y \in \mathcal{N}'(X), Y \neq \emptyset)$. (Consequently, $A <_{\mathcal{N}'} B \rightarrow B \neq \emptyset$.)

Then:

(1) Setting $\mathcal{N}'(X) := \mathcal{N}'_<(X)$, $\mathcal{N}'(X)$ is a proper filter, and UC', DC', RBC', hold for $\mathcal{N}'$.

(2) Setting $< := <_{\mathcal{N}'}$, $<$ will be transitive, cycle-free, and satisfy (B1)–(B4).

(3) The operations are inverse: $\mathcal{N}'(X) = \mathcal{N}'_{<_{\mathcal{N}'}}(X)$ and $< \; = \; <_{\mathcal{N}'_<}$.

(4) If (B5) holds for $<$, then SRM' holds for $\mathcal{N}'_<$. Conversely, if SRM' holds for $\mathcal{N}'$, then (B5) holds for $<_{\mathcal{N}'}$.

Proof:

Note:

(a) If $A \cap B = \emptyset$, and $A <_{\mathcal{N}'} B$, then $B \in \mathcal{N}'(A \cup B)$.

(b) $B' \subseteq A$, $B \subseteq A'$, $B \in \mathcal{N}'(A)$, $B' \in \mathcal{N}'(A') \rightarrow B \cap B' \neq \emptyset$.

Proof:

(a) By prerequisite, $\exists X, Y (A \subseteq X - Y, Y \subseteq B, Y \in \mathcal{N}'(X))$. $Y \in \mathcal{N}'(X) \rightarrow$

$Y \cup A \in \mathcal{N}'(X) \to_{\text{DC'}} Y \in \mathcal{N}'(Y \cup A)$. $Y \in \mathcal{N}'(Y \cup A)$, $B - Y \in \mathcal{N}'(B\text{--}Y)$ $\to_{\text{RBC'}}$ $B \in \mathcal{N}'(B \cup A)$.

(b) $B \in \mathcal{N}'(A)$, $B \subseteq A' \to A \cap A' \in \mathcal{N}'(A) \to_{\text{DC'}}$ $B \in \mathcal{N}'(A \cap A')$, likewise $B' \in \mathcal{N}'(A \cap A')$, so $B \cap B' \in \mathcal{N}'(A \cap A')$, thus $B \cap B' \neq \emptyset$.

(1)

For $X \neq \emptyset$, $\mathcal{N}'(X)$ is a proper filter: 1. $X \in \mathcal{N}'(X)$ by (B3). 2. Let $B \in \mathcal{N}'(X)$, $B \subseteq B' \subseteq X$, then $X - B' \subseteq X - B < B \subseteq B'$, the result follows from (B1). 3. Let $B, B' \in \mathcal{N}'(X)$, then $X - B < B$, $X - B' < B'$, so $X - (B \cap B') < B \cap B'$ by (B2′). 4. If $\emptyset \in \mathcal{N}'(X)$, then $X < \emptyset$, but $\emptyset < X$ by (B3), a contradiction.

UC′ follows from RBC′.

DC′:

Let $B \in \mathcal{N}'(A)$, $C \subseteq B$, $C \in \mathcal{N}'(A)$, so $A - B < B$, and $A - C < C$, so by $C \subseteq B \subseteq A$ $B - C \subseteq A - C < C$, thus $C \in \mathcal{N}'(B)$.

RBC′:

Let $X \in \mathcal{N}'(A)$, $Y \in \mathcal{N}'(B)$, we have to show $X \cup Y \in \mathcal{N}'(A \cup B)$. By prerequisite, $A - X < X$, $B - Y < Y$, we have to show $(A \cup B) - (X \cup Y) < X \cup Y$. $X \cup Y$, $A - (X \cup Y)$, $B - (Y \cup A)$ are pairwise disjoint, and $(A \cup B) - (X \cup Y)$ $= (A - (X \cup Y)) \cup (B - (Y \cup A))$. By prerequisite, $A - (X \cup Y) < X \cup Y$ and $B - (Y \cup A) < X \cup Y$. But if C, D, E are pairwise disjoint, and $C < E$, $D < E$, then $C \cup D < E : C < E \subseteq E \cup D$, $D < E \subseteq E \cup C \to C \cup D < E$ by (B1) and (B2). Thus, $(A \cup B) - (X \cup Y) < X \cup Y$.

(2)

Transitivity:

Let $A < B$, $B < C$, so $\exists X, Y$ $(A \subseteq X - Y$, $Y \subseteq B$, $Y \in \mathcal{N}'(X)$, $Y \neq \emptyset)$, $\exists X', Y'$ $(B \subseteq X' - Y'$, $Y' \subseteq C$, $Y' \in \mathcal{N}'(X')$, $Y' \neq \emptyset)$. We will show $Y' - (X - Y) \in \mathcal{N}'(X' \cup (X - Y))$ and $Y' - (X - Y) \neq \emptyset$, which proves $A < C$, as $A \subseteq (X' \cup (X - Y)) - (Y' - (X - Y))$, and $Y' - (X - Y) \subseteq C$. Note that, by $Y \subseteq X'$, $\mathcal{N}'(X' \cup (X - Y)) = \mathcal{N}'(X' \cup X)$. First, $Y \in \mathcal{N}'(X)$ $\to_{\text{RBC'}}$ $Y \cup (Y' - X) \in \mathcal{N}'(X \cup (Y' - X)) = \mathcal{N}'(X \cup Y')$. Second, if $Z \in \mathcal{N}'(X \cup Y')$, then $Z \in \mathcal{N}'(X \cup X')$: By RBC′ and $Y' \in \mathcal{N}'(X')$, $X \cup Y' \in \mathcal{N}'(X \cup X')$. Thus, if $Z \in \mathcal{N}'(X \cup Y')$, then $Z \in \mathcal{N}'(X \cup X')$ by UC′. Consequently, $Y \cup (Y' - X) \in \mathcal{N}'(X \cup X')$. Third, $Y' \in \mathcal{N}'(X \cup X')$: $Y \in \mathcal{N}'(X)$, $X' \in \mathcal{N}'(X') \to_{\text{RBC'}}$ $X' = X' \cup Y \in \mathcal{N}'(X \cup X')$. Thus, by $Y' \in \mathcal{N}'(X')$ and UC′, $Y' \in \mathcal{N}'(X \cup X')$. Finally, $Y \cup (Y' - X)$, $Y' \in \mathcal{N}'(X \cup X')$, so $Y' \cap (Y \cup (Y' - X)) \in \mathcal{N}'(X \cup X')$, but, as $Y \cap Y' = \emptyset$, $Y' \cap (Y \cup (Y' - X)) = Y' - X = Y' - (X - Y)$. Finally, suppose $Y' - X = \emptyset$, i.e. $Y' \subseteq X$. We thus have $Y \cap Y' = \emptyset$, $Y' \subseteq X$, $Y \subseteq X'$, $Y \in \mathcal{N}'(X)$, $Y' \in \mathcal{N}'(X')$, a contradiction to (b) above.

Acyclicity: By transitivity, it suffices to show that $A < A$ is impossible. $A < A \to \exists X, Y\ (A \subseteq X{-}Y,\ Y \subseteq A,\ Y \in \mathcal{N}'(X),\ Y \neq \emptyset) \to Y = \emptyset$, contradiction.

(B1′) holds by definition of $<$.

(B2): Let A, B, C be disjoint. If $C < A \cup B$, $B < A \cup C$, then by (a) above, $A \cup B, A \cup C \in \mathcal{N}'(A \cup B \cup C) \to A = (A \cup B) \cap (A \cup C) \in \mathcal{N}'(A \cup B \cup C)$, so $B \cup C < A$. (Note that $A \neq \emptyset : A = \emptyset \to C < B < C$, a contradiction to acyclicity.)

(B3): trivial, as $X \in \mathcal{N}'(X)$.

(B4): $A < B \to \exists X, Y\ (A \subseteq X - Y,\ Y \subseteq B,\ Y \in \mathcal{N}'(X))$. The same X, Y will show $A < B{-}A$.

(3)

Let $X \neq \emptyset$. $B \in \mathcal{N}'(X) \to B \in \mathcal{N}'_{<_{\mathcal{N}'}}(X)$: $B \in \mathcal{N}'(X) \to X - B <_{\mathcal{N}'} B$ (note that $B \neq \emptyset) \to B \in \mathcal{N}'_{<_{\mathcal{N}'}}(X)$.

$B \in \mathcal{N}'_{<_{\mathcal{N}'}}(X) \to B \in \mathcal{N}'(X)$: $B \in \mathcal{N}'_{<_{\mathcal{N}'}}(X) \to B \subseteq X$ and $X - B <_{\mathcal{N}'} B$ $\to$ (by (a) above) $B \in \mathcal{N}'(X)$.

$A < B \to A <_{\mathcal{N}'_<} B$: $A < B \to B \neq \emptyset$ by (B3) and acyclicity, and by (B4) $(A \cup B) - (B - A) = A < B - A \to B - A \in \mathcal{N}'_<(A \cup B)$, thus $B - A \neq \emptyset$, and $A = (A \cup B) - (B - A) <_{\mathcal{N}'_<} B$.

$A <_{\mathcal{N}'_<} B \to A < B$: $A <_{\mathcal{N}'_<} B \to \exists X, Y\ (A \subseteq X - Y, Y \subseteq B, Y \in \mathcal{N}'_<(X)$, $Y \neq \emptyset)$. Thus $Y \subseteq X$, $X - Y < Y$. Thus $A \subseteq X - Y < Y \subseteq B \to A < B$ by (B1).

(4)

Let $X \in \mathcal{N}'_<(A)$, $Y \subseteq A$. Then $A - X < X$. If $Y < A{-}Y$, then $A - Y \in \mathcal{N}'_<(A)$. If $Y - X < X \cap Y$, then $Y - X = Y - (X \cap Y) < X \cap Y$, and $X \cap Y \in \mathcal{N}'_<(Y)$. Let $A - X <_{\mathcal{N}'} X$, $X, Y \subseteq A$. Then, by (a), $X \in \mathcal{N}'(A)$, so by SRM′, $A - Y \in \mathcal{N}'(A)$, thus $Y <_{\mathcal{N}'} A - Y$, or $X \cap Y \in \mathcal{N}'(Y)$, thus $Y - (X \cap Y) = Y - X <_{\mathcal{N}'} X \cap Y$. □

We work now with the original plausibility spaces as defined in Definition 7.3.5.

Proposition 7.3.12

(Friedman/Halpern/Schlechta)

To represent Pl's, where $Pl(X) = \bot$ for some $X \neq \emptyset$, we need now degenerate filters, where $\mathcal{N}'(X) = \mathcal{P}(X)$.

(1) Let D and Pl, and $\leq$ be as in Definition 7.3.5. Define $\mathcal{N}'_{\leq}(X) := \{B \subseteq X : Pl(X - B) < Pl(B)\}$ if $Pl(X) \neq \bot$, and $\mathcal{N}'_{\leq}(X) := \mathcal{P}(X)$ if $Pl(X) = \bot$. Then $\mathcal{N}'(X) := \mathcal{N}'_{\leq}(X)$ is a filter, and UC′, DC′, RBC′, hold for $\mathcal{N}'$.

(2) Let $\mathcal{N}'$ be a coherent system of filters, satisfying UC′, DC′, RBC′. Then there is a Plausibility space D, ordered by some $\leq_{\mathcal{N}'}$, and $Pl : \mathcal{P}(U) \to D$, satisfying (A1)–(A3) s.t. $Pl(A) \leq_{\mathcal{N}'} Pl(B) :\leftrightarrow B \in \mathcal{N}'(A \cup B)$

Proof:

(a)

We first note: If $A \subseteq B$, then $B - A \in \mathcal{N}'(B)$ iff ($(Pl(A) <_{\mathcal{N}'} Pl(B - A)$ or $\mathcal{N}'(B) = \mathcal{P}(B))$). For, if $B - A \in \mathcal{N}'(B)$, then $Pl(A) \leq_{\mathcal{N}'} Pl(B - A)$. If $Pl(B - A) \leq_{\mathcal{N}'} Pl(A)$ holds too, then $A \in \mathcal{N}'(B)$, so $\mathcal{N}'(B) = \mathcal{P}(B)$. The converse is trivial.

(1)

If $Pl(X) \neq \bot$, $\mathcal{N}'(X)$ is a proper filter:

1. $X \in \mathcal{N}'(X)$ by $Pl(\emptyset) = \bot < Pl(X)$.

2. Let $B \in \mathcal{N}'(X)$, $B \subseteq B' \subseteq X$, then $PL(X - B') \leq PL(X - B) < PL(B) \leq Pl(B')$.

3. Let $B, B' \in \mathcal{N}'(X)$, then $PL(X - B) < PL(B)$, $PL(X - B') < PL(B')$, so $PL(X - (B \cap B')) < PL(B \cap B')$ by (A2′).

4. $\emptyset \notin \mathcal{N}'(X) : \emptyset \in \mathcal{N}'(X) \to Pl(X) < Pl(\emptyset)$, contradiction.

UC′ follows from RBC′.

DC′:

If $Pl(A) \neq \bot$: Let $B \in \mathcal{N}'(A)$, $C \subseteq B$, $C \in \mathcal{N}'(A)$, so $PL(A - B) < PL(B)$, and $PL(A - C) < PL(C)$, so by $C \subset B \subseteq A$ $PL(B - C) \leq PL(A - C) < PL(C)$, thus $C \in \mathcal{N}'(B)$. If $Pl(A) = \bot$, then by $B \subseteq A$ $Pl(B) = \bot$, too.

RBC′:

Let $X \in \mathcal{N}'(A)$, $Y \in \mathcal{N}'(B)$, we have to show $X \cup Y \in \mathcal{N}'(A \cup B)$. If $Pl(A), Pl(B) \neq \bot$: By prerequisite, $Pl(A - X) < Pl(X)$, $Pl(B - Y) < Pl(Y)$, we have to show $Pl((A \cup B) - (X \cup Y)) < Pl(X \cup Y)$. $X \cup Y$, $A - (X \cup Y)$, $B - (Y \cup A)$ are pairwise disjoint, and $(A \cup B) - (X \cup Y) = (A - (X \cup Y)) \cup (B - (Y \cup A))$. By prerequisite, $Pl(A - (X \cup Y)) < Pl(X \cup Y)$ and $Pl(B - (Y \cup A)) < Pl(X \cup Y)$. But if C, D, E are pairwise disjoint, and

$Pl(C) < Pl(E)$, $Pl(D) < Pl(E)$, then $Pl(C \cup D) < Pl(E)$: $Pl(C) < Pl(E) \leq Pl(E \cup D)$, $Pl(D) < Pl(E) \leq Pl(E \cup C) \to Pl(C \cup D) < Pl(E)$ by (A1) and (A2). Thus, $Pl((A \cup B) - (X \cup Y)) < Pl(X \cup Y)$. If $Pl(A) = \bot$, $Pl(B) \neq \bot$, then $Pl(Y) \neq \bot$, and $\bot = Pl((A - X) - Y) < Pl(Y)$, $Pl((B - Y) - A) < Pl(Y)$, so by (A2) $Pl((A \cup B) - (X \cup Y)) < Pl(Y)$, so $Pl((A \cup B) - (X \cup Y)) < Pl(X \cup Y)$, and $X \cup Y \in \mathcal{N}'(A \cup B)$. If $Pl(A) = Pl(B) = \bot$, then by (A3) $Pl(A \cup B) = \bot$, and $\mathcal{N}'(A \cup B) = \mathcal{P}(A \cup B)$.

(2)

We have to define the partial order, check that it is transitive, reflexive, and antisymmetric, and verify (A1), (A2), (A3). Let $A \approx B :\leftrightarrow A, B \in \mathcal{N}'(A \cup B)$. Note that $A \approx \emptyset$ iff $\mathcal{N}'(A) = \mathcal{P}(A)$: $A \approx \emptyset \leftrightarrow \emptyset \in \mathcal{N}'(A) \leftrightarrow \mathcal{N}'(A) = \mathcal{P}(A)$. We show that $\approx$ is an equivalence relation. Obviously, $A \approx A$, $A \approx B \to B \approx A$. Moreover, if $A \approx B$, $B \approx C$, then $B, C \in \mathcal{N}'(B \cup C)$, so $A \in \mathcal{N}'(A \cup B) \leftrightarrow A \in \mathcal{N}'(A \cup C)$ by Fact 7.3.8, so $A \in \mathcal{N}'(A \cup C)$. Likewise, $C \in \mathcal{N}'(A \cup C) \leftrightarrow C \in \mathcal{N}'(B \cup C)$, so $C \in \mathcal{N}'(A \cup C)$.

Take now $D :=$ the set of $\approx$ $-$equivalence classes $[A]$ for $A \subseteq U$.

Define $[A] <_{\mathcal{N}'} [B]$ iff $B \in \mathcal{N}'(A \cup B)$, but not $A \in \mathcal{N}'(A \cup B)$. This is well defined: Suppose $A \approx A'$, $B \approx B'$. If $B \in \mathcal{N}'(A \cup B)$, then $B \approx B' \to B \cap B' \in \mathcal{N}'(B \cup B') \to$ (by Fact 7.3.8) $B \cap B' \in \mathcal{N}'(B \cup B) = \mathcal{N}'(B) \to$ (by $B \in \mathcal{N}'(A \cup B)$) $B \cap B' \in \mathcal{N}'(A \cup B) \to$ (by Fact 7.3.8) $B \cap B' \in \mathcal{N}'(A' \cup B')$, so $B' \in \mathcal{N}'(A' \cup B')$.

We check the conditions on $\leq$: (with $[A] \leq [B]$ iff $[A] < [B]$ or $[A] = [B]$, i.e. iff $B \in \mathcal{N}'(A \cup B)$) Reflexivity: $[A] \leq [A]$ is trivial. Antisymmetry: $[A] \leq [B] \leq [A] \to A \approx B \to [A] = [B]$. Transitivity: $[A] < [B] < [C]$ $\to B \in \mathcal{N}'(A \cup B)$, $C \in \mathcal{N}'(B \cup C) \to B \cup C \in \mathcal{N}'(A \cup B \cup C) \to C \in \mathcal{N}'(A \cup B \cup C) \to C \in \mathcal{N}'(A \cup C)$. On the other hand, $A \notin \mathcal{N}'(A \cup C)$. For $A \in \mathcal{N}'(A \cup C)$, $C \in \mathcal{N}'(B \cup C) \to A \cup C \in \mathcal{N}'(A \cup B \cup C) \to A \in \mathcal{N}'(A \cup B \cup C) \to A \in \mathcal{N}'(A \cup B)$, contradiction.

Finally, define $Pl(A) := [A]$. Take $\bot := [\emptyset]$, $T := [U]$. Then $[\emptyset] \leq_{\mathcal{N}'} [A] \leq_{\mathcal{N}'} [U]$, as $A \in \mathcal{N}'(A)$, $U \in \mathcal{N}'(U)$.

It remains to show (A1), (A2), (A3).

(A1) Let $A \subseteq B$. Then $Pl(A) \leq_{\mathcal{N}'} Pl(B)$, by $B \in \mathcal{N}'(A \cup B) = \mathcal{N}'(B)$.

(A2′) We have to show $Pl(A \cap B) > Pl(A - B)$, $Pl(A \cap B') > Pl(A - B')$ $\to Pl(A \cap B \cap B') > Pl(A - (B \cap B'))$. If $B, B \not\subseteq A$, consider $B^* := A \cap B$, $B'^* := A \cap B'$, so without loss of generality $B, B' \subseteq A$. Then $Pl(B) > Pl(A - B)$, $Pl(B') > Pl(A - B')$, so $B, B' \in \mathcal{N}'(A)$, and $\mathcal{N}'(A) \neq \mathcal{P}(A)$, so $B \cap B' \in \mathcal{N}'(A)$, so $Pl(B \cap B') \geq Pl(A - (B \cap B'))$, and by $\mathcal{N}'(A) \neq \mathcal{P}(A)$ $Pl(B \cap B') > Pl(A - (B \cap B'))$.

(A3) $Pl(A) = Pl(B) = \perp \to \emptyset \in \mathcal{N}'(A)$, $\emptyset \in \mathcal{N}'(B) \to \emptyset \in \mathcal{N}'(A \cup B)$ by RBC$'$ $\to \mathcal{N}'(A \cup B) = \mathcal{P}(A \cup B) \to Pl(A \cup B) = \perp$.

□

Remark 7.3.13

We conclude with a short remark on the central property of (minimal) preferential structures (μPR). This property corresponds to the coherence property

(F') $A \subseteq B$, $Y \in \mathcal{N}'(A) \to (B - A) \cup Y \in \mathcal{N}'(B)$.

(F') is a consequence of RBC$'$, and corresponds to the following $<$ $-$property: $A \subseteq B$, $A - Y < Y \to A - Y < (B - A) \cup Y$. □

7.4 Theory revision based on model size

7.4.1 Introduction

Recall the basic definitions and results of theory revision as presented in Section 2.2.10, in particular Definitions 2.2.2, 2.2.3.

As already said, the main idea (and part) of this section was published a long time ago, but, at that time, the author did not really understand what was going on, so he had just pushed on and plowed through. He realized only later, looking at the proof again with the notions of size, distance, etc. in the back of the mind, that this at that time ad hoc construction has a more interesting, more general aspect. Consequently, we do not only summarize the old construction, to see how to base theory revision on model size, but analyse the construction more closely to see the central idea.

The main idea is to group elements by size, so that new groups contain really much bigger elements: Any element in the stronger group has to be bigger than all elements of the weaker groups together. This procedure is, of course, highly context dependent. The more we have elements, the more weaker elements can form "coalitions" to beat stronger elements, and pull them into their group of weaker elements. This is asymmetrical, as stronger elements cannot from coalitions, they have to stand on their own. If you wish, this is the opposite of preferential structures, where we may need many stronger elements to beat one weaker element, but then we can beat as many weak elements of the same kind as we like, our forces do not get used up.

This procedure reflects well the strong property of epistemic entrenchment — and thus of revision — that A or B has the same size as $A \cap B$. Speaking in terms of distance, if C is somewhat excentric around X, the farthest elements do not interest us, only the closest ones, so, in a way, we can form intersections until we have the (Grove-) sphere around X. Below this sphere begins something new, and we slide down to the next sphere. This is not exactly the same, but to some degree, and may help the intuition.

Now to the more technical aspects.

We first introduce pre-EE relations (Definition 7.4.1) on the powerset of some set U, which are (essentially) just total orders compatible with the subset relation. We then give, and this is the central idea of this section, a method to construct epistemic entrenchment relations relative to some fixed, arbitrary set $X \subseteq U$ (Definition 7.4.2 and Proposition 7.4.2). This construction allows to recover all epistemic entrenchment relations — see Proposition 7.4.3. Note that we did not speak about X before, so pre-EE relations are — in theory revision terms — universal for all K, K intervenes only when we make the full epistemic entrenchment relation conrete. The analysis of the construction (Fact 7.4.1, Definition 7.4.3, Fact 7.4.4) shows that there is just one interesting case in the construction of the full epistemic entrenchment relation. We then look at the special case where the pre-EE relation is constructed in a natural way from size, and how this translates into the interesting case just mentioned (Definition 7.4.4, Fact 7.4.5, Definition 7.4.5, Fact 7.4.6): by grouping elements into very different levels of size.

We then develop the logics side, which, however, can be read equivalently on the semantical or simply the algebraic side as well, there is a 1-1 correspondence.

To summarize:

We use a quite general technique to go from elements and their size to a robust ranking of sets by size, and apply this to construct a ranking (epistemic entrenchment) from model size, and thus a revision function. The construction works uniformly for all sets K, and gives thus a semantics to iterated revision, too.

7.4.2 Results

7.4.2.1 Pre-EE relations and epistemic entrenchment relations

We first introduce pre-EE relations, and show how to construct full epistemic entrenchment relations from from pre-EE relations, relative to a given set X (or theory K). We also show that we can recover in this way all entrenchment relations, so pre-EE relations are not more special than entrenchment relations. This part goes up to Proposition 7.4.3.

Definition 7.4.1

Definition of a pre-EE relation for sets:

$\leq$ is called a pre-EE relation on $\mathcal{P}(U)$ iff:

(Pre1) $A \subseteq B \rightarrow A \leq B$,

(Pre2) $\leq$ is transitive,

(Pre3) $\leq$ is total,

(Pre4) $\forall B.B \leq A \rightarrow A = U$.

Note that, in the presence of (Pre1), we can replace (Pre4) by

(Pre4′) $U \leq A \rightarrow A = U$.

Definition of a epistemic entrenchment relation for sets from a pre-EE relation for sets:

Definition 7.4.2

Define $\trianglelefteq := \trianglelefteq_X$ on $\mathcal{P}(U)$ relative to fixed $X \subseteq U$ on $\mathcal{P}(U)$ from a pre-EE relation $\leq$ on $\mathcal{P}(U)$ by $A \trianglelefteq B$ iff

(Def1) $A \subseteq B$ or

(Def2) $X \not\subseteq A$ or

(Def3) $B = A \cap C$, $A \leq C$, $X \subseteq A \cap C$.

Finally, let $\preceq = \preceq_X$ be the transitive closure of $\trianglelefteq$.

Fact 7.4.1

There is a standard way of establishing $A \preceq B$: Let $X \subseteq A, C$, then there is B such that $A \trianglelefteq A \cap B$ by (Def3), and $A \cap B \trianglelefteq C$ by (Def1) iff $A \leq (U - A) \cup C$

Proof:

"$\leftarrow$": As $X \subseteq A, C$, $X \subseteq A \cap C = A \cap ((U - A) \cup C)$, so $A \trianglelefteq_{(\text{Def3})} A \cap ((U - A) \cup C) = A \cap C \trianglelefteq_{(\text{Def1})} C$.

"$\rightarrow$": As $A \cap B \trianglelefteq C$ by (Def1), $A \cap B \subseteq C$, so $B \subseteq (U - A) \cup C$, and $B \leq (U - A) \cup C$. As $A \trianglelefteq A \cap B$ by (Def3), $A \leq B$. So by (Pre2) $A \leq (U - A) \cup C$. □

Proposition 7.4.2

If $\leq$ is a pre-EE relation for $\mathcal{L}$, and $\preceq$ defined as in Definition 7.4.2 for some fixed X, then $\preceq$ satisfies (EE1)–(EE5), of an epistemic entrenchment relation (see Definition 2.2.3).

Thus, given one global pre-EE relation for $\mathcal{L}$, we can easily obtain epistemic entrenchment relations for all $X \subseteq U$.

Proof:

We first show two claims, the proof will then be trivial.

Claim 1: For no $X \subseteq A$, $X \not\subseteq B$ we have $A \preceq B$.

Proof: Suppose the contrary. Let $A = A_1 \trianglelefteq A_2 \trianglelefteq \ldots \trianglelefteq A_n = B$. We have to "leave" X somewhere: There is $A_i \trianglelefteq A_{i+1}$ s.t. $X \subseteq A_i$, $X \not\subseteq A_{i+1}$. Examine the cases of the construction of $\trianglelefteq$. (Def1) cannot be, as $X \subseteq A_i$, $A_i \subseteq A_{i+1}$ implies $X \subseteq A_{i+1}$ (Def2) cannot be, as $X \subseteq A_i$ (Def3) cannot be, as $X \not\subseteq A_{i+1}$. Contradiction. □ (Claim 1)

Claim 2: $\forall B. B \preceq A \rightarrow U = A$

Proof: (Induction on the length of the $\trianglelefteq$–chain) Then in particular $U \preceq A$. $U \trianglelefteq_{(\text{Def1})} A \rightarrow U = A$ $U \trianglelefteq_{(\text{Def2})} A$ cannot be, as $X \subseteq U$ $U \trianglelefteq_{(\text{Def3})} A = U \cap C \rightarrow U \leq C \rightarrow U = C$ by (Pre4$'$). □ (Claim 2)

We prove the proposition:

(EE1) is trivial by definition.

(EE2) by (Def1).

(EE3) We have to prove $A \preceq A \cap B$ or $B \preceq A \cap B$ By (Pre3), $A \leq B$ or

$B \leq A$ Case 1: $X \subseteq A \cap B$. If $A \leq B$, then $A \trianglelefteq A \cap B$ by (Def3). $B \leq A$ analogously. Case 2: $X \not\subseteq A \cap B$. Then $X \not\subseteq A$ or $X \not\subseteq B$, so $A \preceq A \cap B$ or $B \preceq A \cap B$ by (Def2).

(EE4) "$\rightarrow$": $X \not\subseteq A \rightarrow A \trianglelefteq B$ by (Def2) for all B "$\leftarrow$": Let $X \neq \emptyset$, $A \preceq B$ for all B. Suppose $X \subseteq A$. By $X \neq \emptyset$, there is B, $X \not\subseteq B$, and by prerequisite $A \preceq B$, contradicting Claim 1.

(EE5) By Claim 2.

□ (Proposition 7.4.2)

Proposition 7.4.3

Let $\leq_X$ be an epistemic entrenchment relation for a knowledge set X. Then, by definition, $\leq_X$ is a pre-EE relation, and $\preceq$ defined for this $\leq_X$ and X as in Definition 7.4.2 is equal to $\leq_X$.

Proof:

(In a simplification due to D. Makinson.)

"$\preceq \subseteq \leq_X$" : It suffices to prove $\trianglelefteq \subseteq \leq_X$. The cases (Def1) and (Def2) are trivial. (Def3): Let $A \trianglelefteq A \cap B$ by $A \leq_X B$. If $B \leq_X A \cap B$, then $A \leq_X B \leq_X A \cap B$, and we are finished by (EE3).

"$\leq_X \subseteq \preceq$": Let $A \leq_X B$. If $X \not\subseteq A$, then $A \trianglelefteq B$ by (Def2). If $X \subseteq A$, then $X \subseteq B$ by (EE4), so $X \subseteq A \cap B$, thus $A \trianglelefteq_{(Def3)} A \cap B \trianglelefteq_{(Def1)} B$. □

7.4.2.2 Stable sets

We take now a closer look at the construction of an epistemic entrenchment relation from a pre-EE relation as done in Definition 7.4.2. We introduce the notion of a stable set, $\subseteq$ −minimal under the epistemic entrenchment relation. Stable sets mark "thresholds", and this will become particularly clear in the natural construction of a pre-EE relation from point size, starting with Definition 7.4.4. This part ends with Fact 7.4.6 and the subsequent discussion.

Definition 7.4.3

A is stable wrt. the epistemic entrenchment relation $\preceq$ iff there is no $A' \subset A$

s.t. $A \preceq A'$.

Fact 7.4.4

Consider the construction in Definition 7.4.2.

Suppose $X \neq \emptyset$.

(1) If $X \not\subseteq A$, then $A \preceq A'$ for all $A' \subseteq A$, so the only stable set A with $X \not\subseteq A$ is $A = \emptyset$.

(2) Let $X \subseteq A$, $A' \subseteq A$. Then $A \preceq A'$ iff $X \subseteq A'$ and there is a sequence C_i with $X \subseteq C_i$ s.t. $A \leq C_1$, $A \cap C_1 \leq C_2$, $A \cap C_1 \cap C_2 \leq C_3$, etc., and $A' = A \cap C_1 \cap C_2 \cap \ldots \cap C_m$.

Proof:

(1) trivial.

(2) "$\leftarrow$": trivial by (Def3). "$\rightarrow$": Wlog., $A' \subset A$. By construction, there is a sequence $A = A_1 \trianglelefteq A_2 \trianglelefteq \ldots \trianglelefteq A_n = A'$. Let i be the first s.t. $A \subseteq A_i$, $A \not\subseteq A_{i+1}$, then $A_i \trianglelefteq A_{i+1}$ by (Def3), i.e. there is C s.t. $A_{i+1} = A_i \cap C$, $X \subseteq C$, $A_i \leq C$. By $A \subseteq A_i \leq C$, $A \leq C$, so we can re-write the beginning of the sequence by $X \subseteq A$, $X \subseteq C$ (we have $A \trianglelefteq A \cap C$) : $A \trianglelefteq_{(\text{Def3})} A \cap C \trianglelefteq_{(\text{Def1})} A_i \cap C = A_{i+1}$. Note that by $A \not\subseteq A_i \cap C$ and $A \subseteq A_i$ $A \not\subseteq A \cap C$, i.e. $A \cap C \subset A$.

Let $C_1 := C$.

If $A \cap C_1 \subseteq A'$, we can take $C' := C_1 \cup A'$, then $X \subseteq C'$, $A \cap C' = A \cap (C_1 \cup A') = (A \cap C_1) \cup A' = A'$, and $A \leq A_i \leq C_1 \leq C'$, so $A \leq C'$, and we are done, as then $A \preceq A \cap C' = A'$ with $A \leq C'$.

Suppose not, so $A \cap C_1 \not\subseteq A'$. We repeat the procedure for $A \cap C_1$. Let j be the first i s.t. $A \cap C_1 \subseteq A_j$, but $A \cap C_1 \not\subseteq A_{j+1}$, so $A_j \not\subseteq A_{j+1}$. As $A \preceq A_j$ and $X \subseteq A$, $X \subseteq A_j$. $A_j \trianglelefteq A_{j+1}$ has to be by (Def3), and there is C' s.t. $A_{j+1} = A_j \cap C'$, $X \subseteq C'$, $A_j \leq C'$. As $A \cap C_1 \subseteq A_j$, $A \cap C_1 \leq A_j \leq C'$, and $A \cap C_1 \trianglelefteq A \cap C_1 \cap C'$, so we re-write $A \trianglelefteq_{(\text{Def3})} A \cap C_1 \trianglelefteq_{(\text{Def3})} A \cap C_1 \cap C' \trianglelefteq_{(\text{Def1})} A_j \cap C' = A_{j+1} \ldots$

Let $C_2 := C'$.

If $A \cap C_1 \cap C' \subseteq A'$, we take C'' as above, and consider $C'' := C' \cup A'$.

By induction, we can thus re-write $A \preceq A'$ by $A \trianglelefteq_{(\text{Def3})} A \cap C_1 \trianglelefteq_{(\text{Def3})} A \cap C_1 \cap C_2 \trianglelefteq_{(\text{Def3})} \cdots \trianglelefteq_{(\text{Def3})} A'$, with $A' = A \cap C_1 \cap C_2 \cap \ldots \cap C_m$ s.t. $X \subseteq C_i$ and $A \cap C_1 \cap \ldots \cap C_i \leq C_{i+1}$. □

Definition 7.4.4

(Construction of epistemic entrenchment relations from a size σ of elements in the finite case.)

Let U be finite, $\sigma : U \to \Re^+$, i.e. $\forall x \in U.\sigma(x) > 0$. Define $\sigma(A) := \Sigma\{\sigma(x) : x \in A\}$ for $A \subseteq U$. Define $A \leq B$ iff $\sigma(A) \leq \sigma(B)$. By $\sigma(x) > 0$ for all $x \in U$, $\leq$ is a pre-EE relation. Define the epistemic entrenchment relation $\preceq = \preceq_X$ as in Definition 7.4.2 from $\leq$ as just defined, for fixed X.

Fact 7.4.5

A is stable wrt. $\preceq_X$ iff

(1) $A = \emptyset$ or

(2) $X \subseteq A$ and $\forall x \in A - X.(\sigma(x) > \sigma(U - A))$.

Proof:

(1) trivial.

(2)

"$\rightarrow$": Suppose $X \subseteq A$, and $\exists x \in A - X.(\sigma(x) \leq \sigma(U - A))$. Consider $C := (A - \{x\}) \cup (U - A)$. Then $\sigma(A) = \sigma(A - \{x\}) + \sigma(x) \leq \sigma(A - \{x\}) + \sigma(U - A) = \sigma(C)$, so $A \leq C$ and $X \subseteq C$ and $A \cap C = A - \{x\}$, so $A \preceq A - \{x\}$ by (Def3), and A is not stable.

"$\leftarrow$": Suppose that $X \subseteq A$, $\forall x \in A - X.(\sigma(x) > \sigma(U - A))$, but A is not stable. Then there is $A' \subset A$ s.t. $A \preceq A'$, $X \subseteq A$. A look at the proof of Fact 7.4.4 shows that there is a first $A \preceq_{(Def3)} A \cap C$ with $A \cap C \subset A$. Take wlog. as A' this $A \cap C$, so there is C s.t. $A \leq C$, $X \subseteq C$, $A \cap C \subset A$. Thus $\sigma(A - C) + \sigma(A \cap C) = \sigma(A) \leq \sigma(C) = \sigma(A \cap C) + \sigma(C - A)$, thus $\sigma(A - C) \leq \sigma(C - A) \leq \sigma(U - A)$. As $A \cap C \subset A$, and $X \cap (A - C) = \emptyset$, there is $x \in A - C$, $x \notin X$, but $\forall x \in A - X.(\sigma(x) > \sigma(U - A))$, contradiction. □

Definition 7.4.5

$n_A := min\{\sigma(x) : x \in A - X\}$ for A s.t. $X \subseteq A$.

Fact 7.4.6

(1) If A is stable, then there is n_A s.t. $A = X \cup \{x : \sigma(x) \geq n_A\}$.

(2) n is the n_A of some stable A iff $\Sigma\{\sigma(x) : x \in U - X, \sigma(x) < n\} < n$.

Proof:

(1) Set $n_A := min\{\sigma(x) : x \in A - X\}$. Then $A \subseteq X \cup \{x : \sigma(x) \geq n_A\}$. If there were $x \in \{y : \sigma(y) \geq n_A\} - A$, this would contradict Fact 7.4.5.

(2) Let A be stable and $n_A = min\{\sigma(x) : x \in A - X\}$. Then $A = X \cup \{x : \sigma(x) \geq n_A\}$ by (1), and $\sigma(U - A) < n_A$ by Fact 7.4.5, thus $U - A = \{x \in U - X : \sigma(x) < n_A\}$, and $\Sigma\{\sigma(x) : x \in U - X, \sigma(x) < n_A\} = \sigma(U - A) < n_A$. Conversely, let n be s.t. $\Sigma\{\sigma(x) : x \in U - X, \sigma(x) < n\} < n$. Set $A := X \cup \{x : \sigma(x) \geq n\}$. Then $X \subseteq A$ and $\forall x \in A - X.\sigma(x) \geq n > \Sigma\{\sigma(x) : x \in U - X, \sigma(x) < n\} = \sigma(U - A)$. Thus, A is stable by Fact 7.4.5.
□

Intuition:

The top layer has one element, U, the bottom layer consists of all sets A s.t. $X \not\subseteq A$. Directly above the bottom is X. The second from top has all elements, except the smallest ones, but they can form alliances to knock still other elements off. All elements in the third layer are bigger than all those together which were thrown out in the 2. layer. All elements in the forth layer are bigger than all those together which were thrown out in the 2. and 3. layer, and so on. So there is a "qualitative" leap between the elements thrown out in successive layers, they are really much smaller. The layers correspond of course to a ranked order, but if we repeat the same construction in subsets, we will usually get a finer distinction, as the possible alliances are smaller, so it is more difficult for small elements to throw out bigger ones. E.g., in the set $\{1, 2, 3\}$, $\{3\}$ is not stable, as $1 + 2 = 3$. But in $\{2, 3\}$ it is, as $2 < 3$. Smaller sets allow finer distinction than coarser ones — see the remarks on clusters below. Consequently, such distinctions cannot be generated by ranked structures (once the ranking is made, as above, then it is a ranked structure, but it cannot be generated by minimizing elements). We can, however, imitate this coalition forming with copies of models: bigger elements can be knocked off by coalitions of smaller ones, they have to destroy all copies of the bigger one.

7.4.2.3 Revision based on model size

We return now to logic. You can, however, read the rest of this section as purely algabraic — there is no difference, apart from notation.

We define a total order (a pre-EE relation) on the formulas of $\mathcal{L}$, and show then how to assign probability values to formulas in a natural way. The construction will be from probability values we give to suitable model sets, and will be carried over to formulas in a natural way. Figure 7.4.1 will give the intuitive picture, and the reader might consult it to get the main idea.

Let, in the following, $\boldsymbol{L}$ be the Lindenbaum-Tarski algebra for the language $\mathcal{L}$ and the empty theory. (Thus, elements of $\boldsymbol{L}$ have the form $[\phi]$, where ϕ is a formula of $\mathcal{L}$, and $[\phi] = [\psi]$ iff $\vdash \phi \leftrightarrow \psi$. Moreover, $[\phi] \wedge [\psi] := [\phi \wedge \psi]$, $-[\phi] := [\neg\phi]$, and $[\phi] \leq [\psi] :\leftrightarrow [\phi] \wedge [\psi] = [\phi]$.)

We have a first constructive result:

Lemma 7.4.7

Extending the natural ordering on the formulas of $\mathcal{L}$ given by $\boldsymbol{L}$ to a total order, preserving [True] as the only maximal element, will give a pre-EE relation for $\mathcal{L}$, and thus, by Proposition 7.4.2, epistemic entrenchment relations $\leq_K$ for all knowledge sets K of $\mathcal{L}$, where a knowledge set is in the terminology of AGM a deductively closed theory. □

Next, we assign probability values to formulas of $\mathcal{L}$, i.e. each $\phi \in \mathcal{L}$ will have a real value $\nu(\phi)$, and the natural order of the real numbers will order the formulas too. Of course, logically equivalent formulas should be given the same probability. We proceed indirectly, assigning first probabilities to models, and defining the probability of a formula as the sum of the probabilities of its models. The above equivalence condition will then be trivially true. It is easily seen (Proposition 7.4.10), that our construction will give a pre-EE relation $\leq$ for $\mathcal{L}$ as needed to define the epistemic entrenchment relations $\leq_K$. We can improve our result and the equivalence condition to obtain ($\phi \leq \psi$ and $\psi \leq \phi$) iff $\vdash \phi \leftrightarrow \psi$ (Proposition 7.4.13). For this end, we use algebraic closure properties of the reals (Fact 7.4.12). We can thus construct in a natural way a total (and natural) extension of the natural order of the Lindenbaum-Tarski algebra $\boldsymbol{L}$, such that ($[\phi] \leq [\psi]$ and $[\psi] \leq [\phi]$) is equivalent to $[\phi] = [\psi]$. In conclusion, we remark that the whole process can be easily relativized to a fixed theory, by considering only models of that theory.

But first, we need some constructions:

Let $\mathcal{A}$ be the $\sigma-$Algebra (i.e. the $\aleph_1-$complete Boolean algebra) of Lebesgue-measurable sets restricted to subsets of the unit interval $[0, 1)$. Let μ be the usual Lebesgue measure. (The reader unfamiliar with these notions will find definitions and properties in any book on measure and integration theory.)

Definition 7.4.6

Let $< x_i : i \in \omega >$ be a sequence of reals in the open interval (0,1).

Define by induction:

$a_0 := [0, x_0)$, $b_0 := \{0, x_0, 1\}$

Let a_n, b_n be defined ($n \in \omega$). b_n will be a set of $2^{n+1} + 1$ elements, a_n a disjoint union of 2^n nonempty intervals. Let $b_n = \{y_j : j < 2^{n+1} + 2\}$, the y_j in increasing order. Define $a_{n+1} := \bigcup \{ \ [\ y_j, y_j + (y_{j+1} - y_j) * x_{n+1} \ ) : j < 2^{n+1} + 1 \ \}$ and $b_{n+1} := b_n \cup \{y_j + (y_{j+1} - y_j) * x_{n+1}\text{: } j < 2^{n+1} + 1\}$. Finally, set $\overline{a_n} := [0, 1) - a_n$. (See Figure 7.4.1 below.)

Let $\boldsymbol{B}$ be the $\aleph_1-$complete subalgebra of $\mathcal{A}$ generated by $\{a_i : i \in \omega\}$.

Fact 7.4.8

For the a_i thus defined we have:

1) $\mu(a_n) = x_n$,

2) $\mu(\overline{a_n}) = 1 - \mu(a_n)$ (trivial),

3) $\mu(\bigcap\{c_n : n \in X\}) = \Pi\{\mu(c_n) : n \in X\}$ where c_n is either a_n or $\overline{a_n}$ for $X \subseteq \omega$ finite, by the "independence" of the construction. This property is essential to all that follows. □

Let, in the rest of this section, $\mathcal{L} = \{p_i : i \in \omega\}$ be a countable language of propositional calculus.

Definition 7.4.7

a) Define f: $\mathcal{L} \rightarrow \{a_i : i \in \omega\}$ by $f(p_i) := a_i$, i.e. $\mu(f(p_i)) = x_i$.

b) Let $\boldsymbol{M}$ be the set of assignments of truth values to finite subsets of $\mathcal{L}$, $t \in \boldsymbol{M}$, t defined on $\mathcal{L}' \subseteq \mathcal{L}$. (It suffices to consider finite subsets, as standard propositional calculus admits only finite formulas.) Define $g(t) := \bigcap\{a_i\text{: } p_i \in \mathcal{L}', t(p_i) = true\} \cap \bigcap\{\overline{a_i}\text{: } p_i \in \mathcal{L}', t(p_i) = false\}$.

Thus, $\mu(g(t)) = \mu(\bigcap\{a_i\text{: } p_i \in \mathcal{L}', t(p_i) = true\} \cap \bigcap\{\overline{a_i}\text{: } p_i \in \mathcal{L}', t(p_i) = false\}) = \Pi\{x_i\text{: } p_i \in \mathcal{L}', t(p_i) = true\} * \Pi\{1 - x_i\text{: } p_i \in \mathcal{L}', t(p_i) = false\}$, and we have defined for every assignment $t \in \boldsymbol{M}$ a real value $\mu(g(t))$. There is a natural way to extend this function to formulas:

Definition 7.4.8

Let ϕ be a formula with propositional variables $p_i \in \mathcal{L}_\phi \subseteq \mathcal{L}$ finite.

(a) Let $Val(\phi) := \{t \in \boldsymbol{M}$: $dom(t) = \mathcal{L}_\phi$, $t(\phi) = true$, i.e. ϕ is true under $t\}$.

(b) So we can define $\nu(\phi) := \Sigma\ \{\mu(g(t))$: $t \in Val(\phi)\}$. (See Figure 7.4.1.)

Let $\mathcal{L} = \{p, q\}$, $t(p) = true$, $t(q) = false$, $t'(p) = false$, $t'(q) = true$, $\phi = p \leftrightarrow \neg q$

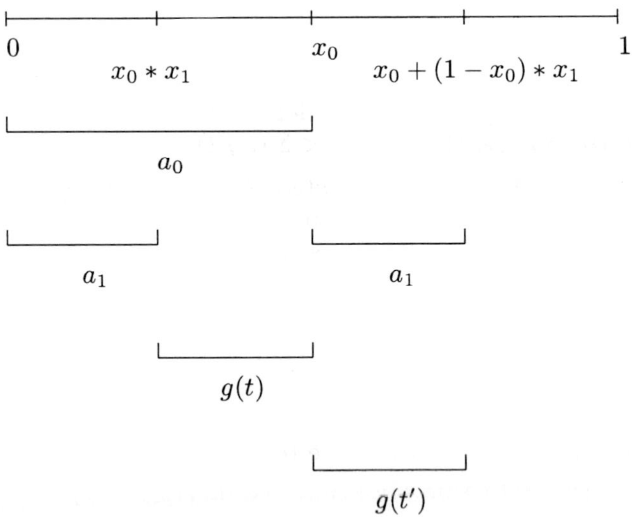

Figure 7.4.1

Thus, $\mu(a_0) = x_0$, $\mu(a_1) = x_1$, $\nu(\phi) = \mu(g(t)) + \mu(g(t')) = x_0 * (1 - x_1) + (1 - x_0) * x_1$.

Our construction has the following properties:

Lemma 7.4.9

(1) $\nu(\phi)$ is independent of $dom(t)$ in the following sense: Let $\mathcal{L}_\phi \subseteq \mathcal{L}' \subseteq \mathcal{L}$ finite. Then $\nu(\phi) := \Sigma\ \{\mu(g(t))\colon\ t \in Val(\phi)\} = \Sigma\ \{\mu(g(t))\colon\ t \in \boldsymbol{M}$, $dom(t) = \mathcal{L}'$, $t(\phi) = true\}$.

(2) By definition of Val and ν, logically equivalent formulas will have the same real value $\nu(\phi)$.

(3) $\vdash \phi \rightarrow \psi$ implies $\nu(\phi) \leq \nu(\psi)$. (To see this, consider $\mathcal{L}' = \mathcal{L}_\phi \cup \mathcal{L}_\psi$, use (1) and the fact, that every assignment which makes ϕ true, will make ψ true too.)

(4) $\nu(\neg\phi) = 1 - \nu(\phi)$. (Use $\nu(true) = \Sigma\ \{\mu(g(t))\colon\ t \in \boldsymbol{M},\ dom(t) = \mathcal{L}'\ finite\} = 1$, $t(\phi) = true \leftrightarrow t(\neg\phi) = false$, and for $t, t' \in \boldsymbol{M}$ with the same domain $t \neq t' \rightarrow g(t) \cap g(t') = \emptyset$.)

(5) Exactly the valid formulas will have real value $\nu(\phi) = 1$. (g: $\boldsymbol{L} \rightarrow \boldsymbol{B}$ (extended suitably to formulas) is an injective homomorphism of Boolean algebras, and use the above arguments.)

(6) $\nu(\phi) \leq \nu(\psi) \leftrightarrow \nu(\neg\psi) \leq \nu(\neg\phi)$ (by 4).

(7) $\nu(\phi \vee \psi) \leq \nu(\phi) \leftrightarrow \vdash \psi \rightarrow \phi$ "$\leftarrow$" by 3) "$\rightarrow$": Suppose $\not\vdash \psi \rightarrow \phi$. Thus $M' := \{t : t \models \phi\} \subset M := \{t : t \models \phi \vee \psi\}$, let $t \in M - M'$. As $x_i \in (0,1)$, $\mu g(t) \neq 0$, thus $\nu(\phi) := \Sigma\{\mu g(t) : t \in M'\} < \Sigma\{\mu g(t) : t \in M\} =: \nu(\phi \vee \psi)$.

(8) We cannot expect $\nu(\phi \wedge \psi) = \nu(\phi) * \nu(\psi)$ or $\nu(\phi \vee \psi) = \nu(\phi) + \nu(\psi)$, just think of $\phi = \psi$. These equations can only be valid if ϕ and ψ are independent. For this reason, we gave first a value to models, which are independent, and then to formulas. □

We have thus proved our main constructive result:

Proposition 7.4.10

Let $p_i : i \in \omega$ be given a probability $x_i \in (0,1)$, then this gives rise naturally to probabilities $\nu(\phi)$ for any formula in $\mathcal{L}$, such that (1) – (6) of Lemma 7.4.9 are valid, and thus to a pre-EE relation $\leq$ for $\mathcal{L}$, i.e. satisfying (Pre1) – (Pre4) of $\leq$ in Definition 7.4.1, and thus the prerequisites of Proposition 7.4.2. □

Fact 7.4.11

Let $0 \leq a \leq b < 1$. Augment the natural order of the reals by setting $x \leq^+ y$ for all $a \leq x, y \leq b$, i.e. "identify" all elements of the interval $[a, b]$. Let ν be defined as in the construction leading to Proposition 7.4.10 and set $\phi \leq \psi$ iff $\nu(\phi) \leq \nu(\psi)$ or $\nu(\psi) \leq^+ \nu(\phi)$. Then $\leq$ is still a pre-EE relation on $\mathcal{L}$.

Proof:

In Definition 7.4.1, (Pre1) and (Pre3) are trivial, (Pre4) holds by $b < 1$. But (Pre2) is simple too: consider, e.g. $x \leq y \leq^+ z$. If $x > z$, then $a \leq z \leq x \leq y \leq b$, and $x \leq^+ z$. □

Example 7.4.1

Consider now $\mathcal{L} := \{A, B, C\}$, and set $\mu f(A) := 1/2$, $\mu f(B) := 1/3$, $\mu f(C) := 1/5$, $a := 5/30$, $b := 10/30$, and identify in the interval $[a, b]$ as described in the above Fact. Then $\nu(A) = 15/30$, $\nu(B) = 10/30$, $\nu(A \wedge B) = 5/30$, $\nu(A \vee C) = 18/30$, $\nu(B \vee C) = 14/30$, $\nu((A \wedge B) \vee C) = 10/30$. By identification, $(A \wedge B) \vee C \leq A \wedge B$, but neither $A \vee C \leq A$ nor $B \vee C \leq B$. Thus, this order is a counterexample as promised above. □

So far, it is quite possible that $\nu(\phi) = \nu(\psi)$, but $\not\vdash \phi \leftrightarrow \psi$. We now make ν injective (modulo $\leftrightarrow$). Thus, we improve our result such that ($\phi \leq \psi$ and $\psi \leq \phi$) iff $\vdash \phi \leftrightarrow \psi$. Choosing the x_i of Definition 7.4.6 above according to the following fact on the reals will do the trick:

Fact 7.4.12

Let $X := \{x_i : i \in \omega\} \subset I \subseteq \Re$, I uncountable be given. Then there is $x' \in I$ s.t. x' is not equal to any real that can be obtained by finite addition, subtraction, multiplication, division from elements of $\boldsymbol{Q} \cup X$ ($\boldsymbol{Q}$ as usual the set of rationals). ($Card(I) > card(\boldsymbol{Q} \cup X) = \aleph_0$ suffices for the proof.) □

We choose the x_i for the above construction of the a_i in Definition 7.4.6 according to this fact.

Suppose that ϕ, ψ are not equivalent, but $\nu(\phi) = \nu(\psi)$. Thus, there is an assignment t s.t. $t(\phi) \neq t(\psi)$. So $\bigcup\{g(t)\colon t \in Val(\phi)\} \neq \bigcup\{g(t)\colon t \in Val(\psi)\}$ (wlog. all t with the same domain $p_0 \ldots p_n$, and n chosen least s.t. the assumption is valid), but $\nu(\phi) = \nu(\psi)$. Thus, $\nu(\phi) = \Sigma\{\Pi\{y_{i,j} : j = 0,n\}\colon i = 0,m\}$, $\nu(\psi) = \Sigma\{\Pi\{y'_{i,j} : j = 0,n\}\colon i = 0,m'\}$, where the $y_{i,j}$, $y'_{i,j}$ are either x_j or $1-x_j$. After multiplication, the equation looks like this: $s_1+\ldots+s_k = t_1+\ldots+t_l$, the s_u and t_u are of the form: 1 or $+/-\ x_{r_1}* \ldots * x_{r_h}$, and each x_j occurs at most once in each summand. After cancelling summands of the same form that occur on both sides of the equation, x_n will still occur in at least one of the summands, as n was chosen least. So, we can solve the equation (linear in x_n) for x_n and have $x_n = f(x_0 \ldots x_{n-1})$, where f is composed of addition, subtraction, multiplication, division — contradicting Fact 7.4.12. As the x_i can be chosen within any distance> 0 from a desired value, choosing x_i according to this fact is no real restriction. We have thus obtained our injectivity result and shown

Proposition 7.4.13

Let $p_i : i \in \omega$ be given a probability $x_i \in (0, 1)$, chosen according to Fact 7.4.12, then this gives rise naturally to probabilities $\nu(\phi)$ for any formula in $\mathcal{L}$, such that (1) – (6) of Lemma 7.4.9 are valid, and ($\phi \leq \psi$ and $\psi \leq \phi$) iff $\vdash \phi \leftrightarrow \psi$. In other words, this defines a total (and natural) extension of the natural order of the Lindenbaum-Tarski algebra $\boldsymbol{L}$, and, in addition, ($[\phi] \leq [\psi]$ and $[\psi] \leq [\phi]$) iff $[\phi] = [\psi]$. □

Chapter 8

Integration

8.1 Introduction

Four questions need an answer when we want to integrate several logical formalisms:

- what do we want to integrate?
- in what form do we want to integrate things?
- on what levels do we want to integrate things?
- how do we do it?

First, we have presented in Chapter 2 a number of types of common-sense reasoning, which we could mostly describe as some kind of generalized modal logic approach, i.e. based on some kind of model choice, with perhaps some pre- or post-processing. It is thus natural to try to integrate several such generalized modal logics. The choice of the reasoning types depends on the problem at hand, there cannot be an universal answer to this question.

Second, we can either choose to take those logics as they are mostly presented, often by some inference relation $\mid\sim$, or we can try to put model choice into the language as a true modal operator. We will take the second approach, and will give below reasons why.

Third, we saw that we can look, e.g. at preferential reasoning on several levels: we can work with "normal" sets, or we can put the relation directly into the language. So, whereas the "usual" level is the abstract one of reasoning with normality, the finer level is the one where we look directly at the relation. On the other side, we can also, higher up, reason about our reasoning, and, as we saw, we need not do much more: Once we code arguments by trees or sequences of formulas, with "gaps" noted as sets or, more abstractly, sizes, we are not very far from the object level reasoning — just trees, whose nodes are formulas, and whose edges are labelled with formulas, with perhaps some size annotated. Thus, we may have the "usual" level of normality, a finer level of the generating structure, and a higher level of reasoning itself.

Fourth, we will see that integration presents some quite subtle problems, but is, once we are aware of these problems, in principle quite straightforward. We will also discuss here some problems with and properties of pure classical and of classical modal logic which present themselves naturally in this context. These questions will be addressed in Section 8.3.

More precisely, we first discuss there some pecularities of classical modal logic, and show also that "empty" modal logic is nothing else than pure propositional logic (Fact 8.3.1). We then show that considering classical operators as modal operators reveals a funny property: usual axiomatizations do not suffice for a unique characterization of the corresponding set operators of models (Example 8.3.1). Finally, we show that composing operators in an overly careless way can lead to exactly the same problems we already saw when neglecting definability preservation: we might work with different things on the level of description and the level of the underlying structure (see Section 8.3.3). Once we take a little care, however, the problem of composition becomes trivial, the result is given in Lemma 8.3.4.

We give a summarizing "answer" to the first question in Section 8.2, where we recall, in the form of a table, the types and and basic concepts as discussed in Chapter 2.

We will turn in a moment to a discussion of "rules vs. object language", i.e. the second of above questions. We feel free to change from formulas to model sets without saying — it should always be clear what is meant.

Before, we would like to underline a very important point, which might be easily overlooked. Take a sentence like "normally, A believes ϕ". "A believes ϕ " can be coded as usual by a subset W_A of the set of worlds W (those A thinks plausible), and in all those worlds ϕ holds. What about "normally"? There seem to be two natural, and very similar, ways, both make "normally" essentially a second order quantifier. The first is to let "normally" range directly over A's beliefs, i.e. over the W_A's. Then in all W_A's $\in N$ and $w \in W_A$ $w \models \phi$ holds, i.e. $w \in W_A \in N \Rightarrow w \models \phi$. The second is to introduce "viewpoints", i.e. we relativize W_A to w's, $W_{A,w}$, and let N range as usual over W, and now $w \in N \Rightarrow (w' \in W_{A,w} \Rightarrow w' \models \phi)$.

In both cases, N does not really range over W, but over the W_A's. A closer analysis of the problem at hand will show whether N (or, in other situations, beliefs, etc.) should range over worlds (elements), or over subsets defined by other operators. Of course, there might be higher levels of nestedness, which will be treated in an analogous way.

8.1.1 Rules or object language?

We think there is a number of reasons to put many things into object language. First, an example: We can either introduce a rule $\alpha \mid\!\sim \beta$ for "normally, if α, then β", or, we can put a modal operator ∇ into the object language, which singles out the normal cases (in the simple variant of a principal filter), and $\alpha \mid\!\sim \beta$ becomes $\vdash \nabla\alpha \rightarrow \beta$. The object language variant has several advantages:

(1) We have to make the "grammatical role" precise: are the normal cases a subset of all cases, is it a set of subsets (a nonprincipal filter), is it a function, which not only assigns to α a subset of α, but also to (all?) subsets α' of α a subset α'' of α', etc.

(2) The language is more expressive, we can have negated formulas, nested formulas, boolean combinations, etc., we can say a lot more than in "flatland".

(3) We obtain contraposition, which seems a sign of quality of a logic: If something goes wrong, we have an indication what happened, it puts things onto the table. If $\alpha \mid\!\sim \beta$ and $\neg\beta$, then we cannot conclude $\neg\alpha$, and we do not really know what to do. The formula $\nabla\alpha \rightarrow \beta$ tells us exactly, what went wrong: Either $\neg\alpha$ was the case, or α, but not $\nabla\alpha$.

Contraposition seems all the more important for defeasible reasoning, as, per definitionem, we can be wrong, so, we will often have to revise, and contraposition is a simple form of revision. We can develop this further: defeasible reasoning formalisms should carry with them their own revision processes — we need a fall back strategy if bold reasoning fails. We think adequate defeasible reasoning should at least have contraposition.

Of course, we have to ask for the price to pay. As a matter of fact, it is very small. The basic algebraic representation results are almost all we need, we can screw them into a representation result for arbitrary object language formulas with a uniform higher abstract nonsense triviality — provided we take some precautions.

We turn to question three.

8.1.2 Various levels of reasoning

Let us carry this reflection one step further. Suppose we put normality into the language, but also the preference relation which generates it. We may then reason on two levels: If a more abstract level suffices, we reason with normality, if we need more detail, we argue with the relation. This is a kind of complexity hiding, so useful in other domains, e.g. computer network organization. The relation explains normality, and this can be used as an argument. If the opponent does not accept normality, we can explain to him our reasons, and he can challenge or acquiesce on the more elementary level. This is like the programming language C, one of whose advantages is, if we understood correctly, that we can do high level programming, as well as handling single bits and registers. Normality hides the underlying complexity of the preference relation, we enter into it only when needed. The double layers give us the finesse when we need it, without drowning us in details when they are not needed, it makes us more flexible.

We see an argument as a concatenation of implications in one or more logics of perhaps differing strengths, "glued" together by perhaps some ad hoc bridges over gaps in the reasoning. The strength of the argument is the width of the gaps and the strengths of the individual implications (which can be stronger than classical logic, to allow for revision). For instance, we read the inheritance diagram $A \to B \to C$ as: normal A's are B's, and normal B's are C's, and, by default, normal A's behave like normal B' s, so they are C' s, too. The problem is, of course, the default reasoning. The default reasoning has to be weaker than normality in above example: If N expresses normality, and N' the default, then the diagram $A \to B \to C$ says: $N(A) \subseteq B$, $N(B) \subseteq C$, $N'(N(A) \subseteq B \to N(A) \subseteq N(B))$ — we prefer situations where normal A's are normal B's in the language of preference. We may have ad hoc bridges over gaps, or a pricipled treatment of such gaps. In the first case, we will try to fill in better reasoning, in the second, this may be all we have at our disposal.

We have already seen in Section 2.2.8 that it seems very difficult to find a definitive theory of argumentation. Consequently, it seems better not to treat the theory of argumentation as a constant, but will allow reasoning about argumentation. This is, of course, a form of metareasoning, just as reasoning about normality is also a form of metareasoning. This is then a further step into the same direction: we gain flexibility by putting more things into our language, the structure of several levels will keep things relatively simple, as we will use the additional complexity only when we need it.

We can summarize: The coarsest argument is the result of the argument. This is explained by the argument(s) which support it — a finer level, and the underlying theory of argumentation, which compares arguments and defines their strength. The implications are explained by their logic and the finer structure underlying them, like the preference relation underlying the notion of normality.

Beyond revision and argumentation, having two or more levels of argumentation gives us also a form of "quick and dirty" reasoning. We can do first high level reasoning, and verify later on the more detailed level whether it was justified. Thus, it opens a way for dynamic reasoning. At the same time, contradictions at a lower level might not be felt at a higher level: The (perhaps contradictory) information $a < b < a$ may have no consequences for a particular $\alpha \mid\!\sim \beta$ in preferential reasoning.

Obviously, the more things we put into the language instead of coding them into a fixed formalism, the more we gain flexibility.

Connection between base and higher concepts

If we put the base concept, like a preference relation, itself into the language, we can define the higher concept from it, as we can define $N(X)$ from $<$, in the style of usual representation constructions. At most, we will have to add some rudimentary set theory or primitive FOL over a fixed finite universe, to have some quantification. The laws about N will follow automatically from the laws about $<$. Usually, we will have more than the laws needed for representation, as we work with a special $<$, which will normally have additional properties.

If more than one higher concept is built on the same base concept, we may have redundant information, and have to be careful that the two sources of information about the base concept do not contradict each other. E.g. revision and counterfactuals are based on distance, and if the distance is to be the same (this is not necessary, but may be the case), they cannot be constructed independently. If we have the base concept in the language, the construction is trivial. In the other case, one higher concept will often allow several different constructions of the base concept (several distances), which generate the same higher concept. Only one of those has to be compatible with the second higher concept.

The extension towards meta-reasoning

To speak about reasoning, we need the reasons for our inferences, their certainty, and the gaps between them. The reasons, e.g. for normality reasoning are size or a relation between models, this can be noted in a simple way, certainty is a set and its size, utility somewhat more complicated, but also within our framework. The gaps are described again by size, or distance.

We work just as in set theory, where a hierarchy is constructed. At each level, we speak about the lower levels. The construction is uniform, but open for upward development. We are arbitrarily flexible, without the problems of reflexive reasoning, as levels are kept separate. We reason first about objects, then reason about our reasoning about objects, etc. In higher levels, we can repair problems of the lower level, by showing weaknesses of the lower argumentation. For example, if an argumentation is attacked, and the attack seems justified, we can try to find a reason: an under- or overestimated distance, or size. Or, on the contrary, we can defend it. Thus, we do not need perfect low level reasoning, instead, we can do repairs on the fly.

8.2 Reasoning types and concepts

We summarize now the different reasoning types and the concepts we have based them on. This is mostly a summary of Chapter 2, more or less in form of a table.

(1) Classical modal logic

(1.1) Semantics: A binary relation between individual models read backwards

(1.2) Keywords: Binary relation between models.

(1.3) The grammatical role: $\Box(X)$ is a set, going backwards.

(2) Theory revision

(2.1) Semantics:

(2.1.1) A relation between formulas (epistemic entrenchment relation).

(2.1.2) A distance between models or model sets read collectively. This defines in a natural way an epistemic entrenchment relation.

(2.1.3) A notion of size of individual models or model sets.

(2.2) Keywords: Binary relation between formulas (= model sets), distance between models, size of models.

(2.3) The grammatical role: $A \mid B \subseteq B$, in the algebraic limit case $A \mid B \subseteq \mathcal{P}(B)$, this can sometimes be reduced again to $A \mid B \subseteq B$. It might be indexed, expressing the viewpoint.

(3) Counterfactuals

(3.1) Semantics:

(3.1.1) A distance between models read individually. The distance may be chosen differently from each viewpoint, it can be unified by admitting copies of models.

(3.1.2) A variant is to neglect only those elements which are "behind" a closer one, where "behind" and "between" are defined from a given distance.

(3.2) Keywords: Distance between models.

(3.3) The grammatical role: As for revision: $A \uparrow B \subseteq B$, in the algebraic limit case $A \uparrow B \subseteq \mathcal{P}(B)$, this can sometimes be reduced again to $A \uparrow B \subseteq B$. It might be indexed, expressing the viewpoint.

(4) Normality

(4.1) Semantics:

(4.1.1) A binary relation of preference between models, resulting in coherent filter systems of various strengths.
Questions:

(4.1.1.1) Properties of the relation.

(4.1.1.2) Strengths of the coherence properties.

(4.1.2) A variant is to consider the center of sets, defined by (sums of) distances. A typical element is then an element which is not marginal. A slight modification is to cut the whole set into clusters and to consider centers of the individual clusters. Clustering might depend on the viewpoint, from farther away, we may put more things together.

(4.1.3) A notion of size of model sets, coded by filters.
Questions:

(4.1.3.1) Relativity of size ("from afar, distinctions blur").

(4.1.3.2) Comparison: "how much bigger than small is big?", etc.

(4.2) Keywords: Binary relation between models, distance between models, clusters, size of subsets (perhaps relativized), (weak) filters, reasoning about size.

(4.3) The grammatical role: $N(X) \subseteq X$, in the case of nonprincipal filters $N(X) \subseteq \mathcal{P}(X)$. $X < B$ or $X \sim B$ expressing size relation. Rules like $X < Y$, $X' < Y \to X \cup X' < Y$.

(5) Update

(5.1) Semantics: The concept of natural inertia is coded by a distance between states, and evaluated individually. If necessary, sums of distances (and other entities) are considered.

(5.2) Keywords: Distance between models, sums.

(5.3) The grammatical role: Depending on situation, e.g. $< A * B * C > \subseteq C$ — where $*$ is a suitable operation.

(6) Certainty of information

(6.1) Semantics:

(6.1.1) Certainty can be measured by an epistemic entrenchment relation, which is definable via a distance. This works (by definition of epistemic entrenchment relations) only for information beyond classical certainty.

(6.1.2) As for normality, we can measure the deviant cases by a relation or an abstract notion of size.

(6.1.3) For information transfer, we can take the distance of transfer, e.g. in analogical reasoning, we can consider the distance between source and destination, or, when both sets are almost the same, the similarity of the two sets, measured by the relative size of the symmetric difference.

(6.2) Keywords: Binary relation between model sets, size of model sets, distance between model sets, similarity of model sets.

(6.3) The grammatical role: Certainty can be a set (of neglected cases), a distance of information transfer $d(X, Y)$, a ranking wrt. a base set (epistemic entrenchment relations).

(7) Utility of information (and quality of an answer)

(7.1) Semantics: Utility will be a product of the size of the model set and individual utility. For each model, we have for (some or all) formulas a utility value for thinking the model satisfies the formula. The utility of ϕ wrt. the question ψ is then the sum of the individual utilities for ψ of all $\phi-$models. Sums and products are calculated in some more or less rough way.

(7.2) Keywords: Size of model sets, values of formulas for individual models, sums, products.

(7.3) The grammatical role: (X, Y) is, e.g. a size, the benefit/prize for thinking that $X \subseteq Y$.

(8) Approximation (of sets)

(8.1) Semantics: Simple sets. They are either singled out as such by some predicate, or, e.g. as convex sets wrt. some distance (or composed by a small number of convex sets), e.g. the Hamming distance. Simple sets can be used to approximate an other set from the inside, the outside, or mixed, the difference can be measured, one can take the best candidate if this exists, etc.

(8.2) Keywords: Simple sets, distance, size of sets.

(8.3) The grammatical role: $\mathcal{Y} \subseteq \mathcal{P}(U)$ is the set of simple sets. Grading can be achieved by counting the simple components of a set.

(9) Reasoning by interpolation

(9.1) Semantics: We take extreme cases (measured by a distance) of a set, reason with them (i.e. with complete theories), and interpolate the result to the rest. This is the dual to considering centers. It is a somewhat more cautious approach than that working with centers, as we consider perhaps more elements than in the former case.

(9.2) Keywords: Distance between models.

(9.3) The grammatical role: Like $N(X)$ above.

(10) Defaults

(10.1) Semantics: Defaults of the type : ϕ do not only say that as many elements as possible satisfy ϕ, but also that for any ψ, as many elements satisfy $\psi \wedge \phi$ as possible. A default is an operator working on a whole family of sets. We can exclude artificial sets, or, admit only admissible ones as destination of the information transfer.

Defaults make as many as possible true, we may have subideal cases, where not all defaults hold, but as many as possible.

(10.2) Keywords: Admissible sets, operator on family of sets, subideal cases.

(10.3) The grammatical role: A single default (or a set of defaults, if several defaults are treated simultaneously) generates a partial order on admissible sets, where the order is determined by the number of defaults which hold for each element of the set. The result is the disjunction of the best candidates.

(11) Analogy and induction (and general transfer of information)

(11.1) Semantics: We have a more general information transfer, and weakening in the confidence or certainty over longer distances of transfer should be possible. Cumulativity gives a natural limit to the transfer, this need not always exist. Transfer may only be authorized from and to admissible sets (reference classes for the origin). Perhaps not all information will be transferred. If we suspect a common mechanism, this should influence our reasoning: The common mechanism causes in principle some property, but other influences may interfere in its manifestation. So, in the background, we may have a theory of causation or update.

(11.2) Keywords: Certainty, distance between sets, admissible sets, admissible information, limits of transfer, theory of causation.

(11.3) The grammatical role: In a complicated case, we have the following structure: For an initial set X, we have a set $\mathcal{Y}$ of information that can be transferred from X, a set $\mathcal{X}'$ of possible destinations, and for each pair $< Y, X' >$ a certainty C (which may be a distance between X and X', etc.).

(12) Inheritance and argumentation

(12.1) Semantics: The steps of argumentation are done in one or several logics. The gaps are bridged by more or less ad hoc reasoning (e.g. inference from $N(X)$ to X), which determines the strength of the argument. The choice of the reference classes can be done by — among other things — specificity (closeness or size of set difference), which can in turn be determined by other arguments. A fully satisfactory theory of argumentation seems difficult to obtain, and is better put into object language itself, to permit reasoning about it. Dynamic reasoning can be seen as argumentation with progressively finer or more reliable information. The starting point of reasoning should not be important, as the measure of reliability or graduation should automatically lead to the best information we have.

(12.2) Keywords: Choice of reference class, admissible sets, specificity, distance between sets, independence of arguments.

(12.3) The grammatical role: This does not apply here, but only individually.

Summary of the main keywords we found:

(1) Binary relations between

  (1.1) individual models,

  (1.2) model sets, e.g. similarity, questions of transitivity, etc.,

(2) (sometimes relativized: seen from far away, distances may shrink) distances between

  (2.1) individual models resulting, e.g. in: clusters, simple sets

  (2.2) model sets

(3) (sometimes relativized: seen from far away, differences may shrink) size of

  (3.1) individual models,

  (3.2) model sets and subsets, e.g. concepts of (weak) filters,

  (3.3) reasoning about size, e.g. sums and products, e.g. sums may be defined via unions or sequences,

(4) values of formulas for individual models,

(5) admissible sets and choice of reference class

  (5.1) admissible information for transfer,

(6) operators on families of sets

  (6.1) subideal cases.

And, somewhat apart:

(7) theory of causation.

8.3 Formal aspects

The following introductory Sections 8.3.1 and 8.3.2 are side remarks, and may be omitted in a first reading. They contain remarks on the (almost nonexistent) expressive power of pure modal logic, and on the fact that classical negation and disjunction are somewhat underdetermined.

8.3.1 Classical modal logic

Classical modal logics goes backwards

Traditional modal logic looks backwards: The operator $\Box$ associates to ϕ the formula $\Box\phi$, which describes all models m, s.t. in all m' reachable from m, ϕ holds. We go backwards, against the sense of the relation, from ϕ to $\Box\phi$. In preferential logic, we go forward, from ϕ to $N(\phi)$. Intuitively, we think it is more natural to go in the sense of the relation. But there is still another point: If we write $\psi \to \Box\phi$, then this is implicitly quantified universally, and thus individually: in all models m, where ψ holds, $\Box\phi$ holds. This is NOT some global property of $\Box\phi$, but defined by validity for every m s.t. $m \models \Box\phi$ — consequently, $\Box$ will be monotone, in the sense: If $\psi \to \psi'$, and $\psi' \to \Box\phi$, then $\psi \to \Box\phi$, too. (The dual operator $\Diamond$ does the same, only existentially.) Thus, the difference between individual (local) and global evaluation is somewhat swept under the carpet, and decided implicitly — which is unfortunate. These are, of course, trivialities, but one should perhaps write them down nonetheless.

(When we try to do, e.g. preferential models backwards, as for traditional modal logic, we will usually have an operator $\mathcal{P}(M_{\mathcal{L}}) \to \mathcal{P}\mathcal{P}(M_{\mathcal{L}})$, as many different ϕ' may have the same $N(\phi)$.)

When we look now at the basic laws of modal logic:

$$\frac{\phi}{\Box\phi}$$

and

$$\Box\phi \wedge \Box(\phi \to \psi) \to \Box\psi$$

we see that the first says that $\Box$ describes a subset of the model set — if something holds everywhere, it will hold in the subset described by $\Box$ — and that the second essentially expresses closure under logical consequence (which will hold for any logic defined by model sets). As a matter of fact, it also expresses an important property, which we may call monotony, and which reads in set terms: $\Box(A \cup B) \supseteq (\Box A) \cup (\Box B)$. Note that this corresponds more to individual distances as in counterfactuals, than to collective ones as in revision.

As classical modal logic is not in the center of our interest, we just briefly conclude with a property which is connected to our main interests.

Coding propositional logic by modal operators

To demonstrate that pure classical modal logic without any additional conditions is nothing more than propositional logic, we show how to express additional propositional variables by modal structure.

Fact 8.3.1

We can code arbitrarily many new propositional variables in modal structures.

Proof:

(Sketch) We work with one propositional variable p and construct a binary tree of $p-$ and $\neg p-$models by induction. xRy will hold iff x is a successor of y.

The root, level 0: We take a $p-$model (this is arbitrary).

Construction of level $n+1$ from level n : For every point y on level n, we make two successors, a $p-$ and a $\neg p-$model.

We stop the construction at some height k, which depends on the number of propositional variables we want to code. Level n will code n propositional variables.

Properties:

(1) At level n $(n > 0)$ $\diamond \ldots \diamond p$ (n times $\diamond$) holds. This will not hold at lower levels.

(2) Each node has different modal properties. Those from different levels can be distinguished by property (1). Consider now two nodes, x and y on the same level n. They are on different paths from the root. But two different paths correspond to different modal formulas. Suppose the first difference of the two paths is on level m. Say, the path to x goes through a $p-$model, the path to y through a $\neg p-$model. If $m = n$, then $x \models p$, $y \models \neg p$. If $l := n - m > 0$, then $x \models \diamond \ldots \diamond p$, and $y \models \diamond \ldots \diamond \neg p$, with l many $\diamond$ in both cases.

(3) Consequently, in this structure, each node on the interesting level has different logical properties, which cannot be equivalent, as our structure is a legal modal structure, and equivalences in pure modal logic would have

to hold in our structure, too. So, each node corresponds to a different propositional model in a classical propositional language with k variables. A natural correspondence is, e.g.: $x \models p_i$ iff $x \models \Diamond \ldots \Diamond p$, with $k-i$ $\Diamond$, i.e., if the path to x goes at level 1 through a $p-$model, then $x \models p_1$, if it goes through a $\neg p-$model, then $x \models \neg p_1$, etc. More generally, p_i is translated by $\Diamond \ldots \Diamond p$, with $k-i$ $\Diamond$, and $\neg p_i$ by $\Diamond \ldots \Diamond \neg p$.

□

8.3.2 Classical propositional operators have no unique interpretation

In our view, the propositional operators $\neg$, $\vee$, etc. are modal operators. It is therefore natural to try two things: First, to make a completeness proof for classical propositional logic along the same lines as we do for other logics. Second, to try to show that the axioms of classical propositional calculus characterize the model set operators -, $\cap$, $\cup$, etc. in a unique way.

Both are not true. The first fails, because we use in the logical parts of the proofs for other logics precisely soundness and completeness for classical logic — which we would try to demonstrate here. The second fails, as -, $\cap$, etc. clearly obey these laws, but, as we shall see now, they are not the only ones.

We make the idea precise, then prove a few facts, and give a counterexample, which shows that we have some freedom with $\neg$ and $\vee$.

The idea: A propositional formula is interpreted semantically as a set expression with operators $\cap$, $\cup$, etc. An axiom is interpreted as saying that the corresponding set expression describes the whole universe, i.e. it holds everywhere, and the rule Modus Ponens ϕ, $(\phi \rightarrow \psi) \Rightarrow \psi$ as saying that $\phi \cap \phi \rightarrow \psi \subseteq \psi$ in shorthand. (We do not distinguish between an expression and its interpretation, this should not cause any problems.) This interpretation seems natural.

We now show some facts we can deduce from the axioms and rules of propositional calculus so interpreted, i.e. where axioms describe the universe, and Modus Ponens is monotone wrt. set inclusion. Thus, we make free use of all theorems of propositional calculus.

We use the following axiomatization of propositional calculus:

Definition 8.3.1

(A1) $\phi \rightarrow (\psi \rightarrow \phi)$,

(A2) $(\phi \rightarrow (\psi \rightarrow \sigma)) \rightarrow ((\phi \rightarrow \psi) \rightarrow (\phi \rightarrow \sigma))$,

(A3) $(\neg\phi \rightarrow \neg\psi) \rightarrow (\psi \rightarrow \phi)$,

($\wedge$) $\phi \wedge \psi \leftrightarrow \neg(\phi \rightarrow \neg\psi)$,

($\vee$) $\phi \vee \psi \leftrightarrow \neg\phi \rightarrow \psi$,

Modus Ponens.

Fact 8.3.2

From these axioms we conclude:

(1) $\phi \cap (\phi \rightarrow \psi) \subseteq \psi$,

(2) $T \vdash \phi \Rightarrow \bigcap T \subseteq \phi$,

(3) $T \vdash \phi \rightarrow \psi \Rightarrow \bigcap T \cap \phi \subseteq \psi$,

(4) $\vdash \phi \leftrightarrow \psi \Rightarrow \phi = \psi$,

(5) $\phi \wedge \psi = \phi \cap \psi$,

(6) $\phi \cup \psi \subseteq \phi \vee \psi$ (but not the converse, as we will see!),

(7) $\neg\neg\phi = \phi$ (but not that $\neg$ has to be interpreted as the set-complement, as we will see!).

Proof:

(1) By Modus Ponens.

(2) By induction on the complexity of the proof.

(3) By the deduction theorem.

(4) By (3).

(5) $\phi, \psi \vdash \phi \wedge \psi$, so $\phi \cap \psi \subseteq \phi \wedge \psi$ by (2). By $\vdash \phi \wedge \psi \rightarrow \phi$ and (3) $\phi \wedge \psi \subseteq \phi$, likewise for $\phi \wedge \psi \subseteq \psi$.

(6) By $\phi \vdash \phi \vee \psi$, $\psi \vdash \phi \vee \psi$, and (2).

(7) By $\vdash \neg\neg\phi \leftrightarrow \phi$ and (4).

□

We turn to the counterexample.

Example 8.3.1

Let $U := U' \cup \{*\}$, $\mathcal{X} := (\mathcal{P}(U') \cup \{U\}) - \{U'\}$. Thus, $\mathcal{X}$ is "almost" $\mathcal{P}(U')$, only U' is replaced by $U := U' \cup \{*\}$, so U is the only element of $\mathcal{X}$ to contain $*$.

Define for $X \in \mathcal{X}$:

$$\neg X := \begin{cases} U \text{ if } X = \emptyset \\ \emptyset \text{ if } X = U \\ U' - X \text{ if } \emptyset \subset X \subset U' \end{cases}$$

Note that $\mathcal{X}$ is closed under $\cap$ and $\neg$, and that $\neg$ is not always the complement.

Define

$X \wedge Y := X \cap Y$ (as we are obliged by Fact 8.3.2, (5)),

$X \vee Y := \neg(\neg X \wedge \neg Y)$ — thus, $\vee$ can be more than $\cup$,

$X \rightarrow Y := \neg X \vee Y$,

$X \leftrightarrow Y := (X \rightarrow Y) \wedge (Y \rightarrow X)$.

As $\mathcal{X}$ is closed under $\neg$ and $\wedge$, it is closed under the other operators, too.

We now show that the operators so defined satisfy the axioms and rule of classical propositional calculus. The proofs are tedious, but straightforward.

Fact 8.3.3

In Example 8.3.1, the following hold:

(1) $\neg\neg X = X$,

(2) $(X \vee Y) \vee Z = X \vee (Y \vee Z)$,

(3) $X \wedge Y = \neg(\neg X \vee \neg Y)$,

(4) $X \wedge Y \rightarrow Z = X \rightarrow (Y \rightarrow Z)$,

(5)

$$X \vee Y := \begin{cases} X \cup Y \text{ if } X \cup Y \subset U' \\ \text{and} \\ U \text{ if } U' \subseteq X \cup Y \end{cases}$$

(5a) $X \vee Y = U$ iff $U' \subseteq X \cup Y$,

(6) $X \rightarrow Y = U$ iff $U' \subseteq \neg X \cup Y$,

(7) $X \vee \neg X = U$,

(8) $X \wedge (Y \vee Z) = (X \wedge Y) \vee (X \wedge Z)$,

(9) $\emptyset \vee X = X$,

(10) $X \cap (X \rightarrow Y) \subseteq Y$ (Modus Ponens),

(11) $X \wedge Y \rightarrow X = U$,

(12) $X \rightarrow (Y \rightarrow X) = U$ (Axiom 1),

(13) $X \rightarrow Y = U$ iff $X \subseteq Y$,

(14) $(\neg X \rightarrow \neg Y) = Y \rightarrow X$,

(15) $(\neg X \rightarrow \neg Y) \rightarrow (Y \rightarrow X) = U$ (Axiom 3),

(16) $X \vee (Y \wedge Z) = (X \vee Y) \wedge (X \vee Z)$,

(17) $\neg X \vee (X \wedge \neg Y) = \neg X \vee \neg Y$,

(18) $(X \rightarrow Y) \rightarrow (X \rightarrow Z) = X \wedge Y \rightarrow Z$,

(19) Axiom 2 holds,

(20) $X \wedge Y = \neg(X \rightarrow \neg Y)$ (Axioms for $\wedge$),

(21) $X \vee Y = \neg X \rightarrow Y$ (Axioms for $\vee$).

Proof:

(1) trivial by examining the cases.

(2) $(X \vee Y) \vee Z = \neg(\neg(X \vee Y) \cap \neg Z) = \neg(\neg\neg(\neg X \cap \neg Y) \cap \neg Z) =_{(1)} \neg(\neg X \cap \neg Y \cap \neg Z)$. The other part works the same way.

(3) $\neg(\neg X \vee \neg Y) = \neg\neg(\neg\neg X \wedge \neg\neg Y) = X \wedge Y$

(4) $X \wedge Y \rightarrow Z = \neg(X \wedge Y) \vee Z =_{(3)} \neg\neg(\neg X \vee \neg Y) \vee Z = \neg X \vee \neg Y \vee Z$. $X \rightarrow (Y \rightarrow Z) = \neg X \vee (Y \rightarrow Z) = \neg X \vee (\neg Y \vee Z)$, finish with (2).

(5) No other cases are possible, as $* \in X \cup Y$ implies $X \cup Y = U$, and thus $X = U$ or $Y = U$, as U is the only $A \in \mathcal{X}$ with $* \in A$.

Case 1: $U' \subseteq X \cup Y$: If $X = U$, $\neg X \cap \neg Y = \emptyset \rightarrow X \vee Y = U$. If $Y = U$, the same way. If $X = \emptyset$, then $Y = U$, if $Y = \emptyset$, then $X = U$. If $\emptyset \subset X, Y \subset U'$, then $\neg X = U'$–X, $\neg Y = U'$–Y, and, as $U' \subseteq X \cup Y$, $\neg X \cap \neg Y = \emptyset$, so $X \vee Y = U$.

Case 2: $X \cup Y \subset U'$: If $X = \emptyset$, then $\neg X = U$, and $X \vee Y = \neg\neg Y = Y = X \cup Y$, likewise for $Y = \emptyset$. Suppose now $X, Y \neq \emptyset$. Then $\neg X = U'$–X, $\neg Y = U'$–Y, so $\neg X \cap \neg Y \neq \emptyset$, so $\neg(\neg X \cap \neg Y) = U' - (\neg X \cap \neg Y) = X \cup Y$.

(5a) $X \cup Y \subset U' \rightarrow X \vee Y = X \cup Y \subseteq U' \neq U$.

(6) By (5a).

(7) $U' \subseteq X \cup \neg X$ and (5).

(8) Case 1: $U' \subseteq Y \cup Z$, so $Y \vee Z = U$.

Case 1.1: $X \subset U'$: The left hand side $= X$, the right hand side $= (X \cap Y) \cup (X \cap Z) = X \cap (Y \cup Z) = X$.

Case 1.2: $X = U$: Trivial.

Case 2: $Y \cup Z \subset U'$, so $Y \vee Z = Y \cup Z$. Thus $(X \cap Y) \cup (X \cap Z) \subset U'$, and on the left $X \cap (Y \cup Z)$, on the right $(X \cap Y) \cup (X \cap Z)$.

(9) $\emptyset \vee X = \neg(\neg\emptyset \cap \neg X) = \neg(U \cap \neg X) = \neg\neg X = X$.

(10) $X \cap (X \rightarrow Y) = X \cap (\neg X \vee Y) =_{(8)} (X \cap \neg X) \vee (X \cap Y) = \emptyset \vee (X \cap Y)$ $=_{(9)} X \cap Y \subseteq Y$.

(11) $X \wedge Y \rightarrow X = \neg(X \wedge Y) \vee X = \neg X \vee \neg Y \vee X = U$.

(12) $X \rightarrow (Y \rightarrow X) =_{(4)} X \wedge Y \rightarrow X = U$.

(13) We use (6), and show $X \subseteq Y$ iff $U' \subseteq \neg X \cup Y$. The following four cases are trivial: $Y = U, Y = \emptyset, X = \emptyset, X = U$. In the other cases, $\emptyset \subset X, Y \subset U'$, and $\neg$ for X and Y behaves like the usual complement.

(14) $\neg X \rightarrow \neg Y = \neg\neg X \vee \neg Y = \neg Y \vee X = Y \rightarrow X$.

(15) By (13) and (14).

(16) Case 1: $X \vee Y = U$, so $U' \subseteq X \cup Y$.

Case 1.1: $X \vee Z = U$, so $U' \subseteq X \cup Z$, so $U' \subseteq X \cup (Y \cap Z)$.

Case 1.2: $X \vee Z \neq U$, so $X \cup Z \subset U'$, $X \vee Z = X \cup Z$, and $X \cup (Y \cap Z) \subset U'$, so $X \vee (Y \cap Z) = X \cup (Y \cap Z) = (X \cup Y) \cap (X \cup Z)$. But $U' \subseteq X \cup Y$, and $X \cup Z \subset U'$, so $(X \cup Y) \cap (X \cup Z) = X \cup Z = X \vee Z = (X \vee Z) \cap U = (X \vee Z) \wedge (X \vee Y)$.

Case 2: $X \vee Y \neq U$, $X \vee Z \neq U$, so $X \cup Y \subset U'$, $X \cup Z \subset U'$, so $X \cup (Y \cap Z) \subset U'$, and $X \vee Y = X \cup Y$, $X \vee Z = X \cup Z$, and $X \vee (Y \wedge Z) = X \cup (Y \cap Z)$, and usual distributivity shows the rest.

(17) $\neg X \vee (X \wedge \neg Y) = (\neg X \vee X) \wedge (\neg X \vee \neg Y) = U \cap (\neg X \vee \neg Y) = \neg X \vee \neg Y$.

(18) $X \wedge Y \rightarrow Z = \neg X \vee \neg Y \vee Z$. $(X \rightarrow Y) \rightarrow (X \rightarrow Z) = \neg(X \rightarrow Y) \vee \neg X \vee Z$ $= \neg(\neg X \vee Y) \vee \neg X \vee Z = (X \wedge \neg Y) \vee \neg X \vee Z =_{(17)} \neg X \vee \neg Y \vee Z$.

(19) By (18) and (4).

(20) $\neg(X \rightarrow \neg Y) = \neg(\neg X \vee \neg Y) = X \wedge Y$.

(21) $\neg X \rightarrow Y = \neg\neg X \vee Y = X \vee Y$.

□

8.3.3 Combining individual completeness results

We would like to assemble now one big completeness result from individual completeness results.

Before we give sufficient — but often too strong — conditions for the procedure to work, we describe the problem, and difficulties which can arise in some situations. The situation is a little complicated, as we will try to integrate two translations with nested operators — different or not — and boolean combinations thereof, to end up in a homogenous language and characterization.

Suppose we have separate completeness proofs perhaps for "flat" situations only. Now, we want to combine them to a completeness result for arbitrary combinations of new operators. Let, e.g. for preferential structures an algebraic characterization be given: $\mu(X) \subseteq X$, $X \subseteq Y \rightarrow \mu(Y) \cap X \subseteq \mu(X)$. We have translated this into logic by the conditions (SC), (PR), etc. — under the caveat of definability preservation.

It is important to note that this translation is a two-step process. First, we establish a correspondence between a relation and a resulting choice function. Second, we establish a correspondence between a choice function and a logic.

In the second step, we had encountered the problem of definability preservation, which was due to the fact that things seemed logically the same, but were not so algebraically. We will see in a moment that there is a similar problem already in the first step of the combination.

Before we address this problem, we return to our general framework.

Our aim was to create a language of "generalized modal logics". For instance, $\phi \mid\sim \psi$ is to be translated into $\vdash N(\phi) \rightarrow \psi$, where $N(.)$ is the central part. Look now at the basic condition $X \subseteq Y \rightarrow \mu(Y) \cap X \subseteq \mu(X)$. It is tempting, but insufficient, to translate this into the formula $(\phi \rightarrow \psi)$ $\rightarrow (N(\psi) \wedge \phi \rightarrow N(\phi))$. Here is an example why such translations are in general insufficient: Let m be a fixed model, let $\not\vdash \phi \rightarrow \psi$, but "by chance" $m \models \phi \rightarrow \psi$. Then $m \models N(\psi) \wedge \phi \rightarrow N(\phi)$ would have to hold, too, which, of course, can be wrong in the intended translation — $N(\phi)$ and $N(\psi)$ need not have anything to do with each other. We need to express that the prerequisite $\phi \rightarrow \psi$ has to hold everywhere for the conclusion to hold, too (the conclusion will then automatically hold everywhere). For this purpose, we introduce a new modal operator $\spadesuit$ to tie local to global evaluation, with the meaning $\vdash$ (or, everywhere): $m \models \spadesuit\phi$ iff for all m' in the structure $m' \models \phi$, equivalently, iff $T \vdash \phi$.

The algebraic conditions we consider are simple set theoretical properties, they are boolean combinations of inclusions, $A \subseteq B$, etc., see Section 1.6.2. The translation is now done as follows:

$A \subseteq B$	$\spadesuit(\phi \to \psi)$
$\cap$	$\wedge$
$\cup$	$\vee$
$\boldsymbol{C}$	$\neg$
etc.	

We have solved this problem, and turn to the next.

We want to treat boolean combinations, and nested formulas. For instance, $N(N(\phi) \wedge \neg\psi)$ should be defined. Thus, we impose the following conditions on the domain $\mathcal{Y} \subseteq \mathcal{P}(X)$ — where X is intended to be the set of models of some language $\mathcal{L}$:

The algebraic operators $\mathcal{O}$ (e.g. μ), go from $\mathcal{Y}$ to $\mathcal{Y}$, $\mathcal{O} : \mathcal{Y} \to \mathcal{Y}$ — otherwise, we could not apply one to the result of the other. Moreover, $\mathcal{Y}$ has to be closed under the finite boolean operators $-$, $\cap$, $\cup$.

As algebraic operators do not care about syntactic reformulations, we add for each new modal symbol a corresponding statement, e.g. $\spadesuit(\phi \leftrightarrow \psi) \to \spadesuit(N(\phi) \leftrightarrow N(\psi))$.

We have solved (part of) the nestedness problem. The result of applying one modal operator to one or more formulas is again equivalent to a formula, and applying any modal operator to equivalent formulas gives equivalent results, it does not matter whether we work with level 1 formulas, or nested ones (counting the new operators), so results from "flatland" carry over to arbitrarily nested and mixed formulas.

We come now to the perhaps most subtle part of the translation — which we will solve with some overkill, just as we solved problems concerning the second translation by imposing (the overkill of) definability preservation.

We see the problem best with an example. E.g. in preferential structures, we had worked with copies. μ was defined by the condition that there is at least one copy which is minimal. So, there is a problem: $\mu(X)$ was won by looking at all copies of $x \in X$, but $y \in \mu(X)$ was won by looking at only some copies of y. So the algebraic μ and the relation operator differ. Y as argument and as result are not the same when looking at the relation, they are so when looking at the $\mu-$operator. When we worked with one operator, we had taken this implicitly into account, e.g. for the transitive case, we had worked with this property. It is not obvious what will happen when we combine several different operators. Consider the following example: We want to apply revision after normality, i.e. express $\phi * N(\psi)$, in algebraic terms, we consider $X \mid \mu(Y)$. If $\mu(Y) = Z$, this should give the same result as $X \mid Z$. Let now μ be defined by a preference relation $\prec$, and $\mid$ by distance d. Suppose $z \in Z$ has an infinite descending chain of ever closer copies, seen from X. Suppose that $z \in \mu(Y)$, but there is just one minimal copy of z. Then combining d with $\prec$ gives a result which is different from working directly with d and Z, the latter might, e.g., result in $X \mid Z = \emptyset$, the former in $X \mid \mu(Y) = \{z\}$. Thus, the underlying structures might have a finer discernation than the resulting set operators, just as the set operators might have a finer discernation than the logical operator. We solved the latter problem by imposing definability preservation (or admitting a small number of exceptions), we solve the former in a similar fashion: we impose that we work with one copy only in all representation constructions.

We summarize now the procedure and the result:

Lemma 8.3.4

Let $\mathcal{S}_1, \ldots, \mathcal{S}_n$ be structures which result in operators $\mathcal{O}_1, \ldots, \mathcal{O}_n$, each individually characterized by the conditions $\mathcal{C}_1, \ldots, \mathcal{C}_n$, and let $\boldsymbol{C}_1, \ldots, \boldsymbol{C}_n$ be their translations into logic as described above ($\subseteq$ becomes $\spadesuit(. \rightarrow .)$, etc., the algebraic $\mathcal{O}_i$ become the logical $\boldsymbol{O}_i$) with $\spadesuit(\phi \leftrightarrow \phi') \rightarrow \spadesuit(\boldsymbol{O}_i(\phi) \leftrightarrow \boldsymbol{O}_i(\phi'))$ added for all $\mathcal{O}_i$ (which are for simplicity here unary).

Let $\boldsymbol{D}$ be the set of formula definable sets of (propositional) models for a fixed language $\mathcal{L}$, let $\mathcal{O}_i : \boldsymbol{D}^n \rightarrow \boldsymbol{D}$, and let the structures work with one copy each. Then the conditions $\boldsymbol{C}_i$ characterize the modal operators $\mathcal{O}_i$ based on the structures $\mathcal{S}_i$ in the full object language, i.e. with nestedness and all finite boolean combinations.

(Of course, the structures have to be independent here, if, e.g. the same distance defines revision and update, they cannot be chosen independently.)

Proof:

(Trivial.)

Soundness:

Let the algebraic operators $\mathcal{O}_i$ be determined by the structures $\mathcal{S}_i$, and translated into the modal operators $\boldsymbol{O}_i$. Let, by prerequisite, the algebraic condition $\mathcal{C}$ hold for $\mathcal{O}_i$, and let $\boldsymbol{C}$ be its logical translation. We have to show that $\boldsymbol{C}$ holds. Let $\phi_1, \ldots, \phi_n$ be its arguments, and $[\phi] := \{m : m \models \phi\}$. By definability preservation, each ϕ_j can be replaced by some classical (base)

ϕ'_j, with $[\phi_j] = [\phi'_j]$ and $\boldsymbol{C}(\phi_1, \ldots, \phi_n)$ holds iff $\boldsymbol{C}(\phi'_1, \ldots, \phi'_n)$ does, as we have added the corresponding conditions $\spadesuit(\phi \leftrightarrow \phi') \rightarrow \spadesuit(\boldsymbol{O}_i(\phi) \leftrightarrow \boldsymbol{O}_i(\phi'))$. Moreover, $\mathcal{C}([\phi'_1], \ldots, [\phi'_n])$ holds iff $\boldsymbol{C}(\phi'_1, \ldots, \phi'_n)$ does. By prerequisite, $\mathcal{C}([\phi'_1], \ldots, [\phi'_n])$ holds, as the ϕ'_i are base formulas. By the 1-copy property (applied perhaps repeatedly) $[\phi_j]$ as result of the algebraic operators is the same as it is as result of the structural operators.

Completeness:

Let the logical conditions hold. Then they hold in particular for base formulas. So the corresponding algebraic conditions hold, and, by the individual characterizations, there are 1-copy structures which correspond and are definability preserving. As they are 1-copy and definability preserving, they give exactly the same results for arbitrary formulas, too, as they correspond to base formulas.

□

Chapter 9

Conclusion and outlook

In the author's opinion, the most important systematic problems we have encountered are perhaps:

(1) Domain closure problems and their relation to possible characterization.

- Sometimes, simplifications are possible for formulas, but not for full theories, e.g. in
  - the KLM counterexample,
  - the limit versions of preferential structures,
  - the limit versions of distance based structures.
- Domain closure can allow simplified descriptions, impossible in poorer domains.
  - Closure under union allows to express the semi-transitivity of smooth structures in one step, without union, we need more complicated conditions.
  - Sufficiently rich domains may allow "reflection" of situations otherwise hidden by closer cases. This can allow finite characterizations, otherwise impossible.

(2) The importance of definability preservation for characterizations.

- Lack of definability preservation necessitates "small amounts of exceptions", or approximate solutions. This, however, is not always expressible by standard ways of characterization.

- "Too much" definability preservation can trivialize the limit versions.
- Similar problems can arise when composing several different structures (see Chapter 8), but are hidden in traditional modal logics.

These subjects should be treated in a more systematic way in future research. We have only scratched the surface, and did not see a general and precise pattern.

Despite the shortcomings and problems left open, there are some techniques the author thinks are useful, and merit attention. Among these are:

- to split representation into an algebraic and a logical part,
- to use choice functions for preferential structures, and trees to encode transitivity, to use suitable hulls $H(U)$ — sets to be to be avoided — in the construction of smooth structures,
- to use topological constructions to obtain positive and negative results for preferential (and similar) structures,
- to use approximations to solve definability preservation problems, even if we have to sacrify standard ways of characterization,
- to reduce the more complicated limit version to the much more simple minimal version,
- to use limited observability of behavior to show lack of finite representation,
- to use exception sets which are small by topological measures but big by cardinality, to show absence of any usual representation,
  - the proof technique to construct bad, but almost good logics (by closure of small model sets) seems to be promising to solve other problems, too.

On the conceptual side, one could perhaps retain:

- the reduction of several types of common-sense reasoning to a small number of underlying semantical notions seems promising, and allows versatile and not too complicated integrated systems,
- multi-layer logics which permit great detail, but can hide the complexity when it is not needed.

Bibliography

[AGM85] C. Alchourron, P. Gärdenfors, D. Makinson, "On the logic of theory change: partial meet contraction and revision functions", Journal of Symbolic Logic, Vol. 50, pp. 510–530, 1985

[ALS99] L. Audibert, C. Lhoussaine, K. Schlechta: "Distance based revision of preferential logics", Logic Journal of the Interest Group in Pure and Applied Logics, Vol. 7, No. 4, pp. 429–446, 1999

[BB94] S. Ben-David, R. Ben-Eliyahu: "A modal logic for subjective default reasoning", Proceedings LICS-94, 1994

[BFH95] C. Boutilier, N. Friedman, J. Halpern: "Belief revision with unreliable observations", AAAI 98, pp. 127–134

[BGLS02] V. Biazzo, A. Gilio, T. Lukasiewicz, G. Sanfilippo, "Probabilistic logic under coherence, model-theoretic probabilistic logic, and default reasoning in system P", Journal of Applied Non-Classical Logics, Vol. 12(2), pp. 189–213, 2002

[BLS99] S. Berger, D. Lehmann, K. Schlechta: "Preferred history semantics for iterated updates", Journal of Logic and Computation, Vol. 9, No. 6, pp. 817–833, 1999,

[Bra87] M. E. Bratman, "Intentions, plans, and practical reason", Harvard Univ. Press, Cambridge, 1987

[BS85] G. Bossu, P. Siegel, "Saturation, nonmonotonic reasoning and the closed-world assumption", Artificial Intelligence, Vol. 25, pp. 13–63, 1985

[CP00] S. Chopra, R. Parikh, "Relevance sensitive belief structures", Annals of Mathematics and Artificial Intelligence, Vol. 28, No. 1–4, pp. 259–285, 2000

[DS99] J. Dix, K. Schlechta: "Explaining updates by minimal sums", 19th. Intern. Conf. on Foundations of Software Technology, 13–15 Dec. 1999, IIT Campus, Chennai, India

[Far02] J. Farkas, "Theorie der einfachen Ungleichungen", Crelles Journal für die Reine und Angewandte Mathematik, Vol. 124, pp. 1–27, 1902

[FH98] N. Friedman, J. Halpern: "Plausibility measures and default reasoning", IBM Almaden Research Center Tech.Rept. 1995, to appear in Journal of the ACM

[FL94] M. Freund, D. Lehmann: "Nonmonotonic reasoning: from finitary relations to infinitary inference operations", Studia Logica, Vol. 53, pp. 161–201, 1994

[FLM90] M. Freund, D. Lehmann, D. Makinson, "Canonical extensions to the infinite case of finitary nonmonotonic inference relations". In: G. Brewka, H. Freitag (eds.), Proceedings of the Workshop on Nonmonotonic Reasoning, GMD, 1989, Gesellschaft für Mathematik und Datenverarbeitung, D-5205 Sankt Augustin, Germany, pp. 133–138

[Fre93] M. Freund, "Injective models and disjunctive relations", Journal of Logic and Computation, Vol. 3, No. 3, pp. 231–247, 1993

[FRS01] L. Forget, V. Risch, P. Siegel, "Preferential logics are X-logics", Journal of Logic and Computation, Vol. 11, No. 1, pp. 71–83, 2001

[Han69] B. Hansson, "An analysis of some deontic logics", Nous, Vol. 3, pp. 373–398. Reprinted in R. Hilpinen ed., Deontic Logic: Introductory and Systematic Readings. Reidel, Dordrecht 1971, pp. 121–147

[Imi87] T. Imielinski, "Results on translating defaults to circumscription", Artificial Intelligence, Vol. 32, pp. 131–146, 1987

[KLM90] S. Kraus, D. Lehmann, M. Magidor, "Nonmonotonic reasoning, preferential models and cumulative logics", Artificial Intelligence, Vol. 44 (1–2), pp. 167–207, 1990

[KM90] H. Katsuno, A. O. Mendelzon, "On the difference between updating a knowledge base and revising it", Univ. of Toronto Tech. Rept., KRR-TR-90-6

[Leh92a] D. Lehmann, "Plausibility logic", Proceedings CSL91

[Leh92b] D. Lehmann, "Plausibility logic", Tech.Rept. TR-92-3, Feb. 1992, Hebrew University, Jerusalem 91904, Israel

[Lew73] D. Lewis: "Counterfactuals", Blackwell, Oxford, 1973

[LM92] D. Lehmann, M. Magidor, "What does a conditional knowledge base entail?", Artificial Intelligence, Vol. 55(1), pp. 1–60, 1992

[LMS01] D. Lehmann, M. Magidor, K. Schlechta: "Distance semantics for belief revision", Journal of Symbolic Logic, Vol. 66, No. 1, pp. 295–317, 2001

[Luk02] T. Lukasiewicz, "Probabilistic default reasoning with conditional constraints", Annals of Mathematics and Artificial Intelligence, Vol. 34(1–3), pp. 35–88, 2002

[Luk04a] T. Lukasiewicz, "Weak nonmonotonic probabilistic logics", To appear in: Proceedings of the 9th International Conference on Principles of Knowledge Representation and Reasoning (KR2004), Whistler, Canada, June 2004, AAAI Press, 2004

[Luk04b] T. Lukasiewicz, "Nonmonotonic probabilistic reasoning under variable-strength inheritance with overriding", To appear in: Synthese

[Mak94] D. Makinson: "General patterns in nonmonotonic reasoning", in D. Gabbay, C. Hogger, Robinson (eds.), Handbook of logic in artificial intelligence and logic programming, Vol. III: Nonmonotonic and uncertain reasoning, Oxford University Press, 1994, pp. 35–110

[Mak03] D. Makinson: "Bridges between classical and nonmonotonic logic", Logic Journal of the IGPL, Vol. 11(1), pp. 69–96, 2003

[Sch91-1] K. Schlechta: "Theory revision and probability", Notre Dame Journal of Formal Logic Vol. 32, No. 2, pp. 307–319, 1991

[Sch91-2] K. Schlechta, "Results on infinite extensions", Journal of Applied Non-Classical Logics, Vol. 1, No. 1, pp. 65–72, 1991

[Sch91-3] K. Schlechta, "Some results on theory revision", in: A. Fuhrmann, M. Morreau (eds.), The Logic of Theory Change, Springer, Berlin, 1991, pp. 72–92

[Sch92] K. Schlechta: "Some results on classical preferential models", Journal of Logic and Computation, Vol. 2, No. 6, pp. 675–686, 1992

[Sch95-1] K. Schlechta: "Defaults as generalized quantifiers", Journal of Logic and Computation, Vol. 5, No. 4, pp. 473–494, 1995

[Sch95-2] K. Schlechta: "Logic, topology, and integration", Journal of Automated Reasoning, Vol. 14, pp. 353–381, 1995

[Sch96-1] K. Schlechta: "Some completeness results for stoppered and ranked classical preferential models", Journal of Logic and Computation, Vol. 6, No. 4, pp. 599–622, 1996

[Sch96-2] K. Schlechta: "A two-stage approach to first order default reasoning", Fundamenta Informaticae, Vol. 28, No. 3–4, pp. 377–402, 1996

[Sch96-3] K. Schlechta: "Completeness and incompleteness for plausibility logic", Journal of Logic, Language and Information, Vol. 5:2, pp. 177–192, 1996

[Sch97-2] K. Schlechta: "Nonmonotonic logics - basic concepts, results, and techniques" Springer Lecture Notes series, LNAI 1187, Jan. 1997,

[Sch97-4] K. Schlechta: "Filters and partial orders", Journal of the Interest Group in Pure and Applied Logics, Vol. 5, No. 5, pp. 753–772, 1997

[Sch99] K. Schlechta: "A topological construction of a non-smooth model of cumulativity" Journal of Logic and Computation, Vol. 9, No. 4, pp. 457–462, 1999

[Sch00-1] K. Schlechta: "New techniques and completeness results for preferential structures", Journal of Symbolic Logic, Vol. 65, No. 2, pp. 719–746, 2000

[Sch00-2] K. Schlechta: "Unrestricted preferential structures", Journal of Logic and Computation, Vol. 10, No. 4, pp. 573–581, 2000

[SD01] K. Schlechta, J. Dix: "Explaining updates by minimal sums", Theoretical Computer Science, Vol. 266, pp. 819–838, 2001

[SFBMS00] K. Schlechta, E. Formenti, J.-M. Batty, J. F. Marsillo, S. Sadok: "Comments on: belief revision with unreliable observations", LIM Research Report 2000-362

[SGMRT00] K. Schlechta, L. Gourmelen, S. Motre, O. Rolland, B. Tahar: "A new approach to preferential structures", Fundamenta Informaticae, Vol. 42, No. 3–4, pp. 391–410, 2000

[Sho87b] Y. Shoham: "A semantical approach to nonmonotonic logics". In Proc. Logics in Computer Science, pp. 275–279, Ithaca, N.Y., 1987, and In Proceed. IJCAI 87, pp. 388–392

[SLM96] K. Schlechta, D. Lehmann, M. Magidor: "Distance Semantics for Belief Revision", in Proceedings of: Theoretical Aspects of Rationality and Knowledge, Tark VI, 1996, ed. Y.Shoham, Morgan Kaufmann, San Francisco, 1996, pp. 137–145

[SM94] K. Schlechta, D. Makinson: "Local and global metrics for the semantics of counterfactual conditionals", Journal of Applied Non-Classical Logics, Vol. 4, No. 2, pp. 129–140, 1994, also: LIM Research Report RR 37, 09/94

[Sno94] P. Snow, "The emergence of ordered belief from initial ignorance". In: Proc. of the 12th National Conf. on Artificial Intelligence (AAAI94), Seattle, pp. 281–286, 1994

Index